Waste Resources: Recycling and Management

Waste Resources: Recycling and Management

Edited by **Adele Cullen**

NY RESEARCH
P R E S S

New York

Published by NY Research Press,
23 West, 55th Street, Suite 816,
New York, NY 10019, USA
www.nyresearchpress.com

Waste Resources: Recycling and Management
Edited by Adele Cullen

International Standard Book Number: 978-1-63238-515-4 (Hardback)

The publisher's policy is to use permanent paper from mills that operate a sustainable forestry policy. Furthermore, the publisher ensures that the text paper and cover boards used have met acceptable environmental accreditation standards.

Trademark Notice: Registered trademark of products or corporate names are used only for explanation and identification without intent to infringe.

Printed in the United States of America.

Contents

                Crumb Rubber Modified Asphalt   160
                Magdy Abdelrahman, Mohyeldin Ragab and Daniel Bergerson

Chapter 27   **Not Too Little, Not Too Much and Shortcut: A Review on the Effectualness of**
                Per Capita Pollutant Discharge Indicators   171
                Yoshiaki Tsuzuki

Chapter 28   **Effect of Ruthenium Oxide/Titanium Mesh Anode Microstructure on**
                Electrooxidation of Pharmaceutical Effluent   177
                Vahidhabanu S, Abilash John Stephen, Ananthakumar S and Ramesh Babu B

Chapter 29   **Evaluation of Energy Conservation with Utilization of Marble Waste in**
                Geotechnical Engineering   182
                Nazile Ural

Chapter 30   **E-Waste Trading Impact on Public Health and Ecosystem Services in**
                Developing Countries   186
                Ahsan Shamim, Ali Mursheda K and Islam Rafiq

Chapter 31   **Degradation of C.I. Reactive Dyes (Yellow 17 and Blue 4) by Electrooxidation**   198
                Vahidhabanu S and Ramesh Babu B

Chapter 32   **Biogas Potential from the Treatment of Solid Waste of Dairy Cattle: Case Study**
                at Bangka Botanical Garden Pangkalpinang   205
                Fianda Revina Widyastuti, Purwanto and Hadiyanto

Chapter 33   **Household Waste Management in High-Rise Residential Building in Dhaka,**
                Bangladesh: Users' Perspective   209
                Tahmina Ahsan and Atiq Uz Zaman

Chapter 34   **Transaction Costs (Tcs) Framework to Understand the Concerns of Building**
                Energy Efficiency (BEE) Investment in Hong Kong   216
                Queena K Qian, Steffen Lehmann, Abd Ghani Bin Khalid and Edwin HW Chan

Chapter 35   **Pollution of Freshwater *Coelatura* species (*Mollusca: Bivalvia: Unionidae*) with**
                Heavy Metals and its impact on the Ecosystem of the River Nile in Egypt   223
                Faiza M El Assal, Salwa F Sabet, Kohar G Varjabedian and Mona F Fol

Chapter 36   **Plastics: Issues Challenges and Remediation**   234
                Vipin Koushal, Raman Sharma, Meenakshi Sharm, Ratika Sharma and
                Vivek Sharma

                Permissions

                List of Contributors

Preface

Every book is initially just a concept; it takes months of research and hard work to give it the final shape in which the readers receive it. In its early stages, this book also went through rigorous reviewing. The notable contributions made by experts from across the globe were first molded into patterned chapters and then arranged in a sensibly sequential manner to bring out the best results.

The rise in population and industrialization has also resulted in increased amounts of wastage. Some of these also result in environmental and health hazards. Waste resources are the byproducts of different processes such as industrial waste, farmland waste, kitchen waste, human and animal waste. Tackling this enormous amount of waste requires special techniques, methods and classification of these resources into different categories. Different approaches, evaluations, methodologies and advanced studies revolving around waste recycling and management have been presented in this text. It encompasses contributions of experts from across the globe that discuss the most vital concepts and emerging trends with a global perspective. From theories to research to practical applications, case studies related to all contemporary topics of relevance have been included in this book. It will provide comprehensive knowledge to the readers. This book will serve as a valuable source of reference for environmentalists, ecologists, conservationists, engineers, academicians and students.

It has been my immense pleasure to be a part of this project and to contribute my years of learning in such a meaningful form. I would like to take this opportunity to thank all the people who have been associated with the completion of this book at any step.

Editor

Reuse Resource of Food Processing Sludge-Derived Fuel Incinerated Ash

Ing-Jia Chiou[1]*, Jun-Pin Su[2], Ching-Ho Chen[3] and I-Tsung Wu[1]

[1]*Department of Environmental Technology and Management, Taoyuan Innovation Institute of Technology, No. 414, Sec. 3, Jhongshan E. Rd., Jhongli, Taoyuan 320, Taiwan*
[2]*Department of Natural Resource, Chinese Culture University, 55, Hwa-Kang Road, Yang-Ming-Shan, Taipei , 11114, Taiwan*
[3]*Department of Social and Regional Development, National Taipei University of Education , No. 134, Sec. 2, Heping E. Rd., Taipei City 106, Taiwan*

Abstract

The heat value of food processing sludge is similar to that of bituminous coal, thus is suitable as biofuel; however, the problem of incinerated ash disposal after combustion should be address. This study evaluated the applicability of food processing sludge-derived fuel incinerated ash (FA) to pozzolanic material and soil improvement, and proposed reuse strategies. When applied to pozzolanic material, the addition of FA reduced the hydration heat of fresh incinerated ash cement paste (FACP) significantly (85.96~91.23%), and prolonged the initial setting times (87.88~134.85%) and final setting times (87.88~134.85%) of FACP significantly. When the FA addition was 10% and 20% respectively, the pozzolanic strength activity index (SAI) was greater than 75% until the hardened FACP was cured for 28 days and 90 days respectively. When applied in soil improvement, the final seed germination of Chinese cabbage and water spinach in the original soil (ash content 0%) and improved soil (ash content 20%) was 98% and 90% respectively. There was no significant effect on the growth rate of Chinese cabbage and water spinach.

Keywords: Food processing sludge-derived fuel incinerated ash (FA); Pozzolanic material; Soil improvement; Germination

Introduction

According to the statistics of Industrial Development Bureau of ROC, about 12 million MT industrial waste could be reused directly in 2012, about 88% was used in construction engineering. The fresh concrete uses pozzolanic materials of fly ash, furnace slag and silica fume. The cement content can be reduced, and the durability and working performance can be improved. The pozzolanic material has become an important material for enhancing the concrete performance and quality. In addition, about 90,000 MT industrial waste was reused for soil improvement in 2012, meaning the industrial waste can be applied to soil improvement. According to Taiwan Ready-Mixed Concrete Industry Association, the consumption of ready-mixed concrete has been stable in Taiwan in recent years, namely 40 million m^3/year in 2012. If each cubic meter concrete is mixed with 10% mineral admixtures (e.g. coal ash and slag), the concrete industry has a large demand for mineral admixtures.

The concrete can use pozzolanic material to partially replace cement or aggregate, not only to improve the concrete performance, but also to reuse the waste. A great variety of pozzolanic materials is used extensively, such as blast furnace slag [1], silicomanganese slag [2], pulp sludge incinerated ash [3], sugarcane incinerated ash [4], palm oil fuel incinerated ash , rice husk ash [5,6], waste glass, ceramics, metakaolin, swelling clay [7-9], cattle manure incinerated ash, coal ash [10], semiconductor industry sludge [11], and TFT-LCD waste glass [12]. Some studies have used industrial wastes, such as soda-lime glass [13], industrial sludge, ocean sludge, fly ash [14-16], waste glass, recycled concrete aggregate and slag, to make artificial aggregate and lightweight aggregate [17,18]. The incinerated ash of the derived fuel made from the mixture of pulp sludge and textile sludge is suitable as the filler of controlled low-strength materials (CLSM) [19]. To sum up, the incinerated ash of different industries mostly has pozzolanic effect, applicable to cement, aggregate and concrete materials.

As the heat value of food processing sludge is similar to that of bituminous coal, it is suitable as biofuel. After the food processing sludge-derived fuel with high heat value is combusted, there is problem of FA disposal. Therefore, this study uses the properties of FA,

rheological properties and mechanical properties of FACP and the seed germination to evaluate the feasibility of FA to pozzolanic material and soil improvement, and proposes reuse strategies. The purpose to handle the disposal of FA and achieve "zero waste".

Methodology

Experimental materials

This study used the dewatered sludge from a food processing plant that produces dairy products and beverage in Taoyuan City. The food processing sludge-derived fuel was produced by preprocessing and extrusion at normal temperature. The mixture of derived fuel was mixed with 10% lime ($Ca(OH)_2$) in the extrusion process, so as to reduce the emission of contaminants (e.g. SOx, NOx). As the sludge-derived fuel was used for boilers, where the flame temperature can reach over 900°C [20], the combustion was maintained at this temperature for 2 h. When the FA was cooled, crushed and sieved (<#100 sieve), the fineness was 250 m^2/kg (Blaine) (Table 1). The cement met the specification of CNS 61 "Portland cement" with a fineness of 300 m^2/kg (Blaine) (Table 1).

The soil samples were collected from a farmland in Taoyuan City without heavy metal pollution. The soil was crushed and sieved to prepare the original soil. The fineness modulus of the sieved original soil was 3.0. The original soil was mixed with 20% incinerated ash thoroughly to prepare the improved soil.

Methods

---

*Corresponding author: Ing-Jia Chiou, Department of Environmental Technology and Management, Taoyuan Innovation Institute of Technology, No. 414, Sec. 3, Jhongshan E. Rd., Jhongli, Taoyuan 320, Taiwan
E-mail: cij@ tiit.edu.tw

There were four weight ratios of cement (PC) in this study, 100%, 90%, 80% and 70%. The weight ratio of FA addition was 0%, 10%, 20% and 30% respectively. They were mixed to make FACP (FACP numbers are PC100FA0, PC90FA10, PC80FA20 and PC70FA30). The water to cement ratio was determined at standard flow (i.e. 110 ± 5%) to make 1 cubic inch FACP. The specimen was cured in 25°C saturated limewater for 3 to 120 days. The impact of FA addition level on the pozzolanic effect of FACP was discussed.

This study used original soil (with 0% FA) and improved soil (with 20% FA), as well as Chinese cabbage and water spinach seeds. The soil was placed in well drained plastic containers (60 cm×20 cm), tested according to the standard germination rate specified by International Seed Testing Association (ISTA). Experiments were conducted in replicates. Each plastic container was divided into 5 blocks, and each block was planted with 20 seeds, meaning each container was planted with 100 seeds (Figure 1). The data were recorded on a daily basis for 30 days to observe the seed germination and growth rate of Chinese cabbage and water spinach.

The chemical composition of FA (Table 1) was mainly Fe_2O_3, CaO and P_2O_5, accounting for 27.35%, 22.70% and 18.63% respectively, and then SiO_2, Al_2O_3, SO_3, K_2O, MgO and Na_2O, accounting for 11.07%, 9.03%, 6.91%, 1.77%, 1.73% and 0.82% respectively. The CaO content in FA was 22.7%, resulted from the addition of 10% lime $(Ca(OH)_2)$ to the food processing sludge-derived fuel. The CaO in FA was much lower than Portland cement, meaning the activity of FA was lower than cement. According to the SEM image, the FA was irregular particles (Figure 2). The water soluble chloride ion (Cl-) content in FA was 0.071%, apparently higher than CNS 3090 specifications. The Cl- content in the reinforced concrete was required to be lower than 0.024%, meaning the FA is inapplicable to reinforced concrete. According to the XRD crystalline phase spectrum, the $CaSO_3$ and CaO of FA had high peak strength, followed by $AlPO_4$ and $Ca(OH)_2$. The incinerated ash pH was 10.21, and was alkaline.

Results and Discussion

Application on cement materials

Rheological properties of fresh cement pastes: The rheological properties of fresh FACP were evaluated by hydration heat, setting time and consistency. The flow value of fresh FACP was controlled within

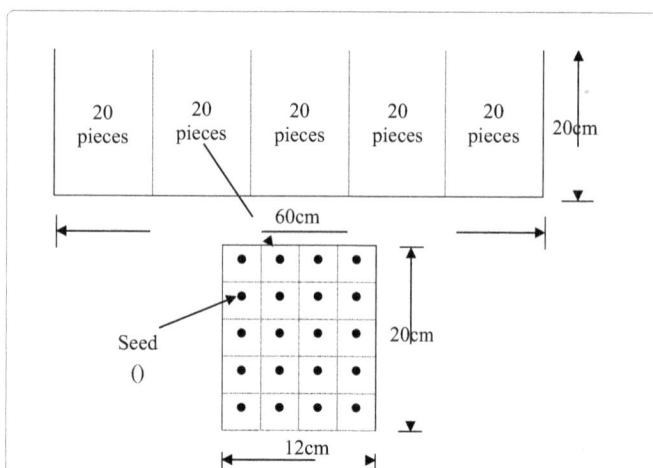

Figure 1: Schematic diagram of germination rate test.

the standard flow range (100-115%), so as to obtain the optimal mixing water consumption. At the standard flow (103~104%), the water-cement ratio corresponding to mix proportions PC100FA0, PC90FA10, PC80FA20 and PC70FA30 was 0.36, 0.38, 0.40 and 0.44 respectively (Table 2). In other words, for each addition of 10% FA, the mixing water for fresh FACP was increased by 6.5%, meaning the FA has high water absorption.

The initial setting time of mix proportions PC100FA0, PC90FA10, PC80FA20 and PC70FA30 was 293 min, 471 min, 580 min and 705 min respectively, which was longer than the initial setting time of mix proportion PC100FA0 by 60.75%, 97.95% and 140.61% respectively. The final setting time was 330 min, 620 min, 661 min and 775 min respectively, which was longer than the final setting time of mix proportion PC100FA0 by 87.88%, 100.30% and 134.85% respectively (Table 2). For each addition of 10% incinerated ash, the initial setting time and final setting time of fresh FACP were prolonged by 49.89% and 53.84% respectively. The hydration heat of mix proportions PC100FA0, PC90FA10, PC80FA20 and PC70FA30 was increased by 5.7°C, 0.8°C, 0.6°C and 0.5°C respectively, which was lower than the hydration heat of mix proportion PC100FA0 by 85.96%, 89.47% and 91.23% respectively (Figure 3). The addition of FA reduced the hydration heat of FACP significantly (85.96~91.23%), and prolonged the initial setting time (87.88~134.85%) and final setting time (87.88~134.85%) of FACP significantly. This phenomenon was due to the fact that CaO and fineness of FA are apparently lower than cement.

Properties of hardened cement pastes: The FA addition level was 0%, 10%, 20% and 30% respectively in this study. The SAI of the hardened FACP should be 75% of control group at the curing age of 28 days, so as to evaluate whether it can be used as pozzolanic material. The compressive strength on Day 3 of mix proportions PC100FA0, PC90FA10, PC80FA20 and PC70FA30 was 344 kgf/cm², 246 kgf/cm², 206 kgf/cm² and 148 kgf/cm² respectively. The compressive strength on Day 28 was 514 kgf/cm², 406 kgf/cm², 349 kgf/cm² and 289 kgf/cm² respectively. The compressive strength on Day 120 was 708 kgf/cm², 618 kgf/cm², 547 kgf/cm² and 450 kgf/cm² respectively. The compressive strength on Day 3 of the four hardened FACPs was 93.22%, 77.85%, 75.86% and 66.07% of compressive strength on Day 28 respectively. The mix proportions PC90FA10 and PC80FA20 should be cured for 28 days and 90 days to meet the requirement of SAI>75%. The mix proportion PC70FA30 did not meet the requirement of SAI>75% at the curing age of 120 days (Figure 4). This phenomenon was resulted from the low hydration heat and low CaO content of FA.

The main crystalline phase species of FACP were $Ca(OH)_2$, $CaCO_3$ and SiO_2. The peak strength of $Ca(OH)_2$ was apparent (Figure 5), namely, the peak strength of $Ca(OH)_2$ in the cement paste decreased slightly as the addition level of incinerated ash increases. It was observed that the peak strength of hardened FACP was reduced slightly as the pozzolanic reaction process consumes $Ca(OH)_2$.

According to the SEM image of hardened cement paste, at the curing age of 7 days, the hydration product of mix proportion PC100FA0 increased gradually, in plate-like structure. The mix proportions PC90FA10, PC80FA20 and PC70FA30 had more particles and pores, especially the mix proportion PC70FA30. At the curing age of 60 days, the mix proportion PC100FA0 had obvious hydration product. The hydration products of mix proportions PC100FA0, PC90FA10 and PC80FA20 increased gradually. The mix proportion PC70FA30 had a little hydration product, mostly particles and pores. At the curing age of 120 days, the mix proportions PC100FA0, PC90FA10, PC80FA20 and PC70FA30 had apparent hydration products (Figure 6).

Oxides (wt.%)	SiO$_2$	Al$_2$O$_3$	CaO	MgO	Fe$_2$O$_3$	Na$_2$O	K$_2$O	SO$_3$	P$_2$O$_5$	Fineness (m²/kg)
PC	20.5	6.50	62.5	1.90	3.20	-	-	2.20	-	300
FA	11.07	9.03	22.7	1.73	27.4	0.82	1.77	6.91	18.63	

PC: Portland cement; FA: Food processing sludge-derived fuel incinerated ash
Table 1: Composition of raw materials.

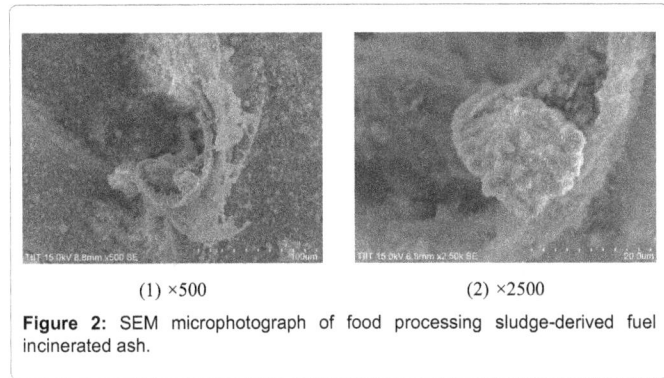
Figure 2: SEM microphotograph of food processing sludge-derived fuel incinerated ash.

(1) ×500   (2) ×2500

Properties	PC100FA0	PC90FA10	PC80FA20	PC70FA30
Flow ability (%)	103	104	104	104
Water to cement ratio	0.36	0.38	0.40	0.44
Initial setting (min)	293	471	580	705
Final setting (min)	330	620	661	775

PC: Portland cement; FA: Food processing sludge-derived fuel incinerated ash
Table 2: Properties of fresh cement paste.

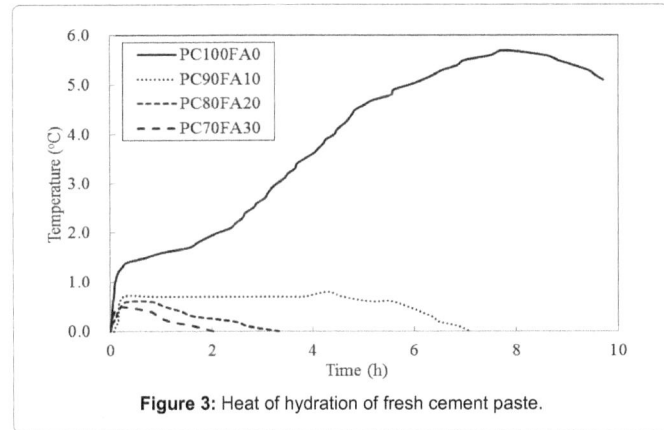
Figure 3: Heat of hydration of fresh cement paste.

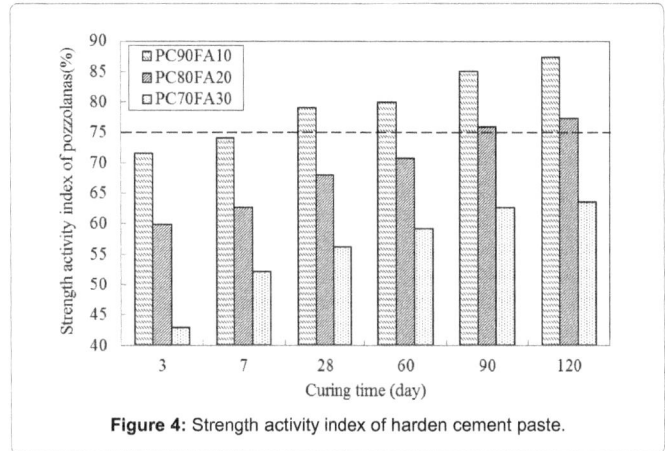
Figure 4: Strength activity index of harden cement paste.

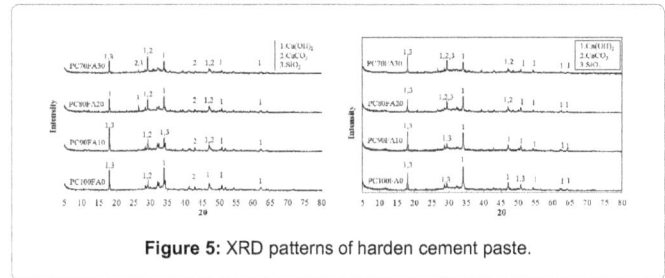
Figure 5: XRD patterns of harden cement paste.

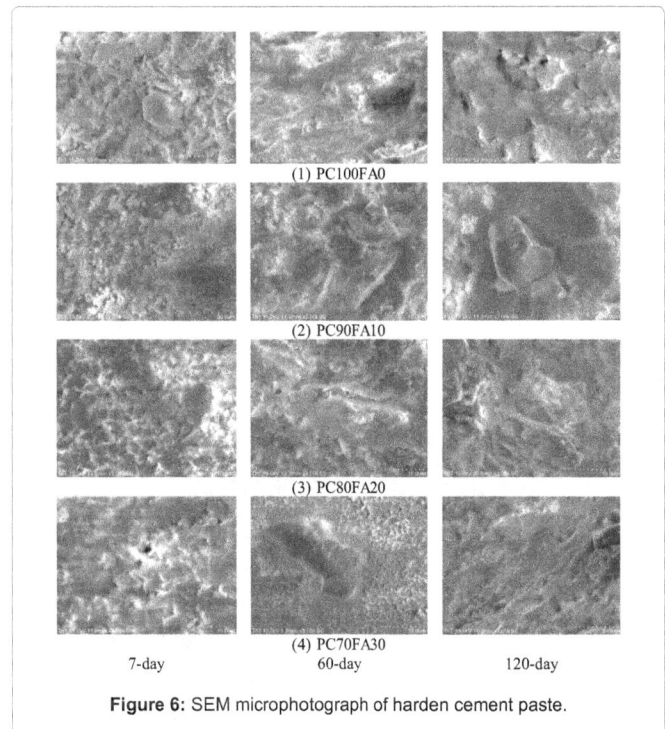
Figure 6: SEM microphotograph of harden cement paste.

Application on soil improvement

Seed germination: According to the standard germination test specified by ISTA, the Chinese cabbage and water spinach were planted in the original soil and improved soil respectively, observed on a daily basis for 30 days. The seed germination of Chinese cabbage in the original soil on Day 1, Day 3, Day 5, Day 10 and Day 15 was 0%, 47%, 75%, 92% and 98% respectively, and that in the improved soil was 0%, 81%, 93%, 98% and 98% respectively. This suggested that the seed germination of Chinese cabbage planted in the original soil and improved soil reached its maximum on Day 14 and Day 7 respectively (98%)(Figures 7 and 8). In terms of water spinach, the seed germination in the original soil on

Day 1, Day 3, Day 5, Day 10 and Day 15 was 0%, 6%, 32%, 83% and 90% respectively, and that in the improved soil was 0%, 5%, 35%, 75% and 90% respectively, suggesting that the seed germination of water spinach planted in the original soil and improved soil reached its maximum on Day 13 (90%)(Figures 7 and 9). The addition of FA could increase the seed germination of Chinese cabbage in the early stage (3~5 days), but it had no significant effect on water spinach.

Growth rate of vegetables

The growth rate of the Chinese cabbage planted in the original soil and improved soil increased gradually on Day 2, and reached its maximum on Day 5. It then decreased gradually, and later increased slightly after 15 days (Figure 10a). The water spinach planted in the original soil and improved soil grew significantly on Day 3. The maximum growth rate was reached in the original soil on Day 4, then decreased gradually, and increased slightly on Day 20 (Figure 10b). The growth rate reached its maximum in the improved soil on Day 7, then decreased gradually, and increased slightly on Day 20. Overall, the addition of FA had no significant effect on the growth rate of Chinese cabbage and water spinach.

Conclusions

(1)The FA is in irregular particle shape, so it has high water absorption and low activity. Its water soluble chloride ion content is as high as 0.071%, so it is inapplicable to reinforced concrete.

(2)The addition of FA reduced the hydration heat of fresh FACP greatly by 85.96~91.23%. Each addition of 10% FA prolonged the initial setting time and final setting time of fresh FACP by about 49.89% and 53.84% respectively.

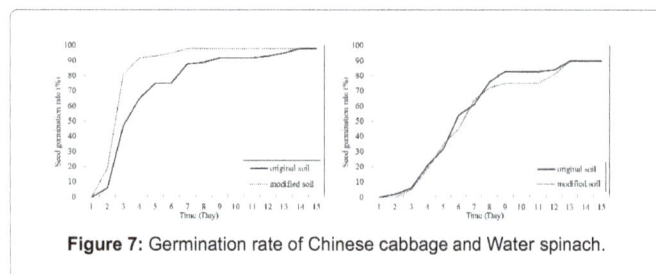

Figure 7: Germination rate of Chinese cabbage and Water spinach.

Figure 8: Growth process of Chinese cabbage.

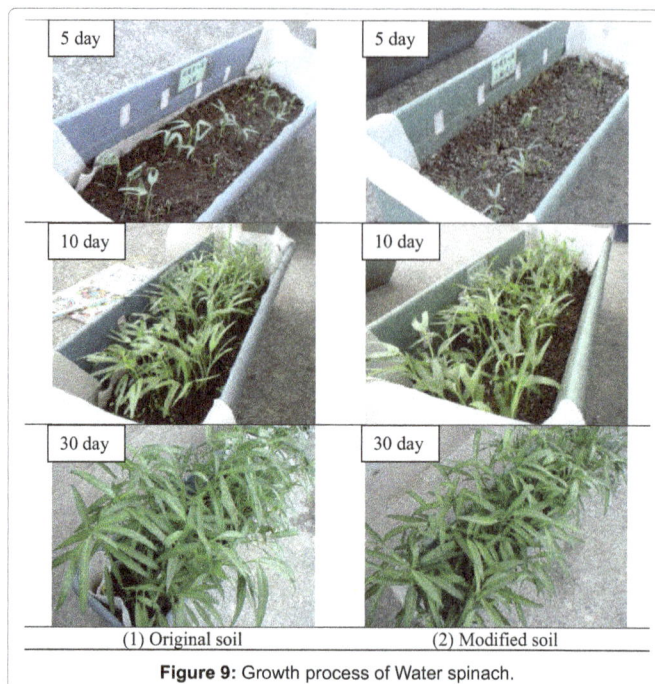

Figure 9: Growth process of Water spinach.

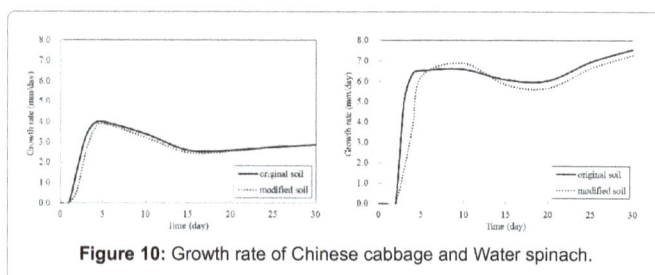

Figure 10: Growth rate of Chinese cabbage and Water spinach.

(3)When the addition level of FA was 10% and 20% respectively, the hardened FACP should be cured for 28 days and 90 days respectively, so that the SAI could be greater than 75%. Therefore, the addition level of FA was recommended as 10%.

(4)The final seed germination of Chinese cabbage and water spinach in the original soil and improved soil was 98% and 90% respectively. The addition of FA could increase the seed germination of Chinese cabbage in the early stage (3~5 days), but it had no significant effect on water spinach. The original soil and improved soil had no significant effect on the growth rate of Chinese cabbage and water spinach.

Acknowledgment

The authors would like to thank the Ministry of Science and Technology (MOST) of the Republic of China, Taiwan, for financially supporting this research under Contract No. MOST102-2221-E-253-002.

References

1. Lumley SJ, Gollop SR, Moir KG, Tayor WFH (1996) Degrees of reaction of the slag in some blends with portland cements. Cem Concr Res 26: 139-151.

2. Frias M, Sanchez de Rojas IM, Santamaria J, Rodriguez C (2006) Recycling of silicomanganese slag as pozzolanic material in Portland cements: Basic and engineering properties, Cem Concr Res 36: 487-491.

3. Garcia R, Vigil de la Villa R, Vegas I, Frias M, Sanchez de Rojas IM (2008) The pozzolanic properties of paper sludge waste. Constr Build Mater 22: 1484-1490.

4. Chusilp N, Jaturapitakkul C, Kiattikomol K (2009) Utilization of bagasse ash as

a pozzolanic material in concrete. Constr Build Mater 23: 3352-3358.

5. Kroehong W, Sinsiri T, Jaturapitakku C (2011) Effect of Palm Oil Fuel Ash Fineness on Packing Effect and Pozzolanic Reaction of Blended Cement Paste. Procedia Engineering 14: 361-369.

6. Sata V, Tangpagasit J, Jaturapitakkul C, Chindaprasirt P (2012) Effect of W/B ratios on pozzolanic reaction of biomass ashes in Portland cement matrix. Cement Concrete Comp 34: 94-100.

7. Lee G, Ling CT, Wong LY, Poon SC (2011) Effects of crushed glass cullet sizes casting methods and pozzolanic materials on ASR of concrete blocks. Constr Build Mater 25: 2611-2618.

8. Pereira-de-Oliveira LA, Castro-Gomes JP, Santos PMS (2012) The potential pozzolanic activity of glass and red-clay ceramic waste as cement mortars components. Constr Build Mater 31: 197-203.

9. Al-Sibahy A, Edwards R (2012) Thermal behavior of novel lightweight concrete at ambient and elevated temperatures: Experimental, modelling and parametric studies. Constr Build Mater 31: 174-187.

10. Zhou S, Zhang X, Chen X (2012) Pozzolanic activity of feedlot biomass (cattle manure) ash, Constr Build Mater 28: 493-498.

11. Lee TC, Liu FJ (2009) Recovery of hazardous semiconductor-industry sludge as a useful resource. J Hazard Mater 165: 359-365.

12. Lin KL, Huang WJ, Shie JL, Lee TC, Wang KS, et al. (2009) The utilization of thin film transistor liquid crystal display waste glass as a pozzolanic material. J Hazard Mater 163: 916-921.

13. Terro MJ (2006) Properties of concrete made with recycled crushed glass at elevated temperatures. Build Environ 41: 633-639.

14. Tay JH, Hong SY, Show KY (2000) Reuse Of Industrial Sludge As Pelletized Aggregate For Concrete. J Environ Manage 126: 279-287.

15. Aursen KL, White TJ, Cresswell DJF, Wainwright PJ, Barton JR (2006) Recycling of an industrial sludge and marine clay as light-weight aggregates. J Environ Manage 80: 208-213.

16. González-Corrochano B, Alonso-Azcárate J, Rodas M (2009) Characterization of Lightweight Aggregates Manufactured from Washing Aggregate Sludge and Fly Ash, Resour Conserv Recy 53: 571-581.

17. Ismail ZZ, Al-Hashmi EA (2009) Recycling of waste glass as a partial replacement for fine aggregate in concrete. Waste Manag 29: 655-659.

18. Maier PL, Durham SA (2012) Beneficial use of recycled materials in concrete mixtures. Constr Build Mater 29: 428-437.

19. Chiou IJ, Chen CH (2013) Reuse of Incinerated Ash from Industrial Sludge-derived Fuel. Constr Build Mater 49: 233-239.

20. Chiou IJ, Wu IT2 (2014) Evaluating the manufacturability and combustion behaviors of sludge-derived fuel briquettes. Waste Manag 34: 1847-1852.

2

Technologies Used in the Wastewater Treatment for Nutrient Removal

Margarida Marchetto*

Department of the Sanitary and Environmental Engineering, Federal University of Mato Grosso, Brazil

Abstract

In this paper will be showed some techniques employed effectively in wastewater treatment for nutrient removal. The results concerning biological phosphorus and nitrogen removal, from an anaerobic expanded bed reactor (AEBR), treating domestic wastewater, in an Intermittent Aeration Reactor (IAR) followed by dissolved air flotation (DAF); aiming to remove chemical oxygen demand (COD), nitrogen and phosphorus on the same unit. The DAF was used for separation of suspended solids. Conditions to obtain nitrification were assessed, denitrification and phosphorus removal in a reactor operated with alternate aerobic and anaerobic periods. The intermittent aeration system was operated with hydraulic detention time (θh) of 8 hour and 6 hours. The system with intermittent aeration (θh:6hr) showed stable results and presented average COD removal of 92% (90 to 94%) and average phosphorus removal of 90% (82 to 96%). For hour detention time, the average orthophosphate (PO_4^{-3}-P) removal was: 84% (60 to 94%) for raw samples and 94% (60 to 98%) for filtered samples (filter: 0.45 μm). Neither the use of the external carbon source nor pH correction was necessary.

Keywords: Biological phosphorus removal; Intermittent aeration activated sludge reactor; Dissolved air flotation

Introduction

In many developing countries, water scarcity and pollution pose a critical challenge in the area of environmental management. In urban areas, it is becoming difficult for the authorities to manage this question. The possibilities of wastewater reuse in agriculture, industry, urban uses, and environment, including groundwater recharge enhancement are discussed worldwide.

Water use has more than tripled globally lately. According to UNEP [1] these problems may be attributed to many factors as increasing population. Inadequate water management is accelerating the depletion of surface water and groundwater resources. Water quality has been degraded by domestic and industrial pollution sources as well as non-point sources. In some places, water is withdrawn from the water resources, which become polluted owing to a lack of sanitation infrastructure and services. Population growth in urban areas is of particular concern for developing countries and water issues are expected to grow.

The capacity building policy-making, institutional strengthening, financial mechanisms, and awareness rising and stakeholder participation are vital to implement these strategies for wastewater reuse.

In sanitary sewer, the phosphorus appears, mainly, as organic phosphorus, polyphosphate and orthophosphate [2]. The organic phosphorus comes from human and animal excretions, as well as from food remains. When organics passes through biological decomposition, they generate orthophosphates. In addition, polyphosphates may also derive from detergents [2]. That fraction has been suffering expressive growth due to increasing consumption of detergents.

The biological phosphorus removal (removal in "excess") is obtained by the selection of bacteria capable to store polyphosphate. The selection is made to expose the bacteria alternately to the anaerobic and aerobic conditions [3]. Under anaerobic condition, bacteria that accumulate phosphorus in excess, use the derived energy of the polyphosphates hydrolyses, to fix organic substratum, that are stored in the form poli-b-hidroxibutirato (PHB) or poli-b-hidroxivalerato (PHV) [3].

The removal of the phosphorus is more efficient in alternate conditions of aerobic and anaerobic [4,5]. Also it was verified, that the phosphorus removal is associated, among other factors, to the contact of the sludge recirculation with the raw affluent, in the initial zone predominantly anaerobic [6].

The technology of reactors with intermittent aeration, in continuous hydraulic permanent flow, constitutes an alternative to the conventional activated sludge. Specific conditions for aerobic and anaerobic activities are provided by these reactors [7].

Choi et al. [8] used synthetic wastewater to evaluate phosphorus and organic matter removal, in a single reactor, combining aerobic and anaerobic conditions. The removal efficiency was about 92%, for a total-P, 90% for PO_4^{-3}-P, while a great difference of phosphorus removal was observed with relationship to N/P ratio. Rittmann and Langeland [9] and Campos [6] observed that combination of nitrification and denitrification processes in same treatment system might result in an increase of phosphate removal efficiency.

For Kerrn-Jespersen et al. [10] the bacteria that store polyphosphate can be classified in two groups: a group which is capable to use oxygen or nitrate as electrons receiver, and another which is capable to just use oxygen. In agreement with the mentioned authors, the differences observed in the phosphorus removal rates, under aerobic and anoxic condition, are due to the fact that, in aerobic condition, both groups of bacteria are capable to absorb phosphorus from the medium. In anoxic condition, the bacteria that can use nitrate as electrons receiver has capacity to store phosphorus, what explains the phosphorus removal, in aerobic condition, it is larger than the removal under anoxic condition.

***Corresponding author:** Margarida Marchetto, Department of the Sanitary and Environmental Engineering, Federal University of Mato Grosso, Brazil
E-mail: m_marchetto@ufmt.br

Methods

There are different technologies and methods for wastewater nutrient removal (Table 1). The author has developed a fully operational laboratory scale one system (Figure 1) [7], being the first one an Anaerobic Expanded Bed Reactor (AEBR), built in steel carbon, with 14.9 m of total height and useful volume of 32.0 m³. The reactor was composed of two regions: reaction, and decantation. The reaction region presented cylindrical format with 12.0 m height, 1.5 m of base and 2.5 m of superior border diameter (operated with average θ_h of 3.2 hours) [11,12]. The second unit is an Intermittent Aeration Reactor (IAR) - with controlled dissolved oxygen (DO) concentration - seeking to the nitrification/denitrification, and subsequently, a unit of DAF. Both IAR and DAF units were in bench scale with continuous flow, being fed for only part of the of expanded bed reactor effluent, (range 11.5 to 15.5 L/hr).

An effluent portion of the anaerobic reactor flowed by the unit of IAR in which DO concentration was controlled. The aerated reactor had a useful volume of 0.092 m³, (width: 0.25 m, length: 0.82 m and height of 0.50 m). The effluent of the aerobic reactor was directed to a subsequent system of DAF, presenting width: 0.25 m, length: 0.10 m and height of 0.5 m with useful area about 0.025 m². The pressure was maintained in the strip of 456 ± 10 kPa in the saturation camera and the air amount was supplied with temperature and recirculation flow function. Part of the accumulated sludge in the flotation unit surface was returned to the beginning of the aeration system and the sludge excess was discarded by the system. The clarified liquid (effluent) was returned to feed the saturation camera, that, after having saturated with air, it was directed to the DAF. In the IAR, the total hydraulic detention time was maintained in 8 h, for 194 days, and of 6 h in the of 41 days, with an average flow of 15.3 and 11.5 L / h, respectively.

The affluent and effluent characteristics were monitored in each unit: pH, DO, temperature, Biochemical Oxygen demand (BOD), (COD), orthophosphate (PO_4^{-3}-P), total solids (TS) and total suspended solids (TSS), total Kjeldahl nitrogen (TKN), ammonia-nitrogen (NH_3-N), nitrite-nitrogen (NO_2^- -N), nitrate-nitrogen (NO_3^- -N) in the " liquor ". Besides those, were also monitored: MPN of denitrifying and nitrifying bacteria, Total organic carbon (TOC) and metals. All determinations were followed the rules of Standard Methods.

Discussion

System constituted by anaerobic reactor, followed by IAR associated with DAF, is a good alternative to promote biological phosphorus removal and also nitrogen and COD. That configuration constitutes quite attractive, capable to promote high a degree of domestic wastewater nutrient removal, without needing external carbon source and alkalinity addition. The intermittent aeration is a resource that results in operational cost reduction for activated sludge reactors fed with anaerobic reactor effluent. Operational changes in existing reactors can bring economy and in certain cases even efficiency increase in the N and P removal. Marchetto [7] observed that anaerobic reactor in series with IAR associated with DAF, presented a good performance in relation to the COD and phosphorus removal, mainly for a total treatment time (θh) of 8:00 and of 6:00 hours. The system can be an economical alternative for the biological phosphorus removal due to the short detention period needed.

The AEBR effluent presented, on average COD of samples without filtering: 183 mg/L, COD of the filtered samples: 76 mg/L; NTK: 26 mg/L; ammoniacal-N: 16 mg/L.

Working with θh on 6:00h was obtained PO_4^{-3}-P removal of 89% and 94% for the effluent without filtering and filtrates, respectively. In addition, for θh of 8:00h was obtained maximum removal of PO_4^{-3}-P, about 95%. Sasaki Kousei et al. [13], aiming at the simultaneous nitrogen and phosphorus removal, of a sanitary sewer, in system of activated sludge aerated intermittently (aerobic-anoxic), used θh on 20:00 h with the residues temperature varying among 9 to 33°C. The TSS concentration of the mixed liquor varied between 3100 to 4300 mg/L and the reason of sludge recirculation among 150 to 200%. During the aerobic period, the DO concentration was maintained in 2.5 mg/L. The pilot installation was shown stable and with high BOD removal efficiency: (98%); KTN: (92%) and total phosphorus: (85%). Sasaki Kousei et al. [13] almost used θh three times superior to the used in this research;

	Preliminary	Primary	Secondary	Tertiary and Advanced
Purpose	Removal of large solids and grit particles	Removal of suspended solid	Biological treatment and removal of common biodegradable organic pollutant	Removal of specific pollutants, such as nitrogen orphosphorous, color, odor, etc.
Sample Technologies	Screening, Settling	Screening, Sedimentation	Percolating/trickling filter, activated sludge Anaerobic treatment Waste stabilization ponds (oxidation ponds)	Sand filtration, Membrane bioreactor, Reverse osmosis, Ozone treatment Chemical coagulation Activated carbon. wetlands

(Asano and Levine [18] and Campos et al. [19])

Table 1: Wastewater Treatment Process.

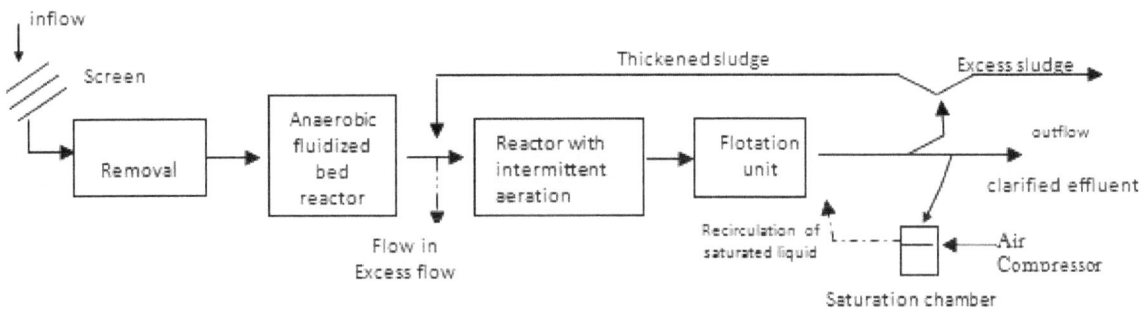

Figure 1: Schematic flowchart of the experimental system.

even so, they obtained inferior phosphorus removal.

In relation to the PO_4^{-3}–P concentration the results showed that, even for the period without aeration, the phosphorus residual of samples without filtering presented medium value of 1.5 mg/L, even for the period without aeration, the PO_4^{-3}–P residual concentration was of 4 mg/L, for raw samples and, 2.5 mg/L for the filtered samples. At the end of the period without aeration, that situation corresponds to the smallest removal efficiency of the (PO_4^{-3} -P), (80%) of the samples without filtering and (85%) of the samples filtered. The cycles were varied from 2:00 h, with aeration and 4 h, without aeration. Ikemoto et al. [14] observed that denitrification was dominant as a result of the competition between bacteria phosphate accumulative and bacteria sulfate reductors, in the organic matter degradation. In anaerobic conditions, the release of phosphate and the sulfate reduction happened simultaneously. The accumulative bacteria of phosphate competed with the bacteria sulfate reductors for organic acids, as propionate. However, the phosphate accumulative bacteria used the acetate produced by the sulfate redactors' bacteria.

It is interesting to observe when the DO concentration increased, the phosphate concentration decreased. It was verified that the values of the concentration of PO_4^{-3} -P began rising, in the corresponding time of 240 min, after 2:00 without the DO supply, (redox potential: -190 mV). This fact shows clearly that, after a certain period without the DO supply, the organisms tend to liberate phosphorus, unlike when it happens in aerobic phase, when phosphorus decrease gradation is verified along the period. It can be inferred that, only when the interior of the flake becomes anaerobic, it should happen the phosphate liberation for the liquid middle.

Finger and Cybis [15] observed that using two distinct conditions (anaerobic-aerobic) was efficient in the removal of phosphorus (81%); however, effluent presented high nitrate concentrations. Sorm et al. [16] also compared the rates of phosphorus removal in anoxic and aerobic conditions. The phosphorus removal rate in anoxic condition was harmed when concentration relatively high, of nitrogen in the nitrate form around 25 mg/L.

For Maurer and Gujer [17] the release of phosphate and substratum accumulation quickly biodegradable only happened under strict anaerobic condition (when there was not nitrate, nitrite or presence of oxygen); however, in agreement with the results of the present researches obtained with intermittent aeration followed by DAF, it was verified that in the periods without aeration, even with the presence of nitrate in the reactor with intermittent aeration, phosphate release was observed.

The organic matter absorbed by the bacteria is used by both purposes, what becomes an advantage, when the organic matter concentration is low, as it happens when the wastewater goes by a pre-treatment system. The anaerobic pre-treatment can reduce the competition significantly for organic substratum between the bacteria poly-P and the denitrifying.

Final Thoughts

The system constituted by (AEBR), followed by (IAR) associated with (DAF) showed a better alternative to promote biological phosphorus removal and also nitrogen and COD. That configuration constitutes quite attractive, capable to promote a high degree of domestic wastewater nutrient removal, without needing external carbon source and alkalinity addition. The intermittent aeration is a resource that results in operational cost reduction for activated sludge

reactors fed with effluent of anaerobic reactor. Operational changes in existing reactors can bring economy and in certain cases even efficiency increase in the N and P removal. This system can be an economical alternative for the biological phosphorus removal due to the short detention period needed. It was obtained efficiency of COD removal of 92% and of 95% for the samples filtered and without filtering, respectively.

The maximum removal of PO_4^{-3}–P in the system, of 95% and 96% of the raw samples and filtered samples, with average removal of 84%, for the raw samples and 88% of the filtered samples.

It can be verified that besides the alternate conditions of anaerobic and aerobic, the phosphate removal is directly related with the sludge recirculation, confirming the several authors' report.

Until 2005 the São Paulo state designed and constructed two wastewater treatment plants with an intermittent system for cities having population around 200,000 inhabitants. In this case, it was employed total hydraulic detention time (with and without aeration) in the range of 12 to 14 hr. Currently is necessary to do a survey to verify the number of existing plants.

Reuse of greywater has also been widely disseminated. Treatment technologies must be easily applied, low-cost implementation, operation and maintenance, aiming at environmental restoration of water bodies and improve the living conditions of the population. Thus, the use of recycled water and rainwater are alternative solutions to increase the supply of water, obeying specific standards and laws.

The reuse of gray water produced in homes, are alternatives to meet the needs in water supply, especially in activities that do not require noble quality standards as in domestic toilet discharges, washing floors and gardens.

Acknowledgment

This research was developed at the Department of Hydraulics & Sanitation, EESC/USP–Sao Carlos School of Engineering, Sao Paulo University. The author thanks the PRONEX (Program of Support for Nucleuses of Excellence CNPq e FINEP) and PROSAB (Program of Basic Sanitation FINEP e CEF) for the financial support. The author also would like to thank the professors Dr José R. Campos and Dr Marco A. P. Reali by guidance and collaboration.

References

1. UNEP (2013) Water and Wastewater Reuse. An Environmentally Sound Approach for Sustainable Urban Water Management.

2. Carol (1990) Health and Aesthetic Aspects of Water Quality. In Pontius F. Water quality and Treatment. A Handbook of Community Water Supplies/ AWWA (4th edn).

3. Kerrn-Jespersen JP, Henze M, Strube R (1994) Biological Phosphorus Release and Uptake Under Alternating Anaerobic and Anoxic Conditions in a Fixed-film Reactor. Water Research 28: 1253-1255.

4. Osada T, Haga KE, Harada Y (1991) Removal of Nitrogen and Phosphorus from Swine Wastewater by the Activated Sludge units with the Intermittent Aeration Process. Water Research 25: 1377-1388.

5. Van Haandel, Marais O (1999) Behavior of Activated Sludge System-Theory and Application for Design and Operation ABES. Campina Grande Federal University of Paraíba. 448.

6. Campos JR (1989) Removal of COD and Nitrogen in a System of Three Fixed Film Bioreactors in series. Habilitation Thesis - School of Engineering of São Carlos-USP. 295.

7. Marchetto M (2001) Nutrient removal from effluents Reactor Anaerobic Reactors Using Microaerado and Intermittent Aeration followed by Dissolved Air Flotation Doctoral Thesis. São Carlos School of Eng -USP.

8. Choi E, Lee HS, Lee JW (1996) Another Carbon Source for BNR System.

Water Science and Techenology 34: 363-369.

9. Rittmann BE, Langeland WE (1985) Simultaneous Denitrification with Nitrification in Single- Channel Oxidation Ditches. Journal WPCF 57: 300-308.

10. Kerrn-Jespernen JP, Henze M (1993) Biological Phosphorus Uptake under Anoxic and Aerobic Conditions. Water Research 27: 617-624.

11. Mendonça N (1999) Material Characterization and Study Support Match-Bed Reactor Anaerobic Expanded Used in Sewage Treatment. Dissertation SHS-EESC/USP.

12. Pereira JAR (2000) Design, Construction and Operation of Anaerobic Expanded Bed, Real Scale for Sewage Treatment. Doctoral Thesis - EESC-USP 339.

13. Sasaki Kousei (1996) Development of 2-Reactor Intermittent-Aeration Activated Sludge Process for Simultaneous Removal of Nitrogen and Phosphorus. Water Science and Technology 34: 111-118.

14. Ikemoto YR, Matsui S, Kiomori T, Hamilton BEJ (1996) Symbiosis and Competition among Sulfate Reduction, Filamentous Sulfur, Denitrification, and Poly-P Accumulation Bacteria in the Anaerobic-oxic Activated Sludge of a Municipal Plant. Water Scence and Technology 34:119-128.

15. Finger JLE, Cybis FL (1999) Biological Phosphorus Removal in Sequential Batch Reactors In: 20 ° Brazilian Congress of Sanitary and Environmental Engineering ABES Rio de Janeiro.

16. Sorm (1996) Phosphate Uptake under Anoxic Conditions and Fixed-Film Nitrification in Nutrient Removal Activated Sludge System. Water Science and Technology 30: 1573-1596.

17. Maurer ME, Gujer W (1994) Prediction of Performance of Enhanced Biological Phosphorus Removal Plants. Water Science and Technology 30: 333-343.

18. Asano T, Levine A (1998) Wastewater Reclamation, Recycling and Reuse: Introduction. In: Asano T (ed), Wastewater Reclamation and Reuse, CRC Press, Boca Raton, Florida, USA.1-55.

19. Campos JR, Reali M, Dombroski S, Marchetto M, Lima F (1996) Physico-chemical treatment of Flotation of Effluents of anaerobic Ballasts, In: XXV Congresso Inter-Ingenieria Y Sanitary Environmental, Mexico. Proceedings electronics AIDIS.

3

Potentials and Evaluation of Preventive Measures - A Case Study for Germany

Henning Wilts* and Bettina Rademacher

Wuppertal Institute for Climate, Environment,Energy GmbH, Research Group Material Flows and Resource Management, Germany

Abstract

The European Waste Framework Directive has defined waste prevention as top of the waste hierarchy meaning nothing less than a fundamental change of the sociotechnical system of waste infrastructures with all its economic, legal, social and cultural elements. Based on an empirical analysis of more than 300 waste prevention measures this paper assesses which prevention effects can realistically be achieved by applying the measures described in the German waste prevention programme or in those of other EU member states. Taking into account waste streams like packaging, food waste, bulky waste and production waste the results show that waste generation is not an unavoidable evil but can be significantly reduced at current level of technology.

Keywords: Waste prevention; Food waste; Production waste, Packaging waste, Eco-innovations

Introduction

Waste prevention as highest priority of the waste hierarchy – as confirmed by the revised Waste Framework Directive 2008/98/EG – is more than just a simple amendment of the way waste is dealt with. This definition means nothing less than a fundamental change of the socio-technical system of waste infrastructures with all its economic, legal, social and cultural elements [1]. Essentially, this is also associated with the transition from end-of-pipe technologies to an integrated resource management [2]. Considering the dimension and the complexity of such a task, it is not surprising that up to now waste prevention as a policy approach played only a minor role within the European Union [3]. Therefore, the Waste Framework Directive obliges member states to develop national waste prevention programmes as a policy instrument.

In Germany, a research project that was funded by the Federal Ministry of the Environment and the Federal Environment Agency developed scientific and technical foundations for a national waste prevention programme [4].

The project has collected and analysed a great number of measures within the public sector in Germany, which already assist in preventing waste generation. Thereby, the analysis focused on public measures but still considered legal framework conditions and economic incentives for private prevention measures. The German case studies are supplemented by specific measures from other countries or measures derived from literature in order to build a foundatio for the national prevention programme. On the basis of these results, generic instruments have been developed as possible elements of the German waste prevention programme in a second research project with the involvement of the federal states [5].

Accounting for these activities, a trivial question soon comes up: Which prevention effects can realistically be achieved by applying the measures described in the waste prevention programme of the federal government or other member states? Isn't the generation of waste unavoidable to a great extent? [6].

A current study on behalf of the European Commission estimates the waste prevention potential at 4%, given that all waste legislation guidelines are adopted by the year 2020. But why does such a low number result, which tempts to consider waste prevention rather a minor addition instead of a top priority in the waste hierarchy? In this regard, the study states the following.

Experience with cleaner production centres in Germany has shown, that some 8% of waste generation can be prevented by supporting the enterprises through audit, consulting and financing schemes. It is assumed that this 8% are a typical waste prevention potential for all waste types. It is further assumed that half of this potential can be activated by the new waste prevention measures till 2020 [7].

This assumption can be discussed in many respects, which is in the nature of assumptions. Nevertheless it is tried hereafter to demonstrate potentials that could be obtained through intelligent waste prevention concepts on the basis of individual waste streams. Naturally, even these results can't always be transferred 1:1. Still they make clear, that especially comprehensive approaches through the value chain can enable the successful reconsideration of waste generation, which meanwhile already has become accepted as a necessary evil.

Potentials and Successes in Prevention

In principle, statements concerning prevention potentials and successes are confronted with the general problem of trying to measure something which has not yet accrued. At the same time, the generation of waste correlates strongly with economic growth, so that decreasing amounts of waste often rather indicate economic crises than successes in prevention.

Meanwhile, a variety of studies exist for waste streams like packaging, food waste, bulky waste and production waste, which on the one hand present information about the potential of waste prevention, and on the other hand also suggest specific measures which could be associated with a specific prevention success. These waste streams

---

***Corresponding author:** Henning Wilts, Wuppertal Institute for Climate, Environment,Energy GmbH, Research Group Material Flows and Resource Management, Germany
E-mail: henning.wilts@wupperinst.org

have been chosen for the purpose of this paper because they show the broad range of different kinds of waste especially with regard to their heterogenity and economic incentives for recovery.

Packaging waste

For some time packaging wastes, especially plastic bags, have been focused by the public debate about waste prevention. For example, a current press release says: "The issue of plastic bags is crucial in the debate on sustainable consumption, the need for a close-loop recycling society and the need to act against litter. "The foreground isn't so much occupied by the actual quantitative or ecological significance of this waste stream, but rather by its perception as symbol of a throwaway society [8]. Of the overall amount of waste arising in German households, which is about 40 Million tons, packaging has only a share of about 7% [9]. At the same time, the average usage time of a plastic bag amounts to 25 minutes [10]. Similarly, packaging is also often thrown away immediately after opening.

Particularly interesting here is taking a look at Great Britain, where the prevention of packaging wastes is a central point of the Waste and Resources Action Programme (WRAP). As part of the campagne ‚Courtauld Commitment', which was initiated in 2005, the non-governmental organization cooperates with grocers, brand owners and producers in order to reduce the impacts of the food industry on climate and environment. The first phase of the Courtald Commitment was already joined by more than 40 larger retailers, brand owners, manufacturers and producers, which in total represented 92% of food trade in Great Britain (the programme is divided into three integrative phases, whereas phase 3 commenced in 2013). These made a commitment, amongst other things, to avoid the increase of packaging waste generation completely until the end of 2008, and to achieve an absolute decrease until 2010 [11]. In the course of the project, packaging wastes were aimed to be reduced through design optimization primarily. For this reason, new packaging strategies for implementation within the whole supply chain have been developed. The applied measures for innovation and improvement of packaged products include [12]:

• Decoating

• Weight reduction

• Use of recycled materials

• New conception of product use, e.g. refillable bottles, design for recyclability

• Reduction of food wastes

• Support of increased collection for reuse

• Import of large quantities (e.g. wine for bottling in Great Britain)

• Improved supply chain and transport efficiencies

Table 1 shows the saving potentials for packaging calculated by WRAP. These could be achieved without new technical developments, but solely by application of the best packaging solution currently available for various products.

Since 2008, the average amount of packaging of every grocery purchased in Gerat Britain has decreased by approximately 4% (however, an absolute reduction has been missed so far, considering a significant increase in overall food sales) [14]. Still a series of specific case studies clarifies the ecological and economic saving potential in the area of packaging waste, which can be attained through intelligent concepts.

Cadbury: It is a leading member of the resource efficiency initiative ‚Seasonal Confectionery Working Group' (SCWG) established by the industry. In 2009, the company took further steps towards reducing its environmental impacts by reducing the packaging of their Easter Eggs. In cooperation with WRAP, a 25% reduction in packaging of their medium eggs has been achieved, which in turn reduced material consumption by 220 tonnes of plastic, 250 tons of cardboard and 90 tons of transport and display packaging. Concerning the large eggs, packaging has also been reduced by 30%, which saved 108 tons of plastic, 65 tons of cardboard and 44 tons of corrugated cardboard. In 2008, the Cadbury, Eco Eggs'-series received an award for best packaging at the, Green Awards'.

Apetito: It is a supplier of high-qualitative frozen food and catering solutions, supplying a wide range of community meals to local authorities and franchise dealers who distribute to individuals in their own home through their meals on wheels service. Together with their employees, the company supports WRAP's Love Food Hate Waste inititiative for the reduction of food wastes by using food optimally in the workspace and by sharing information with employees via their company newsletter ‚Team News'. In this way, cost savings have been achieved by less waste, labelling, freezing, usage of long lasting products, portioning and the use-up of leftovers. Further cost reductions resulted from improved protection of groceries, storing capacities and transport. In total, apetito saves approximately 112 tons of cardboard per year and thereby avoids about 230 tons of CO_2 [12].

Food waste

The prevention of food wastes can be seen as one of the most promising while also most urgent areas of waste prevention. In Germany alone, food waste represents a quarter of all waste arising in households [9]. According to a report „Global Food Losses and Food Waste" published by the FAO, about 1,2 billion tons of food wastes are produced annually. In the European Union, losses along the food chain amount to 280-300 € per capita and year; a quantity which makes up the overall food supply in sub-Saharan Africa. Taking into account the high resource intensity of the food sector, the European Commission has defined the goal of reducing the production of food wastes by 50% until 2020 within the framework of the „Roadmap for a resource efficient Europe". As a matter of fact, a variety of research projects has already proven this goal to be achievable [4,5]. However, there is

	Product Category	Packaging Weight (in tons)	Saving Potential (in tons)
1	wine (bottles)	472,296	120,000
2	convenience foods (frozen)	30,678	19,660
3	soft drinks (cans)	47,725	16,903
4	beer (cans)	46,728	15,545
5	pizza (frozen)	20,344	13,568
6	beer (bottles)	87,470	13,254
7	whisky (bottles)	58,448	12,758
8	ready-made sauces (jars)	106,752	12,152
9	fruit juice (cartons)	51,144	10,283
10	pet food (cans)	80,971	9,212
11	soft drinks (plastic bottles)	74,218	8,833
12	ketchup	21,222	7,720
13	milk (cartons)	12,876	7,433
14	eggs (tablet and box)	12,854	6,697
15	vodka (bottles)	27,048	6,273
	TOTAL	**1,150,829**	**280,291**

Table 1: Waste prevention potentials obtained through the best design solution available on the UK market in 2004 [13]

a need to realize this goal on the basis of systemic approaches, taking into account the whole food chain, including agriculture, processors, dealers and consumers.

In Great Britain, a study with the goal of quantifying prevention potentials of food waste revealed that about two thirds of the total 8.3 million tons of food waste would have been avoidable (disposed of groceries which would have been enjoyable earlier). Furthermore, about half of the remaining 3 million tons have been potentially avoidable (groceries which some people consume and others don't, e.g. potato peelings). Besides this, the study comes to the conclusion that in case of complete prevention of the share of clearly avoidable wastes, a monetary saving of 12 billion pounds could have been achieved. Additionally, a saving ot 20 million tons carbon dioxide per year would have been the result, about 2.4% of greenhouse gas emissions caused by consumption in Great Britain. For Germany, the study "Ermittlung der weggeworfenen Lebensmittelmengen und Vorschläge zur Verminderung der Wegwerfrate bei Lebensmitteln in Deutschland" calculates a total amount of 10,970,000 tons disposed of groceries per year, taking into account every stage of the of the value chain [15]. Hereby the the largest part is made up by households, which are mainly responsible for food waste with 61%. Bulk consumers and industry contribute with 17% each, retail trade with 5%. In households, amongst other things the small share of total consumption expenses for food and the permanent availability of groceries are mentioned as causes for a decreasing appreciation of groceries.

The detection of the above mentioned potentials with regard to waste prevention forms the basis of the WRAP campaign, Love Food Hate Waste', which was initiated in 2007 and targets the reduction of food wastes in private households. For this purpose, the programme cooperates with traders and manufacturers on the one hand, to support those in developing individual respective campaigns. On the other hand, it aims at raising the attention of individuals in order to increase their sensitivity towards the issue of food waste. In this way, the british supermarket chains Sainsbury's and Morrison's, for example, introduced an improved labelling system for best-before dates and installed packaging sizes which enable modern households to be more flexible in the purchase and consumption of groceries [16]. At the same time, ‚Love Food Hate Waste' supplies consumers with practical advice and incentives for using their groceries in the best possible way. This e.g. regards easily acquirable habits of waste reduction, which besides the reduction of environmental impacts also result in significant cost savings for consumers. The acquired habits include the preparation of shopping lists and meal planning as well as freezing products with limited shelf life, appropriate product storage and the creative use-up of leftovers.

During a renewed calculation of food waste amounts in british households in 2011, WRAP could already record a decrease [17]. By means of a comparison of statistical purchase data from the years 2006 to 2009, the reduction of potentially avoidable or unavoidable waste is estimated at 73,000 tons and 77,000 tons respectively. Concerning the overall traceable reduction of 1.1 million tons (13% of food waste in households), this implies a decrease of avoidable food wastes by 950,000 tons. These numbers are further confirmed by two studies, which both quantified the share of food wastes of household wastes in England. For this purpose, information from the WasteDataFlow as well as existing compositinal analyses have been consulted. The first study was conducted in 2006/07, the second one in 2008 [18]. The estimations from this studies indicate a significant reduction of food wastes in households within the investigated period. Taking the

lowest determined number, the decline was 13.1% in comparison to the number of 2006/07 [19].

In order to be able to document changes in behaviour of consumers, WRAP additionaly collected questionnaires giving information about the behaviour of private households regarding three measures: The checkup of stocks before shopping, the planning of meals over several days and the preparation of shopping lists. On this occasion, an increase by 3 to 5 percentage points of all three behaviour patterns could be registered until 2010. Moreover, the understanding of best-before dates was improved, which could also have contributed to the prevention of wastes. In a consumer survey of the "Food and Drink Federation" with more than 1000 respondents in the same year, more than half of the respondents reported to dispose less groceries than the year before.

Bulky waste and used electric and electronic equipment

Bulky waste currently accounts for about 6% of the waste which is generated in households [9]. For used electronic equipment, there is no current data, but in 2005 their proportion amounted to a little less than 1% [20]. While percentage may appear small, it is still significant in view of a number of hazardous chemicals and precious resources used in the production process of these devices. With a view to the issues waste and circular economy, directive 2008/98/EG (EU Waste Framework Directive, short: WFD) requires member states to step up their efforts, especially in terms of reuse, according to the new 5-stage waste hierarchy. Article 11 obliges the member states to take suitable measures for the promotion of reuse and reparature of products, while at the same time mentioning possible measures, i.a. explicitly the installation and support of reparature and reuse networks.

In reality, reparature and reuse of used products has become considerably less important over the last decades. A few reasons for this are the increased complexity of mostly electronic products just as the increasingly shorter innovation cycles, which lead to a rapid depreciation in value of products. Besides, it becomes clear however that the reuse of products is additionally complicated by a conscious deterioration of product qualities (keyword planned obsolescence) [21].

In spite of this, various european regions have succeeded in making relevant amounts of wastes reusable and in placing the reuse of products in public awareness. These examples prove, that reuse is possible despite the prevailing given circumstances!

In Flanders, for example, reusable items are further used via the Kringloop network under the brand "De Kringwinkel". The network essentially consists of the flemish umbrella organization Komosie, reuse centres and De Kringwinkel shops. In total, the organizations operate about 118 shops (state: 2011) in which second hand products are sold. The range of products covers all potentially reusable products (WEEE, clothing, furniture, etc.) collected from households. Cooperations between the reuse centres, waste management and Recupel - the collective collection system of Waste electric and electronic equipment – enable a structured access to products. A reuse quota of 47% (4.41 kg per inhabitant) of collected products is already achieved, which amounts to an increase of more than 1,000% since 1994. Until 2015, the aim is to achieve a reuse quota of 5 kg per inhabitant and employment of 3,000 full-time employees.

An further impressive example are the so called Repair Cafés, which first took place in 2009 in the Netherlands. Repair Cafés are events in which reparature experts and consumers with broken items come together and repair those together over coffee and tea. The organization of the events, just as the engagement of the experts is based on voluntary activities, participation is generally free. On average, 25 products are repaired each Repair Café and month. Currently there are 21 Repair Cafés in Germany, in which 6,300 products are repaired annually. In total, about 70% of devices and products which are brought along by the participants can be repaired [22].

Commercial waste

The overall number of waste generated in Germany amounted to about 387 Million tons in the year 2011 [23]. Waste from industry and production accounted for 15.1% of this number. Operational cost savings through prevention of wastes are a much - discussed topic in Germany, at the latest since the requirements of the 5.1.3 BImSchG in the 1990ies. Within the framework of an investigation of environmental management systems in Germany, almost 80% of the participating enterprises already reported to have exploited potentials in the area of waste prevention. On a scale between 1=high significance and 6=low significance, effects in this area were even awarded with the highest significance. However it can be observed that high waste prevention potentials are still undetected, primarily in cooperation along the value chain [5]. Here, many intersections arise with the issue of material efficiency in the industrial sector. Every prevented material input finally results in a prevention of wastes – be that during raw material extraction, in the industry itself or at the consumption stage.

A study of the Fraunhofer ISI reports that according to assessment of the enterprises themselves, within the current technical standards (!) approximately 7% of material consumption could be saved. In the automobile sector, this estimation even amounts to 10% [24]. By optimizing efficiency consulting of enterprises with regard to waste prevention, cost saving potentials could be traced, especially in the manufacturing industry, which on the one hand lead to increased resource efficiency and on the other hand offer a monetary incentive for businesses to realize regarding measures. Efficiency consulting of this kind could take up existing programmes, amongst others that of the German Material Efficiency Agency (demea), while it is also supposed to obtain more realization potential by additional training and mediation of 'bridge qualifications' to consultants. It is assumed that the realization of the recommended measures, with regard to technical progress until 2016, could save one fifth of all raw materials used in production. This relates to quantitative raw materials savings of 300 million tons per year [25].

An extension of this promising measure whose effects are also significant but haven't been quantifiable up until now is the extension of existing web-based consultation offers with the aspect of purchasing low-waste and low-emission raw materials. The consideration of environmental and waste criteria in the purchase of raw materials could be specifically promoted by governmental bodies. In this way, enterprises could be sensitised for the environmental impacts outside of their own business as well as expand their knowledge about possibilities of a low-waste material purchase. *Waste-minimizing cooperations in value chains* are another potential approach of waste prevention. In real supply chains, interface problems are often "bridged" in a way that leads to waste generation. One example of this is the conscious delivery of logistical supersets to guarantee 100%-availability at the point of sale. By means of a systematic cooperation of all partners of a value chain, potentials for the reduction of material losses resulting from interface arrangements, unnecessary or wrong specifications or logistical requirements could be identified which wouldn't be manageable for a single actor. Especially the addition of waste issues to optimized logistical planning within the supply chain cooperation could lead to relevant waste prevention potentials, according to expert opinions. In product segments which feature high rates of remittances, such as fashion, magazines or even groceries, saving potentials are estimated to be a double-digit percentage.

A study named "Study on the design of a program for increasing material efficiency in SMEs" and conducted by author D. Little in cooperation with the Fraunhofer Institute and the Wuppertal Institute shows, that waste prevention through material efficiency can also involve significant monetary saving potentials.

The *manufacture of fabricated metal products*, for example, is done mainly with production techniques that have been used for decades. Even in the future, no major changes are expected in this regard. Improvements of material efficiency could be achieved through material know-how and calculation methods which lead to a more material efficient design of products and manufacturing processes, such as e.g. automated manufacturing and continuous quality control during operation. New tool materials can reduce wear and tear and the dry process of machining production can save a subsequent cleaning process including auxiliaries and thermal heating. A high number of small enterprises with high shares of material cost and relatively high saving potentials make the manufacture of fabricated metal products appear to be a promising branch for the increase of material efficiency. The potential is estimated at 800 to 1,500 million € in 2012-15.

Within the framwork of a study named "Study on reinforced waste prevention in the commercial sector" conducted by the Bavarian State Ministry for the Environment, Public Health and Consumer Protection, measure concepts of company-specific approaches for reduction and utilization of waste have been introduced in cooperation with pilot enterprises of different sectors [26]. For the sector of machine tools, metal products and metal working, saving potentials of 131.2 kg prevented waste per employee and year resulted. Additionally, 87.2 kg per employee could be recycled. With 499,000 employees in whole Germany, a theoretical prevention potential of 65,468 tons per year and a recycling potential of 43,512 tons per year can be calculated for this sector only, assuming that the measures from the pilot enterprises were realized in every german business [27].

In total, the nine sectors which were analysed within the framework of the study exhibit saving potenials between 6.4 billion €/a and 13 billion €/a through the increase of material efficiency. For the manufacturing industry, the estimate is about 27 billion €/a (for autonomous material specific and technical progress) and up to almost 60 billion €/a within the years 2012-2015 as policy-inducible potential [28].

Total Potentials and Approaches of Priorisation

In the overall debate about waste prevention, it is striking up to now that in the face of a variety of suggested measures or realized individual projects, no clear concept seems to exist so far concerning the priorisation of a starting point. Often, focal points seem to be determined primarily through public discussion of individual waste streams (cell phones, plastic bags, old clothes..). With a view to limited public funds, approaches which determine multi-sectoral prevention potentials for all waste streams seem necessary.

A first useful point of reference from the perspective of waste prevention is the consideration of ecological backpacks. These specify the amount of natural resources used for different consumption goods. "The more nature has been put into a consumption good, the heavier is its ecological backpack". Essential is the fact that in order to calculate the weight declarations of the ecological backpack, the whole life cycle of a consumption good is taken into consideration [29]:

- From raw material production and manufacturing (including extraction, production of pre-products, transport and sale)

- To utilization (including all consumption, transports and reparatures)

- Up to recycling or reuse.

This consideration focusses on resource input, which is extracted from nature. Because of the fact that material doesn't "vanish", it is obvious, that every input into the system has to end up as waste at some point. The consideration of ecological backpacks is therefore a well suitable indicator to carry out a first priorisation which products waste prevention should concentrate on. It gives information about the consumption of natural resources of a product. The more resources a product requires, the heavier its ecological backpack gets. This includes resources used along the whole life cycle of a product, from raw material extraction to use phase and disposal or recycling. Table 2 gives a selected number of examples of products with their specific ecological backpacks.

Another, product-related approach is the question, for which products optimizations of product design and production processes can result in environmental reliefs. In the course of the discussion about an extension of the ecodesign guideline to questions of energy consumption, a study presented by BioIS identified products and product categories for which the ecodesign guideline could deduce potentials for ecological improvements. For this purpose, 60 product categories in total have been investigated for all of their environmental impacts. Table 3 shows the product categories for which a particularly high ecodesign-potential has been determined according to a review of existing life cycle assessments.

Conclusions

The analysis of the herein considered waste streams makes clear that there are still high unexploited potentials in the prevention of waste under existing technical and institutional circumstances. Even for the automobile sector, which is generally regarded to be very cost-sensitive, practicians deem a saving potential of more than 10% of used materials realistic, which would become visible in waste prevention

of mainly enterprises. The area of food waste also shows that simple measures could already result in more than 10% savings.

Taking into account the area of reuse of waste electronics, successful reuse networks can be found which already feature quotas more than a factor 10 above the german average. A particular case in Germany is certainly packaging waste, which already demonstrated incentives for reducing packaging because of high license fees during the monopoly stage of 'Der Grüne Punkt'. This example also clarifies that economic incentive instruments show potentials, especially for enterprises, to contribute to waste prevention. Even more tragic, against this background, appears the current situation on the german incineration market, where spot prices around 30 € per ton certainly don't induce many companies to invest in waste preventing technologies or processes at the moment. Figure 1 gives an overview over the price differences between spot and contract prices for commercial waste as well as municipal waste disposal prices for incineration in different areas of Germany.

Furthermore, the consideration shows that the initial question "How much waste could be prevented?" can only be answered for short time periods. The consulted studies allow the conclusion, that about 10-15% of all wastes within a year could be prevented – even if drastic legal specifications (principally conceivable would be e.g. the prohibition of particularly waste intensive products via the ecodesign guideline analogous to light bulbs) or financial incentives (e.g. an additional tax on waste incineration as it already existits in various EU member states) were omitted.

However, changes in product design, in used raw materials or useful life will only be effective medium-to-long term. On the other hand, systemic approaches such as the "share economy" or the "leasing society" as frontrunners of a fundamental change in consumptions patterns and a new relationship between wealth and product possession present potentials that make at least a halving of all generated waste seem realistic [30,31]. Further research will be needed to analyze long term effects of specific innovations, processes or technologies with regard to their waste prevention potentials. Especially an integrated assessment of environmental and ecologic saving potentials will be necessary in order to trigger further eco-innovations and to allow an efficient allocation of benefits between the different actors in the value chain.

Product	Ecological Backpack
bedstead	666 kg
sofa (3 seats)	694 kg
desk	272 kg
combination fridge/freezer	2381 kg
washing machine	1215 kg
LCD TV	2666 kg
laptop	743 kg
cell phone	44 kg
fleece jacket	9.1 kg
aluminium foil (20 meters)	4.8 kg
recycled paper (100 sheet)	15 kg
plastic foil (20 meters)	0.34 kg
vacuum cleaner	84 kg
DVD player	1928 kg
jeans	6.8 kg
t-shirt	16 kg

Table 2: Ecological backpacks of selected examples [29].

Product Categories	
AV devices	Motors (ICE)
Batteries	Hand tools
Computer related	Furniture
Office equipment	Household goods
Power tools	Paper packaging
Detectors	Hygiene papers
Other electronics	Paper goods, tablets and related products
Gardening tools	Personal road transport
Heating	Forwarding
Household devices	Motorbikes, bicycles
Print	Rail transport

Table 3: Most Important Product Categories for the Evaluation of the Ecodesign Instrument as Policy Approach [7].

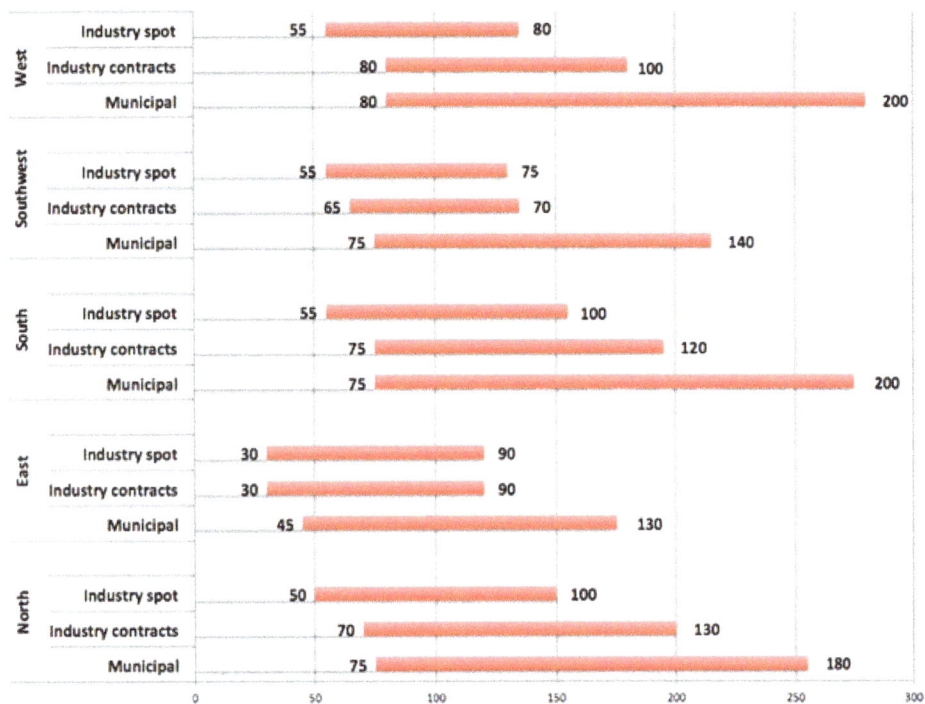

Figure 1: Prices and Fees in Incinerators in Germany, Nov./ Dec. 2012 [22].

References

1. Berkhout F, Smith A, Stirling A (2003) Socio-technological regimes and transition contexts. SPRU Electronic Working Paper No. 106. Brighton.

2. ISWA Working Group on Recycling and Waste Minimization (2011) ISWA Key Issue Paper on Waste Prevention Waste Minimization and Resource Management. Vienna.

3. Gentil EC, Gallo D, Christensen TH (2011) Environmental evaluation of municipal waste prevention. Waste Management. 31: 2371-2379.

4. Dehoust G, Küppers P, Bringezu S, Wilts H (2010) Development of scientific and technical foundations for a national waste prevention programme. Environmental Research Programme of the Federal Ministry for the Environment, Nature Conservation and Nuclear Safety. On behalf of the Federal Environment Agency.

5. Dehoust G, Jepsen D, Knappe F, Wilts H (2012) Substantive implementation of Article 29 of Directive 2008/98/EC. Scientific technical foundation for a national waste prevention program. Final Report. Environmental Research Programme of the Federal Ministry for the Environment, Nature Conservation and Nuclear Safety.On behalf oft he Federal Environment Agency.

6. Wilts H, Dehoust G, Jepsen D, Knappe F (2013) Eco-innovations for waste prevention-Best practices, drivers and barriers. Science of the Total Environment. 461: 823–829.

7. BIO Intelligence Service (2011) Implementing EU Waste Legislation for Green Growth. Final Report. On behalf of the European Commission DG ENV. Paris.

8. ACR Plus (2013) Plastic bags: an analysis of policy options and instruments.

9. Destatis (2013) Waste Environment. Fachserie 19. Reihe 1. Wiesbaden.

10. Robin Wood (2010) Colorful and dangerous - Robin Wood. Issue 104: 8-9

11. WRAP (2013) Courtauld Commitment 2.

12. WRAP (2012) Courtauld Commitment 2. Signatory case studies and quotes.

13. WRAP (2007) Understanding Food Waste. Key findings of our recent research on the nature, scale and causes of household food waste.

14. WRAP (2013) Courtauld Commitment 1.

15. Kranert M, Hafner G, Barabosz J, Schuller H, Leverenz D, et al. (2012) Ermittlung der weggeworfenen Lebensmittelmengen und Vorschläge zur Verminderung der Wegwerfrate bei Lebensmitteln in Deutschland. Institut für Siedlungswasserbau, Wassergüte- und Abfallwirtschaft. Stuttgart.

16. WRAP (2010) Love Food Hate Waste. An Introduction.

17. WRAP (2011) New estimates for household food and drink waste in the UK.

18. DEFRA (2008) Review of Municipal Waste Component Analyses. Resource Futures for Defra.

19. WRAP (2010) Evaluation of Courtauld Food Waste Target – Phase 1.

20. Schridde S, Kreiß C (2013) Geplante Obsoleszenz. Report on behalf of the Bundesfraktion Bündnis 90/ Die Grünen. Berlin.

21. Destatis (2007) Waste Environment.Fachserie 19. Reihe 1. Wiesbaden.

22. Wilts H, Gries N (2013) Reuse – One Step Beyond. Feasibility Study Wuppertal.

23. Destatis (2013) waste arise in Germany at 387 million tonnes.

24. Schroter M, Lerch C, Jäger A (2011) Material efficiency in production: savings and dissemination of concepts for saving materials in manufacturing.. Final Report to the an das Federal Ministry of Economics and Technology (BMWi). Fraunhofer Institute for Systems and Innovation Research (ISI). Karlsruhe.

25. BMU (2012) Federal Ministry for the Environment, Nature Conservation and Nuclear Safety. GreenTech made in Germany 3.0. Umwelttechnologie-Atlas für Deutschland. Berlin.

26. StMUGV (2005) Bavarian state ministry of the environment, health and consumer protection.

27. Destatis (2013) Federal Office of Statistics. May 2013: 0.6% more employees in manufacturing.

28. Baron R, Alberti K, Gerber J, Jochem E, Bradke H, et al. (2011) Study on the design of a program for the improvement of material efficiency in mittelständischen company.

29. Federal Ministry of Education and Research (BMBF) (2012) Commodity expedition. The ecological rucksack.

30. Heinrichs H, Grunenberg H. (2013) Sharing Economy: Towards a New Culture of Consumption? Centre for Sustainability Management Luneburg.

31. Fischer S, Steger S, Jordan ND, O'Brien M, Schepelmann P (2012) Leasing Society: Study. Brussels.

32. WRAP (2009) Household Food and Drink Waste in the UK. Final Report. Banbury.

33. WRAP (2011) The Courtauld Commitment Phase 2. First Year Progress Report. Banbury.

4

Effects of Hydrothermal Process on the Nutrient Release of Sewage Sludge

Xiao Han Sun[1]*, Hiroaki Sumida[2] and Kunio Yoshikawa[1]

[1]Department of Environmental Science and Technology, Tokyo Institute of Technology, Yokohama 226-8502, Japan
[2]Laboratory of Soil Science, Department of Chemistry and Life Science, Nihon University, Fujisawa 252-8510, Japan

Abstract

Hydrothermal treatment has demonstrated the ability of improving the dehydration and drying performances of sewage sludge, as well as shown its suitability for producing fuel. On the other hand, because of the abundant nutrient matters in sewage sludge, the produced liquid may be used as the liquid organic fertilizer. In this work, the effect of the hydrothermal treatment on the nutrient behavior in sewage sludge was investigated. The effects of the reaction temperature (180-240°C), and the reaction time (30-90 min) were investigated, and both of solid and liquid products were analyzed individually. The results showed that 40%-70% of nitrogen, 50%-70% of potassium and 10%-15% of phosphorus in sewage sludge could be dissolved into the liquid product, and that the solubilization was highly influenced by the temperature and the reaction time during the hydrothermal process. The hydrothermal treatment can effectively transport nutrient components in sewage sludge into the liquid product.

Keywords: Sewage sludge; Hydrothermal treatment; Nutrient

Introduction

In recent years, huge amount of sewage sludge production has become a serious problem in many countries. Especially in developing countries, for instance, China, the sewage sludge production is increasing together with the increase of wastewater discharge, causing very serious environmental pollutions. Thus, the sludge pollution problem should be solved as soon as possible.

Sewage sludge is a kind of biomass, which comes from waste water treatment, its moisture content could reach above 80% and also the organic matter content is high. Thus, it is very easy to rot, and the odor problem is also very prominent. With the increasing production of sewage sludge, the disposal method is also gathering interests. At present, the methods for sludge disposal include landfill, incineration, and agricultural use and so on. These traditional sewage sludge disposal methods, however, cannot meet the treatment need, because even with these methods the sludge is still a serious risk to human health and the environment. For the method of landfill, the heavily polluted leachate will pollute the groundwater; and the landfill gas, which is mainly methane, has a hidden peril of causing explosion and fire. In addition, the choice of suitable sludge landfill sites is also another awkward problem. On the other hand, if the sewage sludge is used for incineration, the operating cost is very expensive due to its high moisture content and the emissions of toxic air pollutants such as NO_x, SO_2 and dioxins are also problems. Originally, sewage sludge is a good agricultural fertilizer. Its nitrogen, phosphorus and potassium content are much higher than manure. However, unfortunately, sewage sludge also contains parasites, bacteria and so on, coupled with unpleasant odor, therefore, it is more unwelcomed as fertilizer. The accumulation of heavy metals in soils and subsequent accumulation along the food chain are the potential threat to animal and human health. Therefore, all of these issues motivate the development of more economic and environmentally friendly approaches to the disposal of sewage sludge and simultaneously utilize its valuable components as soon as possible. Converting the sewage sludge into valuable products can not only alleviate the disposal problem, but also can bring economic benefits, so it is of great interest and should be intensely investigated.

The hydrothermal treatment employing high pressure (around 2 MPa) saturated steam to convert wastes into usable products is a new applicable technology to sewage sludge. It has been already applied to sewage sludge to improve their dewaterability [1]. It is proved that at certain temperature and pressure, hydrothermal treatment will rupture the cell wall and membranes of organics in sewage sludge, and will improve the dewaterability of the sludge at the same time [2]. Added with the fact that the hydrothermal process uses water as reaction medium, high moisture content waste can be directly processed without the need for the predrying process, and the hot water can serve as a solvent, reactant and even a catalyst for the raw material [3]. After the treatment, the solid product is always used as a kind of RDF (Refuse Derived Fuel) or reused for agricultural fertilization after anaerobic digestion. However, the liquid phase, which is rich in dissolved organic compounds, is always treated as wastewater or discarded. In addition, on one hand, the expenditure for treating this so-called wastewater is not inexpensive; on the other hand, there are actually plenty of nutrients, such as nitrogen, phosphorus and potassium and so on in the sewage sludge. During the hydrothermal processes, accompanied with the destruction of bacterial cells, a certain amount of these nutrients will also dissolve into the liquid phase. Thus treating the liquid byproduct as wastewater seems regretful. If this so-called wastewater can be recycled and utilized, it could not only solve the wastewater treatment problem of sewage treatment, but also could achieve a huge economical benefit.

Basing on all the merits of the hydrothermal treatment, many researches are not only limited on the manufacture of fuel, but also focused on using this technology to produce other products. In recent years, using the hydrothermal treatment for converting organic wastes into more valuable substances has been investigated. According to the investigation about microalgae, an additional benefit

*Corresponding author: Xiao Han SUN, Department of Environmental Science and Technology, Tokyo Institute of Technology, Yokohama 226-8502, Japan
E-mail: xiaohansun1984@gmail.com

of the hydrothermal processing routes had been found, which has the potential to recycle liquid byproduct rich in nutrients, as well as other mineral matter and polar organics [4]. The hydrothermal treatment was also demonstrated to be able to produce high yield of amino acid from biomass waste by controlling reaction temperature and reaction time range [5]. Fish meat and silk fibroin can reportedly be converted into organic acids and amino acids by the hydrothermal treatment [6,7]. As more detailed investigation, hydrothermal conversions of cellulose and disaccharides [8-11] were also studied and had been found readily convertible into glucose and low-molecular-weight carboxylic acids [12].

In the application of the hydrothermal treatment to sewage sludge, most of the previous researches have been focused on the improvement of the dewatering performance for solid fuel production [13], or enhancement of the anaerobic digestion [14]. Nevertheless, due to the complicate components of the sludge, few researches are focused on the nutrient release during the hydrothermal process. Considering the abundant nutrient concentration in the hydrothermally treated liquid residue, it is also possible to make use of it as fertilizer. Therefore, the hydrothermal effect on the nutrient release of sewage sludge is gaining interest. The investigation on the nutrient solubilization is of great significance.

The objective of this study is to investigate the impact of hydrothermal treatment on the solubilization of the nutrient component in sewage sludge and the characteristics of the liquid products. The influences of process variables such as reaction temperature and reaction time have been studied. The influence of process variables on the yield and quality of the nutrient is discussed including the carbon balance and the nitrogen partitioning between the product phases. Since the level of macronutrients as well as micronutrients and heavy metals released from sewage sludge also depend on the process variables, their concentrations in the liquid product and the release quantity at various treatment parameters is tested to investigate their variation trend during the hydrothermal process. The results are expected to be favorable for optimizing the hydrothermal treatment process for

treating sewage sludge and creating a new component for constructing the systematic hydrothermal theory to use sewage sludge as a resource in a maximum range, and suggest a new theoretical basis for reusing the byproduct from hydrothermal process.

Experimental Materials and Methods

Material

In this research, the sewage sludge was obtained from a wastewater treatment facility located in Shimane city of Japan. The nutritional value along with other properties of the sludge is provided in Tables 1 and 2.

Methods

Hydrothermal treatment: In this research, a bench-scale hydrothermal reactor with 0.5 L capacity was utilized. The schematic view of the facility is shown in Figure 1. The reactor is a batch type (MMJ-500, Japan) which is equipped with an automated stirrer, a pressure sensor and a temperature controller. 60 g of sludge (as received based) mixed with 60 ml of distilled water, was introduced into the reactor without any pretreatment. After sealing the reactor, the air inside the reactor was purged by inert gas (argon) to prevent combustion during the treatment. Initial pressure inside was set to near atmospheric. Then, the reactor was heated to target temperature (180, 200, 220 and 240°C) with the average heating rate of 7°C/min and the constant stirring speed of 100 rpm. After reaching the target temperature, the mixture was further kept in the reactor for a certain period of time (30, 60 and 90 min) called as the holding time. Once the holding time is completed, the reactor was cooled down (< 90°C) and depressurized. Then the treated mixture was taken out and was subjected to centrifugation 3000 rpm, 30 min for solid and liquid separation. The solid phase was oven-dried (105°C for 24 hours), cooled in desiccators and subjected to chemical analyses. The liquid phase was filtered through sterile analytical filter units (with a membrane of 0.2 µm pore size) before it was used for analytical measurements.

pH	Moisture (%)	N(%)db	C(%)db
6.28	85.94	7.2	39.95

Table 1: Characteristics of sewage sludge.

Raw	unit	P	K	S	Na	Mg	Ca	Fe	Al	Mn	Pb	Zn	Cu	Cr
db	mg/g	13.88	1.43	7.60	0.35	1.4	7.44	20.69	8.71	0.16	0.02	0.33	-	-

Table 2: Nutrient and heavy metal of sewage sludge.

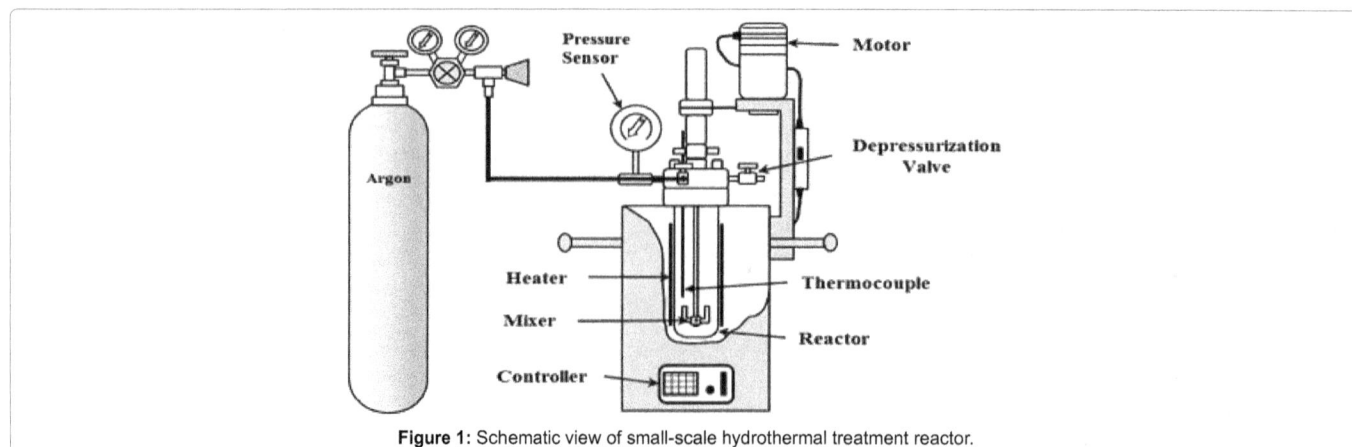

Figure 1: Schematic view of small-scale hydrothermal treatment reactor.

Analyses: The total carbon (Total-C) and the total nitrogen (Total-N) contents in solid were measured using an automatic high sensitive analyzer (Sumigraph NC-220F, SCAS, Japan). The macro and micro nutrients were analyzed after pretreatment of subsample of solid in HClO4 and HNO3 solution by a DigiPREP block digestion system (SCP SCIENCE, Canada). The phosphorus (P), sulfur (S) and other heavy metal were determined using the ICP emission spectroscopy (ICPE-9000, SHIMADZU), while potassium (K), calcium (Ca), sodium (Na) and magnesium (Mg) were analyzed by the atomic absorption spectrophotometer (180-50, HITACHI).

The total-C (TC) and inorganic carbon (IC) in liquid phase was determined by the total organic carbon analyzer (TOC-5000, SHIMADZU), the total organic carbon (TOC) is represented by the difference between TC and IC. The Total-N was measured using the Kjeldahl method; it involved acid digestion of the sludge followed by distillation and measurement of the released ammonia. The ammoniacal nitrogen (NH_4^+-N) was analyzed by employing distillation procedure. The nitrate and nitrite were detected in negligible amount, and therefore omitted in this research. The difference between the Total-N and NH_4^+-N was then taken as the organic nitrogen (Org.-N). The potential of the hydrogen ion (pH) and Electrical Conductivity (EC) values were measured using glass pH and EC electrodes (HORIBA, JAPAN).

In order to show the result of the hydrothermal treatment to the solid fraction, the solubilization ratio of the solid phase and the chemical elements had been defined as the ratio of the initial sample content minus the treated sample content divided by the initial sample content. It can also be defined as follows:

$$\text{Solubilization ratio} = \frac{\text{content in initial sample - content in treated sample}}{\text{content in initial sample}} \times 100$$

(1)

Results and Discussion

Effects of hydrothermal treatment on the solid solubilization

The effects of the reaction temperature and the holding time on the reaction are demonstrated in Figure 2, which showed the solid solubilization ratio at various reaction temperatures ranging from 180°C to 240°C and the holding time from 30 min to 90 min.

The reaction temperature positively affected the solid solubilization ratio for all the examined conditions, while the holding time only showed the positive effect on the solubilization ratio at the reaction temperature below 200°C. The solubilization ratio gradually increased with the increase of the reaction temperature. It also can be inferred that with the dissolution of the solid component, the organic matter in the solid phase also dissolved correspondingly. When the reaction temperature was 180°C, in the case of 30 min holding time, only about 32% of the solid is dissolved; however, by increasing the holding time to 40%, the solid solubilization ratio reached to about 40%. On the other hand, when the reaction temperature was higher than 200°C, the holding time had a minor influence.

It is well studied that the main composition of sewage sludge is complex organic compounds, and Barlindhaug and Odegaard [15] reported that carbohydrates were easier to be degraded but more difficult to be solubilized than proteins, which indicates that not all organic compounds react in the same way. In order to understand better the behavior of each kind of compounds, the forms and concentrations

of carbon, nitrogen and other elements in the liquid phase should be investigated in the future.

Effects of hydrothermal treatment on the nutrient components of solid phase

Effects of hydrothermal treatment on the solubilization of macronutrient: In this section, the solubilization of nitrogen (N), phosphorus (P), and potassium (K) and carbon were investigated. N, P, and K are the three major macro nutrients for fertilizer production. Therefore, the investigation of the effectiveness of the hydrothermal treatment for the dissolutions of the three components in the solid phase is essential. On the other hand, carbon component also plays an important role during the hydrothermal treatment since the main content of sewage sludge is organic component.

Figure 3 shows the effect of the reaction temperature and the holding time on the nitrogen solubilization ratio in the solid phase. In general, a certain amount of nitrogen in the solid phase (from 40% to 70%) was dissolved during the hydrothermal treatment and similar trends with those of the solid solubilization were observed for all examined conditions. The nitrogen solubilization ratio significantly increased with the temperature increase while it was only affected by the holding time at the temperature below 200°C. At 180°C, by increasing the holding time from 30 min to 60 min, the nitrogen solubilization ratio increased from 42% to 51%; but when the holding time was increased to 90 min, there was no obvious effect. When the reaction temperature was elevated to higher than 200°C, it is shown that the reaction temperature played the dominate role.

The nitrogen content in sewage sludge mainly comes from the protein contained in the microorganisms and debris inside sewage sludge. Before the treatment, the nitrogen is present in the form of

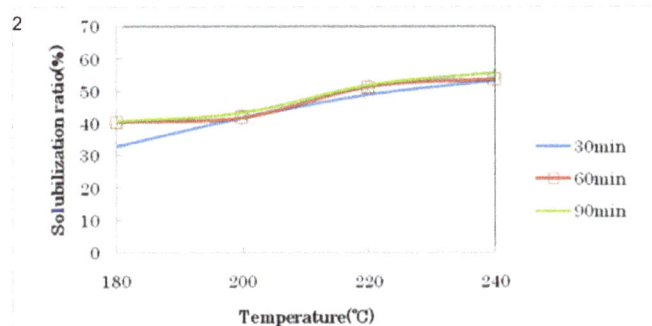

Figure 2: Solid solubilization ratio as functions of the reaction temperature and the holding time.

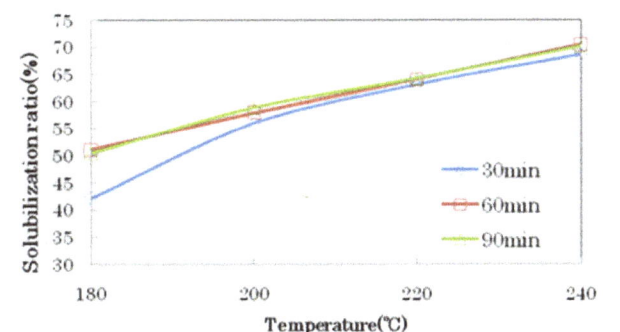

Figure 3: Nitrogen solubilization ratio as functions of the reaction temperature and the holding time.

macro molecular organic nitrogen in the solid phase of the sludge. As the reaction temperature and the pressure in the reactor increased, the organizational structure of sewage sludge starts to break up simultaneously with the nitrogen solubilization into the liquid phase. At the temperature of 180°C, the macromolecular protein in the sludge dissolved into the liquid phase gradually. In addition at this temperature, the extension of the holding time from 30 to 60 min made more protein dissolve, but further extension to 90 min created insignificant effect on the solubilization. As the holding time increases, a large amount of organic compounds get into the liquid phase and formed a high organic concentration until the equilibrium state is reached (at the holding time around 60 min). Higher reaction temperature and the pressure seem to able to transfer more nitrogen content from the solid phase to the liquid phase, but considering the material recovery and the operating cost, a relatively lower reaction temperature (lower than 200°C) and a shorter holding time (30 min) seems more suitable for large-scale production.

Figure 4 shows the effect of the hydrothermal treatment on the carbon content in the solid phase. Similar to nitrogen, the solubilization of carbon also increased with increasing the reaction temperature. However, for all the conditions tested, the holding time of 60 min and 90 min showed no difference. It is observable that at 180°C, the solubilization ratio increased with the increase of the holding time from 30 min to 60 min and above. At the temperature higher than 200°C, the reaction temperature becomes the dominant factor over the holding time. This phenomenon might be explained by the presence of extracellular polymers (ECP), whose main components are polysaccharide, protein and DNA that are significant in sewage sludge [16]. Under the relatively lower reaction temperature (180°C), the sludge structure break up was followed by the gradual ECP solution into the liquid phase and the longer holding time enhanced the solubilization. Mok and Antal [17] reported that when biomass was heated in hot-compressed water, solvolysis of hemicellulose and lignin began to occur at 190°C and all of the hemicellulose and much of the lignin dissolved in water at 220°C. Therefore, it can be inferred that during this process, accompanied with the polysaccharide dissolve, a certain amount of celluloses, hemicelluloses or lignin in the solid phase also dissolved gradually. Some previous studies had been demonstrated that for the biomass conversion process in hot-compressed water, the temperature was the most critical parameter and the holding time has little influence on the solubilization [1,18]; however, in this study, it seems that the holding time, in a relatively low reaction temperature, also plays an important role. At the temperature above 200°C, the holding time showed little effect. It indicates that all the micro-molecule soluble organic matters (mainly saccharide) has been fully

dissolved into the liquid, and the further observed solubilization might be related to the macromolecules (cellulose, lignin and so on), which is much more difficult to dissolve.

Figure 5 shows the solubilization of phosphorus (P) and potassium (K). The solubilization of phosphorus does not show a significant increase by increasing the reaction temperature as well as extending the holding time. Similar to those of nitrogen and carbon, potassium solubilization was significantly affected by the reaction temperature but not by the holding time. In addition, it is obvious that the amount of dissolved potassium far exceeds the amount of dissolved phosphorus. From 180°C to 240°C, the potassium solubilization ratio increased by about 20%, from about 50% to about 70%. The solubilization of phosphorus is only around 10% to 15% at all conditions, and from its high initial content together with the presence of high Fe initial content (Table 2), it can be inferred that phosphorus was added as flocculants to the sludge, is mainly in the form of insoluble $FePO_4$ precipitate. In addition, the small amount of phosphorus solubilization mainly comes from the DNA content in the ECP component. As for potassium, because its concentration in the sludge is not high, and it is very easy to dissolve in water, so when high temperature and pressure were applied, it is released simultaneously with the broke up of the complex sludge flocculation network and the breakdown of organic matter.

Effects of hydrothermal treatment on the micronutrient: The contents of micronutrients and heavy metal in the solid phase are also determined before and after the hydrothermal treatment. The results are shown in Table 3.

In Table 3, the release ratio of the micronutrient and heavy metal from the solid phase is provided. With exception of Na and Pb, most of the elements follow the trend that the higher the reaction temperature is, the greater the release is. The holding time only showed a minor influence on the solubilization.

On the other hand, for the elements of Na and Pb, the holding time showed very significant effects. For Na, when the reaction temperature increased from 180°C to 240°C, the solubilization ratio increased at most by only about 8% (the holding time of 90 min); however, for each reaction temperature, obvious decreasing trend can be seen when the holding time increased. By increasing the holding time from 30 min to 90 min, the solubilization of Na can decrease around 15%. It is implied that increasing the holding time is not beneficial for the releasing of Na. On the other hand, the solubilization of Pb showed significant decreasing trend with the increase of both the reaction temperature and the holding time. From 180°C to 240°C, the solubilization ratio decreased around 45% (with 60 min holding time),and the highest solubilization ratio decrease reached about 53% when the holding time was extended from 30 min to 90 min at the temperature of 180°C. It is

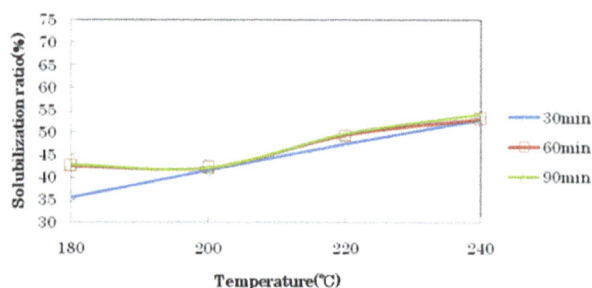

Figure 4: Carbon solubilization ratio as functions of the reaction temperature and the holding time.

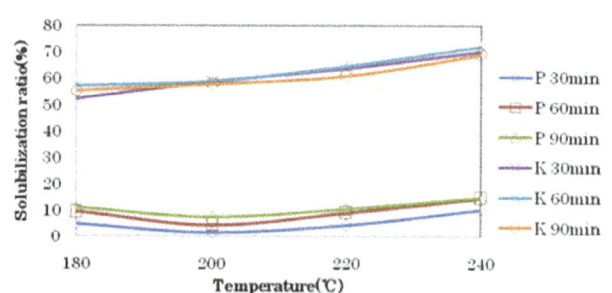

Figure 5: Phosphorus (P) and potassium (K) solubilization ratios as functions of the reaction temperature and the holding time.

Element	Unit	180°C			200°C			220°C			240°C		
		30min	60min	90min	30min	60min	90min	30min	60min	90min	30min	60min	90min
Na		62.21	58.51	45.74	62.16	57.67	48.44	65.76	56.43	50.75	69.98	57.99	52.19
Mg		8.08	9.11	10.82	10.54	12.32	9.16	4.62	5.03	5.87	5.13	4.32	4.39
Ca		5.14	7.92	6.78	3.52	3.65	6.19	10.11	11.12	10.98	7.80	7.68	12.88
Fe		4.16	4.53	5.27	5.76	6.59	5.35	3.53	3.63	5.78	7.71	8.73	9.04
Al	%	11.53	16.74	12.21	14.09	17.75	14.08	19.09	18.79	16.34	24.28	20.05	22.89
Mn		8.67	9.52	9.23	5.06	6.46	9.81	8.41	11.64	7.73	8.75	13.40	15.15
Pb		82.89	74.11	29.66	71.06	53.93	29.97	68.80	30.22	26.17	68.83	29.58	17.82
Zn		0	7.72	3.28	0.92	5.18	5.11	5.25	7.75	3.39	5.78	9.23	11.55
S		44.08	51.49	54.22	52.87	54.91	56.51	59.77	60.86	61.83	62.81	66.33	68.96

Table 3: Solubilization of the inorganic material and heavy metal.

shown that at a lower reaction temperature and a shorter holding time, most of Pb will be released from the solid phase. When the reaction temperature and the holding time increased, it can be seen that some complex reactions were happened and that the release of Pb amount was decreased. There are many different forms of metals in sludge, including those forms of exchangeable, bound to carbonates, bound to iron and manganese oxides, bound to organic matter and residual. The exchangeable fraction is likely to be affected by changes in water ionic composition, sorption and desorption processes. The carbonate fraction is sensitive to changes in pH, while the reducible fraction, which consists of iron and manganese oxides, is thermodynamically unstable under anoxic conditions [19]. At the lower temperature and shorter holding time, the release of Na and Pb are huge, indicating that these two elements most likely appear in the form of exchangeable, and thus are easy to dissolve. On the other hand, at a higher temperature and at a longer holding time, accompanying the hydrolysis of protein and carbohydrate, some organic monomers are produced. The monomers reacted with these dissolved metal ions, generating more complex water-insoluble precipitate, thus the solubilization ratio of Na and Pb was significantly reduced.

However, the solubilization ratios of other elements showed very low levels entirely. Generally, in addition to Al and S, the solubilization ratios of other measured elements are less than 20%, most of which are less than 10%. As for Al, although the effect of the holding time is not obvious, the reaction temperature showed a linear relationship. In addition, for S, the solubilization was improved from 44% to around 70% during the treatment process; therefore, by adjusting the treatment parameters, most of the S content can be transported from the solid phase to the liquid phase.

It is reported that heavy metals, incorporated in the sludge flocs, can only be transported from the flocs to the aqueous phase by diffusion [19]. For a relatively lower temperature, the rate of extraction as the function of ions diffusivity is promoted [20]. Conformational changes of the sludge flocs enhance further the mobility of the metal ions. Moreover, a large part of heavy metals is adsorbed to the EPS, which shows a lot of potential binding sites including carboxylates, amines and thiols. The degradation of these structures leads also to the release of the adsorbed metals [19]. However, in this study, the heavy metal contents are too low, and thus, the release ratio can only be seen as a reference.

Effects of hydrothermal treatment on the liquid phase

pH and EC value: After the hydrothermal treatment, the liquid phase became not transparent anymore and turned brown to a large extend after the treatment which confirms visually that chemical reaction happened. At the same time, not only the unpleasant odor

disappeared completely, the liquid product even gives off a coffee like smell, which suggested the caramelizing reaction or maillard reaction as was discussed in our previous work [21]. During the hydrothermal process, because of the hydrolysis of the organic components, a lot of reducing sugar and amino acids were produced. Under high temperature and high pressure, the C- in the open-chain carbonyl radicals in reducing sugar are attacked by the lone pair electrons in N in the amino group nucleophillically to lose H_2O and closed chain to form new kind of substances [22], which caused the change of both the color and odor of the liquid product.

The pH and EC values were measured as soon as possible after each experiment. The results are shown in Tables 4 and 5. For the case of pH values, when the reaction temperature was 180°C, with extending the holding time, the solution's pH value showed a decreasing trend. However, an obvious difference was observed at the temperature of 200°C with a longer holding time. In the first 60 min, the same as the trend of 180°C, the pH value decreased to lower than 6, and then when the holding time was extended to 90 min, the pH value showed an increasing trend. The pH decrease indicates the destruction or transformation of organic matters to organic acids in the liquid product. Under a comparative low temperature ($\leq$ 200°C), it is obvious that chemical reactions happened in the dissolved organic matters. With extending the holding time, more organic acid is produced, and this reaction is exhibited by the decrease of the pH value. Then pH increase could be due to the organic acid decomposition or acidic compounds volatilization. The EC value also increased with the reaction temperature increase, which can be attributed to some dissolved macromolecular organic compounds decomposed into small and inorganic molecules, and as we had discussed before, metals, released into the liquid phase caused the EC increase.

Yields of TOC, Org-N and ammonium: We have found that both the reaction temperature and the holding time are very important parameters for the nitrogen solubilization. A large number of previous studies have shown that during the hydrothermal treatment process, part of the protein will degrade to inorganic matters due to decomposition. Some lectures also pointed that while the protein dissolves, it is also hydrolyzed to form multipeptide, dipeptide and amino acid [13]. By the hydrothermal treatment, more than 40% of nitrogen components are dissolved into the liquid phase; to take advantage of this liquid, it is necessary to know the characteristics of the dissolved nitrogen in the liquid. In order to know the effect of the operating parameters of the hydrothermal treatment on the protein degradation, Org-N and NH_4^+-N concentrations were also measured. The degradation of protein was characterized by the percentage of Org-N and NH_4^+-N in the total nitrogen.

The result is presented in Table 6, The NH_4^+-N ratio increased

pH	30 min	60 min	90 min
180	7.05	6.68	6.41
200	6.08	5.89	6.28
220	6.78	7.12	7.49
240	8.17	8.31	8.41

Table 4: The pH value of liquid products after each hydrothermal process.

EC(ms/cm)	30 min	60 min	90 min
180	7.49	8.13	8.94
200	9.17	9.89	10.19
220	10.29	11.16	11.42
240	11.43	11.99	12.57

Table 5: The EC value of liquid products after each hydrothermal process.

Parameter	unit	180°C			200°C			220°C			240°C		
	%	30 min	60 min	90 min	30 min	60 min	90 min	30 min	60 min	90 min	30 min	60 min	90 min
Or-N		73.08	72.20	64.09	64.52	62.39	61.25	59.87	56.80	55.75	53.90	50.26	49.22
NH4+-N		26.92	27.80	35.91	35.48	37.61	38.75	40.13	43.20	44.25	46.10	49.74	50.78

Table 6: The ratio of different kind of nitrogen in liquid phase.

and Org-N decreased with increasing the reaction temperature, demonstrating that a certain amount of Org-N in the liquid phase was decomposed to NH_4^+-N with increasing the reaction temperature. At the temperature of 180°C, in the first 60 min holding time, the NH_4^+-N ratio did not show obvious change, demonstrating that during this process, the main reaction is dissolution; when the holding time was extended to 90 min, the NH_4^+-N ratio increased by about 8%, and the Org-N ratio decreased. However, when the reaction temperature was increased to 200°C, it seems that even though the holding time was extended, the NH_4^+-N ratio maintained constant levels in this temperature. At the temperatures of 220°C and 240°C, by changing the holding time from 30 min to 90 min, the NH_4^+-N ratio increased by 4.12% and 4.68%, respectively, which can also be considered to be maintained at the constant level. The NH_4^+-N ratio change showed almost a linear relationship with the reaction temperatures, demonstrating that the degradation of organic nitrogen is influenced by the reaction temperature. For the case of protein decomposition, at a low reaction temperature, protein was decomposed into amino acid first, and then, when more energy was applied, the destruction of amino acids was broken up, and the decarboxylation and the deamination happened [23]. Then the amino acid further hydrolyzed to form organic acid, NH_4^+ and CO_2 [22]. At a high reaction temperature, since no obvious effect can be seen from extending the holding time, the dissolution and the decomposition can be completed in less than 30 min.

Referring to Table 5, it can be inferred that the pH change, not only depends on the production of NH_4^+-N, but also depends on the carbon form extremely since even though a certain amount of ammonia was produced, the total content is not high enough to affect the pH value. Particularly, in the low reaction temperature region, with the decomposition of protein, ammonia content increased, but the liquid phase showed a certain degree of acidity, and, the reaction temperature of 200°C can be seen as an inflection point. The pH value showed a trend of decreasing first, and then increasing later by increasing the reaction temperature. When the reaction temperature is higher than 220°C, the liquid phase showed alkalinity and the pH value reaches higher than 8. Therefore, the huge amount of carbon content in the liquid phase played the dominant role for both the physical and chemical characteristics of the liquid phase.

To identify the carbon morphology change in the liquid phase,

the TC and IC concentrations of the liquid phase was measured with a TOC analyzer. TOC was calculated by subtracting Inorganic Carbon (IC) from the Total Carbon (TC). The IC values were less than 3% of the TC values for all samples examined. The TOC yield of liquid phase is defined by the following equation:

$$\text{TOC yield} = \frac{\text{TOC concentration} \times \text{liquid phase volume}}{\text{dissolved carbon}} \quad (2)$$

Where the unit of the TOC concentration is [mg/l], the liquid phase volume is [l], the dissolved carbon is [mg], respectively.

The TOC yield is shown in Figure 6. It can be seen that even though the carbon solubilization ratio increased with increasing the reaction temperature, visually reverse trends were observed for the yields of total carbon and TOC with regard to the holding time at the temperature of 180°C and 200°C: an increased holding time results in a larger carbon solubilization ratio but a lower yield of TOC. The highest decreasing ratio of 17% happened at 200°C from 30 min to 90 min. Moreover, at the temperatures of 220°C and 240°C, there are no obvious differences between different holding times. This result suggested that during the hydrothermal treatment, in different treatment conditions, the carbon decomposition rates are also different. Generally speaking, the TOC yield decreased with increasing the reaction temperature, and for all the situations tested, the results for 30 min and 60 min holding times are almost identical; but at 180°C and 200°C, when the holding time was extended to 90 min, the TOC ratio showed obvious decrease, which means the formation of gaseous product.

The trend of 180°C and 200°C showed that the dissolved organic carbon in the solution, decomposed to gaseous products gradually, and most of the gaseous products were produced between 60 min and 90 min. As can be seen from Table 5, the conclusion showed that during the first 60 min holding time, the dissolution and the organic acid formation are main processes. That is because at temperatures lower than 200°C, only exopolymers were affected by the hydrothermal treatment, carbohydrates and also a few proteins were dissolved. As carbohydrates are located in exopolymers whereas proteins are mainly located inside the cells, carbohydrates decomposition was superior to proteins decomposition [24]. Therefore, the organic content from exopolymers, dissolved into the liquid phase first, then, with the holding time increase, polysaccharide decomposed into organic acid,

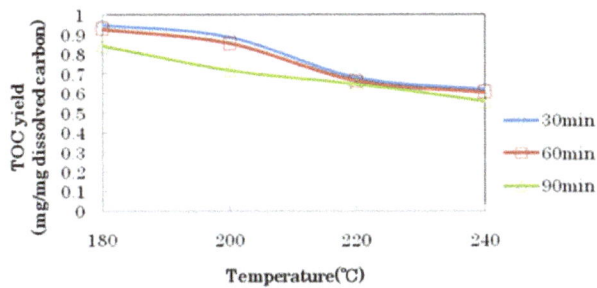

Figure 6: TOC yield as functions of the reaction temperature and the holding time.

and with a further extension of the holding time, the organic acid was decomposed into gases and the organic nitrogen also decomposed into inorganic nitrogen.

The formation of gaseous products from biomass in hot-compressed water mainly results from the decarboxylation and the fragmentation reactions of the intermediates or liquid products [25]. At a comparatively higher temperature, it seems that the decomposition of organic substance happened in a very short time, because at 220°C, no obvious difference in the TOC content was shown by extending the holding time. That is because under a higher reaction temperature, with more energy was applied, the pressure in the reactor also became higher, which caused the reaction acceleration. In addition, at this temperature, the pH value of the solution showed obvious increase trend, expressing the decomposition of the organic acid. By referring to Figure 6 and Table 5, this assumption could be demonstrated very well.

It is found that the hydrothermal treatment of sewage sludge at 190°C and 2 MPa can considerably enhance the dehydration performance of the slurry like product, and then the water content can be reduced to lower than 60% by the mechanical dehydration [25]. The fragmentation of ECP caused the sludge dewaterability improvement, so it can be seen that at 180°C to 200°C, nearly all of the carbohydrates dissolved into the liquid phase accompanied with the fragmentation of ECP and decomposed gradually, which caused the dewaterability improvement; when the temperature is higher than 200°C, the TOC product kept consistent, expressing that all of the carbohydrates were decomposed into gaseous product, leaving only complicate organic matters which is difficult to decompose.

Cost Effectiveness of Hydrothermal Process for Sewage Sludge Treatment

The cost effectiveness of the hydrothermal process has been analyzed based on the Japanese tipping fee for sewage sludge treatment (100US$/ton) and the expected selling price of the liquid fertilizer produced (100US$/ton) for the 30tons/day capacity plant. The detail assumptions and calculations are shown below. From this analysis, we can expect that the total expenditure of the plant will be 0.7 MUS$/year, while the total income of the plant will be 2.0 MUS$/year. This analysis clearly shows the cost effectiveness of liquid fertilizer production from sewage sludge employing the hydrothermal process.

- Treatment capacity of sewage: 30 tons/day
- Investment for full treatment plant: 2 MUS$
- Moisture content of sewage sludge: 80%
- Annual operation period: 330 days

- Daily operation: 24 hours/day
- Daily solid fuel production (dry base): 6 tons/day (30 tons/day x 0.2)
- Daily liquid fertilizer production: 30 tons/day
- Boiler fuel consumption: 100% of produced solid fuel will be utilized as a boiler fuel
- Maintenance cost per annum (5% of the capital cost): 0.1 M US$/year
- Labor and utility costs per annum: 0.2 MUS$/year
- Capital cost (5 years depreciation) per annum: 0.4 MUS$/year
- Total expenditure per annum: 0.7 MUS$/year
- Tipping fee income per annum: 1.0 MUS$/year (100US$/ton x 30tons/day x 330days/year)
- Fertilizer sales income per annum: 1.0 MUS$/year (100US$/ton x 30tons/day x 330days/year)
- Total income: 2.0 MUS$/year

Conclusions

The present paper studied the application of moderate temperature hydrothermal treatment (180, 200, 220, 240°C) as a new treatment method of nutrient recycling from sewage sludge. It is observable that the hydrothermal treatments are effective to solubilize sludge. For the temperature lower than 200°C, the holding time was found to be more important factor for the solid solubilization, but when the temperature was higher than 200°C, the solid solubilization was found to increase linearly with the reaction temperature and the effect of the holding time was insignificant. Nitrogen, potassium and sulfur solubilization linearly increased as the reaction temperature was increased and they reached the maximum values of about 70%, while only less than 20% of phosphorus was dissolved. In addition, with different treatment parameters, different kinds of organic and inorganic nutrients are also produced; protein and saccharine components will be decomposed into amino acid and other organic acid first and finally into ammonia and CO_2, with different reaction rates. Considering the large quantity of nutrient in the liquid product, the hydrothermal treatment also suggested a possibility of producing more valuable products from sewage sludge with a lower cost.

References

1. Neyens E, Baeyens J (2003) A review of thermal sludge pre-treatment processes to improve dewaterability. Journal of Hazardous Materials 1: 51–67.

2. Rui Xun, Wei Wang, Wei Qiao, Keqing Yin (2008) Status of Urban sludge treatment and hydrothermal reduction technology of enhanced dewatering. Environmental Sanitation Engineering 16: 28-32.

3. Mark Crocker (2010) Thermo chemical Conversion of Biomass to Liquid Fuels and Chemicals, RSC Publishing Cambridge.

4. Ross AB, Biller P, Kubacki ML, Li H, Lea-Langton A, et al. (2010) Hydrothermal processing of microalgae using alkali and organic acids. Fuel 89: 2234–2243.

5. Cheng H, Zhu X, Zhu C, Qian J, Zhu N, et al. (2008) Hydrolysis technology of biomass waste to produce amino acids in sub-critical water. Bioresource Technology 99: 3337–3341.

6. Yoshida H, Terashima M, Takahashi Y (1999) Production of organic acids and amino acids from fish meat by sub-critical water hydrolysis. Biotechnology Progress 15: 1090–1094.

7. Kang K, Chun B (2004) Behavior of hydrothermal decomposition of silk fibroin to amino acids in near-critical water. Korean J. Chem. Eng. 21: 654–659.

8. Park JH, Park SD (2002) Kinetics of cellobiose decomposition under subcritical

and supercritical water in continuous flow system. Korean J Chem Eng 19: 960-966.

9. Sasaki M, Kabyemela B, Malaluan R, Hirose S, Takeda N, et al. (1998) Cellulose hydrolysis in subcritical and supercritical water. Journal of Supercritical Fluids 13: 261–268.

10. Sasaki M, Fang Z, Fukushima Y, Adschiri T, Arai K (2000) Dissolution and hydrolysis of cellulose in subcritical and supercritical water. Ind Eng Chem Res 39: 2883–2890.

11. Oomori T, Khajavi SH, Kimura Y, Adachi S, Matsuno R (2004) Hydrolysis of disaccharides containing glucose residue in subcritical water. Biochemical Engineering Journal 18: 143–147.

12. Quitain AT, Faisal M, Kang K, Daimon H, Fujie K (2002) Low-molecular weight carboxylic acids produced from hydrothermal treatment of organic wastes. J Hazard Mater 93: 209–220.

13. Jiang Z, Meng D, Mu H, Kunio Y (2010) Study on the hydrothermal drying technology of sewage sludge. Science China Technological Sciences 53: 160-163.

14. Wilson CA, Novak JT (2009) Hydrolysis of macromolecular components of primary and secondary wastewater sludge by thermal hydrolytic pretreatment. Water research 43: 4489-4498.

15. Barlindhaug J, Odegaard H (1996) Thermal hydrolysis for the production of carbon source for denitrification. Water Sci Technol 34 : 371–378.

16. Urbain V, Block JC, Manem J (1993) Bioflocculation in activated sludge an analytical approach. Water Res 27: 829–838.

17. Mok WSL, Antal MJ (1992) Uncatalyzedsolvolysis of whole biomass hemicellulose by hot-compressed water. Ind EngChem Res 31: 1157–1161.

18. Brooks RB (1970) Heat treatment of sewage sludge. Water Pollut Control 69: 221–231.

19. Dewil R, Baeyens J, Appels L (2007) Enhancing the use of waste activated sludgeas bio-fuel through selectively reducing its heavy metal content. J Hazard Mater. 144: 703–707.

20. Veeken AH, Hamelers HV (1999) Removal of heavy metals from sewage sludge by extraction with organic acids. Water Sci Technol 40: 129-136.

21. Namioka T, MorohashiY, Yoshikawa K (2011) Mechanisms of malodor reduction in dewatered sewage sludge by means of the hydrothermal torrefaction. Journal of Environment and Engineering 6: 119-130.

22. REN L (2006) Impact of hydrothermal process on the nutrient ingredients of restaurant garbage. Journal of Environmental Sciences 18: 1012-1019.

23. Toor SS, Rosendahl L, Rudolf A (2011) Hydrothermal liquefaction of biomass A review of subcritical water technologies. Energy 36: 2328-2342.

24. Bougrier C, Delgen JP, Carrere H (2008) Effects of thermal treatments on five different waste activated sludge samples solubilisation, physical properties and anaerobic digestion. Chemical Engineering Journal 139: 236-244.

25. Xu C, Lancaster J (2008) Conversion of secondary pulp/paper sludge powder to liquid oil products for energy recovery by direct liquefaction in hot-compressed water. Water Research 42: 1571-1582.

5

State of The Art Strategies for Successful ANAMMOX Startup and Development: A Review

Suneethi S[1]*, Sri Shalini S[1] and Kurian Joseph[2]
Research Scholar, Centre for Environmental Studies, Anna University, India

Abstract

ANAMMOX (Anaerobic ammonium oxidation) bacteria, leading proponent of autotrophic ammonia removal, requires dedicated enrichment and cultivation techniques due to its slow growth rate and low biomass yield. Sensitivity to inhibitory concentrations of NO_2^--N, O_2 etc., often present in industrial effluents makes ANAMMOX startup difficult to achieve. In this paper, significant parameters for ANAMMOX startup and development such as source of seed, type of reactors used, one and two stage ANAMMOX process, operational strategies and experimental conditions promoting long term enrichment, cultivation and quantitative analysis with biomass retention are discoursed in detail. Key chemical and molecular signatures including NO_2^--N, Alkalinity, O_2, Polymerase chain reaction (PCR), Fluorescence in situ hybridization (FISH), Scanning electron microscopy (SEM), inhibitors and affinity factors for ANAMMOX activity are deliberated. Summary of state of the art on ANAMMOX enrichment, recommendations for future research with specific contributions of this paper to scientific community is brought out in conclusion.

Keywords: ANAMMOX; Enrichment; Startup; Growth rate; NH_4^+-N removal

Introduction

Anaerobic ammonium oxidizers (ANAMMOX), the often overlooked key player in Nitrogen cycle, are involved in highly complex autotrophic microbial interactions to remove NH_4^+-N from nitrogen rich wastewaters. In this process NH_4^+-N is directly converted into dinitrogen gas using NO_2^--N as electron acceptor under anaerobic conditions. The ANAMMOX reaction was first observed in an anoxic denitrifying fluidized bed reactor (FBR) in Gist-Brocades, Delft, Netherlands, treating the effluent from a methanogenic reactor [1]. An ammonium loading rate of $0.4gNH_4^+$-N/L/d was removed in this system. The stoichiometry of the ANAMMOX reaction is represented by the following Equation (1).

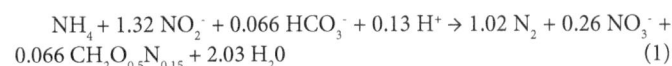

$$NH_4 + 1.32\ NO_2^- + 0.066\ HCO_3^- + 0.13\ H^+ \rightarrow 1.02\ N_2 + 0.26\ NO_3^- + 0.066\ CH_2O_{0.5}N_{0.15} + 2.03\ H_2O \tag{1}$$

The ANAMMOX reaction is carried out by members of deeply branched *Planctomycetes*, such as '*Candidatus brocadia anammoxidans, Candidatus kuenenia stuttgartiensis, Candidatus scalindua, Candidatus anammoxoglobus* and *Candidatus jettenia*' which makes use of NH_4^+-N as the electron donor (energy source) and NO_2^--N as the electron acceptor. These autotrophs utilize dissolved CO_2 or HCO_3^- for cell biosynthesis [2-5].

Due to the notorious slow growth rate and low biomass yield (0.13g dry weight/g NH_4^+-N oxidized) of ANAMMOX bacteria, dedicated enrichment and cultivation techniques are required [6-9]. The average doubling time of ANAMMOX bacteria is reported to be 14d [10]. Doubling times as low as 5.3 d and 8.9d were achieved by Park et al. (2010) [11], whilst 11d and 14d were reported by Strous et al. (1998) [12], Li et al. (2009) [9], Strous et al. (2006) [10] from ANAMMOX enriched laboratory scale reactors. The doubling time from the full-scale studies by Van der star et al. (2007) [13] were about 8.3 to 11 d, while Third et al. (2005) [8] achieved 12d. The main bottlenecks of the ANAMMOX process could be ascribed to the low yield (0.14gVSS/g NH_4^+-N) and slow growth rate of ANAMMOX bacteria (0.003 h^{-1}; $0.072d^{-1}$ at 32°C) and difficulty in isolating ANAMMOX bacteria in pure cultures [7,12]. Besides these rate limiting factors, presence of NO_2^--N, oxygen at critical concentrations can reversibly/irreversibly inhibit ANAMMOX process [1,12,14].

The low growth rate combined with the sensitivity of the microorganisms to inhibitory concentrations of some compounds that are often present in industrial effluents makes the startup of the ANAMMOX process very difficult to achieve. The startup period can be minimized by selecting an appropriate seed biomass and running a suitable reactor configuration that promotes long term enrichment, cultivation and quantitative analysis with biomass retention. Proper control and monitoring of key parameters with no compromise on NH_4^+-N removal rates is essential. Some of the important parameters that play a significant role during ANAMMOX enrichment are the source of seed, type of reactors used, operational strategy and experimental conditions. In this paper, the different ANAMMOX enrichment procedure from seeds of different origin is evaluated and the key parameters (both chemical and molecular) that enable both quick growth of ANAMMOX bacteria and affect its growth rate are identified.

Key Parameters for ANAMMOX Enrichment

Source of seed

ANAMMOX bacteria could be developed and applied to startup new reactors from obtaining enriched seed from already operational ANAMMOX reactors which is already containing significant composition of ANAMMOX populations. The long startup period could be reduced and a number of ANAMMOX operational reactors could be initiated simultaneously. Full-scale ANAMMOX reactors are

*Corresponding author: Suneethi S, Research Scholar, Centre for Environmental Studies, Anna University, Chennai- 600 025, India
E-mail: sundar.suneethi@gmail.com

operational in Netherlands, Austria, and Germany and in other parts of Europe and USA. ANAMMOX process application can be developed by giving attention to application of enriched seed biomass for startup, since ANAMMOX process consumed long startup period (> 200d) [15]. But as of date, it has not been possible to achieve a pure culture of any ANAMMOX bacteria. It could be attributed to the fact that the activities or presence of other organisms such as AOBs and NOBs are required for them to grow in mixed cultures. ANAMMOX process could also be initiated from non enriched seeds that are availed locally when enriched seed is unavailable and/or exotic enriched seeds could be deemed unsuitable for local conditions.

Inoculation of enriched ANAMMOX seed can accelerate the startup operation within 2 months, while inoculation from non enriched locally available seed may take longer than 4 months [6,16]. Date et al. (2009) [16] carried out ANAMMOX enrichment using sewage, digester and nitrifying sludge as seed in non-woven fabric carrier for immobilization. Besides the simultaneous removal of NH_4^+-N and NO_2^--N and generation of NO_3^--N indicative of ANAMMOX activity, it was also reported that NH_4^+-$N_{removed}$: NO_2^--$N_{removed}$: NO_3^--$N_{produced}$ was 1:1.02:0.23, 1:1.19:0.25 and 1:1.31:0.33 for sewage, digester and nitrifying sludge respectively. The evaluation of efficacy of ANAMMOX process from both enriched and non enriched seeds are depicted in Table 1.

Experimental condition

Gas sparging: Sparging the feed tank and ANAMMOX reactor with inert gases such as Ar/CO_2 or N_2/CO_2 with 95/5% composition enables to maintain strict anaerobic condition for ANAMMOX enrichment. Oxygen leakage that could occur from recirculation operations would require strict deoxygenation steps by sparging. In any case, the gas mixture should contain about 5% CO_2 in order to buffer the medium. Although there are reports of ANAMMOX enrichment under aerobic conditions [17-19], anaerobic conditions at least during the first periods of enrichment is advisable due to the high susceptibility of ANAMMOX bacteria to even microaerobic conditions [17].

Light : Light is considered to be inhibitive to ANAMMOX bacteria since it leads to the undesirable growth of phototrophic algal growth [20]. Studies were conducted by Uyanik et al. (2011) [20] which compared the operation of ANAMMOX reactors with and without control over light penetration. The ANAMMOX reactors that were operated with control over light penetration achieved ANAMMOX enrichment within 50d of startup compared to 100d taken by ANAMMOX reactors that was exposed to light. The reactors that were covered and kept in dark room favored no development of photosynthetic O_2 production and algal development.

Addition of NO_3^--N, N_2H_4 and NH_2OH: During ANAMMOX startup from non enriched seeds, NO_3^--N concentration of about (10–50mg/L) is usually added [20]. Under anaerobic condition, the organic load found in the seed biomass and those produced from initial decomposition could drive the process towards denitrification against the favor of ANAMMOX bacteria. This is due to the autotrophic nature of ANAMMOX bacteria. The addition of NO_3^--N into the reactors was used to feed heterotrophic denitrification bacteria in order for them to consume the organic compounds. It has been reported that addition of NO_3^--N to the feed was effective on removal of organic load due to bacterial decomposition at the start of ANAMMOX enrichment [20]. In the startup phase, NO_3^--N is added to establish oxidative conditions, preventing the growth of other anaerobically respiring microorganisms such as sulfate reducers and methanogens. In response to the NO_2^--N consumption, its concentration in the feed is gradually increased [21]. The idea of supplying NO_3^--N was that, since the inoculum contained an active denitrifying population, NO_2^--N would be produced in the culture in low amounts by NO_3^--N reduction, using remaining organic compounds in the inoculum. The low amount of NO_2^--N production should be sufficient for any ANAMMOX bacteria present in the sludge. After the ANAMMOX reaction is clearly observed, NO_3^--N addition should be discontinued because of both production of nitrate by ANAMMOX reaction and lack of interest in enrichment of denitrification bacteria.

From the works of Third et al. [8], addition of 0.1mM final reactor concentration N_2H_4/NH_2OH to the ANAMMOX reactor was also reported to kick start the ANAMMOX process. The spiking of N_2H_4 and NH_2OH was attempted to take advantage of the cyclic nature of the ANAMMOX mechanism, since cells needed to invest reducing power to start their catabolism by producing NH_2OH from nitrite. This initial energy barrier can be overcome by the direct addition of NH_2OH or N_2H_4 [8,12].

Shear stress: Thorough mixing condition inside the ANAMMOX reactor is important so as to avoid accumulation of NO_2^--N that could inhibit ANAMMOX activity. The speed of stirrer which is used to mix contents in ANAMMOX reactors influences the stability of ANAMMOX

Sl No	Origin of seed	Highest Nitrogen Loading rate (kg N/m³/d)	Highest Nitrogen Removal Rate (kg N/m³/d)	Specific Nitrogen Removal Rate (g N/g VSS/d)	Reference
	Enriched ANAMMOX seed	0.28	0.08	0.13	Padin et al. (2009)
		1.2	0.75	0.18	Van Dongen et al. (2001)
		0.662	0.582	-	Guven et al. (2004)
		2.0	1.78	1.15	Dapena-Mora et al. (2004a)
		10.7	8.9	-	Sliekers et al. (2003)
	Activated sludge	2.6	2.4	0.30	Fux et al. (2002)
		1.6	1.57	0.92	Lopez et al. (2008)
		0.08	0.072	0.35	Wang et al. (2009)
	Anaerobic sludge digester	0.231	0.216	0.43	Bagchi et al. (2010)
	Anaerobic/Aerobic granular sludge	2.5	2.5	-	Yang et al. (2006)
	Anaerobic granular sludge	125-137.1	74.3-76.7	-	Tang et al. (2011)
	Full-scale UASB	0.015	0.009	0.64	Tran et al. (2006)
	Denitrifying reactor	58.5	26	1.6	Tsushima et al. (2007)
		1	1.8	180	Strous et al. (1997)
		1.2	1.5	150	

Table 1: Nitrogen removal officiency achieved by ANAMMOX process from both enriched and non enriched seeds.

granules [22]. The stirring speed in the range of 60–250rpm was tested by Arrojo et al. [22]. It was reported that upto 180rpm (input power between 0.003 and $0.09kW/m^3$) there was no significant effect on the performance of ANAMMOX process. ANAMMOX activity was reduced by 40% when the stirring speed was increased to 250rpm (input power $0.23kW/m^3$). The biomass retention worsened due to the breakage of the granules and floatation caused by nitrite accumulation. This caused a loss of the system efficiency due to a combination of cellular lysis and granules breakage. The increase of stirrer speed which involves an increase of shear stress was found to provoke the changes on the properties of biomass aggregates and on the system efficiency [22]. Due to rise in shear stress the rupture of the granules or even the detachment of weak patches of biomass from the surface of the granules could result in increase of suspended solids washout of the reactor. This is due to the biomass flotation that is closely related to the NO_2^--N accumulation in the system [15].

Reactor Configurations

High NH_4^+-N removal rates could be obtained using ANAMMOX process in two ways: two reactors in series, with a partial nitrification reactor as a first step, and a separate unit for the anaerobic oxidation of NH_4^+-N as a second step. With this configuration, the two biological processes can be controlled separately [23]. The second option was to use biofilm systems where classical nitrification is developed by the ammonium oxidizers in the outer aerobic layers, and anaerobic oxidation takes place in the deeper zones of the biofilm [23]. Application of ANAMMOX process or coupling partial nitrification with ANAMMOX seemed promising. It could result in 60% savings in O_2 generation, 100% savings of external carbon source addition, less sludge production and CO_2 emission, with a total reduction in treatment cost by 90% [14,24-26].

One stage ANAMMOX process

Direct application of ANAMMOX process was adopted by Xu et al. [27] to treat NH_4^+-N rich leachate using Sequencing batch biofilm reactors (SBBR). The system was started up in 58d and stabilized in 33d, with DO of 1.2–1.4 mg/L, under alternate periods of aeration and anoxic condition. The leachate was used by spiking it with NH_4Cl to about 450 mg/L prior to feeding as influent to SBBR. The organic load was in the range of 1876 ± 547mg/L of COD and 1048 ± 436mg/L of BOD_5. NLR was optimized to 300mg/L/d, with pH around 7.3 to 7.8 without addition of alkali or acid. It was proposed that the repeated alteration between aeration and anoxic period made the acidity generated in the aeration phase neutralized in time by the alkalinity produced in the anoxic phase. The ratio of NH_4^+-N/ NO_2^--N was in the range of 1.058 to 1.074 in the aeration phase, and 0.558 to 0.776 in the anoxic phase, as compared to the theoretical value of 0.758 in ANAMMOX reaction [12]. It was proven that anoxic condition favored ANAMMOX activity when weighed against oxic condition.

While Guo and Qi [28] treated aged landfill leachate in an Upflow anaerobic sludge blanket (UASB) ANAMMOX bioreactor (HRT 24h) and achieved about 80% total nitrogen removal efficiency from influent containing 900mg TN/L and 88% NH_4^+-N removal from an influent of 350mg NH_4^+-N/L. During the study period for >200d, the mean COD removal was 24% from an influent of 1000mg/L. While alkalinity concentrations of both the influent and effluent during the steady phase of ANAMMOX activity were 1g/L and pH of influent and effluent were 8.3. This study indicated that alkalinity and pH could also be used to monitor the ANAMMOX activity. The ratio of NO_2^--N/ NH_4^+-N was in the range of 0.96 to 1.49, as compared to the stoichiometric value of 1.24 [12].

Application of ANAMMOX process to remove NH_4^+-N (1545mg/L) from wastewaters was performed by Sliekers et al. [29] in a gas lift reactor. The highest NLR reached during the ANAMMOX process was $10.3kgN/m^3/d$ with Ar/CO_2 sparged at flow of 200 mL/min. The ANAMMOX process was maintained in a gas lift reactor by maintaining anoxic condition during ANAMMOX process. NRR of 8.9 $kgN/m^3/d$ was achieved for the ANAMMOX process.

For one stage ANAMMOX reactors, besides the NO_2^--N limitation, O_2 consumption by AOB plays a role in process design. O_2 transfer was indeed indicated as the limiting process for a lab-scale air-lift [29] and for a lab scale moving bed reactor [8]. The O_2 limitation could have originated from the slow diffusion into the biofilm or from the gas–liquid transfer. It was reported by Van der Star et al. [13] that O_2 penetration is limiting the rotating disk contactor and the moving bed reactor with conversions of 2.5 and 1.2 kg $N/m^3/d$, respectively. For the other reactors, gas–liquid oxygen transfer is potentially limiting as well. With a superficial gas velocity of 0.025 m/s the oxygen transfer is approximately 15 kg $O_2/m^3/d$ (equivalent to a conversion of 8 kg-N/m^3/d [13,29].

Two stage ANAMMOX process

Application of coupling partial nitrification with ANAMMOX process was adopted by Liang and Liu [30] while treating landfill leachate (NH_4^+-N 1500 to 2500mg/L). An integrated Partial nitration – ANAMMOX reactor – Underground soil infiltration system was applied. ANAMMOX operation was performed by upflow fixed bed biofilm reactor and achieved 67% NH_4^+-N and 77% NO_2^--N removal within 97d. The effluent of the partial nitritation process yielded a suitable influent for ANAMMOX process, by yielding 50% partial conversion of NH_4^+-N to NO_2^--N (ratio 1:1.3) favoring anaerobic ammonium oxidation. Nearly 60% of NH_4^+-N removal was achieved by the end of ANAMMOX process, and 97% removal was obtained at the end of the combined treatment train. From the initial COD of 1100-2500mg/L, nearly 89% COD was removed with the final effluent showing 30-250mg/L, where nearly 32% COD was removed by ANAMMOX process. The main limitation of the process could be ascribed to the low yield and slow growth rate of ANAMMOX bacteria resulting in slow removal of NO_3^--N (half the time taken for aerobic nitrification) [7,8,12].

The investigation of the aquatic humic substances (AHS) degradation by ANAMMOX process was conducted by Liang et al. [31] where the initial partial nitritation reactor was run for 166d continuously using raw leachate, with NH_4^+-N of 1430 to 2720mg/L and COD of 1170 to 2600mg/L. Upon removal of VFA and acquiring the proper mixture of NO_2^--N to NH_4^+-N ratio, this effluent with NH_4^+-N of 506 to 885mg/L and COD 303 to 954mg/L was further treated in ANAMMOX reactor. The pretreatment in partial nitritation enabled removal of biodegradable organics from raw leachate, resulting in higher content of AHS in the feed to ANAMMOX reactor (228mg/L). The effluent from ANAMMOX reactor is reduced to 91mg/L. The Dissolved organic carbon (DOC) was also reduced from 288 to 136mg/L in the ANAMMOX reactor.

Furukawa et al. [32] successfully demonstrated partial nitritation using nitrifying activated sludge entrapped in a polyethylene glycol (PEG) gel carrier, as a pretreatment to ANAMMOX process for treating supernatant of anaerobically digested sludge. The partial nitritation reactor was operated at a NLR of 3.0 kgN/m³/d and suspended solids (SS) concentration of 2000 to 3000 mg/L that resulted in effluent with a NO_2^--N/NH_4^+-N ratio between 1.0 and 1.4 that was suitable for ANAMMOX process. The ANAMMOX reactor achieved TN removal rates of 4.0 kgN/m³/d under an applied nitrogen loading rate of 5.3 kgN/m³/d. The authors reported that entrapping ANAMMOX bacteria in the gel carrier prevented inhibition from influent COD and SS concentration. The mean C/N ratio was 0.84 g TOC/g NH_4^+-N with no observed autotrophic ammonium oxidation inhibition.

Application of two stage partial nitration and ANAMMOX process was also adopted by Fux et al. [33] for treating supernatant of anaerobically digested sludge. The partial nitration was carried out in a continuous stirred tank reactor (CSTR) of working volume 2L resulting in 58% conversion of NH_4^+-N to NO_2^--N. The ANAMMOX process was conducted in a Sequencing batch reactor (SBR) of working volume 1.6L with nitrogen removal rate of 2.4 kgN/m³/d, which constituted nearly 90% of the influent to ANAMMOX reactor. The nitrogen load to the ANAMMOX reactor was gradually increased based on the NO_2^--N concentration in the effluent, while the nitrogen removal was dependent on the ratio of NO_2^--N/NH_4^+-N in the influent.

Gali et al. [34] carried out studies to produce the influent for ANAMMOX process by partial nitrification process in SBR. The influent was the supernatant of anaerobically digested sludge of NH_4^+-N concentration of 700 to 800mg N/L with low HCO_3^-/ NH_4^+-N ratio favoring partial nitrification. Complete nitrification required 2mol of HCO_3^- for each mol of NH_4^+-N, whereas the HCO_3^-/ NH_4^+-N ratio was 1 for the influent, which bodes well for partial nitrification. The SBR was operated in 4 stages such as aerobic filling (5min), mixing (210min), settling (20min) and drawing (5min). By the end of 5 months the partial nitrification process yielded 50% NH_4^+-N removal efficiency with specific ammonium uptake rate of 42mg NH_4^+-N/g VSS /h.

Partial nitration–ANAMMOX process was applied for treating livestock manure by Yamamoto et al. [35]. The partial nitration reactor was maintained for 32d under NLR of 1.6kgN/m³/d resulted in mean conversion efficiency of 51%. The partial nitration reactor achieved 1.65kgN/m³/d of maximum NO_2^-- N production rate under NLR of 2.58 kgN/m³/d. The ANAMMOX process was performed in a UASB reactor that achieved nearly 2.0 kgN/m³/d under a NLR of 2.2 kgN/ m³/d. Another partial nitration–ANAMMOX study was conducted by Yamamoto et al. [35] for treating livestock manure effluent of NH_4^+-N concentration in the range of 2000 to 4000 mg/L. The partial nitration reactor was maintained at NLR of 1.0kgN/m³/d for over 4 months which was followed by the ANAMMOX reactor. The conversion efficiency of NH_4^+-N to NO_2^--N and NH_4^+-N to NO_3^--N were estimated to be 58% and <5% respectively. The ANAMMOX process yielded consistent nitrogen removal rate of 0.22 kgN/m³/d.

Leachate pretreatment prior to ANAMMOX reactor using SBR was studied by Ganigue et al. [36]. By adopting a step feed strategy, with alteration between aerobic and anoxic condition, leachate of 5000 mg NH_4^+-N/L was partially nitrified and obtained in the required ratio of 1:1.32. The effluent of 1500 to 2000 mg NH_4^+-N/L and 2000 to 3000 mg NO_2^--N/L was obtained. The reactor was operated for 450d with NLR around 0.85 kgN/m³/d to 1.2 kgN/m³/d. The suitable mixture of NH_4^+-N and NO_2^--N obtained was used as feed for ANAMMOX reactor. The presence of low concentrations of biodegradable organic matter (4357 ± 692 mg COD/L) in the leachate was used for denitrification. Supplementation of HCO_3^- externally to favor partial nitration was adopted, thereby increasing NH_4^+-N conversion and NO_2^--N formation.

Type of Reactor Operation

It is important to comprehend the physical, chemical and the biological needs in the form of symbiotic relationships with key partners (AOBs, NOBs and ANAMMOX) in the mixed microbial community [37]. Hence the widespread strategy to obtain dense enrichments is by using batch cultures and different reactor types, both conventional and advanced configurations to facilitate the growth and development of ANAMMOX bacteria.

The slow growing ANAMMOX organism cannot be cultivated using conventional microbiological techniques [6]. Biological reactors are proven best for cultivating ANAMMOX bacteria [4,6,12,15,26,38-41]. Batch cultures were applied for enrichment purposes to confirm a wide array of samples from different seed origins, to reduce the startup duration needed to reduce trial and error in biological reactors and to decrease the number of reactors for simultaneous enrichment [37]. To apply the ANAMMOX process, the choice of reactor type is very important. It should be suited for long term enrichment, cultivation and quantitative analysis [12]. Full-scale application of ANAMMOX process can be achieved by choosing the appropriate seed sludge and an adequate reactor configuration for ANAMMOX enrichment [42].

ANAMMOX cultivation and enrichment in batch cultures

Applying enrichment in batch studies will enable obtaining successful inocula for starting up ANAMMOX biological reactors. The physiological pH and temperature ranges to be maintained are 6.7 to 8.3 and 20°C to 43°C [14]. Well mixing is vital to maintain redox potential in the denitrification zone and sulfide formation has to be avoided. Sludge retention is important owing to the slow growth rates of the bacteria [6-8]. Replenishment of both Nitrogen supplements (NH_4^+-N and NO_2^--N) and nutrients (enrichment medium) need to be performed to avoid substrate limitation.

In the batch study of ANAMMOX enrichment carried out by Sanchez – Melsio et al. [37], seed biomass were taken from natural environments and treatment plants, which included sediment samples from marine, brackish and freshwater system, seeds from WWTP digesters and from anoxic SBR systems. Enrichments were carried out in Erlenmeyer flasks, with 100mL sample and 300mL enrichment medium [43]. The enrichment flasks contained essential nutrients along with the presence of trace metals. NH_4^+-N and NO_2^--N were added as supplements step wise in order to increase the concentrations from 20mg/L. The enrichments were maintained in shaking incubator in the dark at 37°C, with pH in the range of 6.7 to 8.3. The presence and activity of ANAMMOX was confirmed through the changes in nitrogen compound consumption and by application of molecular analyses. Kawagoshi et al. [44] performed ANAMMOX enrichment from marine sediments. *Candidatus Scalindua wagneri* was observed in the enrichment reactor that was operated at NLR of 0.4kgN/m³/d. Mulder et al. [1] conducted batch studies with seed originating from FBR, with the NH_4^+-N conversion capacity of 2.7mg NH_4^+-N/g VSS/d. ANAMMOX activity was proved based on nitrogen and redox balances. It was concluded that NH_4^+-N conversion was NO_3^--N dependent [1].

While Bagchi et al. [45] carried out ANAMMOX enrichment from a non enriched locally procured seed in a completely stirred tank reactor (CSTR) with HRT of 2 d. Anaerobic digester sludge from a local STP with TSS and VSS of 38500 mg/L and 29800 mg/L was used as seed. The biomass was initially acclimatized in batch cultures incubated with synthetic medium and amoxicillin (250 mg/L). After 65d of incubation, the sludge with VSS of 78000mg/L was used as seed for

CSTR. The synthetic wastewater of total nitrogen concentration (NH_4^+-N+NO_2^--N) of 115 ± 5 mg/L was fed into CSTR. The NLR was increased as the ANAMMOX activity developed from 0.057 ± 0.003 kgN/m³/d to 0.225 ± 0.014 kgN/m³/d. The presence of AOB was identified on the surface and on the exterior layers of granules and the NOB and ANAMMOX bacteria inside. The NO_2^--N/NH_4^+-N was adjusted to enable ANAMMOX activity sustenance even in the presence of mixed community. The highest NRR obtained was 0.216 kgN/m³/d and highest specific NRR was 0.434g N/g VSS/ d. It was also reported that the ANAMMOX activity in this system was not inhibited by NH_3 toxicity and the pH variations.

Batch tests were conducted by Chen et al. (2010) [46] in 500 mL flasks, for ANAMMOX enrichment, with temperature (35°C), pH (7–8) control and continuous mixing (150 rpm). The anoxic condition was maintained by flushing with a gas mixture of 95% Ar and 5% CO_2. The ANAMMOX enrichment was monitored by analyzing the initial and final substrate concentration for mass balance calculation. The initial NH_4^+-N and NO_2^--N concentrations were 60mg/L (each) and final substrate concentrations were 17 mg/L for NH_4^+-N and 0 mg/L of NO_2^--N with effluent NO_3^--N of 9.7 mg/L. The maximum Specific ANAMMOX activity (SAA) was determined to be 1.8 gN/g VSS/d. Enrichment of ANAMMOX using immobilized microbial consortia was carried out by Pathak et al. (2007) [47]. It was performed at low nitrogen level (<3 mg/L) at low temperature (20°C) in a laboratory scale upflow anoxic reactors. Nitrogen removal efficacy >92 % with the total nitrogen in the effluent <0.2 mg/L was achieved upon operating for 300d.

ANAMMOX process is not inhibited by NH_4^+-N or the byproduct NO_3^--N up to 1g of N/L [14]. The ANAMMOX process is completely inhibited by NO_2^--N>0.1g of N/L. The NO_2^--N inhibition could be overcome by addition of trace amounts of either of the ANAMMOX intermediates 1.4mg of N/L of N_2H_4 or 0.7 mg of N/L of NH_2OH, and enable restoration of the ANAMMOX activity [14]. NO_2^--N toxicity due to increase in substrate concentrations is addressed by recirculation of the effluent, thereby protecting the ANAMMOX sludge [46,48]. High NO_2^--N concentrations above 70mg/L for 'Candidatus Brocadia Anammoxidans' and above 180mg/L 'Candidatus Kuenenia stuttgartiensis' for prolonged periods is harmful to the process, so is O_2 concentration higher than 0.06 mg/L [24]. Complete irreversible inhibition of this process when CH_3OH concentration is ≥ 40 mg/L [23,49]. Since there was an existing competition for dominance between ANAMMOX bacteria, denitrifiers and AOBs, it might be a problem while treating wastewaters with high organic (Carbon) and nitrogen content [40].

ANAMMOX cultivation and enrichment in continuous systems

The development and growth of ANAMMOX bacteria in a number of conventional biological reactors such as FBR, SBR, UASB etc., has been undertaken to evaluate the apt configuration for ANAMMOX enrichment and application in treating NH_4^+-N rich wastewaters.

Fluidized Bed Reactor (FBR): The ANAMMOX activity was first reported in the denitrifying FBR treating effluent from a methanogenic reactor by Mulder et al. [1]. The FBR of 23L volume was operated at 36°C and pH 7 with HRT 4.2h. The anoxic liquid was recirculated to keep the bed fluidized of 255L/h flow with superficial liquid velocity 30 to 34m/h. Sand particles were used as biocarrier with biofilm concentration of 150 to 300 mg VSS/g carrier (14g VSS/L). During the initial 420d NH_4^+-N concentrations was similar to effluent concentrations (125 mg/L).

From 420–560d the NH_4^+-N concentration started to decrease (from 50 mg/L to <5 mg/L). Highest NH_4^+-N removal rate of 0.4 kgN/m³/d was reported. Gas composition analysis (v/v) performed during the ANAMMOX process showed 68 to 72% of N_2, 15 to 18% of CH_4, 13 to 18% of CO_2, while N_2O was below the detection limit of 65 ppm.

Sequencing Batch Reactor (SBR): SBR was considered for ANAMMOX enrichment as it held the following strong points such as simplicity, efficient biomass retention, homogenous distribution of substrates, products and biomass aggregates in the reactor, reliable operation for long period (>1 year), stable conditions under substrate limiting environment [4,12]. Using SBR, ANAMMOX process was successfully started in 4 months by Chamchoi and Nitisoravut [6] from sludge obtained from UASB, activated sludge and anaerobic digester. The initial biomass concentration in MLSS was 1500mg/L. The synthetic medium composition used was according to Van de Graaf et al. [43], with regular addition of NH_4^+-N and NO_2^--N (1:1.5) at neutral pH (7.7 to 8.4). The ratio of NH_4^+-N to NO_2^--N in the feed was higher than the stoichiometric ratio of 1:1.32 [12]. The reactors were operated under anaerobic conditions, with reaction (5 to 7h), settling (0.5h) and discharge periods (0.25h). The reactor was kept at anaerobic condition with Ar/CO_2 (95/ 5%) gas flushing. During the first 5 to 7 weeks after seeding, the effluent NH_4^+-N concentrations (75mg/L) were higher than influent (30mg/L). Significant removal of NH_4^+-N and NO_2^--N (80 and 100%) along with small amounts of NO_3^--N produced was observed in the initial three months of operation, indicative of ANAMMOX process [4]. Complete NO_2^--N removal was obtained in four months. The ratios obtained were 1:1.5, 1:1.53 and 1:1.5 for the three reactors containing seeds from UASB, activated sludge and anaerobic digester.

According to Liao et al. [42], granulation of ANAMMOX biomass is favored upon using SBR. Anaerobic methanogenic granular sludge was used (MLSS 72800mg/L and MLVSS 63400mg/L) to startup SBR (7L volume). About 1L of anaerobic granular sludge and 5L of enrichment medium [43] were added to SBR with N_2 gas sparging for fluidization of biomass with liquid and to maintain anaerobiosis of the reactor. The ANAMMOX activity showed improvement when the VSS concentrations reduced from 8.913 to 4.554mg/L and VSS/SS ratio also reduced from 94 to 84%.

Direct application of ANAMMOX process was adopted by Xu et al. [27] to treat NH_4^+-N rich leachate using Sequencing batch biofilm reactors (SBBR). The system was started up in 58d and stabilized in 33d, with DO level of 1.2–1.4mg/L, with alternate periods of aeration and anoxic condition. The leachate was used by spiking it with NH_4Cl and regulated to about 450mg/L with distilled water prior to feeding as influent to SBBR. The organic load was in the range of 1876±547mg/L of COD and 1048±436mg/L of BOD_5. Nitrogen loading rate (NLR) was optimized to 300mg/L/d, with pH around 7.3 to 7.8 without addition of alkali or acid. It was proposed that the repeated alteration between aeration and anoxic period made the acidity generated in the aeration phase neutralized in time by the alkalinity produced in the anoxic phase. The ratio of NH_4^+-N /NO_2^--N was in the range of 1.058–1.074 in the aeration phase, and 0.558–0.776 in the anoxic phase, as compared to the theoretical value of 0.758 in ANAMMOX reaction [12]. It was proven that anoxic condition favored ANAMMOX activity when weighed against oxic condition.

Rotating biological contactor (RBC): Liu et al. [39] developed ANAMMOX in rotating biological contactor (RBC) within 100d by increasing the NLR in the influent feed with gradually shortened HRT. Influent concentration of NH_4^+-N and NO_2^--N was raised from 100 to 350mg/L, with initial HRT of 1d and 99% and 97% of NH_4^+-N and NO_2

-N removal efficiency was observed. Upon varying the HRT during optimization, after 100d of operation, mean NH_4^+-N removal of 99%, NO_2^--N removal 100%, TN removal 84% was achieved at 350mg/L of NH_4^+-N and NO_2^--N with HRT 4h. Nearly 70% of nitrogen was removed in a nitrifying RBC for NH_4^+-N rich leachate (100 to 400mg/L) from a hazardous waste landfill [26]. With a pretreated inflow of leachate from RBC at 30-170m³/d, the system was designed to nitrify 30kg NH_4^+-N/d (3.7gN/m²/d). The nitrogen elimination without organic carbon was attributed to ANAMMOX process, where the NO_2^--N produced in the aerobic biofilm layer by *Nitrosomonas* near the surface, is diffused and reduced by anaerobic ammonium oxidation in the lower anoxic layer by the ANAMMOX bacteria. Thus spontaneously both aerobic and anaerobic ammonia oxidation was developed on the biofilm [50].

Similarly in another study on RBC [51] NH_4^+-N conversion efficiency of 79%, TN removal efficacy of 70% and COD removal of 94% was obtained at 0.69kgN/m³/d and 0.34kg/m³/d of nitrogen and COD loading rates, respectively. The ANAMMOX reactors used were capably mixed with mechanical stirrer and uncommonly with recirculation of the produced N_2 gas as well. Anoxic conditions were maintained by N_2/Ar/CO_2 sparging [4,6,15]. In these biofilm studies, long term establishment of AOB and ANAMMOX communities occurred with limited scope for NOBs [50].

Gas Lift reactor: Dapena-Mora et al. [15] used both SBR and Gas lift reactor under anaerobic conditions with steady increment in NH_4^+-N and NO_2^--N under maximum nitrogen load that could be treated by ANAMMOX. The SBR was operated in 330 min, 20 min, 10 min of reaction, settling and discharge phases. The mean ratio of NH_4^+-N utilization to generation of NO_2^--N was 1:0.26 and 1:0.2 in the gas lift and SBR as compared to 1:0.04 by Chamchoi and Nitisoravut [6]. Sliekers et al. [29] also used Gas lift reactor for ANAMMOX and CANON processes. The NH_4^+-N removal by ANAMMOX and CANON process attained were 8.9 and 1.5 kgN/m³/d respectively. The reactor was initiated from an enriched seed from already operational ANAMMOX SBR. The gas lift reactor was operated as ANAMMOX mode and then followed by CANON mode. In the ANAMMOX mode, the NLR was 10.7 kgN/m³/d at 6.7 h HRT that achieved nearly 80% NH_4^+-N removal. During the CANON mode of operation operated at 10h HRT, feed and effluent NH_4^+-N concentrations were 1545 and 899 mg/L, respectively.

Upflow Anaerobic Sludge Blanket Reactor (UASB): Besides SBR, Gas lift reactor and RBC, ANAMMOX bacteria was also developed using UASB reactor too [38]. Granular sludge from anaerobic digester (18.6g VSS/L) was used as potential seed source. Recycling of sludge at 3Q ratios was applied to accelerate mixing and induce dilution of influent thereby reducing the NO_2^--N toxicity to ANAMMOX bacteria. Onset of NH_4^+-N reduction was detected after 200d of sludge cultivation. The effluent NH_4^+-N concentration exhibited steadfast reduction with NO_2^--N consumption and low production of NO_3^--N as byproduct. The color of the sludge turned from black to reddish brown, which is a well-documented characteristic of ANAMMOX activity [6,7,38,41]. After 11 months of operation, the stoichiometric ratio of the ANAMMOX reaction was maintained at 1:1.32 (NH_4^+-N: NO_2^--N) with 60% of NH_4^+-N conversion.

In another study by Ahn et al. [40], ANAMMOX process was developed to treat piggery wastewater (56g COD/L and 5g TN/L), using UASB reactor at 35°C. The reactor was operated with 5d HRT, with NLR of 0.43 kg NH_4^+-N/ m³/ d. To enable ANAMMOX process, Nitrogen loading rate was 0.36 kg NH_4^+-N/m³/d and 0.5 kg NO_2^--N/m³/d, using granular sludge of 18.6 g VS/L as seed biomass. The UASB was operated

in semi continuous mode. At the end of the study, nitrogen conversion of 0.6 to 0.7g TN/L/d was achieved and the color of the biomass at the bottom of the reactor changed from dark gray to reddish brown, along with granulation of the ANAMMOX sludge in the lower part of reactor. The reddish brown color characteristic of the ANAMMOX enrichment was mainly attributed to the increase in cytochrome content [43]. Nearly 47 % of NH_4^+-N removal and 83% NO_2^--N removal by ANAMMOX process was reported with the mean specific nitrogen removal to be 0.064 to 0.08g TN/gVSS/d. UASB reactor was also used by Ni et al. [52], for quick ANAMMOX enrichment from enriched ANAMMOX seed as compared to Ahn et al. who used a locally available seed for startup. The NLR achieved by Ni et al. [52] was 0.28 kgN/m³/d to 1 kgN/m³/d with HRT<1d.

Upflow Biofilter (UBF): Chen et al. (2010b) [48] successfully started up ANAMMOX process in an Upflow Biofilter with effluent recirculation at high loading rates (34.5kgN/m³/d). The high loading rates in ANAMMOX reactor was dependent on two parameters such as, the quantity and activity of functional biomass in the reactor. The seed used was taken from full-scale WWTP treating monosodium glutamate effluents of SS (33.3g/L) and VSS (14.6 g/L). The reactor was fed with synthetic water and trace metals solution [43]. The ANAMMOX reactor was started with (30 to 50 mg/L) of NH_4^+-N and (50 to 70 mg/L) of NO_2^--N with HRT of 9.6h. HRT and substrate concentration was adjusted to increase the NLR, with the highest substrate concentrations of 976 mg/L of NH_4^+-N and 1280 mg/L of NO_2^--N. The recirculation was adopted in UBF to decrease the influent substrates in the ratio of 1:1, 2:1 and 4:1, especially till the effluent NO_2^--N concentrations was below 50mg/L. The UBF operation resulted in granulation and biofilm formation of ANAMMOX bacteria. Tang et al. [25] also used UBF that was inoculated with anaerobic granular sludge of SS (51.2 g/L) and VSS (43.5 g/L) originating from an UASB reactor treating paper mill wastewater. The UBF was operated at a fixed HRT of 9.1h and upon occurrence of ANAMMOX activity with continuous removal of NH_4^+-N and NO_2^--N, the NLR was raised stepwise by adjusting the NH_4^+-N and NO_2^--N concentrations. The inhibitory effects of high pH and NH_3 on ANAMMOX bacteria were addressed in this study, since it affects the stabilization of the ANAMMOX process. The buffer concentration was increased from 0.5 g $KHCO_3$/L to 1.25 g $KHCO_3$/L to effectively reduce the pH variation, and to enhance the nitrogen removal performance of the reactor.

Membrane Bioreactor (MBR): Development of advanced biological system, such as MBR for startup of ANAMMOX process was considered as a better alternative for a quicker and stable system when compared to the other biological systems. Main limitation of SBR and other biofilm bioreactors which is the dependence of biomass retention on biomass settling was overcome by MBR. Cultivation of ANAMMOX bacteria as biofilms or cell aggregates in conventional biological systems can be beneficial from an applied, operation point of view. But based on microbiological, physiological and biochemical perspective, development of ANAMMOX bacteria as free cells is favored to avoid mass transfer and nutrient diffusion limitation experienced in biofilm bioreactors. Improvement of growth rate amongst free cells has been reported by Kartal et al. [53], with *K. stuttgartiensis* exhibiting doubling time of 11–20d in SBR compared to 7d in MBR. Successful cultivation of ANAMMOX bacteria with complete biomass retention, operated at high NLRs with production of ANAMMOX bacterial suspension as free cells or aggregates with high growth rate was achieved using MBR [7,41].

In a study conducted by Wang et al. [4], with mixed activated sludge as seed source (MLSS–2.23g/L ; MLVSS–1.52g/L) and synthetic wastewater of composition prescribed by Van de Graaf et al. [43], the MBR was operated with <0.05mg/L of DO concentration to enable ANAMMOX growth and metabolism. Within 2 months of operation, successful enrichment of ANAMMOX population as the dominant climax community was achieved. Trigo et al. [7] developed ANAMMOX as aggregates in Membrane Sequence Batch Reactor (MSBR). Complete mixing was performed using mechanical stirrer from 40 to 160 rpm. The reactor was operated in 6h cycle comprising of 330 min of reaction, 9 min settling and 18 min discharge with permeate backwash for 3 min. At the end of 375d of operation, 1.22 mole of NO_2^--N and 0.22 mole of NO_3^--N were consumed for 1.0 mole of NH_4^+-N consumed.

ANAMMOX bacteria were cultivated as free cells in MBR in the study conducted by Van der Star et al. [41]. Enrichment medium and reduction of calcium and magnesium concentrations along with addition of yeast extract triggered the growth of ANAMMOX as free cells, reducing the appearance of flocs. Enriched granular ANAMMOX seed from a full-scale ANAMMOX reactor was used in this study. The reactor was operated for >250d. At the end of the study, the NO_2^--N:NH_4^+-N conversion ratio obtained was 1.1 to 1.3, and NO_3^--N:NH_4^+-N ratio was 0.10 to 0.25. In another study by Suneethi and Joseph [54] where 96 % NH_4^+-N removal efficiency was achieved with an influent Nitrogen loading rate of 5 kg NH_4^+-N/m³/d within 129d. The performance in various biological systems applying ANAMMOX process is compared in Table 2.

Key Signatures for Anammox Activity/Enrichment

The prominent signatures applied for indicating the activity or enrichment of ANAMMOX bacteria includes both chemical and molecular nature. The details of which are entailed in 5.1 and 5.2.

Chemical signatures

pH: Changes in pH reversibly affect ANAMMOX activity. Optimal pH favorable for efficient ANAMMOX activity is in the range of 7 to 8 [12]. Instances of failures in pH control have resulted in unsuccessful attempts in ANAMMOX cultivation [12]. This usually leads to changes in concentrations of NH_4^+-N, NO_2^--N, NO_3^--N, which should be immediately addressed. Since increase in the growth of AOB and NOB instead of ANAMMOX will result in decrease in autotrophic ammonia removal by ANAMMOX activity.

N_2H_4, NH_2OH: The concentrations of N_2H_4 and NH_2OH in AnMBR [54] were around 0.03 to 0.001 mg/L (average of 0.011 ± 0.008 mg/L) and 0.08 to 0.33 mg/L (average of 0.055 ± 0.107 mg/L) respectively, confirming the AOB and ANAMMOX activity in the system [8,55]. The N_2H_4 generated from substrates NH_4^+-N and NO_2^--N during ANAMMOX activity with NO as a direct precursor [56]. Changes in N_2H_4 concentration is usually corresponded to changes in NO_2^--N concentration. This could be attributed to the fact that when NO_2^--N accumulation occurs inhibiting ANAMMOX activity, increased concentrations of N_2H_4 could be noticed [56]. Similarly the changes in the concentration of NH_2OH coincided with the change in feed NH_4^+-N concentration. NH_2OH was reported to be an intermediate of ammonia oxidation carried out by AOB, with increase in NH_2OH meant an inhibition of HAO enzyme activity [56]. Since NH_4^+-N is a common substrate for both AOB and ANAMMOX, change in feed NH_4^+-N concentrations might have led to NH_2OH buildup resulting in comparable tendencies of NH_2OH with NLR [56].

N_2O, NO_2, NO: In ANAMMOX enrichment units, production of N_2O, NO_2, and NO occurs. The changes in gas concentrations corresponded to subtle changes in N_2H_4 and NH_2OH, with NO classified to be one of the intermediate in ANAMMOX reaction [56]. NO could be produced from NH_2OH and NO_2^--N by AOB and ANAMMOX activity, respectively, while N_2O generated from NO by AOB activity [57]. NO produced can also revert to NO_2^--N and then NO_3^--N by NOB activity. Denitrification process was also claimed to produce NO and N_2O as well [57]. Not with standing such complexity in the sources of NO and N_2O concentrations, the ultimate harmless N_2 gas production could be resulted only by ANAMMOX and/or denitrification process.

SI No	Type of Reactor	Working Volume (L)	Duration of operation (d)	HRT (d)	Influent Nitrogen concentration (mg/L)	Nitrogen removal efficiency (%)	Reference
	FBR	2.25	84	0.9-2	70 – 840 (NO_2^--N) 70 – 840 (NH_4^+-N)	83 – 85	Strous et al. (1997)
			150	0.1 – 11	70 – 840 (NO_2^--N) 1100 – 2100 (NH_4^+-N)	81 – 99	
	SBR	7	150	-	50 – 70 (NO_2^--N) 40 – 60 (NH_4^+-N)	100 80	Chamchoi and Nitisoravut (2007)
	Gas lift	7	200	1	1100 (NO_2^--N) 900 (NH_4^+-N)	>99 88	Dapena-Mora et al. (2004a)
	SBR	1		0.625	375 (NO_2^--N) 375 (NH_4^+-N)	100 78	
	Gas lift	1.8	-	0.42	1370 (NO_2^--N) 1360 (NH_4^+-N)	83	Sliekers et al. (2003)
	RBC	1.7	100	0.167	350 (NO_2^--N) 350 (NH_4^+-N)	100 99	Liu et al. (2008)
	UASB	6	325	3.5	90 (total inorganic nitrogen)	60	Tran et al. (2006)
	UBF	8	236	0.06 - 0.2	331.5 – 1280 (NO_2^--N) 204.6 – 976 (NH_4^+-N)	98.8	Chen et al. (2010a)
	MBR	4.8	60	2	50 (NO_2^--N) 50 (NH_4^+-N)	90 90	Wang et al. (2009)
	MBR	8	>250	2	552 (NO_2^--N) 552 (NH_4^+-N)	-	Van der Star et al. (2008)
	MSBR	5	375	1	390 (NO_2^--N) 390 (NH_4^+-N)	90 90	Trigo et al. (2006)
	AnMBR	15	129	2	100 (NO_2^--N) 10000 (NH_4^+-N)	96%	Suneethi and Joseph (2011)

Table 2: ANAMMOX process performance in conventional and advanced biological reactors.

Alongside these crucial elements, factors such as pH, HNO_2 and DO concentrations were also speculated to play a role in the NO, N_2O and NO_x gas emissions [57]. The production of N_2O and NO by AOBs at low O_2 conditions was reported, with especially a pure culture of *Nitrosomonas* generating N_2O and NO only in the absence of NO_2^--N consumer such as *Nitrobacter* [58].

NH_3 and HNO_2: The possibility of NOB inhibition by NH_3 was supported by studies conducted by Anthonisen et al. [59]. Occurrence of unionized NH_3 and HNO_2 is dependent on the pH and temperature in the biological system and leads to inhibition of NO_2^--N conversion [60]. Presence of NOB (Nitrite Oxidizing Bacteria) inhibition is evident when NH_3 concentration is above 0.1 mg/L [32,59,60]. Likewise all nitrifying bacteria showed inhibition above 0.2mg/L of HNO_2 concentration as reported by other studies [32,59].

NH_4^+-N, NO_2^-, NO_3^-: The concentration of nitrite during the startup is of crucial importance for growth: a too low amount will result in substrate limitation and thus slower growth, while concentrations above 50–150 mg N/L can already lead to inhibition [14,50,61]. The stoichiometric ratio for NH_4^+-N removed: NO_2^--N converted: NO_3^--N produced indicating the ANAMMOX process was 1:1.32:0.26 [12]. During the buildup of Nitrogen Loading rate, low influent NO_2^--N/NH_4^+-N ratio could be observed with the overall obtained ratio was 1:0.84:0.02 [62]. Wyffels et al. [63] reported 0.20mol of production of NO_3^--N per mol of NH_4^+-N oxidized for ANAMMOX MBR system. Low degree but significant occurrence of ANAMMOX process, of approximately 1:1.15 for NO_2^--N consumption to NH_4^+-N consumption, was reported by Wang et al. [4] and about 1.05 by Wyffels et al. [63]. Denitrification process could also affect the effluent NO_2^--N to NH_4^+-N molar ratio [64]. The molar ratio of NO_2^--N conversion to NH_4^+-N removed of 1.22 while the NO_3^--N production to NH_4^+-N oxidized ratio of 0.22 was obtained by Trigo et al. [7].

Alkalinity and dissolved oxygen: Feed alkalinity along with DO concentrations are critical controlling parameters in a single-stage biological process for nitrogen removal. The control of alkalinity and dissolved oxygen (DO) concentrations in the feed to maintain an alkalinity to ammonia ratio of <8 and DO of <0.06 mg O/mg N/d, respectively, was necessary for inhibiting nitratation and enhancing partial nitration and ANAMMOX [45].

To achieve a biological nitrogen removal in a single-stage, the activity of NOB has to be inhibited without affecting the activities of AOB and ANAMMOX bacteria. These three groups of microorganisms are closely interlinked with common electron donor and acceptors. From the works of Gong et al. [19] and Paredes et al. [23] it has been reported that by regulating the concentration of DO and NO_2^--N, partial control of NOB activity could be achieved. As NOB competes for DO and NO_2^--N with AOB and ANAMMOX bacteria, respectively in the absence of NO_2^--N in wastewater, NOB depends directly on AOB for their source of electron donor. By limiting DO concentration, AOB consume available DO for NO_2^--N production. Hence, under this condition, NOB experiences two-way limitations, initially in terms of electron donor (NO_2^--N) and later in terms of electron acceptor (O_2). AOB, NOB and ANAMMOX bacteria which are chemolithotrophic microorganisms also require inorganic carbon source for their cell growth [65]. By controlling bicarbonate alkalinity, the process of elimination of NOB can be further fine-tuned [45].

Molecular signatures

The common methodological approaches adopted to detect, identify and confirm ANAMMOX bacteria or their activity includes:

(i) Detection of the single ladderane lipids as biomarkers, which were reported to be unique structures, found in the intracytoplasmic organelle like structure of ANAMMOX bacteria by HPLC – Atmospheric Pressure Ionization – MS and GC – MS [66]. The compound specific stable carbon isotope ratios can also be monitored by GC – IR – MS system [67].

(ii) Chemical analyses of the Nitrogen compounds to detect the change in the concentrations of NH_4^+-N, NO_2^--N and NO_3^--N owing to ANAMMOX activity [60] along with determination of N_2H_4 and NH_2OH [67].

(iii) Analysis of gas composition of N_2, N_2O, NH_3, CO_2, CH_4 by GC – TCD [40,41], or only the gas composition analysis of N_2O carried out in real-time using gas-filter correlation (Teledyne API 320E).

(iv) Estimation of NO (nitric oxide) using a chemiluminescence detector (CLD) [41], or determination of NO and NO_2 by chemiluminescence method (Ecophysics CLD 64 monitor).

(v) Application of molecular techniques such as Polymerase Chain Reaction (PCR) or Fluorescence *In Situ* Hybridization (FISH), which are based on the nucleic acid analysis for identification. A number of specific sequences and primers were developed to amplify the 16S rRNA from the environmental and enrichment samples using the PCR based approach or by FISH analysis, as detailed in 5.2.2 [4,11,37,69].

(vi) Application of ISR FISH to assess the precursor rRNA concentration in a cell, the intergenic spacer region (ISR) between 16S rRNA gene and 23S rRNA gene by targeting using fluorescently labeled oligonucleotide probes [67].

(vii) Combination of Fluorescence *In Situ* Hybridization – Micro Autoradiography (FISH-MAR) that can directly link the uptake of radioactively labeled substrate (Eg: $NaH_{14}CO_3$) with uncultured organisms such as ANAMMOX [67].

(viii) Quantification of changes in microbial population of a mixed culture of nitrifiers, denitrifiers, NOBs, AOBs and ANAMMOX using quinone profiles, since this profile is usually represented as the mole fraction of each quinone type that is specific for a microbial community [70].

(ix) Observation of ANAMMOX cells using confocal micrographs or scanning electron microscope (SEM). The observations from SEM images are subject to confirmation using molecular analyses [4,615,22].

(x) Trace experiments with labeled $[_{15}N]NH_4^+$-N reacts uniquely, in a 1:1 ratio with unlabeled $[_{14}N]NO_2^-$-N to $N_2(_{14}N_{15}N)$ in the ANAMMOX reaction.

SEM observation of ANAMMOX seed biomass: Scanning electron microscope images were used to visualize the seed and membrane morphology in AnMBR by Suneethi and Joseph [54]. The seed cultivated in the AnMBR and the biomass on the membrane surface were mostly spherical and elliptical bacterial clusters with rough surface, interspersed with abundant aggregates of inorganic origin. This observation together with monitoring of the nitrogen transformations revealed the presence of ANAMMOX activity. There was also presence of few filamentous and short rod bacteria, indicating the harmonious coexistence of ANAMMOX bacteria with other microbial populations like AOBs, NOBs and denitrifiers in AnMBR [4,22]. The SEM photographs of the cultivated ANAMMOX sludge in AnMBR indicated the various bacterial morphologies found in the sludge, though the cause of filamentous bacteria formation was unclear.

Biomass growth occurred as granules with an irregular cauliflower appearance, which was reported by Trigo et al. [7]. Dapena-Mora et al. [71] described the ANAMMOX biomass growth in SBR-ANAMMOX system as granular with an irregular cauliflower appearance. Presence of co–existence of cocci bacteria such as filamentous and short rod bacteria with spherical shaped bacteria in the seed sludges from activated sludge and anaerobic digestion sludge was reported by Chamchoi and Nitisoravut [6]. Negative role played by filamentous bacteria in system performance in UASB seed sludge was reported by Chamchoi and Nitisoravut [6]. Frequent appearance and development of *Chlorobi*– like filamentous bacteria in ANAMMOX reactors, was also reported by Li et al. [9]. The relation between the filamentous bacteria and ANAMMOX seed could be typically linked to unconfirmed metabolic connection with the involvement of the filamentous bacteria in bestowing structural integrity to the ANAMMOX in both granular and biofilm phase [11,69].

The SEM observations of the enriched ANAMMOX seeds were reported by Chamchoi and Nitisoravut [6] as spherical flocs of ANAMMOX sludge with smooth surfaces along with cocci and filamentous type bacteria. The seed sludge from UASB exhibited both filamentous and spherical shaped bacteria while the seed sludge from activated sludge and anaerobic digestion sludge, the main types of bacteria were spherical and short rod-shaped bacteria. The spherical cells were presumed to be ANAMMOX that coexisted harmoniously with AOBs and NOBs as observed from various bacterial morphologies. The cause for occurrence of filamentous bacteria was inconclusive in their study.

ANAMMOX identification: Usually the color of the ANAMMOX enriched sludge is reported to be reddish brown, which is a well-documented characteristic of ANAMMOX activity [6,7,38,41]. The recent detection methods and different molecular techniques available for the ANAMMOX organisms are fluorescence in situ hybridization (FISH), 16S rRNA or functional gene analysis, membrane lipids and tracer experiments with $[_{15}N]$ labeled ammonia [3,72-78]. FISH can give both quantitative and qualitative figures of ANAMMOX bacterial population. In situ hybridizations with DNA probes for the

detection of ANAMMOX bacteria are performed with fluorescent labeled compounds and fluorescent bacterial cells, which are detected by epifluorescent microscopy, confocal laser scanning microscopy or flow cytometry. The number of FISH probes targeting the specific ANAMMOX organisms is given in the Table 3. FISH signal intensity is directly proportional to the precursor rRNA concentrations in ANAMMOX cells and could be used as a direct measure of the growth rate (ribosome turnover rate) of the ANAMMOX organisms. The advantages of analyzing with FISH technique are reliable, with reduced misinterpret artifacts, conveys about the spatial location of bacteria and density in a limited region without the destruction of the sample. But if large amount of inert material was present and when the ANAMMOX bacterial cell was low in the sample (i.e. low numbers of rRNA) then the FISH technique may not be applied since the detection will not be possible in microscopy. In that case, the 16S rRNA or functional gene-based approach i.e. PCR amplification with 16S rRNA gene-targeted primers and phylogenetic analysis of the product is an excellent technique. Some ANAMMOX specific FISH probes are used as PCR primers for the specific amplification of the 16S rRNA genes of ANAMMOX organisms. The different PCR primer pairs for the ANAMMOX identification along with its PCR conditions are listed in the Table 4 are used for analyzing the entire group of ANAMMOX organisms. Disadvantages of PCR are low DNA extraction yield and the production of artifacts after PCR. Selection of an appropriate DNA extraction method should be carried out [76]. Real time quantitative PCR (RT-qPCR) and competitive PCR (cPCR) could also be used to find the ANAMMOX bacterial mass based on 16S rRNA or functional genes. qPCR is widely used as it is cheaper when compared to Competitive PCR which has lower accuracy. When compared to FISH, qPCR has higher throughput, more reliability and more sensitive quantification [9,74]. Functional genes like Hydrazine oxidoreductase (hzo) gene can be used as a biomarker in ANAMMOX detection, since it is reported to be an intermediate in ANAMMOX process to dehydrogenate N_2H_4 to convert it into N_2. The different primers targeting the hzo gene is given in the Table 4 along with their operating conditions and target length.

Inhibitor/Stimulator for ANAMMOX activity: The effects on the ANAMMOX activity due to addition of various inhibitors/stimulators are summarized in the Table 5. According to Strous et al. [14], 1gN/L for NH_4^+-N and NO_3^--N respectively, has no effect. Whereas 100 mg/L

SI No	Probe	Specificity	Sequence	Target site	Formamide concentration (%)	Wash buffer NaCl (mM)	Reference
	Pla46	Planctomycetales	GACTTGCATGCCTAATCC	46-63[a]	30	112	Neef et al. 1998; Kartal et al. 2007
	Pla886		GCCTTGCGACCATACTCCC	-	35	-	Neef et al. 1998
	Amx368	All ANAMMOX microorganisms	CCTTTCGGGCATTGCGAA	368-385[a]	15	338	Schmid et al. 2003; Kartal et al. 2007
	Amx820	Candidatus "Brocadia Anammoxidans"	AAAACCCCTCTACTTAGTGCCC	820-841[a]	25 / 40 / 25	159 / 56 / 56	Egli et al. 2001; Schmid et al. 2000; Mobarry et al. 1996
	Amx432		GTTAACTCCCGACAGTGG	-	40	-	Schmid et al. 2000
	Amx997		TTTCAGGTTTCTACTTCTACC	-	20	-	Schmid et al. 2000
	Amx1240		TTTAGCATCCCTTTGTACCAACC	-	60	14	Egli et al. 2001
	Apr 820	Candidatus "Anammoxoglobus propionicus"	AAACCCCTCTACCGAGTGCCC	820-840[a]	40	56	Kartal et al. 2007
	Kst1273	Candidatus "Kuenenia stuttgartiensis"	TCGGCTTTATAGGTTTCGCA	-	25	159	Schmid et al. 2000; Egli et al. 2001

[a](16S rRNA position E.Coli numbering)

Table 3: Probes used in FISH technique for ANAMMOX identification

SI No	Probe	Specificity	Sequence	Condition (°C)	Formamide concentration (%)	Wash buffer NaCl (mm)	Reference
	Pla46F	Planctomycetales	GACTTGCATGCCTAATCC	58 46-63	25	159	Neef et al. 1998; Schmid et al. 2005
	Pla886	Isosphaera, Gemmata, Pirellula, Planctomyces	GCCTTGCGACCATACTCCC	-	30	112	Neef et al. 1998; Schmid et al. 2005
	First run PCR PLA46F& PLA886R Second run PCR P338F & P518R		-	56 63	-	-	Pynaert et al. 2003
	PLA40F & P518R P338F, P338-IIF & P518R	Planctomycetales	-	60 53	-	-	Pynaert et al. 2003
	Pla46F 1390R		GGATTAGGCATGCAAGTC ACGGGCGGTGTGTACAA	59	-	-	Li et al. 2009
	Pla46F Amx667R	16S rRNA ANAMMOX	GGATTAGGCATGCAAGTC ACCAGAAGTTCCACTCTC	57 56	-	-	Neef et al. 1998; Yapsakli et al. 2011
	Pla46F Amx368R	All ANAMMOX microorganisms	GACTTGCATGCCTAATCC CCTTTCGGGCATTGCGAA	59 (c300) 52 (c323)	-	-	Schmid et al. 2003; Li et al. 2010
	Amx368F/R		CCTTTCGGGCATTGCGAA	56	15	338	Schmid et al. 2003; Kartal et al. 2007
	Amx809F		GCCGTAAACGATGGGCACT	c809-826	-	-	Tsushima et al. 2007
	Amx818F	ANAMMOX bacteria	ATGGGCACTMRGTAGAGGGGTTT	c818–839	-	-	Tsushima et al. 2007
	Amx1066R		AACGTCTCACGACACGAGCTG	c1047–1066	-	-	Tsushima et al. 2007
	Amx808F Amx1040R	AnAOB 16S rRNA gene	ARCYGTAAACGATGGGCACTAA CAGCCATGCAACACCTGTRATA	60	-	-	Li et al. 2009
	Amx694F Amx960R		GGGGAGAGTGGAACTTCGG GCTCGCACAAGCGGTGGAGC	c694–713 c960–979	-	-	Jetten et al. 1999
	Brod541F Brod1260R	ANAMMOX bacteria	GAGCACGTAGGTGGGTTTGT GGATTCGCTTACCTCTCGG	60 (c720)	-	-	Li et al. 2010
	Brod541F Amx820R		GAGCACGTAGGTGGGTTTGT AAAACCCCTCTACTTAGTGCCC	59 (c280)	-	-	Li et al. 2010
	Pla46F Amx820R		GACTTGCATGCCTAATCC AAAACCCCTCTACTTAGTGCCC	59 (c780) 52 (c775)	-	-	Schmid et al. 2000; Li et al. 2010
	Amx368F Amx820R	Candidatus "Brocadia Anammoxidans"/ Candidatus Kuenenia	TTCGCAATGCCCGAAAGG AAAACCCCTCTACTTAGTGCCC	c368–385 c820–841	-	-	Schmid et al. 2003; Schmid et al. 2000
	Amx820 F/R		AAAACCCCTCTACTTAGTGCCC	56	25 40	159 56	Egli et al. 2001; Schmid et al. 2000
	Amx1900		CATCTCCGGCTTAACAA	-	30	112	Schmid et al. 2000; Schmid et al. 2001
	Kst0288		GCGCAAAGAAATCAAACTGG	-	10	450	Schmid et al. 2001
	Kst0193		CAGACCGGACGTATAAAAG	-	10	450	Schmid et al. 2001
	Kst0077		TTTGGGCCACACTCTGTT	-	10	450	Schmid et al. 2001
	Kst0031		ATAGAAGCCTTTTGCGCG	-	10	450	Schmid et al. 2001
	Kst1275	Candidatus "Kuenenia stuttgartiensis"	TCGGCTTTATAGGTTTCGCA	-	25	159	Schmid et al. 2000
	Kst0157		GTTCCGATTGCTCGAAAC	-	25	159	Schmid et al. 2001, Schmid et al. 2005
	Ban0389		GGATCAAATTGCTACCCG	-	10	450	Schmid et al. 2001
	Ban0222		GCTTAGAATCTTCTGAGGG	-	10	450	Schmid et al. 2001
	Ban0108		TTTGGGCCCGCAATCTCA	-	10	450	Schmid et al. 2001
	Ban0071		CCCTACCACAAACCTCGT	-	10	450	Schmid et al. 2001
	Amx1240		TTAGCATCCCTTTGTACCAACC	-	60	14	Schmid et al. 2000
	Amx1154	Candidatus "Brocadia Anammoxidans"	TCTTGACGACAGCAGTCT	-	20	225	Schmid et al. 2000
	Amx1015		GATACCGTTCGTCGCCCT	-	60	14	Schmid et al. 2000
	Amx0997		TTTCAGGTTTCTACTTCTACC	-	20	225	Schmid et al. 2000
	Amx0613		CCGCCATTCTTCCGTTAAGCGG	-	40	56	Schmid et al. 2000
	Amx0432		CTTAACTCCCGACAGTGG	-	40	56	Schmid et al. 2000
	Amx0223		GACATTGACCCCTCTCTG	-	40	56	Schmid et al. 2000
	Amx0156		CGGTAGCCCAATTGCTT	-	40	56	Schmid et al. 2000
	Ban0162		CGGTAGCCCCAATTGCTT	-	40	56	Schmid et al. 2000
	BS820R	Candidatus "Scalindua wagneri / sorokinii"	TAATTCCCTCTACTTAGTGCCC	56	40	56	Schmid et al. 2000

Scabr1114R	Candidatus "Scalindua brodae"	CCCGCTGGTAACTAAAAACAAG	56	20	225	Schmid et al. 2003; Schmid et al. 2005
Sca1309R	Candidatus "Scalindua"	TGGAGGCGAATTTCAGCCTCC	56	5	675	Schmid et al. 2003; Schmid et al. 2005
Targeting hzo gene						
hzocl1F1 hzocl1R2	AnAOB hzo gene	TGYAAGACYTGYCAYTGG ACTCCAGATRTGCTGACC	50C (^{c}470)	-	-	Schmid et al. 2008
hzocl1F2 hzocl1R2		TGYAAGACYTGYCAYTGGG ACTCCAGATRTGCTGACC	53C (^{c}470)	-	-	Schmid et al. 2008
Ana-hzo 1F Ana-hzo 2R		TGTGCATGGTCAATTGAAAG ACCTCTTCWGCAGGTGCAT	53 (^{c}1000)	-	-	Li et al. 2010
hzoF1h hzoR1		TGTGCATGGTCAATTGAAAG CAACCTCTTCWGCAGGTGCATG	53 (^{c}1000)	-	-	Li et al. 2010
Ana-hzo1f Ana-hzo2r		TGTGCATGGTCAATTGAAAG ACCTCTTCWGCAGGTGCAT	53	-	-	Li et al. 2010

^{c}PCR product length (-bp)

Table 4: Probes used in PCR for ANAMMOX identification.

SI No	ANAMMOX activity inhibitor /stimulator	Mode of action	Concentration or period tested	Effect	Reference
	Ammonium	-	1gN/L	No effect	Strous et al. 1999; Bettazi et al. 2010
			55mM	50% inhibition	Dapena-Mora et al. 2007; Bettazi et al. 2010
			>70 mgN/L	Free ammonia inhibition	Jung et al. 2007; Bettazi et al. 2010
			13-90 mg NH_4^+-N/L	Negative effect	Waki et al. 2007; Bettazi et al. 2010
			90 mg NH_4^+-N /L	No effect	Bettazi et al. 2010
	Nitrate	-	1gN/L	No effect	Strous et al. 1999; Bettazi et al. 2010
			45 mM	50% inhibition	Dapena-Mora et al. 2007; Bettazi et al. 2010
			57 mg NO_3^--N/L	No effect	Bettazi et al. 2010
	Nitrite	-	100 mgNO_2^- N/L	Complete inhibition	Strous et al. 1999; Bettazi et al. 2010
			>13.2 mM	No activity	Egli et al. 2001; Bettazi et al. 2010
			25 mM	50% inhibition	Dapena-Mora et al. 2007; Bettazi et al. 2010
			70 mgNO_2^- N/L	Activity decrease	Jung et al. 2007; Bettazi et al. 2010
			>100 mgNO_2^- N/L	Inhibition	Lopez et al. 2008; Bettazi et al. 2010
			60 mgNO_2^- N/L (spiked)	Activity decrease	Bettazi et al. 2010
			>30 mgNO_2^--N/L (long exposure)	Activity decrease	Bettazi et al. 2010
	No biomass	None	0 mg/L	No activity	Jetten et al. 1999
	Sterilization at 121³C	Denaturation	20-120 min	No activity	Jetten et al. 1999
	Gamma irradiation	Inactivation	60 min	No activity	Jetten et al. 1999
	Penicillin V	Inhibition of cell wall synthesis of bacteria	0-100 mg/L	None	Jetten et al. 1999
	Penicillin G	-	0-1000 mg/L	None	Jetten et al. 1999
	Bromoethane sulfonic acid	Inhibition of methanogenesis	0-20 mM	None	Jetten et al. 1999
	Na_2SO_4	Stimulation of sulphate reduction	20 mM	None	Jetten et al. 1999
	Na_2MoO_4	Inhibition of sulphate reduction	20 mM	None	Jetten et al. 1999
	Chloramphenicol	Inhibition of protein synthesis	0-400 mg/L	None	Jetten et al. 1999
	Hydrazine	Inhibition of NH2OH oxidation	0-3 mM	Activation	Jetten et al. 1999
	Acetone	Solvent for N-serve	10 mM	None	Jetten et al. 1999
	N-serve	Inhibition of nitrification	0-50 mg/L	None	Jetten et al. 1999
	Allylthiourea	Inhibition of nitrification	0-10 mM	None	Jetten et al. 1999
	Acetylene	Inhibition of nitrification and denitrification	6 mM	Inhibition	Jetten et al. 1999
	2,4-Dinitrophenol	Uncoupler	0-400 mg/L	Inhibition	Jetten et al. 1999
	Carbonyl cyanide m-chlorophenylhydrazone	Uncoupler	0-40 mg/L	Inhibition	Jetten et al. 1999
	$HgCl_2$	Cell damage	0-300 mg/L	Inhibition	Jetten et al. 1999
	Oxygen	Oxidative stress	0-0.2 mM	Inhibition	Jetten et al. 1999

Phosphate	Chelating agent	<1 mM	None	Jetten *et al.* 1999; Dapena-Mora *et al.* 2007
		>2 mM	Inhibition	Jetten *et al.* 1999
		5 or 50 mM	Loss of activity	Van de Graaf *et al.* 1996; Dapena-Mora *et al.* 2007
		21 mM	50% inhibition	Dapena-Mora *et al.* 2007
Sulphide	-	1 or 5 mM	Increase	Van de Graaf *et al.* 1996; Dapena-Mora *et al.* 2007
		0.3 mM	50 % inhibition	Dapena-Mora *et al.* 2007
Chloride	-	50 mM	No effect	Van de Graaf *et al.* 1996; Dapena-Mora *et al.* 2007
		200 mM	50% inhibition	Dapena-Mora *et al.* 2007
Acetate	-	1 or 5 mM	Increase	Van de Graaf *et al.* 1996; Dapena-Mora *et al.* 2007
		39 mM	50% inhibition	Dapena-Mora *et al.* 2007
		50 mM	70% of inhibition	Dapena-Mora *et al.* 2007
Oxygen	-	0.5%	Reversibly inhibited	Van Dongen *et al.* 2001
		0.06 mg/L	Reversibly Inhibition	Paredes *et al.* 2007
		1 μM (>18 % oxygen saturation)	Irreversibly Inhibition	Zhang *et al.* 2008
Organic matter	-	300 mg COD/L	Inactivation	Chamchoi *et al.* 2008
Methanol	-	<1 mM	Inhibition	Guven *et al.* 2004 Kartal *et al.* 2004
Ethanol	-	<1 mM	Inhibition	Guven *et al.* 2004; Kartal *et al.* 2004

Table 5: Inhibitor/Stimulator for ANAMMOX activity.

of NO_2^--N cause complete inhibition. Biomass plays an important role in the ANAMMOX process; if no biomass available then the activity was not observed [79]. When subjecting the biomass with gamma radiation or sterilization, no change in the NH_4^+-N or NO_2^--N was noticed showing no activity until the 60-120 min. Around 10-20 mM concentration of the compounds such as Penicillin V, Penicillin G, Bromoethane sulfonic acid, Sodium sulphate, Sodium molybdate, Chloramphenicol, Acetone and Allylthiourea has no effect on the ANAMMOX activity [79]. But with chemical compounds such as Methanol, Ethanol and sulphide even with very low concentrations of 1mM has inhibition over ANAMMOX activity. The Oxygen from 0.2 to 200mM has an oxidative stress on the ANAMMOX bacteria [21,80]. 50 mM of acetate resulted in 70% of inhibition in the ANAMMOX process [15]. Phosphate concentrations higher than 180 mg/L and NH_3 with higher concentration inhibit ANAMMOX activity [23]. Trace amounts of either of the ANAMMOX intermediates N_2H_4 (>1.4 mgN/L) and NH_2OH (>0.7 mgN/L) can activate the ANAMMOX process [14,18,45,81].

Conclusions

Some of the challenges facing successful ANAMMOX startup and development include slow growth rate, operational difficulty and long startup time. With dedicated enrichment and cultivation techniques the sensitivity of ANAMMOX bacteria to inhibitory concentrations of NO_2^--N, Alkalinity, O_2 etc., could be minimized and ANAMMOX bacteria could be successfully developed to yield sustained NH_4^+-N removal. With the key parameters such as the source of seed, type of reactors used, operational strategy, experimental conditions and molecular signatures such as PCR, FISH, SEM and the inhibitors and affinity factors being monitored and optimized, ANAMMOX startup and development could be deemed successful.

Acknowledgement

The support of the University Grants Commission Research Fellowship for meritorious scholars in Sciences to the corresponding author is gratefully acknowledged.

References

1. Mulder A, Graaf AAV, Robertson LA, Kuenen JG (1995) Anaerobic ammonium oxidation discovered in a denitrifying fluidized bed reactor. FEMS Microbiology Reviews 16: 177–184.

2. Berge ND, Reinhart DR. Townsend TG (2005) The Fate of Nitrogen in Bioreactor Landfills. Critical Reviews in Environmental Science and Technology 35: 365–399.

3. Pynaert K, Smets BF, Wyffels S, Beheydt D, Siciliano SD, et al. (2003) Characterization of an autotrophic nitrogen-removing biofilm from a highly loaded lab-scale rotating biological contactor. Appl Environ Microbiol. 69: 3626–3635.

4. Wang T, Zhang H, Yang F, Liu S, Fu Z, et al. (2009) Startup of the ANAMMOX process from the conventional activated sludge in a membrane bioreactor. Bioresource Technology 100: 2501–2506.

5. Kimura Y, Isaka K, Kazama F (2011) Effects of inorganic carbon limitation on anaerobic ammonium oxidation (Anammox) activity. Bioresource technology 102: 4390–4394.

6. Chamchoi N, Nitisoravut S (2007) ANAMMOX enrichment from different conventional sludges. Chemosphere 66: 2225–2232.

7. Trigo C, Campos JL, Garridio JM, Mendez R (2006) Startup of the ANAMMOX process in a membrane bioreactor. Journal of Biotechnology 126: 475–487.

8. Third KA, Paxman J, Schmid M, Strous M, Jetten MSM, et al. (2005) Enrichment of ANAMMOX from Activated Sludge and its Application in the CANON process. Microb Ecol 459: 236–244.

9. Li X, Du B, Fu H, Wang R, Shi J, et al. (2009) The bacterial diversity in an anaerobic ammonium-oxidizing (Anammox) reactor community. Syst Appl Microbiol 32: 278–289.

10. Strous M, Pelletier E, Mangenot S, Rattei T, Lehner A, et al. (2006) Deciphering the evolution and metabolism of an Anammox bacterium from a community genome. Nature, 440: 790–794.

11. Park H, Rosenthal A, Jezek ., Ramalingam K, Fillos J, et al. (2010) Impact of inocula and growth mode on the molecular microbial ecology of anaerobic ammonia oxidation (Anammox) bioreactor communities. Water Res 44: 5005–5013.

12. Strous M, Heijnen JJ, Kuenen JG, Jetten MSM (1998) The sequencing batch reactor as a powerful tool for the study of slowly growing anaerobic ammonium-oxidizing microorganisms. Applied Microbiology and Biotechnology 50: 598–596.

13. Van der Star WRL, Abma WR, Blommers D, Mulder JW, Tokutomi T, et al. (2007) Startup of reactors for anoxic ammonium oxidation: Experiences from the first full scale Anammox reactor in Rotterdam. Water Res 41: 4149 – 4163.

14. Strous M, Kuenen JG, Jetten MSM (1999) Key physiology of anaerobic ammonium oxidation. Appl Environ Microbiol 65: 3248–3250.

15. Dapena-Mora A, Campos JL, Mosquerra-Corral A, Jetten MSM, Mendez R (2004) Stability of the ANAMMOX process in a gas – lift reactor and a SBR. Journal of Biotechnology 110: 159–170.

16. Date Y, Isaka K, Ikuta H, Sumino T, Kaneko N, et al. (2009) Microbial diversity of Anammox bacteria enriched from different types of seed sludge in an anaerobic continuous-feeding cultivation reactor. J of Biosci Bioeng 107: 281–286.

17. Strous M, van Gerven E, Kuenen JG, Jetten MSM. (1997) Effects of Aerobic and Microaerobic Conditions on Anaerobic Ammonium-Oxidizing (Anammox) Sludge. Appl Environ Microbiol 63: 2446–2448.

18. Xu G, Xu X, Yang F, Liu S, Gao Y (2012) Partial nitrification adjusted by hydroxylamine in aerobic granules under high DO and ambient temperature and subsequent Anammox for low C/N wastewater treatment. Chemical Engineering Journal 213: 338–345.

19. Gong Z, Yang F, Liu S, Bao H, Hu S, et al. (2007) Feasibility of a membrane – aerated biofilm reactor to achieve single – stage autotrophic nitrogen removal based on Anammox. Chemosphere 69: 776–784.

20. Uyanik S, Bekmezci OK, Yurtsever A (2011) Strategies for successful ANAMMOX enrichment at laboratory scale. Clean-Soil, Air, Water 39: 653–657.

21. Kartal B, de Almeida NM, Maaclke W, den Camp O, Jetten MSM, et al. (2013) How to make a living from anaerobic ammonium oxidation. FEMS Microbiol Rev 37: 428–461.

22. Arrojo B, Mosquera-Corral A, Campos JL, Mendez R (2006) Effects of mechanical stress on Anammox granules in a sequencing batch reactor (SBR). J Biotechnol 123: 453–463.

23. Paredes D, Kuschk P, Mbwette TSA, Stange F, Muller RA, et al. (2007) New aspects of Microbial Nitrogen Transformations in the Context of Wastewater treatment – A Review. Engineering and Life Science 7: 13-25.

24. Strous M, Gerven V, Zheng P, Kuenen JG, Jetten MSM (1997) Ammonium removal from concentrated waste streams with the anaerobic ammonium oxidation (Anammox) process in different reactor configurations. Water Research 31: 1955–1962.

25. Tang CJ, Zheng P, Mahmood Q, Chen JW (2009) Startup and inhibition analysis of the Anammox process seeded with anaerobic granular sludge. J Ind Microbiol Biotechnol 36: 1093–1100.

26. Siegrist H, Reithaar S, Lais P(1998) Nitrogen loss in a nitrifying rotating contactor treating ammonium rich leachate without organic carbon. Water Science and Technology, 37: 589–591.

27. Xu Z., Zhaohui Y., Guangming Z., Yong X., Jiuhua D. (2007) Mechanism studies on nitrogen removal when treating ammonium-rich leachate by sequencing batch biofilm reactor. Frontiers of Environmental Science and Engineering in China 1: 43–48.

28. Guo ZH, Qi ZS (2006) Treating leachate mixture with anaerobic ammonium oxidation technology. Journal of Central South University of Technology 6: 663–667.

29. Sliekers AO, Third KA, Abma W, Kuenen JG, Jetten MSM (2003) CANON and Anammox in a gas-lift reactor. FEMS Microbiol Lett 218: 339–344.

30. Liang Z, Liu J (2008) Landfill Leachate Treatment with a novel process: Anaerobic ammonium oxidation (Anammox) combined with soil infiltration system. J Hazard Mater 151: 202–212.

31. Liang Z, Liu JX, Li J (2009) Decomposition and mineralization of aquatic humic substances (AHS) in treating landfill leachate using the Anammox process. Chemosphere 74: 1315–1320.

32. Furukawa K, Inatomi Y, Qiao S, Quan L, Yamamoto T, et al. (2009) Innovative treatment system for digester liquor using ANAMMOX process. Bioresource Technology 100: 5437–5443.

33. Fux C, Boehler M, Huber P, Brunner I, Siegrist H (2002) Biological treatment of ammonium-rich wastewater by partial nitritation and subsequent anaerobic ammonium oxidation (Anammox) in a pilot plant. Journal of Biotechnology 99: 295–306.

34. Gali A, Dosta J, Loosdrecht MV, Alvarez JM (2007) Two ways to achieve an Anammox influent from real reject water treatment at lab-scale: Partial SBR nitrification and SHARON process. Process Biochemistry 42: 715–720.

35. Yamamoto T, Wakamatsu , Qiao S, Hira D, Fujii T, et al. (2011) Partial nitritation and Anammox of a livestock manure digester liquor and analysis of its microbial community. Bioresour Technol 102: 2342–2347.

36. Ganigue R, Gabarro J, Sanchez-Melsio A, Ruscalleda M, Lopez H, et al. (2009) Long-term operation of a partial nitritation pilot plant treating leachate with extremely high ammonium concentration prior to an Anammox process. Bioresour Technol 100: 5624–5632.

37. Sanchez-Melsio A, Caliz J, Balaguer MD, Colprim J, Vila X (2009) Development of batch-culture enrichment coupled to molecular detection for screening of natural and man-made environments in search of Anammox bacteria for N-removal bioreactors systems. Chemosphere 75: 169–179.

38. Tran HT, Park YJ, Cho MK, Kim DJ, Ahn DH (2006) Anaerobic Ammonium Oxidation Process in an Upflow Anaerobic Sludge Reactor with Granular Sludge selected from an anaerobic digester. Biotechnology and Bioprocess Engineering 11: 199–204.

39. Liu S, Yang F, Gong Z, Chen H, Xue Y, et al. (2008) The enrichment of ANAMMOX bacteria in non – woven rotating biological contactor reactor. The 2nd International Conference on Bioinformatics and Biomedical Engineering 3315–3318.

40. Ahn YH, Hwang IS, Min KS (2004) Anammox and partial denitritation in anaerobic nitrogen removal from piggery waste. Water Sci Technol 49: 145–153.

41. Van der Star RLW, Miclea AI, Dongen UGJ0MV, Muyzer G, Picioreanu C, et al. (2008) The Membrane Bioreactor: A Novel tool to grow ANAMMOX Bacteria as free cells. Biotechnol Bioeng 101: 286–294.

42. Liao D, Li X, Yang Q, Zhao Z, Zeng G (2007) Enrichment and granulation of Anammox biomass started up with methanogenic granular sludge. World Journal of Microbiology Biotechnology 23: 1015–1020.

43. van de Graaf AA, Bruijn PD, Robertson LA, Jetten MSM, Kuenen JG (1996) Autotrophic growth of anaerobic ammonium oxidizing microorganisms in a fluidized bed reactor. Microbiology 142: 2187–2196.

44. Kawagoshi Y, Nakamura Y, Kawashima H, Fujisaki K, Fujimoto A, et al. (2009) Enrichment culture of marine anaerobic ammonium oxidation (Anammox) bacteria from sediment of sea-based waste disposal site. J Biosci Bioeng 107: 61–63.

45. Bagchi S, Biswas R., Nandy T (2010) Alkalinity and Dissolved Oxygen as Controlling Parameters for Ammonia Removal through Partial Nitritation and ANAMMOX in a Single-stage Bioreactor. J Ind Microbiol Biotechnol 37: 871–876.

46. Chen J, Ji Q, Zheng P, Chen T, Wang C (2010a) floatation and control of granular sludge in a high-rate anammox reactor. Water Research 1–8.

47. Pathak BK, Kazama F, Tanaka Y (2007) Quantification of anammox populations enriched in an immobilized microbial consortium with low levels of ammonium nitrogen and at low temperature. Journal of Bacteriology 1173–1179.

48. Chen J, Zheng P, Yi Y, Tang C, Mahmood Q (2010) Promoting sludge quantity and activity results in high loading rates in anammox UBF. Bioresour Technol 101: 2700–2705.

49. Schmid MC, Walsh K, Webb RI, Rijpstra WIC, van de Pas-Schoonen K, et al. (2003) Candidatus "Scalindua brodae", sp. nov., Candidatus "Scalindua wagneri", sp. nov., two new species of anaerobic ammonium oxidizing bacteria. Syst Appl Microbiol 26: 529–538.

50. Egli K, Fanger U, Alvarez PJJ, Siegrist H, Van der Meer JR, et al. (2001) Enrichment and characterization of an Anammox bacterium from a rotating biological contactor treating ammonium-rich leachate. Arch Microbiol 175: 198–207.

51. Chen HH, Liu ST, Yang FL, Xue Y, Wang T (2009) The development of simultaneous partial nitrification, ANAMMOX and denitrification (SNAD) process in a single reactor for nitrogen removal. Bioresour Technol 100: 1548–1554.

52. Ni SQ, Gao BY, Wang CC, Lin JG, Sung S (2011) Fast startup, performance and microbial community in a pilot scale Anammox reactor seeded with exotic mature granules. Bioresour Technol 102: 2448–2454.

53. Kartal B, Niftrik LV, Sliekers O, Schmid MC, Schmidt I, et al. (2004) Application, eco-physiology and biodiversity of anaerobic ammonium-oxidizing bacteria. Reviews in Environmental Science and Biotechnology 3 : 255–264.

54. Suneethi S, Joseph K (2011) ANAMMOX process startup and stabilization with an anaerobic seed in Anaerobic Membrane Bioreactor (AnMBR). Bioresour Technol 102: 8860–8867.

55. Shivaraman N, Shivaraman G (2003) ANAMMOX–A novel microbial process for ammonium removal. Current Science 84: 1507–1508.

56. Kartal B, Maalcke WJ, de Almeida NM, Cirpus I, Gloerich J, et al. (2011) Molecular mechanism of anaerobic ammonium oxidation. Nature 479: 127–130.

57. Chandran K, Stein LY, Klotz MG, Van Loosdrecht MCM (2011) Nitrous Oxide Production by Lithotrophic Ammonia-oxidizing Bacteria and Implications for Engineered Nitrogen removal Systems. Biochem Soc Trans 39): 1832–1837.

58. Kester R ., Deboer W, Laanbroek HJ (1997) Production of NO and N_2O by pure cultures of nitrifying and denitrifying bacteria during changes in aeration. Appl Environ Microbiol 63: 3872–3877.

59. Anthonisen AC, Loehr RC, Prakasam TBS, Srinath EG (1976) Inhibition of nitrification by ammonia and nitrous acid. Journal of Water pollution control Fed 48: 835–852.

60. Yamamoto T, Takaki K, Koyama T, Furukawa K (2008) Long-term stability of partial nitritation of swine wastewater digester liquor and its subsequent treatment by ANAMMOX. Bioresour Technol 99: 6419–6425.

61. Dapena-Mora A, Fernandez I, Campos JL, Mosquera-Corral A, Mendez R, et al. (2007) Evaluation of activity and inhibition effects on Anammox process by batch tests based on the nitrogen gas production. Enzyme and Microbial Technology 40: 859–865.

62. Suneethi S, Joseph K (2013) Autotrophic ammonia removal from landfill leachate in Anaerobic Membrane Bioreactor (AnMBR). Environ Technol 34: 3161–3167.

63. Wyffels S, Boeckx P, Pynaert K, Zhang D, Van Cleemput O, et al. (2004) Nitrogen removal from sludge reject water by a two-stage oxygen-limited autotrophic nitrification denitrification process. Water Sci Technol 49: 57–64.

64. Ganigue R, Gabarró J, López H, Ruscalleda M, Balaguer MD, et al. (2010) Combining Partial Nitritation and Heterotrophic Denitritation for the Treatment of Landfill Leachate Previous to an Anammox Reactor. Water Sci Technol 61: 1949–1955.

65. Kuai L, Verstraete W (1998) Ammonium removal by the oxygen – limited autotrophic nitrification – denitrification system. Appl Environ Microbiol 64: 4500–4506.

66. Van Niftrik LA, Fuerst JA, Damste JSS, Kuenen JG, Jetten MSM, et al. (2004) The anammoxosome: an intracytoplasmic compartment in Anammox bacteria. FEMS Microbiol Lett 233: 7–13.

67. Jetten MS, Cirpus IEY, Kartal B, van Niftrik LAMP, van de Pas-Schoonen K, et al. (2005) 1994 – 2004: 10 years of research on the anaerobic oxidation of ammonium. Biochem Soc Trans 33: 119–123.

68. Cho S, Takahashi Y, Fujii N, Yamada Y, Satoh H, et al. (2010) Nitrogen removal performance and microbial community analysis of an anaerobic upflow granular bed ANAMMOX reactor. Chemosphere, 78: 1129–1135.

69. Park H, Rosenthal A, Ramalingam K, Fillos J, Chandran K (2010) Linking community profiles, gene expression and N-removal in Anammox bioreactors treating municipal anaerobic digestion reject water. Environ Sci Technol 44: 6110–6116.

70. Sinha B, Annachhatre AP (2007) Assessment of partial nitritation reactor performance through microbial population shift using quinone profile, FISH and SEM. Bioresour Technol 98: 3602–3610.

71. Dapena-Mora A, Arrojo B, Campos JL, Mosquera-Corral A, Mendez R (2004b) Improvement of the settling properties of Anammox sludge in an SBR. Journal of Chemical Technology and Biotechnology 79: 1412–1420.

72. Kartal B, Rattray J, van Niftrik LA, van de Vossenberg J, Schmid MC, et al. (2007) Candidatus "Anammoxoglobus propionicus" A new propionate oxidizing species of anaerobic ammonium oxidizing bacteria. Syst Appl Microbiol 30: 39–49.

73. Neef A., Amann R, Schlesner H, Schleifer KH (1998) Monitoring a widespread bacterial group: in situ detection of planctomycetes with 16S rRNA-targeted probes. Microbiology-UV 144: 3257–3266.

74. Tsushima I, Kindaichi T, Okabe S (2007) Quantification of anaerobic ammonium-oxidizing bacteria in enrichment cultures by real-time PCR. Water Research 41: 785–794.

75. Schmid M, Twachtmann U, Klein M, Strous M, Juretschko S, et al. (2000) Molecular evidence for genus level diversity of bacteria capable of catalyzing anaerobic ammonium oxidation. Syst Appl Microbiol. 23: 93–106.

76. Schmid MC, Maas B, Dapena A, Pas-schoonen KVD, Vossenberg JVD, et al. (2005) Biomarkers for In Situ Detection of Anaerobic Ammonium-Oxidizing (Anammox) Bacteria. Appl Environ Microbiol 71 : 1677–1684.

77. Schmid MC, Hooper AB, Klotz MG, Woebken D, Lam P, et al. (2008) Environmental detection of octahaem cytochrome c hydroxylamine/hydrazine oxidoreductase genes of aerobic and anaerobic ammonium-oxidizing bacteria. Environ Microbiol 10: 3140–3149.

78. Li M, Hong Y, Klotz MG, Gu JD (2010) A comparison of primer sets for detecting 16S rRNA and hydrazine oxidoreductase genes of anaerobic ammonium-oxidizing bacteria in marine sediments. Appl Microbiol Biotechnol 86: 781 – 790.

79. Jetten MSM, Strous M, Pas-schoonen KTVD, Schalk J, Dongen UGJMV, et al. (1999) The anaerobic oxidation of ammonium. FEMS Microbiology Reviews 22: 421–437.

80. Jung JY, Kang SH, Chung YC, Ahn DH (2007) Factors affecting the activity of Anammox bacteria during Startup in the continuous culture reactor. Water Sci Technol 55: 459–468.

81. Bettazzi E, Simone C, Claudia V, Claudio L (2010) Nitrite Inhibition and Intermediates Effects on Anammox Bacteria: A Batch-scale Experimental Study. Process Biochemistry 45: 573–580.

6

Current Review on the Coagulation/Flocculation of Lignin Containing Wastewater

Abu Zahrim Yaser*, Cassey TL, Hairul MA and Shazwan AS

Chemical Engineering Programme, Faculty of Engineering, Universiti Malaysia Sabah, Jalan UMS, 88400 Kota Kinabalu, Sabah, Malaysia

Abstract

Agro based industries including pulp and paper mill, palm oil mill, textile, dairy parlour etc. are characterised by its water intensive nature. The pollution problems related to the agro-based industries wastewater are colour, odour and toxicity. Lignin is the major colorant in the agro based industry wastewater. This review is dedicated to explore the lignin and its removal from wastewater via coagulation/flocculation. Over the years, several studies have been conducted for lignin removal using coagulation/flocculation. Aluminium based and iron based coagulants are widely used for this purpose. However, their usage could deteriorate the performance of biological post treatment. Among others coagulant, calcium based coagulant is seems to be an alternative. Hence, there is interest in developing the use of calcium based coagulants alone or hybrid calcium-other metals coagulant that might fully dismiss or minimize the adverse impact of the aluminium and iron based coagulants. On the whole it looks for the possibility of calcium based salt as a potential coagulant.

Keywords: Lignin; Coagulation; Agro based industry wastewater; Calcium

Introduction

Agro based industries such as pulp and paper mill, palm oil mill, textile and dairy parlour are characterised by water intensive nature [1]. The industry consumes considerable quantities of water for their processes including cleaning and washing purposes and then end up as coloured wastewater. The coloured wastewater might be due to the presence of plant component i.e. lignin, tannin [1] as well as its biodegradation product e.g. melanoidin [2].

The presence of colored compounds in surface water is aesthetically undesirable and causes disturbance of the aquatic biosphere due to reduction of sunlight penetration and depletion of dissolved oxygen. Some of the colored compounds are toxic and mutagenic and have the potential to release the carcinogenic amines [1,3]. Due to their recalcitrant properties, coloured compounds can also contribute to the failure of biological processes in wastewater treatment plants [4]. In some applications, coloured treated water is also not suitable for water reuse [2]. In addition, colour commonly increases throughout biological treatment systems, which may be due to the organic material being converted into smaller chromophoric units rather than being mineralized [5]. This review will focus on the lignin as the main colorant and its treatment using coagulation/flocculation. Prior to that, the lignin structure and class will be discussed.

Lignin

Lignin is the second most abundant natural raw material and nature's most abundant aromatic (phenolic) polymer [6,7] (as cited in Suhas [8]. Lignin is a natural polymeric product from an enzyme initiated dehydrogenative polymerization of the three primary precursors (Figure 1) [9,10]. In another study, Essington [11] proposed the lignin structure as shown in the Figure 2. There are three classes of lignin (Table 1): softwood, hard wood and grass lignin (as cited in Suhas [8]). Generally, lignin particles has negatively charge surface in water [4].

Coagulation/Flocculation

Coagulation/flocculation is one of the most important treatment steps in wastewater treatment plants. Coagulation is the destabilisation of particles using coagulant(s), which can be classified into three main categories: (1) inorganic-based coagulants, (2) organic-based flocculants and (3) hybrid materials [12]. Coagulation tends to overcome the factors that promote particles stability and form agglomerates or flocs. Flocculation is the process of whereby destabilised particles, or particles formed as a consequence of destabilisation, are induced to come together, make contact, and thereby form large(r) agglomerates [13]. The widely used coagulants for lignin containing wastewater are aluminium and iron based (Table 2). They were chosen since they are effective in removing organic substances.

Since in the normal water condition, the lignin particle is negatively charged, then the proposed mechanism of lignin removal could include [14-16]:

- Charge neutralization (inorganic-based coagulants, organic-based flocculants (e.g. poly-diallyl-dimethyl ammonium chloride), hybrid materials)

- Adsorption-precipitation (inorganic-based coagulants),

- Complex chemical reactions/chelation-precipitation (inorganic-based coagulants)

- Sweep coagulation (inorganic-based coagulants at alkali pH)

- Bridging (organic-based flocculants (e.g. poly-diallyl-dimethyl ammonium chloride), hybrid materials)

- Electrostatic patch (organic-based flocculants (e.g. poly-diallyl-

*Corresponding author: Abu Zahrim Yaser, Chemical Engineering Programme, Faculty of Engineering, Universiti Malaysia Sabah, Jalan UMS, 88400 Kota Kinabalu, Sabah, Malaysia
E-mail: zahrim@ums.edu.my

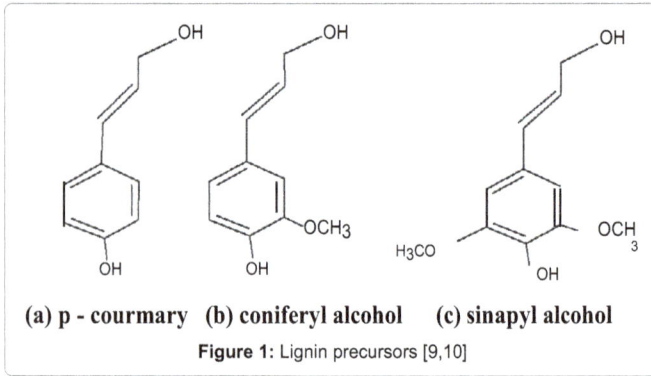

(a) p - courmary   (b) coniferyl alcohol   (c) sinapyl alcohol

Figure 1: Lignin precursors [9,10]

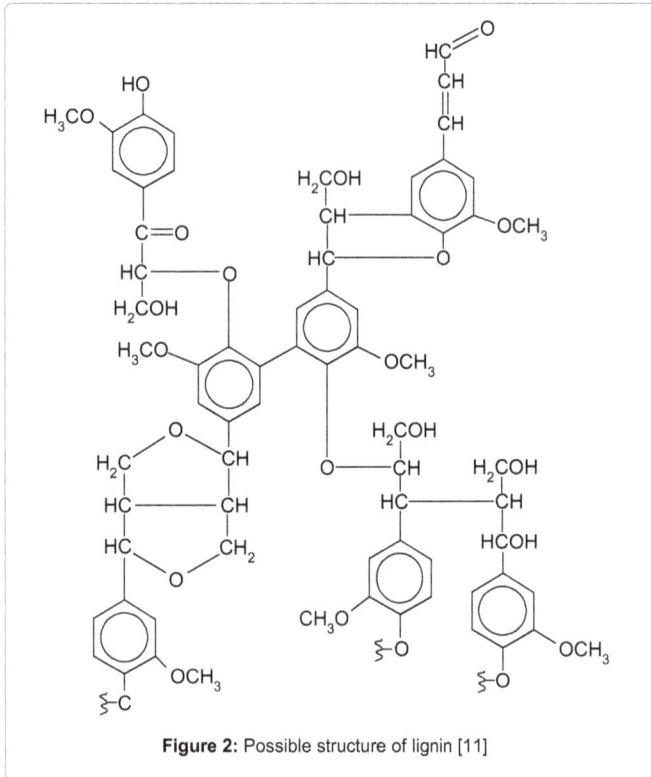

Figure 2: Possible structure of lignin [11]

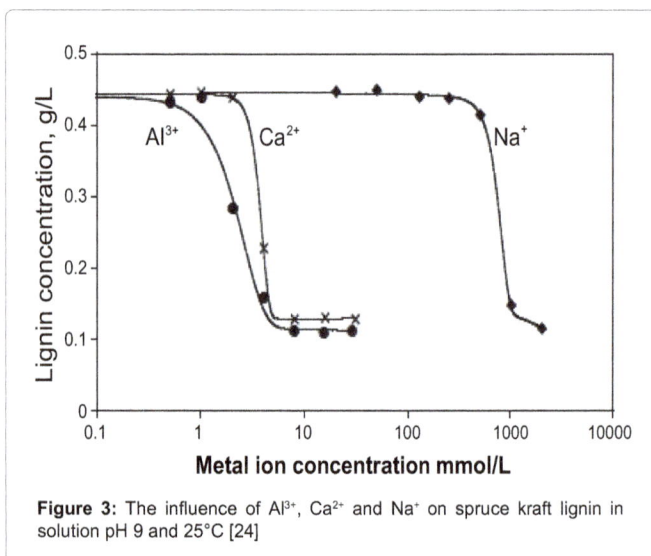

Figure 3: The influence of Al^{3+}, Ca^{2+} and Na^+ on spruce kraft lignin in solution pH 9 and 25°C [24]

dimethyl ammonium chloride), hybrid materials)

The efficiency of coagulation or flocculation process depends on five important factors [12]:

(1) The rate of transport of coagulant or flocculant molecules to the lignin particles in the fluid,

(2) The rate of adsorption of coagulant/flocculant on the surface of lignin particles,

(3) The time scale needed for the coagulant/flocculant layer to reach equilibrium,

(4) The rate of aggregation of lignin particles having adsorbed coagulant/flocculant, and

(5) The frequency of collisions of lignin particles with adsorbed particles to form flocs

Several types of coagulants that can be used for treating lignin containing wastewater are shown in Table 2. Table 2 shows the performance of metal, polymer and hybrid metal-polymer coagulant. The most frequent metal coagulant used in treating agro based wastewater is aluminium sulphate (Table 2).

Ganjidoust [17] investigated aluminium sulphate and synthetic polymers for pulp and paper wastewater treatment. The result shows that the TOC and lignin removal is ~30% and ~80%, respectively. However, treatment using chitosan successfully remove ~90% lignin. Rohella [18] reported that aluminium chloride (10-20 mg/L) unable to remove colour efficiently. However, the application of cationic polyelectrolyte (0.2 mL/L) alone shows that the removal for colour is 82.58%.

Ahmad [19] investigated aluminium sulphate-polyaluminium chloride for palm oil mill effluent treatment. They stated that the COD and TSS removal is 95% and 98%, respectively. Ahmad [20] studied coagulation of pulp and paper mill effluent using aluminium sulphate (~1 g/L) and found that the highest percentage removal of COD, TSS and turbidity is 91%, 99.4% and 99.8%, respectively. Srivastava [21] that utilised poly aluminium chloride (PAC) followed by adsorption with bagasse fly ash (BFA). They reported that at the optimal condition (pH 3; PAC dosage 300 mg/L), it can remove about 80% COD removal and 90% colour [21].

Irfan et al. [22] investigated the performance of different coagulants and flocculants like alum, ferric chloride, aluminium

Class	Other name (lignin)	Unit	Reference
Soft wood	• Guaiacyl • Coniferous	Coniferyl alcohol	
Hard wood	• Dicotyledonous angiosperm	Coniferyl alcohol Sinapyl alcohol	Suhas [8]
Grass lignin	• Annual plant • Monocotyledonous angiosperm	Coniferyl alcohol Sinapyl alcohol p-coumaryl alcohol	

Table 1: Classes of lignin

Industry	Metal	Polymer	Operating Condition			Other	Lignin removal			Reference
			pH	Dosage (mg/L)	Temp (°C)	Removal (%)	Before (mg/L)	After (mg/L)	Removal (%)	
Metal										
Pulp & Paper mill	Aluminium sulphate	–	6	999.6		Turbidity-99.8 TSS-99.4 COD-91	–	–	–	Ahmad [20]
Pulp & Paper mill	Aluminium sulphate	–	7	30		–	–	–	88	Dilek and Bese [29]
Pulp & Paper mill	Aluminium sulphate	–	–	–		–	–	–	80	Ganjidoust [17]
Pulp & Paper mill	Ferric chloride + Aluminium sulphate	–	3	FeCl3-799.97 AlCl3-800.04		COD-18 TSS-49 Color-48	–	–	–	Irfan et al. [22]
Kraft pulp mill (washing water)	Sodium	–	9	22989.80 - 114949.00	25	–	450 - 400	150 -100	66.7 - 75.0	Sundin [22]
	Calcium	–	9	200.39 - 2805.46	25	–	451 - 400	150 -100	66.7 - 75.0	
	Magnesium	–	11	680.54 - 729.15	25	–	452 - 400	100 -50	77.8 - 87.5	
	Aluminium	–	9	134.91 - 1888.71	25	–	453 - 400	120 - 90	73.3 - 77.5	
Palm oil mill (AnPOME)	Calcium lactate		8-8.4	500	–	–	1173 - 1517	_	91	Zahrim et al. [30]
Polymer						_				
Sugar mill	_	PAC	3	300	_	COD-80	_	_	_	Srivastava et al. [21]
Oily	–	Poly-Zinc-Silicate-Sulfate (PZSS) + Anionic Polyacrylamide (APAM)	2	Zn/Si ratio - 1.00-1.50	Ambient	COD-superior TSS- 95 Turbidity- 96.3	–	–	–	Zeng and Park [31]
Paper and pulp	–	Polydadmac + Poly-acrylamide(PAM)	–	Polydadmac-1.20 PAM - 2.00	Ambient	COD-98 TSS-96.8	–	–	71.7	Ariffin [32]
Pulp mill	_	Acrylamide + Starch + 2-methyarcyloyloxyethyl trimethyl ammonium chloride(DMC)	8.35	22.30 and 22.30	Ambient	Turbidity-95.7 Water recovery-72.7	–	–	83.4	Wang [23]
Pulp mill	_	Chitosan + 2-methyarcyloyloxyethyl trimethylammonium chloride(DMC)	7.1	Chitosan - 17.80 DMC- 17.80	Ambient	Turbidity- 99.4 COD-90.7 Water recovery - 89.4	–	–	81.3	Wang [33]
Wood	–	Poly-aluminium Chloride	9.2	125	Ambient	COD-40.9 Color-83.8 LES-92.8	–	–	58.4	Brovkina [34]
Pulp	_	Poly-ethylene	2	350	Ambient	COD-15	_	_	15	Shi [35]
Pulp and paper	–	Natural polymer (chitosan)	–	–	–	TOC-70	–	–	90	Ganjidoust [17]
Metal-Polymer										
Pulp and paper	Aluminium sulphate	Poly-aluminium Chloride (PACl)	6	PACl: 500.00 Alum: 1000.00	Ambient	COD-91 TSS-99.4	–	–	–	Ahmad [36]
Pulp and paper	Ferric chloride	PACl (Poly aluminum chloride) + PAM (Poly acrylamide)	Ferric Chloride: 2 PAC: 3 PAM: 2	Ferric chloride and PACl: 200.00 PAM: 4.00	Ambient	COD - 81 TSS - 95 Colour - Not efficient	–	–	–	Irfan et al. [22]
Pulp and paper	Aluminum chloride + ferrous sulphate	Anionic PAM	Aluminium chloride and Ferrous Sulphate: 6	Aluminium chloride: 800.00 Ferrous sulphate: 800.00 PAM: 4.00		COD-76 TSS-95 Colour-95	–	–	–	Irfan et al. [22]

Palm oil mill	Aluminium sulphate	Poly-aluminium Chloride (PACl)	4.5	Alum : 8000.00 PACl: 600.00	Ambient	COD-95 TSS-98 Residue oil and suspended solids -99	–	–	–	Ahmad [19]
Palm oil mill	Aluminium sulphate + Activated Carbon	Cationic Polyacrylamide	6	Alum: 1700.00	Ambient	COD-85 TSS-99.9 BOD-86.3 Residue oil And suspended solids-95	–	–	–	Ahmad [37]
Pulp and paper	Aluminium sulphate	Syntheticpolymer (HE, PEI, and PAM)	6	–	–	TOC-30	–	–	80	Ganjidoust [17]
Pulp and paper	Aluminium sulphate	Modified natural polymer, starch-g-PAM-g-PDMC [polyacrylamide and poly (2-methacryloyloxyeth-yl) trimethyl ammonium chloride]	8.35	Alum: 871 mg/L, Flocculant dosage: 22.3 mg/L	–	Turbidity: 95.7 Water Recovery Efficiency: 83.4	–	–	72.7	Wang [23]
Kraft paper mill waste water	Aluminium sulphate	Anionic Polyelectrolyte	–	Alum: 300mg/L Anionic Polyelectrolyte : 0.05mg/L	–	Suspended Solid: 91.6 COD: 97	–	–	66.7	Pawels and Bhole [38]
Pulp and Paper	Aluminium Chloride (Alum)	Cationic Polyelectrolyte	–	Alum: 20mg/L Cationic Polyelectrolyte : 0.02mg/L	–	Turbidity: 96.26 COD: 55.65	–	–	82.58	Rohella [18]

Table 2: Coagulants Used to treat lignin containing wastewater

chloride, ferrous sulphate, poly aluminium chloride (PAC), cationic and anionic polyacrylamide polymers in individual form as well as in different combinations. They found that mixed coagulants were found to be more effective in reducing COD, TSS and colour than the individual form. Combination of ferric chloride, PAC and cationic polymer was excellent for reduction of 81% COD and 95% TSS while combination of Aluminium chloride, PAC and Anionic PAM was good in 88% colour reduction [22].

In another study, Wang [23] reported coagulation/flocculation of pulp and paper mill wastewater using aluminum chloride as coagulant and a modified natural polymer, starch-g-PAM-g- PDMC [polyacrylamide and poly (2-methacryloyloxyethyl) trimethyl ammonium chloride], as flocculant aids. They concluded that the coagulation/flocculation was able to reduce the turbidity up to 95.7% with water recovery as much 72.7% [23].

Adverse effect of iron and aluminium based coagulants

From the previous studies, it can be shown that high valence coagulants such as aluminium sulphate and ferric chloride; were being used (Table 2). This probably because as the valence electron increased, the metal ion concentration for coagulation decreases [24] (Figure 3); and this affected the choice of coagulant due to the lower cost. However, these conventional coagulants have been shown to produce less biodegradable wastewater since the coagulation process also remove the amino acids, proteins, and long-chain fatty acids from wastewater [25]. Moreover, it also have been reported that the residual alum and ferric based coagulants (initial concentration for both coagulants is 25 mg/L) inhibit the biological treatment process which is indicated by the reduction of microorganism respiration rate and low organic matter removal [26]. Besides that, the predatory growth in biological wastewater treatment plant also enhanced significantly [26].

The effects of alum addition on the treatment performance of

membrane bioreactor were investigated by Zahid and El-Shafai [27]. The authors reported that the alum doses above 60 mg/L have toxic effect on the autotrophic bacteria with significant reduction in ammonia oxidation and hence the nitrogen removal [27]. Furthermore, the addition of iron and aluminium based coagulants could turn treated wastewater into acidic condition [14,28] in which the neutralization step is required prior to biological treatment.

Potential of Calcium Based Coagulant

Among others coagulant, calcium based coagulant is seems to be an alternative. Hence, there is interest in developing the use of calcium based coagulants alone [39,40] or hybrid calcium-other metals coagulant [41,42] that might fully dismiss or minimize the adverse impact of the aluminium and iron based coagulants. Herewith, we give some examples of the treatment lignin containing wastewater using various calcium based coagulants.

Olive mill wastewater (OMW) generated by the olive-oil extracted industry is a great pollutant contributor in Mediterranean countries. Aktas [43] proposed lime pretreatment proposed to reduce the polluting effect of OMW is an applicable method in practice since lime can easily be purchased anywhere and it is cheaper than other chemicals such as aluminium sulfate, ferric chloride, magnesium sulfate, etc., used for pretreatment of wastewaters [43]. After lime treatment process, chemical oxygen demand (COD) values of the wastewater samples could be reduced by 41.5–46.2% in pressing and centrifugal methods, respectively. The average removal percentage of the other parameters are 29.3–46.9% for total solids, 41.2–53.2% for volatile solids, 74.4– 37.0% for reduced sugar, 94.9–95.8% for oil-grease, 73.5–62.5% for polyphenols, 38.4–32.0% for volatile phenols and 60.5–80.1% for nitrogenous compounds, respectively [43]. The application of $Ca(OH)_2$ for the removal of soluble COD, phenolic and polyphenolic like compounds, and other organic compounds responsible for the olive mill wastewater (OOWW) colour has been studied by Boukhoubza [44]. For 1% lime concentration, the contents of COD and TSS are reduced to around 72%, 73% and 60% for polyphenols [44].

Palm oil mill industry is one of the main agro based industry in Malaysia andIndonesia. Calcium lactate has been used before by Zahrim et al. [30] for the removal of lignin in aerobic palm oil mill effluent (AnPOME) and it shows that 91% of lignin- tannin was removed from the AnPOME which initially had a range concentration of lignin-tannin 1173-1517 mg/L using 500 mg/L of calcium lactate.

Conclusion

Coagulation of lignin particles is important to ensure the effectiveness of wastewater treatment plant. Composite coagulants were found to be more effective in reducing lignin particles than the individual coagulant. Over the years, the treatment of lignin containing wastewater was dominated by aluminium and iron based coagulants. Application of these conventional coagulants should be minimised especially if the next treatment is biological. In this study, the calcium based coagulant has an advantage tend to more biodegradable treated water than the conventional aluminium and iron based coagulants. However, published literatures on this matter are lacking. In addition, the calcium based power of coagulating can be increased by the addition of suitable additive(s).

Acknowledgement

The author would like to thank Ministry of Education, Malaysia for awarding FRGS grant, FRGS/2/2013/TK05/UMS/02/1.

References

1. Anjaneyulu Y, Chary NS, Raj DSS (2005) Decolourization of industrial effluents - available methods and emerging technologies - a review. Rev Environ Sci Biotechnol 4: 245-273.

2. Zahrim AY, Hilal N (2013) Treatment of highly concentrated dye solution by coagulation/flocculation-sand filtration and nanofiltration. Water Resources and Industry 3: 23-34.

3. Neoh CH, Yahya A, Adnan R, Majid ZA, Ibrahim Z (2012) Optimization of decolorization of palm oil mill effluent (POME) by growing cultures of Aspergillus fumigatus using response surface methodology. Environmental Science and Pollution Research: 1-12.

4. Zahrim AY, Mariani R (2014) Diluted Biologically Digested Palm Oil Mill Effluent as a Nutrient Source for Eichornia crassipes. Current Environmental Engineering 1: 45-50.

5. Lewis R, Nothrop S, Chow CWK, Everson A, van Leeuwen JA (2013) Colour formation from pre and post-coagulation treatment of Pinus radiata sulfite pulp millwastewater using nutrient limited aerated stabilisation basins. Separation and Purification Technology 114: 1-10.

6. Gosselink RJA, Jong ED, Guran B, Abacherli A (2004) Co-ordination network for lignin standardisation,production and applications adapted to market requirements (EUROLIGNIN). Industrial Crops and Products 20: 121-129.

7. Lora JH, Glasser WG (2002) Recent Industrial Applications of Lignin: A Sustainable Alternative to Nonrenewable Materials. Journal of Polymers and the Environment 10: 39-48.

8. Suhas, Carrott PJ, Ribeiro Carrott MM (2007) Lignin--from natural adsorbent to activated carbon: a review. Bioresour Technol 98: 2301-2312.

9. Boeriu CG, Bravo D, Gosselink RJA, Dam JEG (2004) Characterization of structure-dependent functional properties of lignin with infrared spectroscopy, Industrial Crops and Products 20: 205 - 218.

10. Chakar FS, Ragauskas AJ (2004) Review of current and future softwood kraft lignin process chemistry. Industrial Crops and Products 20: 131-141.

11. Essington ML (2003) Soil and Water Chemistry: An Integrative Approach.

12. Lee KE, Morad N, Teng TT, Poh BT (2012) Development, characterization and the application of hybrid materials in coagulation/flocculation of wastewater: A review. Chemical Engineering Journal 203: 370-386.

13. Bratby J (2006) Coagulation and Flocculation in Water and Wastewater Treatment. (2ndedtn) IWA Publishing London.

14. Zahrim AY, Tizaoui C, Hilal N (2010) Evaluation of several commercial synthetic polymers as flocculant aids for removal of highly concentrated C.I. Acid Black 210 dye. J Hazard Mater 182: 624-630.

15. Bolto B, Gregory J (2007) Organic polyelectrolytes in water treatment. Water Res 41: 2301-2324.

16. Lee CS, Robinson J, Chong MF (2014) A review on application of flocculants in wastewater treatment. Process Saf Environ Prot.

17. Ganjidoust H, Tatsumi K, Yamagishi T, Gholian RN (1997) Effect of synthetic and natural coagulant on lignin removal from pulp and paper wastewater. Water Science and Technology 35: 291-296.

18. Rohella RS, Choudhury S, Manthan M, Murty JS (2001) Removal of colour and turbidity in pulp and paper mill effluents using polyelectrolytes. Indian J Environ Health 43: 159-163.

19. Ahmad AL, Sumathi S, Hameed BH (2006) Coagulation of residue oil and suspended solid in palm oil mill effluent by chitosan alum and PAC. Chemical Engineering Journal 118: 99-105.

20. Ahmad AL, Wong SS, Teng TT, Zuhairi A (2008) Improvement of alum and PACl coagulation by polyacrylamides (PAMs) for the treatment of pulp and paper mill wastewater. Chemical Engineering Journal 137: 510–517.

21. Srivastava VC, Mall ID, Mishra IM (2005) Treatment of pulp and paper mill wastewaters with poly aluminium chloride and bagasse fly ash Colloids and Surfaces A: Physicochemical and Engineering Aspects 260: 17-28.

22. Irfan M, Butt T, Imtiaz N, Abbas N, Khan RA, et al. (2013) The removal of COD TSS and colour of black liquor by coagulation–flocculation process at optimized pH settling and dosing rate. Arabian Journal of Chemistry.

23. Wang JP, Chen YZ, Wang Y, Yuan SJ, Yu HQ (2011) Optimization of the coagulation-flocculation process for pulp mill wastewater treatment using a combination of uniform design and response surface methodology. Water Res 45: 5633-5640.

24. Sundin J (2000) Precipitation of Kraft Lignin under Alkaline Conditions in Department of Pulp and Paper Chemistry and Technology. Royal Institute of Technology: Stockholm.

25. Dentel SK, Gossett JM (1982) Effect of chemical coagulation on anaerobic digestibility of organic materials. Water Research 16: 707-718.

26. Lees EJ, Noble B, Hewitt R, Parsons SA (2001) The Impact of Residual Coagulant on the Respiration Rate and Sludge Characteristics of an Activated Microbial Biomass. Process Safety and Environmental Protection 79: 283-290.

27. Zahid WM, El-Shafai SA (2012) Impacts of alum addition on the treatment efficiency of cloth-media MBR. Desalination 301: 53-58.

28. Tatsi AA, Zouboulis AI, Matis KA, Samaras P (2003) Coagulation-flocculation pretreatment of sanitary landfill leachates. Chemosphere 53: 737-744.

29. Dilek F, Bese S (2001) Treatment of pulping effluents by using alum and clay - Colour removal and sludge characteristics. African Journal Online 27: 361-366.

30. Zahrim AY, Hillery AH, Yasmi II, Nasimah A, Nurmin B, et al. (2014) Towards recycling of palm oil mill effluent: Coagulation/precipitation of anaerobicallydigested palm oil effluent as a pre-treatment, in Third international conference on recycling and reuse of materials. Kerala, India.

31. Zeng Y, Park J (2009) Characterization and coagulation performance of a novel inorganic polymer coagulant-Poly-zinc-silicate-sulfate Colloids and Surfaces A: Physicochemical and Engineering Aspects 334: 147-154.

32. Ariffin A, Razali MAA, Ahmad Z (2012) PolyDADMAC and polyacrylamide as a hybrid flocculation system in the treatment of pulp and paper mills waste water. Chemical Engineering Journal 179: 107-111.

33. Wang JP, Chen YZ, Yuan SJ, Sheng GP, Yu HQ (2009) Synthesis and characterization of a novel cationic chitosan-based flocculant with a high water-solubility for pulp mill wastewater treatment. Water Res 43: 5267 5275.

34. Brovkina J, Shulga G, Ozolins J (2011) Coagulation of wood pollutants from model wastewater by aluminum salts. Proceedings of the 8th International Scientific and Practical Conference.

35. Shi H, Fatehi P, Xiao H, Ni Y (2011) A combined acidification/PEO flocculation process to improve the lignin removal from the pre-hydrolysis liquor of kraft-based dissolving pulp production process. Bioresour Technol 102: 5177-5182.

36. Ahmad AL, Chong MF, Bhatia S (2008) Population Balance Model (PBM) for flocculation process: Simulation and experimental studies of palm oil mill effluent (POME) pretreatment. Chemical Engineering Journal 140: 86-100.

37. Ahmad AL, Ismail S, Bhatia S (2005) Optimization of coagulation-flocculation process for palm oil mill effluent using response surface methodology. Environ Sci Technol 39: 2828-2834.

38. Pawels R, Bhole AG (1997) Use of synthetic polyelectrolytes in treatment of kraft paper mill wastewater. 39.

39. Devesa-Rey R, Fernandez N, Cruz JM, Moldes AB (2011) Optimization of the dose of calcium lactate as a new coagulant for the coagulation-flocculation of suspended particles in water. Desalination 280: 63-71.

40. Vazquez-Almazan MC, Ventural E, Ricol E, Rodríguez-García ME (2012) Use of calcium sulphate dihydrate as an alternative to the conventional use of aluminium sulphate in the primary treatment of wastewater. Water SA 38: 813-818.

41. Georgiou D, Aivazidis A, Hatiras J, Gimouhopoulos K (2003) Treatment of cotton textile wastewater using lime and ferrous sulfate. Water Res 37: 2248-2250.

42. van Vuuren LRJ, Stander GJ, Henzen MR, Meiring PGJ, van Blerk SHV (1967) Advanced purification of sewage works effluent using a combined system of lime softening and flotation. Water Research 1: 463-474.

43. Aktas ES, Imre S, Ersoy L (2001) Characterization and lime treatment of olive mill wastewater. Water Res 35: 2336-2340.

44. Boukhoubza F, Jail A, Korchi F, Idrissi LL, Hannache H, et al. (2009) Application of lime and calcium hypochlorite in the dephenolisation and discolouration of olive mill wastewater. J Environ Manage 91: 124-132.

7

Estimation of Current and Future Generation of Medical Solid Wastes In Hanoi City, Vietnam

Duc Luong Nguyen[1]*, Xuan Thanh Bui[2] and The Hung Nguyen[3]

[1]Department of Environmental Technology and Management, National University of Civil Engineering (NUCE), Vietnam
[2]Faculty of Environment, Ho Chi Minh City University of Technology (HCMUT), Vietnam
[3]Hanoi Urban Environment One Member Limited Company, Vietnam

Abstract

Management of medical waste is of great importance due to its infectious and hazardous nature that can cause adverse impacts on human health and environment. The objectives of this study were to estimate the current generation of medical solid waste and its existing management practices in Hanoi city, Vietnam. This study also aimed at providing the predictions for future generation of medical solid waste that could serve as scientific basis for planning of medical waste management in Hanoi city. Based on the collected secondary data, the analyses indicated that the generation rate of total medical waste (including normal and hazardous medical waste) is 0.86 kg/bed.day, in which the generation rate of hazardous medical waste is 0.14 kg/bed.day. The major problem associated with existing management practices of medical waste is the treatment and disposal stage. There are no official recycling activities for normal medical waste at present although its legal basis has been setting up in the Medical Waste Management Regulation in 2007 issued by Ministry of Health. With respect to the treatment of hazardous medical waste, incinerators-the major applied technology are being operated inefficiently. For overcoming these obstacles, the local government and relating agencies need to put more effort, in terms of financial and human resources, in facilitating the official recycling activities for normal medical waste and developing more environmentally-friendly alternative treatment technologies for medical waste, towards the gradual replacement of unnecessary incineration. The study predicted that in 2020 and 2030, the quantities of total medical waste generated in Hanoi city would be 30.44 and 46.05 tons/day, respectively which 1.7 and 2.6 times higher than those in 2010. This would be challenging the local government in managing medical waste generated in the future.

Keywords: Medical solid waste management; Normal medical waste; Hazardous medical waste; Prediction; Hanoi city

Introduction

Recently, there has been an increase in the public concern about the management of medical waste on a global basis [1]. Medical wastes are considered as a special category of waste because they pose potential human health and environmental risks, as they contain sharps, human tissues or body parts, discarded plastic materials contaminated with blood, discarded medical equipment, and other infectious materials [2]. About 15–25% (by weight) of medical waste is infectious materials [1]. Despite the fact that current medical waste management practices vary from hospital to hospital, the concerning problems are similar for all hospitals and at all stages of management, including segregation, collection, packaging, storage, transport, treatment and disposal [3]. Improper management of medical wastes could cause environmental pollution, unpleasant odors, and growth of insects, rodents and worms. Subsequently, it may lead to transmission of diseases like typhoid, cholera, and hepatitis through injuries from sharps contaminated with human blood [4]. Therefore, it is important to properly manage medical waste to avoid human health and environmental risks.

In Vietnam, especially in the large cities, the rapidly increasing population as well as the increasing healthcare demand in the recent years have led to the enlarging scale of existing hospitals, in terms of the number of sickbed, from 17.7 sickbeds/10,000 people in 2005 to 22 sickbeds/10,000 people in 2009 [5]. This means that the quantities of medical waste generated from hospitals have significantly increased.

Hanoi–the capital city of Vietnam, is the place where the majority of national-level and largest hospitals located. These hospitals are always in the overloaded status since they must receive a large number of patients from most of provinces in the North of Vietnam. Currently, Hanoi has 16 hospitals with the total of 6,680 sickbeds and 16 medical institutes with the total of 1,030 sickbeds that under the management of Ministry of Health. Hanoi also has 15 hospitals with the total of 3,270 sickbeds that under the management of the other ministries such as Ministry of Construction, Ministry of Transportation, Ministry of Industry and Trade, Ministry of Agriculture and Rural Development, Ministry of Public Security, and Ministry of National Defense. In addition, there are 23 provincial-level hospitals, 16 district-level hospitals, 20 private hospitals, and 300 healthcare centers at communal level located in Hanoi city [6]. With a large number of hospitals and healthcare centers concentrated and the rapidly increasing population using the healthcare services, a large quantity of medical waste, including both normal and hazardous medical waste, being generated daily in Hanoi city. On the other hand, according to Hanoi People's Committee [7], the area of Hanoi city will be expanding in the future with newly developed areas, satellite towns, and districts surrounding the existing centre area. The population of these areas will be increased accordingly. As a result, the quantity of medical waste generated will be increased significantly in the future. This study aims at estimating the current generation and evaluating the existing management practices of medical solid wastes in Hanoi city. This study also provides the

*Corresponding author: Duc Luong Nguyen, Department of Environmental Technology and Management, National University of Civil Engineering (NUCE), 55 Giai Phong Road, Hai Ba Trung District, Hanoi, Vietnam
E-mail: luongnd1@nuce.edu.vn

predictions for future generation of medical solid wastes to the years of 2020 and 2030 in Hanoi city which could be used as scientific basis to support the responsible agencies for the management planning of municipal solid waste in general and medical solid waste in particular.

Materials and Methods

Estimation of current generation and management practices of medical waste

Current generation and management practices of medical solid wastes in hospital in Hanoi city are estimated and evaluated basing on the questionnaire survey data and data collected from the secondary sources.

Estimation of future generation of medical waste

Total medical waste generation in the future in Hanoi city is predicted using the following equation:

M=B.R/1000 (ton/day)

Where:

- M: Total medical waste generated (ton/day)

- B: Number of sickbed (bed). According to Hanoi People's Committee [7], there will be 25 sickbeds/10,000 people and 30 sickbeds/10,000 people in Hanoi city in 2020 and 2030, respectively. The predicted population of Hanoi city in 2020 and 2030 is also taken from [7].

- R: Generation rate of medical waste (kg/bed.day), depending on a number of factors such as the increasing rate of population, social-economic condition, types and scales of hospitals, etc. The generation rates of medical waste used in this study are based on the standard for management planning of solid waste issued by Ministry of Construction as the following [8]:

- National-level hospitals: R=2.2 kg/bed.day

- Provincial-level hospitals: R=1.5 kg/bed.day

- District-level hospitals: R=1.0 kg/bed.day

The proportions of normal and hazardous medical wastes are estimated to account for about 80% and 20% of the total medical waste, respectively [8].

Results and Discussions

Current generation of medical wastes

Generation of medical wastes has been continuously increasing due to the increases of number of hospitals and sickbeds, once using

medical products, and population in the recent years. On average, the generation rate of total medical waste is 0.86 kg/bed.day, in which the generation rate of hazardous medical waste is 0.14 kg/bed.day [5]. Compared to the other studies around the world, the generation rate of medical waste in this study is higher than those reported for Nanjing city, China [9], Turkey [10], and Korea [11]; comparable to the value estimated for Jordan [4]; however, lower than the values reported for North and South American countries and European countries [12], Greece [3], and Taiwan [13]. The generation rate of hazardous medical waste in this study falls into the range (from 0.01 to 0.65 kg/bed.day) reported for developing countries by Diaz et al. [14]. Actually, the generation rates of medical waste depend on the types and scales of hospital (national-, provincial-, and district-level hospitals), types and scales of units in hospitals, technical operation used, and quantities of materials used as shown in Table 1. Generally, the generation rates of medical waste in different units of national-level hospitals are higher than those of provincial- and district-level hospitals.

Generated quantities of normal and hazardous medical waste of the major hospitals in Hanoi city based on the surveyed data are presented in Table 2. The differences in the hospitals' generated medical waste quantities could be attributed to the differences in the number of sickbed, generation rate of medical wastes, and management practices among the surveyed hospitals.

In medical waste generated from hospitals in Hanoi city, the proportion of recyclable materials is relatively high, ~25% of total mass. Medical waste usually has large organic content (about 52% of total mass) and high moisture. The compositions of medical wastes generated from hospitals in Hanoi city [15] are shown in Table 3.

Segregation and collection of medical wastes

Following the Decision No. 43/2007/QĐ-BYT dated 20 Nov 2007 of Ministry of Health on issuing the Medical Waste Management Regulation [16], medical wastes generated in hospitals are segregated and collected into 4 groups: group A (biomedical wastes), group B (sharp wastes), group C (medicine and chemical wastes), and group D (radioactive wastes). Each group is stored in appropriate facilities such as nylon bag, plastic box, metal box, etc. for subsequent transport and treatment and disposal. Medical wastes are classified to normal and hazardous waste. Normal medical wastes (or non-hazardous wastes) are those do not contain infectious, hazardous chemical, radioactive, explosible elements including general waste generated from patient rooms, wastes generated from medical activities (glass bottle, plastic materials, etc.), and wastes generated from offices (paper, nylon bag, etc.). Hazardous medical wastes are those contain harmful elements for people health such as infectious, radioactive, explosible, and flammable elements. The current status of segregation and collection of medical

Units in hospital	Total medical waste (normal & hazardous) (kg/bed.day)			Hazardous medical waste (kg/bed.day)		
	National-level hospitals	Provincial-level hospitals	District-level hospitals	National-level hospitals	Provincial-level hospitals	District-level hospitals
Intensive care and casualty unit	1.08	1.27	1.00	0.30	0.31	0.18
Medical unit	0.64	0.47	0.45	0.04	0.03	0.02
Child unit	0.50	0.41	0.45	0.04	0.05	0.02
Surgical unit	1.01	0.87	0.73	0.26	0.21	0.17
Maternity unit	0.82	0.95	0.74	0.21	0.22	0.17
Eyes and otorhinolaryngology unit	0.66	0.68	0.34	0.12	0.10	0.08
Clinical unit	0.11	0.10	0.08	0.03	0.03	0.03

Table 1: Generation rate of medical waste in different units of hospitals in Hanoi city.

Hospitals	Number of sickbed (bed)	Normal medical waste (kg/month)	Hazardous medical waste (kg/month)
19-8 hospital	450	NA*	1690
National acupuncture hospital	440	832	NA
E hospital	618	NA	3560
Viet Xo hospital	537	3000	NA
National tuberculosis hospital	400	2400	NA
National child hospital	650	NA	4250
National otorhinolaryngology hospital	250	1000	NA
Traditional hospital	200	NA	80
National scalded hospital	300	NA	740
National malaria hospital	33	NA	62
103 army hospital	650	NA	3000
354 army hospital	250	NA	1022
Xanh Pon general hospital	520	1040	NA
Hanoi tuberculosis hospital	110	1500	500
Hanoi endocrine hospital	242	1200	480
Hanoi heart hospital	70	630	NA
Hanoi protuberance hospital	160	NA	500
North Thang Long hospital	359	1538	636
South Thang Long hospital	100	NA	300
Dong Anh general hospital	2150	NA	700
Dong Da general hospital	270	826	500
Ha Dong general hospital	520	NA	1500
Hoe Nhai general hospital	70	125	26
My Đuc general hospital	150	600	30
Soc Son general hospital	200	NA	260
Van Dinh general hospital	230	750	500
Ba Vi district hospital	200	750	300
Phu Xuyen district hospital	130	NA	150
Phuc Tho district hospital	120	987	210
Thuong Tin district hospital	180	NA	420
Son Tay district hospital	420	2632	1375

*: NA = Not available

Table 2: Generated quantity of normal and hazardous medical waste of hospitals in Hanoi city.

Compositions	% by weight
General waste	26.6-40.0
Packing paper, tissues waste	3.0-9.8
Sharp wastes	1.3-2.3
Blood-contaminated wastes	4.6-18.1
Chemical wastes	1.3-13.8
Plastic waste	2.6-3.2
Metal waste	0.6-1.4
Broken glass bottles	1.8-2.6
Medicine waste	0.1-1.6
Others	12.5-26.0

Table 3: Compositions of medical wastes.

wastes generated in surveyed hospitals of Hanoi city is shown in Table 4 which indicated that for several requirements such as waste containing bag in accordance with standard of color, container with coverage lid, etc., the followed proportions were not really high.

Transportation, treatment and disposal of medical wastes

Currently, medical wastes generated from hospitals in Hanoi city are collected and kept at the storage areas of hospitals. For normal medical wastes, they are then transported to the city's centralized treatment areas by Hanoi Urban Environment One Member Limited Company (URENCO) and mixed with municipal solid wastes for final disposal at landfills. There are no official recycling activities for normal medical wastes at present although its legal basis has been setting up in the Medical Waste Management Regulation [16]. One of the major reasons is the lack of investment capital for constructing and operating treatment systems. For hazardous medical wastes, they are combusted in small incinerators located within hospitals or transported for final treatment in incinerators located in the city's centralized treatment areas. However, the proportion of well-operated incinerators is just about 50% of the total number of installed incinerators at present. The major reasons of inefficient operation of incinerators are the lack of finance, lack of trained human resources, high cost for treatment of generated air pollutants, and high cost of combusting fuels. In the future, incinerators should be gradually replaced with more environmental-friendly technologies such as autoclave and microwave oven for treatment of hazardous medical wastes.

Prediction of future generation of medical wastes

The prediction results for the generation of total, normal, and hazardous medical wastes from hospitals in Hanoi city to 2020 and 2030 are shown in Tables 5 and 6, respectively. In 2010, the total, normal, and hazardous medical waste quantities generated in Hanoi city were reported to be 18, 13.5, and 4.5 tons/day, respectively [17]. Therefore, the generated quantities of total, normal, and hazardous medical wastes in 2020 would be about 1.7, 1.8, and 1.4 times higher than those in 2010, respectively. Meanwhile, the generated quantities of total, normal, and hazardous medical wastes in 2030 would be about 2.6, 2.7, and 2.0 times higher than those in 2010, respectively. This would be challenging the local government in managing medical waste generated in the future.

Clearly, the quantities of medical wastes generated in the urban areas are much higher than those in the rural areas, for instance, 3.6 and 4.3 times higher for prediction results in 2020 and 2030, respectively. The reason is that the population and generation rate of medical waste assumed for the urban areas were about 2 times higher than those for the rural areas.

Within the urban areas, the predictions show that the quantities of medical wastes generated in the centre areas are 7.6 and 39.7 times higher than those in the satellite towns and districts in 2020. Meanwhile, in 2030, the quantities of medical wastes generated in the centre areas are predicted to be 4.9 and 42.8 times higher than those in the satellite towns and districts. These imply that in the longer term, the more development of the satellite towns surrounding the city centre, particularly their increasing population is the major cause for the significant increase in the quantity of medical wastes generated in these areas. Whereas, the prediction for the districts shows the lower increasing rate of medical wastes, mainly due to the lower increasing rate of population in these areas.

Conclusions and Recommendations

With a large number of hospitals and healthcare centers concentrated and the rapidly increasing population using the healthcare services, a large quantity of medical waste, including both normal and hazardous medical waste, is being generated daily in Hanoi city. Moreover, with the future enlargement, in terms of area and population, the future generation of medical waste would be significantly increased. In conclusion, the main findings of this study are:

• The generation rate of total medical waste (including normal and hazardous medical waste) is 0.86 kg/bed.day, in which the generation rate of hazardous medical waste is 0.14 kg/bed.day.

Medical Waste Management Regulation	Followed proportion (%)
Waste containing bag in accordance with standard of thickness and volume	66.67
Waste containing bag in accordance with standard of color	30.67
Waste containing bag in accordance with standard of packing	81.33
Sharp waste containing box in accordance with standard	93.90
Container with coverage lid	58.33
Container with label	66.67

Table 4: Current status of segregation and collection of medical wastes.

No	Area	Population* (1,000 people)	Number of sickbed/10,000 people (bed)	Total number of sickbed (bed)	Generation rate of medical waste (kg/bed.day)	Total medical waste (ton/day)	Normal medical waste (ton/day)	Hazardous medical waste (ton/day)
1	Urban areas	4,676.8	25	11,692	-	23.84	19.07	4.77
1.1	Centre areas	3,748.3	25	9,371	2.2	20.62	16.49	4.12
1.1.1	Inner areas	1,727.8	25	4,320	2.2	9.50	7.60	1.90
1.1.2	Newly developed areas	2,020.5	25	5,051	2.2	11.11	8.89	2.22
1.2	Satellite towns	722.2	25	1,806	1.5	2.71	2.17	0.54
1.3	Districts	206.2	25	516	1.0	0.52	0.41	0.10
2	Rural areas	2,642	25	6,605	1.0	6.61	5.28	1.32
3	Hanoi city	7,318.8	25	18,297	-	30.44	24.36	6.09

*: Data taken from 6.   Hanoi Department of Health [6]

Table 5: Prediction of medical waste generation to 2020 in Hanoi city.

No	Area	Population* (1,000 people)	Number of sickbed /10,000 people (bed)	Total number of sickbed (bed)	Generation rate of medical waste (kg/bed.day)	Total medical waste (ton/day)	Normal medical waste (ton/day)	Hazardous medical waste (ton/day)
1	Urban areas	6,218.5	30	18,656	-	37.30	29.84	7.46
1.1	Centre areas	4,606	30	13,818	2.2	30.40	24.32	6.08
1.1.1	Inner areas	1,656	30	4,968	2.2	10.93	8.74	2.19
1.1.2	Newly developed areas	2,950	30	8,850	2.2	19.47	15.58	3.89
1.2	Satellite towns	1,377	30	4,131	1.5	6.20	4.96	1.24
1.3	Districts	235.4	30	706	1.0	0.71	0.56	0.14
2	Rural areas	2,917	30	8,751	1.0	8.75	7.00	1.75
3	Hanoi city	9,135.5	30	27,407	-	46.05	36.84	9.21

*: Data taken from Hanoi Department of Health [6].

Table 6: Prediction of medical waste generation to 2030 in Hanoi city.

• In 2010, the total, normal, and hazardous medical waste quantities generated in Hanoi city were 18, 13.5, and 4.5 tons/day, respectively.

• In 2020, the total, normal, and hazardous medical waste quantities generated in Hanoi city predicted to be 30.44, 24.36, and 6.09 tons/day, respectively.

• In 2030, the total, normal, and hazardous medical waste quantities generated in Hanoi city predicted to be 46.05, 36.84, and 9.21 tons/day, respectively.

Although the Ministry of Health has issued the Medical Waste Management Regulation in 2007 aiming to ensure appropriate handling and processing of medical waste, there is still need to put the regulations into practice, for example, those related to the treatment and disposal of medical waste. For maximizing the health and environmental benefits, the local government and relating agencies should put more effort in the facilitation of the official recycling activities for normal medical wastes. The recycling and reuse of normal medical waste materials are very important in reducing the waste generation as well as disposal cost. Currently, many hospitals in developed countries like USA, UK are operating recycling program to recycle uncontaminated solid waste materials like office paper, cardboard, metal cans and selected glass [18,19]. For successful recycling program, it is important to promote separation of medical waste components at the source. In addition, development and application of more environmentally friendly alternative treatment technologies for medical waste (e.g. microwave sanitation, autoclave, chemical disinfection, pyrolysis, and gasification) should be encouraged, towards the gradual replacement of unnecessary incineration. For example, estimated that microwaving medical waste might be economically competitive compared to the incinerator [20].

References

1. Shinee E, Gombojav E, Nishimura A, Hamajima N, Ito K (2008) Healthcare waste management in the capital city of Mongolia. Waste Manag 28: 435-441.

2. Baveja G, Muralidhar S, Aggarwal P (2000) Medical waste management – an overview. Hospital Today 5: 485–486.

3. Tsakona M, Anagnostopoulou E, Gidarakos E (2007) Hospital waste management and toxicity evaluation: a case study. Waste Manag 27: 912-920.

4. Abdulla F, Abu Qdais H, Rabi A (2008) Site investigation on medical waste management practices in northern Jordan. Waste Manag 28: 450-458.

5. Ministry of Natural Resource and Environment (MONRE) (2011) National report on solid waste management. Medical solid waste.

6. Hanoi Department of Health (2010) Statistic of hospitals in Hanoi city in Vietnamese.

7. Hanoi People's Committee (2011) Report on general planning on social-economic development of Hanoi city to 2020, vision to 2030.

8. Ministry of Construction (2000) Standard for management planning of solid waste in Vietnamese.

9. Yong Z, Gang X, Guanxing W, Tao Z, Dawei J (2009) Medical waste management in China: a case study of Nanjing. Waste Manag 29: 1376-1382.

10. Birpinar ME, Bilgili MS, ErdoĂŸan T (2009) Medical waste management in Turkey: A case study of Istanbul. Waste Manag 29: 445-448.

11. Jang YC, Lee C, Yoon OS, Kim H (2006) Medical waste management in Korea. J Environ Manage 80: 107-115.

12. Hossain MS, Santhanam A, Nik Norulaini NA, Omar AK (2011) Clinical solid waste management practices and its impact on human health and environment--A review. Waste Manag 31: 754-766.

13. Cheng YW, Li KC, Sung FC (2010) Medical waste generation in selected clinical facilities in Taiwan. Waste Manag 30: 1690-1695.

14. Diaz LF, Eggerth LL, Enkhtsetseg Sh, Savage GM (2008) Characteristics of healthcare wastes. Waste Manag 28: 1219-1226.

15. Nguyen TKT (2011) Solid waste management–Hazardous waste. Hanoi Science and Engineering Publishing House.

16. Ministry of Health (2007) The Decision No. 43/2007/QĐ-BYT dated 20 Nov 2007 of Ministry of Health on issuing Medical Waste Management Regulation.

17. Pham ND, Tran HN, Tran TH, Phung TB (2010) Prediction of environmental trend and proposing environmental protection solutions in urbanization process of Hanoi city.

18. Lee BK, Ellenbecker MJ, Moure-Eraso R (2002) Analyses of the recycling potential of medical plastic wastes. Waste Manag 22: 461-470.

19. Tudor TL (2007) Towards the development of a standardized measurement unit for healthcare waste generation. Resources Conservation and Recycling 50: 319–333.

20. Lee BK, Ellenbecker MJ, Moure-Ersaso R (2004) Alternatives for treatment and disposal cost reduction of regulated medical wastes. Waste Manag 24: 143-151.

8

Comparison of Mercury Emissions in USA and China-The Way of Effective Control of Hg from the Power Plant

Haoyi Chen[1], Xiaolong Wang[2], Li Zhong[2], Shisen Xu[2] and Tiancun Xiao[1,2,3]*

[1]*D-102, Guangzhou Boxenergytech Ltd, Guangzhou International Business Incubator, Guanagzhou City, P R China*
[2]*China Huaneng Group Clean Energy Technology Research Institute, China No. 249, Xiaotangshan Industrial Park, Changping District, Beijing, China*
[3]*Inorganic Chemistry Lab, Oxford University, South Parks Road, OX1 3QR, UK*

Abstract

Coal-fired power plants are the largest sources of mercury in China as well as United States, accounting for nearly 50% of industrial mercury releases. USA consumed much more coal than China before 1987, while no Hg control measures had been taken during this period. There has been lots of background Hg emitted into air system, including those from mining, oil & gas extraction, coal combustion, volcanoes and geothermal et.al. However the coal combustion has been considered as the dominant sources. Since 2005, quite a few coal power plants started adopting mercury control measures based on the powder Active Carbon Injection (ACI) technology in USA, metallic Hg emission to air environment has been thus significantly reduced, but leading to Hg content in fly ash is up to 10 ppm, potentially a Hg pollution source if not disposed properly. Although short terms studies have showed little leachate of Hg occurred from the fly dust, there is a potential that the landfilled or dumped dust may be a possible source for the Hg leachate to contaminate underground water in long term, as Hg is mainly trapped as Hg^{2+} or other forms, which are either soluble or minor soluble in water and could go to water environment. Based on this analysis, a more safe and reliable Hg removal technology has been developed and tested.

Keywords: USA; China; Coal power station; Mercury removal

Introduction

Mercury is a heavy metal pollutant with high toxicity, volatility, persistence and bioaccumulation in the environment. Mercury compounds are emitted into the atmosphere from various anthropogenic and natural sources, and are later transported to the surface water and land. It has been shown that mercury content in coal varies between 0.01 and 1.5 g per ton of coal, with world coal consumption in 2011 estimated at 7695 million tons per year therefore coal consumption has been considered as one of the main resources of Hg emission [1].

According to UNEP, current anthropogenic sources are responsible for about 30% of annual emissions of mercury to air [2-6]. Since 2005, the increases in the application of air pollution controls in USA particularly the selective catalytic reduction (SCR), together with more stringent regulations in a number of countries have reduced mercury emissions from coal burning in power plants, and thus offset some part of the emissions arising from the increased coal consumption. In the United States, for example, emissions from coal burning at power plants have reportedly decreased from about 53 tons in 2005 to 27 tons in 2010. However, this is only based on the mercury vapor reduction rate from the coal power station, the reduced Hg was collected, which is in fact still mixed with the fly ash and exists in environment.

In the report of the Global Mercury Assessment 2013 (UNEP, 2013, Global Mercury Assessment 2013: Sources, Emissions, Releases and Environmental Transport. UNEP Chemicals Branch, Geneva, Switzerland), the total anthropogenic emissions of mercury to the atmosphere in 2010 were estimated at 1960 (in the range 1010-4070) tons which is about 30% of the total mercury that was emitted and re-emitted from anthropogenic and natural sources in that year. East and Southeast Asia are responsible for about 40% of global anthropogenic emissions. It is estimated that about 75% of the mercury from this region comes from China, which is about one-third of the global total.

Mercury is dangerous to aquatic and human life [1]. When mercury is deposited in lakes or streams, natural bacteria action converts it to methyl mercury, which makes the mercury available to concentrate in the tissue of fish, wildlife and people who eat the fish. Humans, plants and animals are routinely exposed to mercury and accumulate in Hg containing environment, potentially resulting in a variety of ecological and human health impacts. Human exposure to mercury can result in long-lasting health effects, especially on fetal development during pregnancy. In addition, mercury poisoning has been linked to nervous system disorders, kidney and liver damage, and impaired childhood development.

In a recent bilateral environmental meeting between China and USA in 2006, the director general of US EPA, Dr. Steven Johnson from EPA pointed out that the US Hg contamination may come from China. In Dec 2013, the US EPA director Gina McCarthy said that the Environment in USA West Coastal was being affected by the air pollutant migrated from China. It is assumed that Hg emitted from China migrated in air environment and deposited to the agriculture and the rivers in USA.

This saying has been there for some time, and so far, there have been no direct data on the US and China Hg emission and also no deep understanding on the control way of Hg to effect on agriculture and rivers. In this work, we review the Hg emission from coal power station, and the effectiveness of the Hg control ways, so as to give a more reliable insight into the Hg emission in the two countries.

*Corresponding author: Tiancun Xiao, Inorganic Chemistry Lab, Oxford University, South Parks Road, OX1 3QR, UK
E-mail: xiao.tiancun@chem.ox.ac.uk

Method and Experimental Setup

In this work, the coal consumption and the emission status in the past years have been reviewed with the focus is on the US and China, including coal consumption, the control measures and the final treatment of the Hg containing materials, based on the open available literatures.

We also report a novel Hg sorbent, which has been developed and tested using a fixed bed micro-reactor test system, the mercury is brought into the sorbent-loaded tubular reactor, the metallic Hg content inlet and outlet were analyzed using an Hg analyzer. The unabsorbed Hg in the vent is absorbed with HNO_3-H_2O_2 and $KMnO_4$ solution. The schematic setup of the Hg test is shown in Figure 1.

In this test setup, a cold vapor atomic fluorescence spectrometry mercury vapourmeter MODELIII made by Brooks Rand was used to detect the concentration of gaseous state elemental mercury. It can measure metallic mercury concentration in the gas stream down to ppt level. In order to ensure the measured value and set value perfectly, mercury concentration fluctuation less than ± 2% hourly was required. Gas supply part used the mass flowmeter to control all kinds of gas flow in mixing for simulating flue gas composition. Water vapor and simulated flue gas flows through the quartz tube from inner pipe and outer tube respectively, and was mixed in the fixed bed upper part, then was absorbed by fixed-bed sorbent reactor. After the reaction, gas passed through the mercury measurement instrument which measured the metallic mercury in flue gas. The whole pipe and pipe connection direct connect to mercury using PTEF to protect the surface inside from Hg0 adsorbed in our experimental system. The Hg sorption test was carried out at 140°C, with linear velocity in 0.5-5 m/s.

XRD measurement of the spent sorbent was carried out using X-ray diffraction(XRD) with a Rigaku D/max 2550 diffractometer using CuK_a radiation (l=0.1541 nm) over a 2 theta range from 10° to 80°at a 0.02 step, so as to identify the Hg phase in the sorbent. The Hg pick up capacity is determined by the weight change of the sorbent as well as the penetration curves.

Results and Discussion

Literature results of the re-emitted Hg in the environment

According to the UNEP investigations, current anthropogenic sources are responsible for about 30% of annual emissions of mercury to air [2-6]. Another 10% comes from natural geological sources, and the rest (60%) is from 're-emissions' of previously released mercury that has built up over decades and centuries in surface soils and oceans, although the original source of this reemitted mercury cannot be detected.

However, it has shown that coal power plant accounts for nearly 50% of anthropogenic Hg emission. Taking the coal consumption changes with the time, as shown in Figures 2 and 3, the coal consumption graph from 1965 to 2012 showed that before 1987, US used much more coal than China, and there had no Hg removal measures or action taken at that period, so it is inferred that a fairly amount of the re-emitted mercury in the early stages was from USA, rather than from China.

The coal consumption in China started to exceed USA from 1987, and the exceeded amount is about 100Mt of coal from 1995 to 2004. Thereafter, China was expected to emit more mercury than USA in this period, but the net surplus Hg emission from China is less than the Hg amount emitted from USA before 1987 when considering the coal consumption amount and assumed not Hg removal measures in place.

The coal consumption in China increased sharply from 2004 onwards, and peaked at 1.9 billion tons in 2012 [1,7-9]. The Hg emission from China in this period is the largest in the world. As shown by Jiming Hao et al. [1,10], the Hg emission in 2008 in China may range from 57 tons to 183 tons depending on the confidence interval. However, the migration of metallic Hg is slow and may be most bonded to the local dust and soil, which is easier to contaminate the local environment. Therefore, even the transport distance of the metallic Hg is short, China should pay more attention to the power station locations, as the Hg may mostly deposit in nearby. To prevent the soil from pollution, Hg purification technologies and facilities are urgently needed to install in China, as the coal consumption is soaring since 2005, even the Hg will

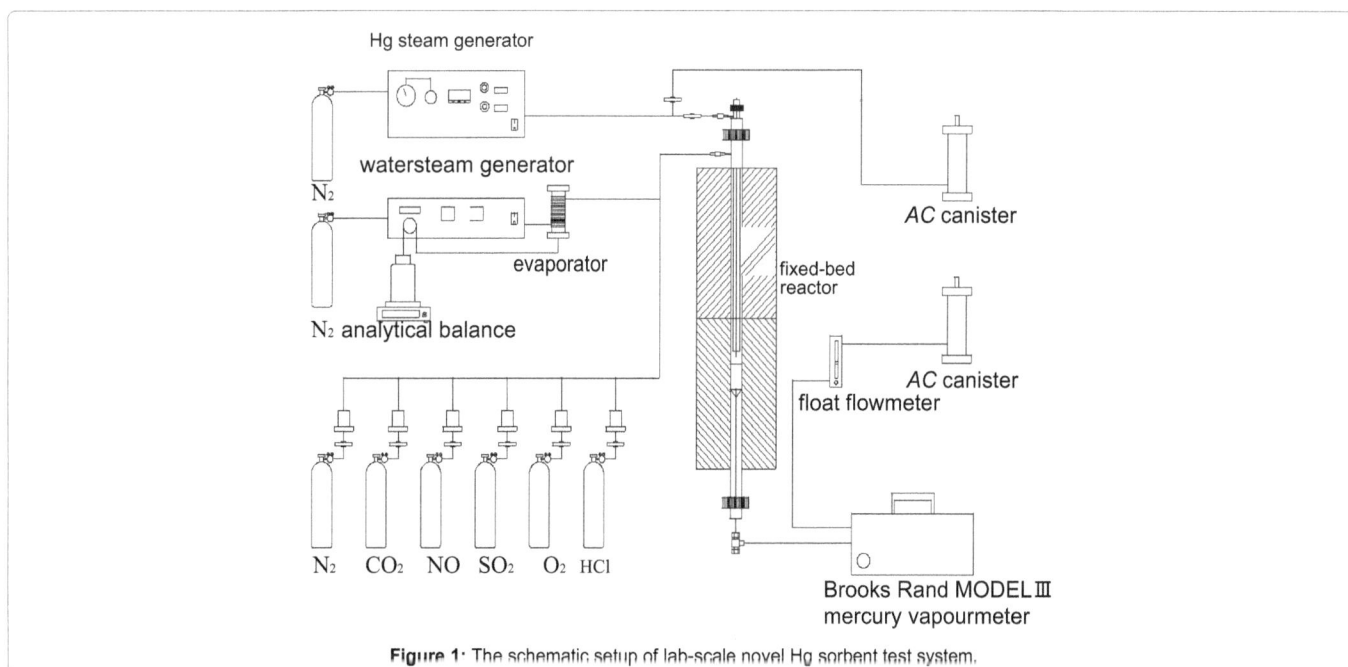

Figure 1: The schematic setup of lab-scale novel Hg sorbent test system.

World Coal Consumption, 1990-2030

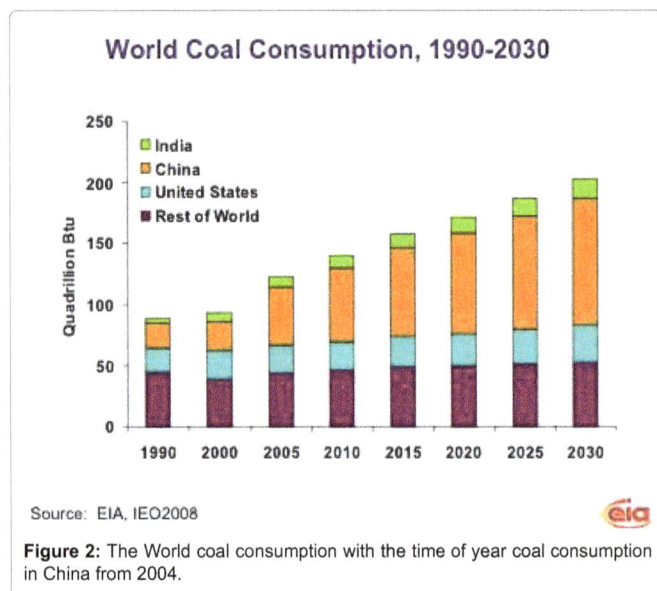

Source: EIA, IEO2008

Figure 2: The World coal consumption with the time of year coal consumption in China from 2004.

US and Chinese Coal Consumption 1965-2012

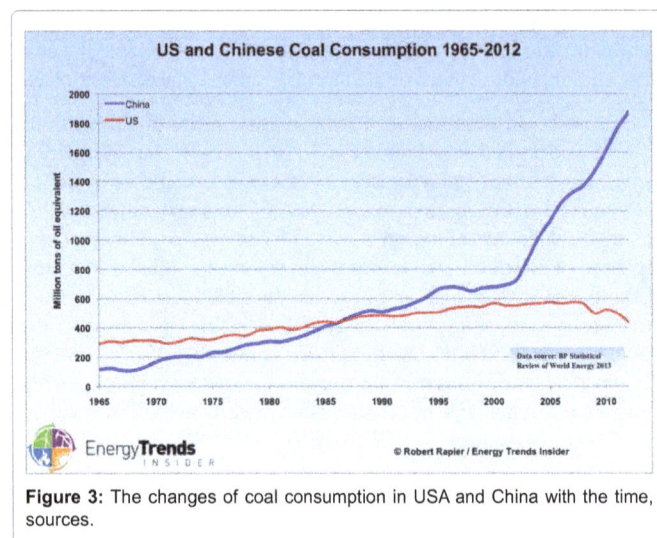

Figure 3: The changes of coal consumption in USA and China with the time, sources.

not migrate to USA, the deposited Hg may contaminate the soil and water in China, which goes into the food chain, and in the end affect the whole world [1,11-15].

Mercury control technology and the remaining issues

The pollution of mercury emission has caused worldwide attention, and many country governments have set out regulations to control Hg emission [16-20]. The US EPA firstly regulated newly built or reconstructed cement plants Hg emission in 2002, and required them to install Hg removal after December 2, 2005. In 2013 U.S. EPA set new limits on mercury emissions for the coal power plants, the Hg emission should be less than 1.36×10^{-3} kg/GWh for all the power stations in 2015. This emission limits correspond to 10 and 4 µg/Nm^3, which are the strictest mercury emission limits. Before that, the mercury removal was mostly removed using the activated carbon injection system, which is shown in Figure 4.

The US is making significant effort in decreasing harmful Hg emissions into the environment through regulations such as the EPA's Clean Air Interstate Rule (CAIR) and the Clean Air Mercury

Rule (CAMR), which placed caps on NOx, SOx and Hg emissions. In March 2011 the EPA stated that all hazardous air pollutants must have emission standards and proposed that for existing sources in the category, that the standards are at least as stringent as the emission reductions achieved by the average of the top 12% best controlled sources for source categories with 30 or more sources. With this new rule, a reduction of mercury from coal emissions of approximately 90% is anticipated continuously to improve.

Typically, activated carbon injection is used to capture oxidized mercury [21-32]. A major limitation with using activated carbon is that in flue gases with low halogen concentrations a large amount of activated carbon needs to be added to the system to effectively control Hg0. Over the last 10 years, most research and development work has focused on the full-scale and slip-stream field testing of activated carbon injection (ACI) and flue gas desulfurization enhancements at nearly 50 U.S. coal-fired power plants. The goal was to demonstrate high levels (50% to 90%) of mercury capture over an extended period of operation, while also reducing the cost of mercury removal. Until 2008, nearly 90 full-scale ACI systems have been ordered by U.S. coal-fired power generators, accounting for over 44 GW of coal-fired elec. generating capacity, and now more coal fired power plants are installing or considering installing the Hg removal facilities using activated carbon injection technology [33-35].

It should be noted that the injected carbon would absorb some of the metallic mercury and may convert it into Hg^{2+} when there exist halogen promoters, such as Br or Cl [36-40]. This is the reason lots of chlorinated or bromide activated carbon has been in the industrial application [41-44]. As shown in Figure 4, the Hg deposited carbon would be collected together with the fly ash. These Hg enriched ashes would be used for cement or land filled. Although there have been some studies on the potential effect of the Hg containing fly ash leachate. However, the leachate experiments were studied for a relatively short period, and so far, there is no quantitative estimation of the Hg in the coal. Here we make a rough estimation in the following.

Generally Hg content is 0.01 and 1.5 g per ton of coal, averaged at 0.17 g per ton [15,16,45,46]. The coal normally contains 5 to 24% of ash, in which average ash content is supposed to be 10%, and 80% of the ash is fly ash. Assume 80% of the mercury is captured by the active carbon, whose content in the collected ash may be neglected, the total mercury from each ton coal combustion in the collected fly ash can be calculated as follows:

The total Hg collected   MHg=0.8×0.17=0.136 g

The total ash collected from each ton of coal=10%×80%×1000=80 Kg of ash The Hg content in the ash 0.136/80000=1.7 ppm.

It should be noted that the Hg content in coal ranges from 0.01 g to 5.0 g in each ton of coal, but in the above calculation, Hg content in coal was set to be 0.17 g, and the Hg content in the resultant collected ash is 1.7 ppm. If Hg content is 1.7 g in coal, the resultant ash may contain up to 17 ppm. Therefore the Hg in the collected ash is in a wide range.

It is clear that the Hg in the collected ash is about 1.7 ppm even with 0.17 g Hg/ton of coal as the fuel, which is significant, thus the collected ash should be carefully disposed. In fact, quite a lot of this kind ash has been landfilled or stored in ash pounds, a short term leachate study may not be able to show the true changes the trace elements changes. There have been few reports on long period leachate study. As shown

How Power Plants Are Reducing Air Emissions

This simplified diagram is illustrative of the operations at a large coal-based electric power plant. It explains the various control technologies in place at many U.S. power plants to reduce emissions to air, land, and water. These technologies are designed to control emissions of nitrogen oxide (NOx), sulfur dioxide (SO2), and particulate matter (PM). In addition, these control technologies capture significant amounts of other air emissions, including mercury. The diagram also illustrates the ways that byproducts of coal combustion are recycled into useful products.

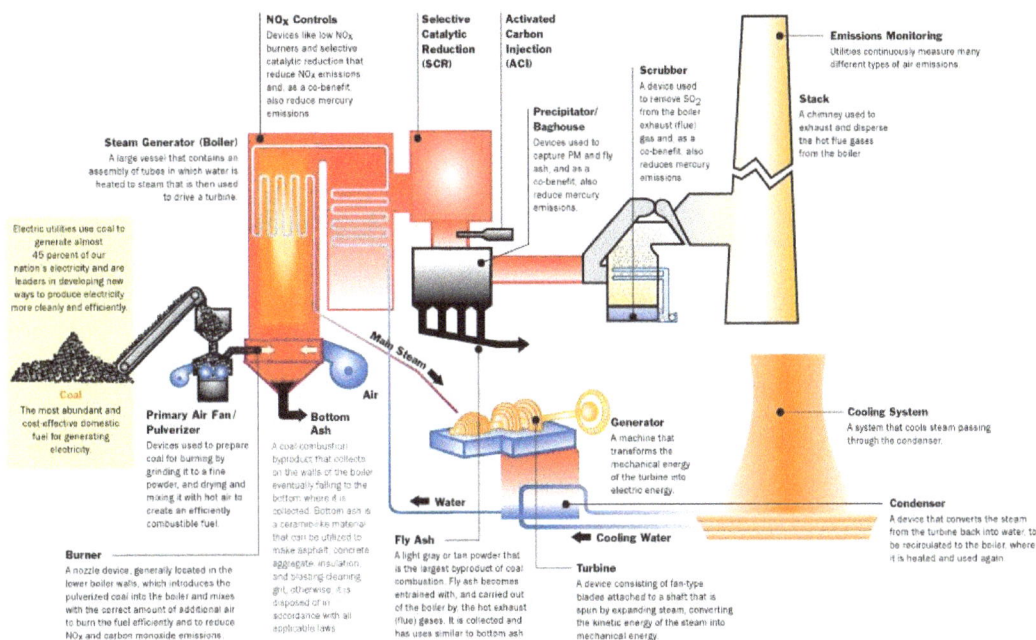

Based on data from the Edison Electric Institute

Figure 4: The typical flue gas purification schematic in US coal power plant.

	Name	Molar Weight (g/mol)	Melting point (°C)	Boiling point (°C)	Decomposition /sublimate temperature (°C)	Density (g/cm³)	Aqueous solubility (g/l at 25°C)
Hg(0)	Elemental mercury	200.59	−38.8	356.7	n.a.	13.534	5.6×10^{-7}
Hg_2Cl_2	Mercurous chloride	472.09	525	n.a.	383	7.15	0.002
$HgCl_2$	Mercuric chloride	271.5	277	302	n.a.	5.43	28.6
Hg_2SO_4	Mercurous sulphate	497.24	n.a.	n.a.	n.a.	7.56	0.51
$HgSO_4$	Mercuric sulphate	296.66	n.a.	n.a.	450	6.47	Decomposes
HgS	Mercury sulfide	232.66	n.a.	446–583	580	8.1	Insoluble
HgO	Mercuric oxide	216.59	n.a.	356	500	11.14	Insoluble
Hg_2Br_2	Mercurous bromide	560.99	405	n.a.	340–350	7.307	3.9×10^{-4}
$HgBr_2$	Mercuric bromide	360.44	237	322	n.a.	6.03	Slightly soluble
Hg_2I_2	Mercurous iodide	654.98	n.a.	n.a.	140	7.7	Slightly soluble
HgI_2	Mercuric iodide	454.4	259	350	n.a.	6.36	0.06
Hg_2F_2	Mercurous fluoride	439.18	n.a.	n.a.	570	8.73	Decomposes
HgF_2	Mercuric fluoride	238.59	645	650	645	8.95	Soluble, reacts
$Hg_2(NO_3)_2$	Mercurous nitrate	525.19	n.a.	n.a.	70 (dihydrate)	4.8 (dihydrat)	Slightly soluble, Reacts
$Hg(NO_3)_2$	Mercuric nitrate	324.7	79	n.a.	n.a.	4.3	Soluble
$Hg(CN)_2$	Mercuric cyanide	252.63	320	n.a.	n.a.	3.996	Soluble

Table 1: Properties of selected mercury compounds [20,47-49]. n.a.: not available.

in Table 1 of the various Hg compounds, $HgCl_2$ or $HgBr_2$ are either soluble or minor soluble in water, hence it may gradually leach out and pollute the environment if they are present in the collected ash.

Based on the above analysis, it is very difficult to conclude that the mercury collected in the injected activated carbon in US power plants has been securely stored and no secondary pollution of the collected active carbons occurs. Because if mercury is emitted from the flue gas in metallic form, it may transport into air environment. However,

when it is enriched as Hg^{2+} in the fly ash, it may increase the risk of the Hg leachate under some special conditions; more attention needs to be paid to the nearby underground water in the ACI Hg collection power plant.

Identification of the Hg species from the spent sorbent

So far in the fly ash collected from the active carbon injection technology, the Hg species in the sorbent is very difficult to be identified, as the amount of Hg is too little to be detected. Here we have developed series Hg sorbent and tested them over a fixed bed test system.

The Hg break curves are shown in Figure 5, the Hg removal ratio is about 99.9999% in the first 17 days. The inlet Hg content is about 16806 ppb, with the Hg conversion about 99.9999%; the outlet Hg content is reduced down to lower than 0.1 ppb in a single pass. The Hg pickup capacity is up to 10 wt% according to the mass change of the sorbent. The spent sorbent was characterized using XRD and the results are shown in Figure 6. It is shown that the Hg is mainly present as $HgCl_2$, HgCl when $CuCl_2$ is present on the sorbent. For the bromided sorbent, Hg exists as $HgBr_2$. The SO_2 sulfur seems to have little influence on the formation of $HgBr_2$.

Because the Hg capacity is up to more 10 wt%, and given the Hg amount in a power plant is limited, it is proposed to install fixed bed Hg removal system in coal power station, e.g., installing a mercury removal sorbent bed in the chimney and let the flus gas flow through the sorbent bed after the dust is removed. The spent sorbent could thus be collected and properly landfilled, so as to avoid the secondary pollution by the leachate, which may be more difficulty to remedy in the future.

Conclusion

Mercury emission from anthropogenic sources has become increasing serious issue and caused environmental pollution. More than 60% of current Hg pollution might be from the re-emission of the Hg vapor discharged to the environment before 1987.

Coal fired power stations contribute to up to 40% of Hg emission worldwide. Before 1987, US consumed much more coal than in China, which may be the major source for Hg emission in the air, as no control measures were taken at that period.

In recent years, many countries started to regulate the Hg emission form coal fired power station and the measure is mainly based on activated carbon or halogen modified activated carbon injection. This can remove up to 90% of Hg from the flue gas, and converted it into Hg^{2+} and deposits over the carbon. However the Hg containing activated carbon is present in the fly ash in which Hg content can be up to 17 ppm, which may be a static source for Hg leachate if the fly ash is not disposed properly.

Novel Hg sorbent has been developed and tested in a fixed bed Hg removal micro-reactor whose Hg pick capacity is up to 10wt%, which can reduce the Hg in the flue gas to 0.1 ppb. The concentrated Hg sorbent has small volume and can be safely disposed easily.

References

1. Wang S, Zhang L, Zhao B, Meng Y, Hao J (2012) Mitigation Potential of Mercury Emissions from Coal-Fired Power Plants in China. Energy Fuels 26: 4635-4642.

2. Forte R, Ryan J, Johnson TP, Kariher PH (2012) The United States Environmental Protection Agency's Mercury Measurement Toolkit: An Introduction. Energy Fuels 26: 4643-4646.

3. Masekoameng KE, Leaner J, Dabrowski J (2010) Trends in anthropogenic mercury emissions estimated for South Africa during 2000-2006. Atmos Environ 44: 3007-3014.

4. Nelson PF, Morrisona AL, Malfroya HJ, Copeb M, Leeb S, et al. (2012) Atmospheric mercury emissions in Australia from anthropogenic, natural and recycled sources. Atmos Environ 62: 291-302.

5. Romanov A, Sloss L, Jozewicz W (2012) Mercury Emissions from the Coal-Fired Energy Generation Sector of the Russian Federation. Energy Fuels 26: 4647-4654.

6. Sloss L (2012) Special Section on Mercury Emissions from Coal. Energy Fuels 26: 4623.

7. Wang SX, Zhang L, Li GH, Wu Y, Hao JM, et al. (2010) Mercury emission and speciation of coal-fired power plants in China. Atmos Chem Phys 10: 1183-1192.

8. Xu W, Wang H, Zhu T, Kuang J, Jing P (2013) Mercury removal from coal combustion flue gas by modified fly ash. J Environ Sci 25: 393-398.

9. Zhang L, Zhuo Y, Chen L, Xu X, Chen C (2008) Mercury emissions from six coal-fired power plants in China. Fuel Process Technol 89: 1033-1040.

10. Zhang L, Wang S, Meng Y, Hao J (2012) Influence of Mercury and Chlorine Content of Coal on Mercury Emissions from Coal-Fired Power Plants in China. Environ Sci Technol 46 6385-6392.

11. Feng L (2009) Study on modes of occurrence of mercury in coals and mercury removal during coal preparation. Prog Environ Sci Technol 2: 2080-2083.

12. Wang LG, Chen CH, Kolker KH (2005) Vapor-phase elemental mercury adsorption by residual carbon separated from fly ash. J Environ Sci 17: 518-20.

13. Wu D, Zhang S, Zhu T (2011) Controlling mercury emission for China's coal fired electricity plants: an economic analysis. Energy Procedia 5: 1439-1454.

14. Zhao YC, Zhang JY, Liu J, Mercedes DS, Patricia AV, et al. (2010) Experimental study on fly ash capture mercury in flue gas. Sci China Technol Sci 53: 976-983.

15. Zheng L, Liu G, Chou CL (2007) The distribution, occurrence and environmental effect of mercury in Chinese coals. Sci Total Environ 384: 374-383.

16. Hower JC, Mastalerz M, Drobniak A, Quick JC, Eble CF, et al. (2005) Mercury content of the Springfield coal Indiana and Kentucky. Int J Coal Geol 63: 205-227.

17. Kay JP, Pavlish BM, Jones ML, Lentz NB, Eutizi JG (2009) Assessment of

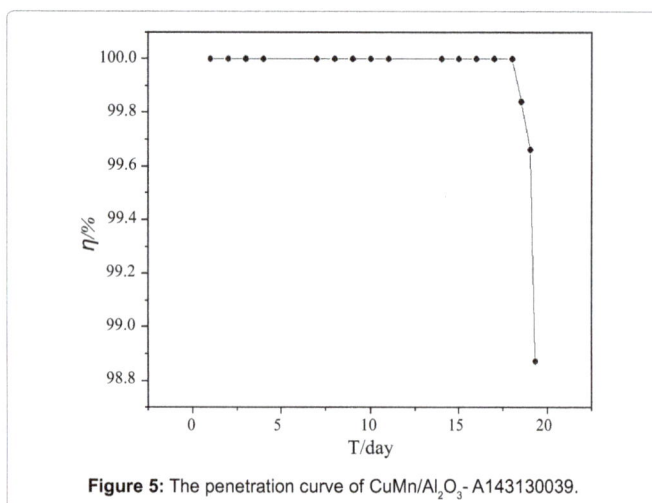

Figure 5: The penetration curve of CuMn/Al$_2$O$_3$- A143130039.

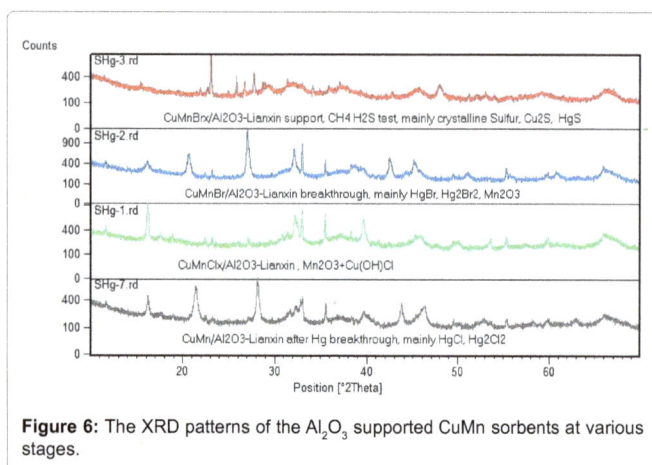

Figure 6: The XRD patterns of the Al$_2$O$_3$ supported CuMn sorbents at various stages.

mercury control options for the San Miguel Electric Cooperative Power Plant. Int Tech Conf Clean Coal Fuel Syst 34: 491-493.

18. Strezov V, Morrison A, Nelson PF (2007) Pyrolytic Mercury Removal from Coal and Its Adverse Effect on Coal Swelling. Energy Fuels 2007 21: 496-500.

19. Winkel H, Meij R (2006) Mercury emissions from coal-fired power stations: The current state of the art in the Netherlands. Sci Total Environ 368: 393 – 396.

20. Zheng Y, Jensen AD, Windelin C, Jensen F (2012) Review of technologies for mercury removal from flue gas from cement production processes. Prog Energy Combust Sci 38: 599-629.

21. Chang R, Bustard CJ (1994) Sorbent injection for flue gas mercury control. Proc Annu Meet-Air Waste Manage Assoc.

22. Chen WC, Lin HY, Yuan CS, Hung CH (2009) Kinetic modeling on the adsorption of vapor-phase mercury chloride on activated carbon by thermogravimetric analysis. J Air Waste Manage Assoc 59: 227 – 235.

23. Chi Y, Yan N, Qu Z, Qiao S, Jia J (2009) The performance of iodine on the removal of elemental mercury from the simulated coal-fired flue gas. J Hazard Mater 166: 776-781.

24. De M, Azargohar R, Dalai AK, Shewchuk SR (2013) Mercury removal by bio-char based modified activated carbons. Fuel 103: 570-578.

25. Diao YF, Ding JD, Yu WX, Zou Y, hao WH (2012) Application of activated carbon fiber coated with cobalt oxide in mercury removal from simulated flue gas. Adv Mater Res 356-360: 547-553.

26. Dombrowski K, Richardson C, Padilla J, Fisher K, Campbell T, et al. (2009) Evaluation of low ash impact sorbent injection technologies for mercury control at a Texas lignite/PRB fired power plant. Fuel Process Technol 90: 1406-1411.

27. Durham M (2004) Sorbent injection making progress. Power 64: 66-67.

28. Fan X, Li C, Zeng G, Gao Z, Chen L, et al. (2010) Removal of Gas-Phase Element Mercury by Activated Carbon Fiber Impregnated with CeO2. Energy Fuels 24: 4250-4254.

29. Feeley TJ, Brickett LA, O'Palko AB (2009) DOE/NETL's mercury control technology R&D program-taking technology from concept to commercial reality. Power Plant Chem 11: 402-411.

30. Felsvang K, Gleiser R, Juip G, Nielsen KK (1994) Activated carbon injection in spray dryer/ESP/FF for mercury and toxics control. Fuel Process Technol 39: 417-430.

31. Ghorishi SB, Keeney RM, Serre SD, Gullett BK, Jozewicz WS (2002) Development of a Cl-Impregnated Activated Carbon for Entrained-Flow Capture of Elemental Mercury. Environ Sci Technol 36: 4454-4459.

32. Hargis RA, O'Dowd WJ, Pennline HW (2001) Pilot-scale research at NETL on sorbent injection for mercury control. Proc Int Tech Conf Coal Util Fuel Syst 26: 469-480.

33. Gleiser R, Felsvang K (1994) Mercury emission reduction using activated carbon with spray dryer flue gas desulfurization. Proc Am Power Conf 56: 452-457.

34. Helfritch DJ, Feldman PL (1998) The pilot scale testing of a circulating fluid bed fine particulate and mercury control device. FACT Am Soc Mech Eng 22: 757-763.

35. Hutson ND (2008) Mercury capture on fly ash and sorbents: the effects of coal properties and combustion conditions. Water Air Soil Pollut: Focus 8: 323-331.

36. Connell DP, Roll DJ, Huber WP (2008) First-year operating experience from the Greenidge Multi-Pollutant Control Project. Proc Int Tech Conf Coal Util Fuel Syst 33: 1179-1190.

37. Diamantopoulou I, Skodras G, Sakellaropoulos GP (2010) Sorption of mercury by activated carbon in the presence of flue gas components. Fuel Process Technol 91: 158-163.

38. Laumb JD, Benson SA, Olson EA (2004) X-ray photoelectron spectroscopy analysis of mercury sorbent surface chemistry. Fuel Process Technol 85: 577-585.

39. Murakami A, Uddin A, Ochiai R, Sasaoka E, Wu S (2010) Study of the Mercury Sorption Mechanism on Activated Carbon in Coal Combustion Flue Gas by the Temperature-Programmed Decomposition Desorption Technique. Energy Fuels 24: 4241-4249.

40. Sidhu S, Varanasi P (2005) Mercury transformation reactions on Ohio coal fly ashes. Proc Annu Int Pittsburgh Coal Conf.

41. Rallo M, Lopez-Anton MA, Meij R, Perry R, Maroto-Valer MM (2010) Study of mercury in by-products from a Dutch co-combustion power station. J Hazard Mater 2010 174: 28-33.

42. Stergarsek A, Horvat M, Frkal P, Stergarsek J (2010) Removal of Hg0 from flue gases in wet FGD by catalytic oxidation with air-An experimental study. Fuel 89: 3167-3177.

43. Vidic RD, Siler DP (2000) Vapor-phase elemental mercury adsorption by activated carbon impregnated with chloride and chelating agents. Carbon 39: 3-14.

44. Zeng H, Jin F, Guo J (2003) Removal of elemental mercury from coal combustion flue gas by chloride-impregnated activated carbon. Fuel 83: 143-146.

45. Dabrowski JM, Ashtona PJ, Murray PJ, Leanerb JJ, Masonc RP (2008) Anthropogenic mercury emissions in South Africa: Coal combustion in power plants. Atmos Environ 42: 6620-6626.

46. Szwed-Lorenz J, Lorenz K (1999) Environmental hazards due to mercury: mercury content in the xylites of Poland's brown coal mines and in conifer wood. Environ Prot Eng 24: 27-30.

47. Bisson TM, MacLean LC, Hu Y, Xu Z (2012) Characterization of Mercury Binding onto a Novel Brominated Biomass Ash Sorbent by X-ray Absorption Spectroscopy. Environ Sci Technol 46: 12186-12193.

48. Li Z, Hwang JY (1997) Mercury distribution in fly ash components. Proc Annu Meet Air Waste Manage Assoc.

49. Zheng Y, Jensen AD, Windelin C, Jensen F (2012) Dynamic measurement of mercury adsorption and oxidation on activated carbon in simulated cement kiln flue gas. Fuel 93: 649-657.

9

High Quality Oil Recovery from Oil Sludge Employing a Pyrolysis Process with Oil Sludge Ash Catalyst

Shuo Cheng[1]*, Aimin Li[2] and Kunio Yoshikawa[1]

[1]Department of Environmental Science and Technology, Tokyo Institute of Technology, Japan
[2]School of Environmental Science and Technology, Dalian University of Technology, China

Abstract

In this study, pyrolysis experiments of two kinds of oil sludge were conducted using a bench-scale fixed bed reactor, which is mainly consisted of a pyrolyzer and a reformer reactor. The effects of the pyrolysis method and the catalysts on the yield and quality of oil products obtained from pyrolysis of oil sludge were investigated. Characterization of the oil products was preliminary judged by comparing their distillation results with the diesel standard and the oil product without catalyst usage. Then, the chemical characterization of a part of the pyrolytic oil was conducted by FT-IR and NMR analysis. In the last step of these experiments, the oil products obtained from pyrolysis of oil sludge with the use of oil field sludge ash and oil tank sludge ash as catalysts were mixed with diesel oil by the volume ratio of one-to-five and one-to-ten. The fuel characteristic of the oil mixtures was identified from the point of the viscosity, the density, the cetane index, the carbon number distribution and the higher heating value (HHV). The results of experiment show that the highest oil product yield was obtained at the one-stage pyrolysis of the oil field sludge and also the oil tank sludge with the oil field sludge ash as the catalyst. According to the results of NMR analysis, the main part of the carbon in the oil product from pyrolysis of the oil sludge with the oil sludge ash was aliphatic carbon. The total ring number of oil products decreased owing to the hydrogenation and ring-open of naphthenic hydrocarbon during the catalytic pyrolysis process. The highest HHV and the cetane index could be obtained when the oil product from the one-stage pyrolysis process of the oil field sludge with the oil field sludge ash was mixed with diesel by the mixing ratio of 1: 10. The results of this study are of practical interest and will help to lead the researchers to more thinking about the oil recovery from oil sludge.

Keywords: Oil sludge; Oil sludge ash; Pyrolysis; Structural parameter

Introduction

Oil sludge, an inevitable by-product, is generated wherever exploitation, transportation and refining processes of petroleum industry. It is a complex mixture of petroleum hydrocarbons and water with solid mineral admixture [1]. In China, the growth of oil demand resulted in annual generation of over 110,000 tons of oil field sludge and 800,000 tons of oil storage tank sludge (oil tank sludge) [2]. Both the oil field sludge and the oil tank sludge comprise abundant of toxic substances from the most carcinogenic polycyclic aromatic to heavy metal, and even to radioactive material. Most of them pose the potential risks for human health and the environment. However, at several sites in China, the amount of uncontrolled disposal of oil sludge to the natural environment had reached or even exceeded 20 times of the assimilative capacity of receiving ecosystem. The major sources of oil sludge in this study include the oil field sludge, which is also called hydrocarbon contaminated soil, and the oil storage tank sludge derived from the tank cleaning process.

Currently, the most common way to dispose the oil sludge is immobilization and landfill, which requires much space and causes a serious threat to soil and groundwater environment. Biodegradation is considered to be a new and emerging method to decompose the organic compounds in the oil sludge into the lower molecular weight organic compounds. However, due to its unsatisfactory performance on the decompose process of asphaltene composition; this technique remains to be developed.

On the other hand, oil sludge is also a potential recycling resource for its high heating value. So the conversion of the stored energy of oil sludge to various fuel sources for power plants or engines has been recognized as an attractive approach. Some chemical recycling methods such as solvent extraction or adding the demulsifier have been tested by many researchers [3-5], and they are indeed feasible in improving the rate of oil recovery. Nevertheless, both of these ways need vast amount of extra organic solution or additive.

Thermal disposal methods: incineration and pyrolysis offer some benefits over these methods as is mentioned above in the aspect of energy recovery and waste reduction [6]. Since the existence of problems of the secondary pollution [7,8] and the high viscosity of fuels [9], the application of incinerating oil sludge directly is limited. With its characteristics to crack high molecular weight organic compounds into lower ones, and to separate the stable emulsion of oil sludge into oil, water, and residue fraction efficiently, pyrolysis treatment were used widely in the field of oily waste disposal. A series of studies on the pyrolysis of oil sludge using TGA [10-15], rotary kiln [1], and fluidized bed [9] have been carried out. In these previous studies, the behavior of thermal conversion and the reaction models of the pyrolysis kinetics of oil sludge were proposed to explain the reaction mechanism. The oil products were reported to be close to diesel oil, however, they also contain certain amount of heavy components, like asphaltene, which affect the qualities of oil products severely [15,16].

*Corresponding author: Shuo Cheng, Department of Environmental Science and Technology, Tokyo Institute of Technology, G5-8, 4259 Nagatsuta, Midori-ku, Yokohama 226-8502, Japan
E-mail: cheng.s.ab@m.titech.ac.jp

To deal with this problem, the catalytic cracking, which means adding a useful catalyst or additive in the cracking process, has some advantages over the non-catalytic cracking. This catalyst adding method has a great potential to shorten the reaction time, lower the reaction temperature and narrow the product distribution [12]. In the previous study, the catalytic cracking of oil sludge in the presence of aluminum compounds (Al, Al_2O_3 and $AlCl_3$); and iron compounds (Fe, Fe_2O_3, $FeSO_4 \cdot 7H_2O$, $FeCl_3$ and $Fe_2(SO4)_3 \cdot nH_2O$) [12] have been reported in detail. Based on the above studies, the addition of Fe_2O_3 and $Fe_2(SO_4)_3 \cdot H_2O$ can improve the qualities of pyrolysis oil products. The effect of sodium compounds (NaOH, NaCl and Na_2CO_3); and potassium compounds (KCl, KOH and K_2CO_3) were tested by Je-LuengShie and Ching-Yuan Chang [13]. They pointed out that KCl and Na_2CO_3 showed better performance in the pyrolysis process of oil tank sludge. Recently, these authors tested the catalytic solid wastes (DAY-zeolite and PVA), fly ash, and oil tank sludge ash as additive, and proved the effectiveness of fly ash and oil tank sludge ash in the improvement of oil quality [15]. However, most of the studies about catalytic cracking of oil sludge were carried out by TGA. Although, TGA can be seen as a simulation of fixed bed reactor, still there is a big difference between them.

In the present study, a series of pyrolysis experiment was conducted using a bench-scale fixed bed reactor, which is mainly consisted of a pyrolyzer and a reformer reactor. The experiment includes two basic pyrolytic methods namely: one-stage catalytic pyrolysis and two-stage catalytic pyrolysis. In the one-stage catalytic pyrolysis, the oil sludge was placed in the pyrolyzer reactor with a catalyst or an additive. In this case, however, the recovery of catalyst could be difficult. Therefore, the two-stage catalytic pyrolysis process is commonly used to treat the waste plastic. By this method, a high recovery rate of a catalyst can be guaranteed due to keeping the catalyst separate from the feedstock. One of the objectives of this study is to compare the pyrolysis effect of these two pyrolysis methods on the product yield and the quality of the pyrolysis oil obtained from two kinds of oil sludge.

Although most of the catalyst can be recovered from the reformer, they would be deactivated by poisoning, like sulfur, after several runs [12,17]. From the commercial point of view, the catalyst or additive should be cheap and easily available material. Therefore, anther objective of this study is to identify the catalytic effect of oil sludge ash, which is considered to be a waste of the disposal process of oil sludge, and comparing it with some common catalysts such as ZSM-5 zeolite, activated Al_2O_3, and a natural zeolite obtained from Indonesia. More detailed information of the catalysts used in this study will be explained in the experimental section.

The effects of the pyrolysis method and the catalysts on the yields and properties of oil products obtained from pyrolysis of oil sludge were investigated in this study. Characterization of the oil products was preliminary judged by comparing their distillation results with the diesel standard. Then, the chemical characterization of a part of the pyrolytic oils was conducted by FT-IR and NMR. In the last step of this experiment, the oil products obtained from pyrolysis of oil sludge using oil field sludge ash and oil tank sludge ash as catalysts were mixed with diesel oil by the volume ratio of one-to-five and one-to-ten. The fuel characteristic of the oil mixtures was analyzed by GC-MS, and identified from the view point of the viscosity, the density, the cetane index, the carbon number distribution and the heating value.

Materials and Methods

Oil sludge

Both the oil field sludge and the oil storage tank sludge used in this study were sampled from Shengli oil field, China. The oil field sludge is finely gritty texture which is wet by crude oil. As to the oil tank sludge, it generally is dark and viscous slurry, and there is almost no particulate material can be found in it. Table 1 shows the results of the proximate analysis, the ultimate analysis, the higher heating value (HHV), the oil content and a part of metal element analysis results of the oil sludge.

The moisture, volatile, and ash were measured according to the standard test method ASTM E871, D1102, and E872. The fixed carbon was calculated by the difference to 100%. There is a considerable difference between the proximate analysis results of the oil field sludge and the oil storage tank sludge. It may be due to the different resource of these two sludge samples. The ultimate analysis was performed using a CHNS/O analyzer. The oxygen was estimated by subtracting the sum of C, H, N, S, and ash constituent from 100. A bomb calorimeter was employed to measure the low heating value of the samples. The oil content and the high heating value of the oil storage tank sludge are more than double of those of the oil field sludge. Shengli crude oil is a relatively high-Ni, high-nitrogen, and intermediate base crude oil. The oil sludge from Shengli oil field and storage tanks inherited these genetic characteristics, which will pose a threat to catalysts. Sludge drying can remove a part of moisture from the samples, but it also cause light hydrocarbon volatilization and leads to a loss of oil recovery. Moreover, the dried sludge became stickier, more dense, and hard to be treated. Therefore, in all experiments of this study, the oil sludge was used as it was received.

Catalysts

The following five catalysts were selected for use in the catalytic cracking:

	Oil field sludge	Oil tank sludge
Proximate analysis		
(Wt% wet basis)		
Moisture	7.63	20.61
Volatile	14.57	41.59
Fixed carbon	8.47	4.62
Ash	76.96	33.18
Ultimate analysis		
(Wt% dat)		
C	16.89	58.97
H	2.32	9.14
N	0.16	1.21
S	0.53	1.91
O (a)	2.6	4.67
H/C (mol/mol)	1.65	1.86
S/C (mol/mol)	0.0118	0.0121
N/C (mol/mol)	0.0082	0.0175
HHV (MJ/kg)	10.45	21.51
Oil content (wt%)	21.65	43.2
Metal element analysis (ppmw)		
Fe	14865.98	955.92
Cu	22.55	2.36
Na	164.35	620.31
Ni	23.44	5.3
V	20.28	18.54

Table 1: Selected properties and composition of the tested oil sludge obtained from Shengli oil field.

1. Oil field sludge ash: a mixture of metallic oxide, amorphous, surface area: 4.74 m^2/g, pore volume: 0.024 cm^3/g, mean pore size: 20.57 nm, mesopore structure. It was prepared by burning the pyrolysis residue of the oil field sludge in a muffle furnace at 600°C for 3 hours.

2. Oil storage tank sludge ash: a mixture of metallic oxide, amorphous, surface area: 135.34 m^2/g, pore volume: 0.415 cm^3/g, mean pore size: 12.28 nm, mesopore structure. The preparation process is the same as the oil field sludge ash.

3. Natural zeolite (obtained from Indonesia): Crystalline aluminosilicate (with impurities), mordenite, SiO$_2$/Al$_2$O$_3$ (mole ratio): 8.12, surface area: 50.54 m^2/g, pore volume: 0.134 cm^3/g, mean pore size: 10.61nm, mesopore structure.

4. Activated aluminium oxide: Aluminium oxide, amorphous, surface area: 319.93 m^2/g, pore volume: 0.371 cm^3/g, mean pore size: 4.64 nm, mesopore and micropore structure.

5. ZSM-5: Crystalline aluminosilicate, mordenit, SiO$_2$/Al$_2$O$_3$ (mole ratio): 80, surface area: 425 m^2/g, pore volume: 0.365 cm^3/g, mean pore size: 4.51 nm, mesopore and micro pore structure.

The chemical components of the oil sludge ashes and the natural zeolite are listed in Table 2. From this table, we can see that the main components of the oil field sludge ash and the oil storage tank ash are SiO$_2$, Al2O$_3$, CaO, Fe$_2$O$_3$, and SO$_3$. Also from this table, we can see that the main components of natural zeolite are SiO$_2$, Al$_2$O$_3$, CaO, Fe$_2$O$_3$, and K$_2$O. The natural zeolite has been proved that it is effective and applicable to recovery the diesel range fuel from waste plastic. It works on breaking the macromolecules into smaller ones. As Table 2 presents, the oil sludge ash has the similar chemical components with natural zeolite. Considering the result of Je-LuengShie [15], we speculate that the adding of oil sludge ash has the ability to affect the yield and quality of the oil product to some extent. However, the ratios of components of the oil sludge ash are quite different from those of natural zeolite.

Experimental apparatus

The pyrolysis and catalytic reforming experiments were carried out using a self-fabricated stainless steel two stage reactor, which is shown in Figure 1. This system consists of the pyrolysis reactor and the catalytic reforming reactor. The inner diameter and the height of the pyrolysis reactor are 43 mm and 284 mm, respectively. As to the reformer, its inner diameter is 38mm, and the height is 210 mm. Both of the reactors were covered with the electric heater, and the reaction temperatures in both the pyrolysis reactor and the reforming reactor

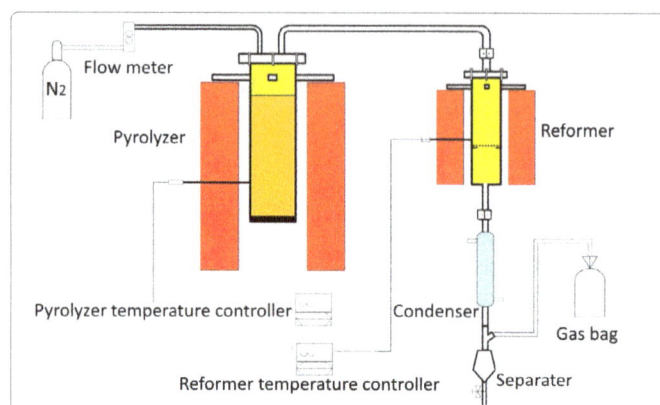

Figure 1: Diagram of the bench-scale pyrolysis and catalytic reforming reactors.

were controlled by a closed loop feedback system, including K-type thermocouples, a controller, and a heater. A double-tube condenser was installed at the outlet side of the reformer, and another side of it was connected to the gas/liquid separator.

Pyrolysis experiment

Pyrolysis experiment was conducted using the stainless steel pyrolyzer coupled with the reforming reactor made of the same material. In the one stage reaction case, 60 g of oil sludge was mixed with 20 g of catalyst, and then the mixture was placed in the pyrolyzer. In the case of the two stage reaction, 60 g of oil sludge and 20 g of catalyst were put into the pyrolyzer and the reformer, respectively. In the reformer side, on the top surface of the catalyst, 40 g of quartz sand were packed to ensure that the catalyst was flatten and separated from the solid particles carried by the pyrolysis gas. To compare with the two stage reaction case, the reformer side of the one stage reaction case was also filled with 40 g of quartz sand. N$_2$ carrier gas with the flow rate of 50 ml/min was passed through the pyrolysis and reforming reactors before the experiments for 1hour, to remove the air in the reactors. After the air in the reactors was replaced with N$_2$, the reformer was heated to 500°C first to dry and activate the catalyst and the quartz sand. Then the pyrolyzer was heated at the rate of 20°C/min until the temperature reached to 500°C, and held at this temperature for 1hour. When the temperature rose to around 390°C, due to the endothermic reaction of the oil sludge, the heating rate dropped obviously. Then the temperature hovered at around 395°C for 2 minutes. After this stagnation, the temperature continued to rise. The quantity of the pyrolysis gas began to increase when the temperature reached to 450°C. The reformed gas was introduced to the condenser in which the reaction mixture was cooled to the room temperature. The condensable (liquid product) and the non-condensable (gas product) parts were separated in the gas-liquid separator and collected in the liquid collector and the gas collector respectively. The products can be classified into: gas, liquid (oil and water), and solid residue. The liquid and solid products were weighted, and the yields of gas were calculated by the difference.

Analysis

The instruments used for the analysis of the oil product and the mixed fuel oil are given below.

Fourier transform infrared spectrometer (FT-IR): FT-IR was considered as a common analysis technique by the researchers of heavy oil and asphaltene field [1,18]. In this study, FT-IR analysis of the pyrolytic oil was conducted by JEOL JIR-SPX200 FT-IR spectrometer.

Table 2: Chemical components of oil field sludge ash, oil tank sludge ash, and natural zeolite.

Wt%	Oil field sludge ash	Oil tank sludge ash	Natural zeolite
SiO$_2$	54.4	25.6	68.76
Al$_2$O$_3$	11.5	40.3	14.4
CaO	9.69	6.79	10.47
Fe$_2$O$_3$	8.29	7.7	4.24
K$_2$O	3.6	1.09	1.08
MgO	1.76	1.37	0.45
Na$_2$O	1.37	2.89	0.31
TiO$_2$	1.36	1.6	-
SrO	0.13	0.17	-
SO$_3$	7.11	11.1	-
Ignition loss	0.519	0.475	0.29

The oil samples were prepared as a cast film between two KBr windows, and the range of wave number was set from 400 to 4000.

Nuclear magnetic resonance (NMR) ^{1}H, ^{13}C: The results of NMR analysis are often used to speculate the molecular structure of organism, and calculate the average structural parameters of crude oil and heavy oil by bringing the chemical shift results into the equation list of Brown-Ladner method [19] or the other method. Our NMR results of oil products were carried out using a BrukerbiospinAvance III NMR spectrometer. The relative peak areas were captured in accordance with the range of: Hα (2.0~4.0 ppm); Hβ (1.0~2.0 ppm); Hγ (0.5~1.0 ppm); HA (9.0~6.0 ppm); CA (190~100 ppm); CP (70~0 ppm). The average structural parameters of oil products were calculated according to the method described [19,20].

Thermogravimetry analysis (TG): A thermo-gravimetry instrument can measure weight changes of a solid and liquid material under the temperature programmed condition in a controlled atmosphere. It is believed to be an feasible way to analyze the distillation characteristics (below 400°C) of fuel oil, heavy oil, and tar oil by using TGA system [21,22]. The SHIMADZU D 50 simultaneous TGA/DTA analyzer was used to define the distillation characteristics of pyrolytic oil products. 20mg of liquid samples was placed in a platinum micro crucible. For controlling liquid flooding and increasing detection precision, a micro cover with a small gas vent was placed on the crucible before measurement. Then the whole crucible assembly was heated in N$_2$ at the flow rate of 50 ml/min. Data collection was carried out from 30°C to 450°C. During the whole measurement process, the heating rate was kept at 5°C/min. The results were compared with a diesel standard, which was got from ENEOS gasoline station, Japan.

Gas chromatography to mass spectroscopy (GC-MS): The GC-MS analysis of the mixed fuel (diesel and oil product) was conducted by employing an Agilent 19091S-433 with HP-5MS Phenyl Methyl Siloxane (30.0 m×250 µm×0.25 µm) column. The carrier gas was He. The temperature was held at 50°C for 3 minutes, then was increased to 200°C at the heating rate of 3°C/min, and then was held for 40 minutes. The chemical compounds found in the injected samples were identified by comparing their spectra with the data base of the GC-MS system.

Results and Discussion

There are five parts in this section. The first part focuses on the effect of catalysts on the yield of pyrolysis products. Part 2 provides the results of distillation characteristics of all the oil products. The ratio of remaining carbon and the mole H/C, O/C, S/C, and N/C are also discussed in this part. Part 3 and Part 4 go further into the chemical characteristics of the selected typical oil samples, such as the oil product from the one stage pyrolysis of oilfield sludge over the oilfield sludge ash, and the oil product of oil tank sludge, which was produced in one stage pyrolysis process with oil tank sludge ash. The FT-IR analysis and NMR analysis are used to indicate some main functional groups and the chemical structure of oil products in these parts. The last part discusses the fuel characteristics of the mixed fuel of oil products with diesel. Some evaluation indexes, like the density, the viscosity, the carbon distribution, the heating value, and the cetane index are provided in this part. The results presented in this section are the mean values of at least three experiment runs and analysis under the same conditions.

Effect of experiment conditions on the yield of pyrolysis products

The liquid product (oil and water) distribution for the oil field

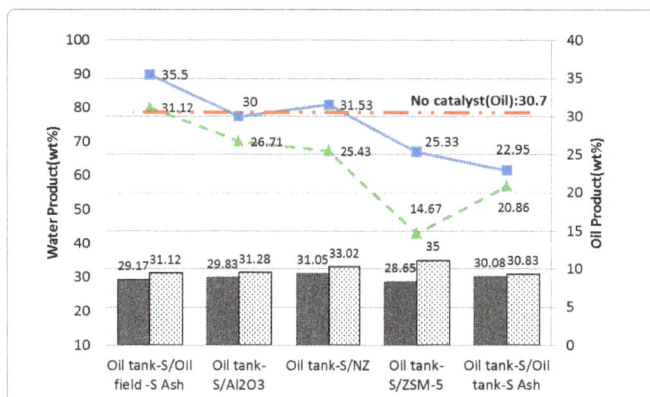

Figure 2: Effect of the catalysts and pyrolysis methods on the oil and water yields of the oil tank sludge.

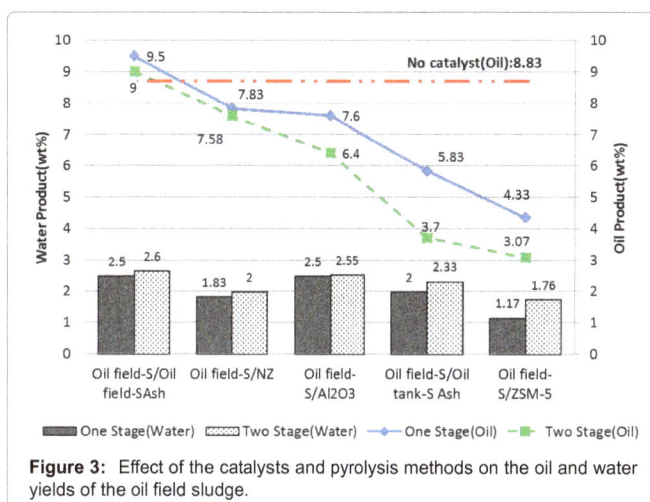

Figure 3: Effect of the catalysts and pyrolysis methods on the oil and water yields of the oil field sludge.

sludge and oil tank sludge are presented in Figures 2 and 3, respectively. The oil product under different experiment conditions are shown by the line graphs. Their Y-axis is on the right side. The bars whose Y-axis is on the left side show the water product yield. On the whole, the oil products of the one stage pyrolysis are higher than the two stage case, whether the feedstock is the oil field sludge or the oil tank sludge. This might be caused by the fully functioning of catalysts in the reformer, which would lead to the increasing of gas product and the decreasing of oil product. However, the opposite occurs in water products.

As seen from Figure 2, the oil product of the oil tank sludge, about 35.5 wt%, in the one stage pyrolysis, with the catalyst of the oil field sludge ash is higher than the oil products with other catalysts(31.53-22.95 wt%), and higher than that without catalyst(30.7 wt%). In Figure 3, The highest oil product yield (9.5 wt%) of the oil field sludge is also obtained at one stage pyrolysis with the oil field sludge ash. According to the main components of the oil sludge ash (Table 2), both the oil field sludge ash and the oil tank sludge ash contained lots of Fe oxide and S oxide. Previous studies [12,23] proved that the presence of Fe element and S element in the catalyst can improve the conversion of oil sludge by inhibiting the agglomeration of metal oxides and increasing the surface area of catalysts. This advantage became more meaningful in the one stage pyrolysis. The oil field sludge ash improved the physical contact between the oil sludge and the catalyst, and resulted in improved oil recovery. That might explain why the one stage pyrolysis with the oil field sludge ash can produce more oil product. We noted then that compared with other catalysts, the oil product yield over the oil tank sludge ash stayed on a lower level, due to their higher amounts of gas products. This tendency is generally similar to the pyrolysis experiments with the ZSM-5 catalyst. Here we can speculate that the oil tank sludge ash has a higher catalytic activity or dispersive capacity than the oil field sludge ash, even though there are only a few differences in their main components.

As to the gas and residue product yields which can be seen from Figures 4 and 5, the gas product yields from the one stage pyrolysis are lower than that from the two stage pyrolysis. This trend once again proves that the catalysts functioned fully in the two stage case, and enhanced the cracking of large molecule products into small molecules such as some non-condensable gas molecules. The residue product yields of the catalytic pyrolysis of the oil field sludge are generally above 80 wt%, owing to the high ash content of the oil field sludge.

Distillation characteristics of oil products

The distillation characteristics of oil products having the boiling range between 30°C to 400°C were determined at the atmospheric pressure using a thermo-gravimetry instrument. Tests revealed the boiling range of all the oil samples were between 90°C and 370°C. The distillation characteristic curves are given in Figures 6 and 7. The ratio

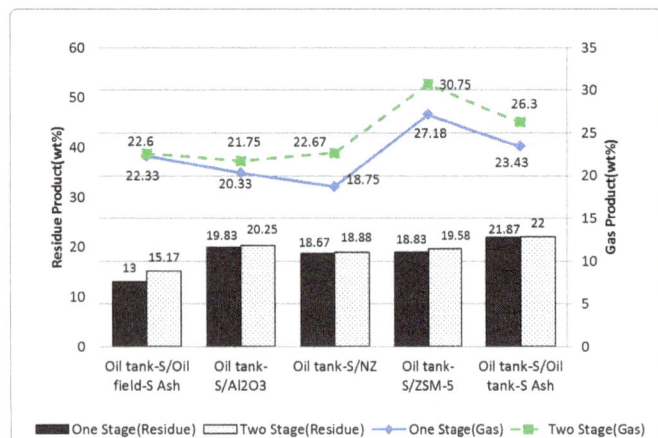

Figure 4: Effect of the catalysts and pyrolysis methods on the gas and residue yields of the oil tank sludge.

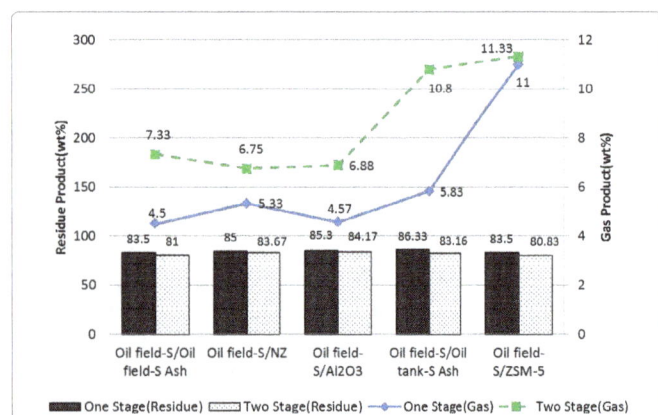

Figure 5: Effect of the catalysts and pyrolysis methods on the gas and residue yields of the oil field sludge.

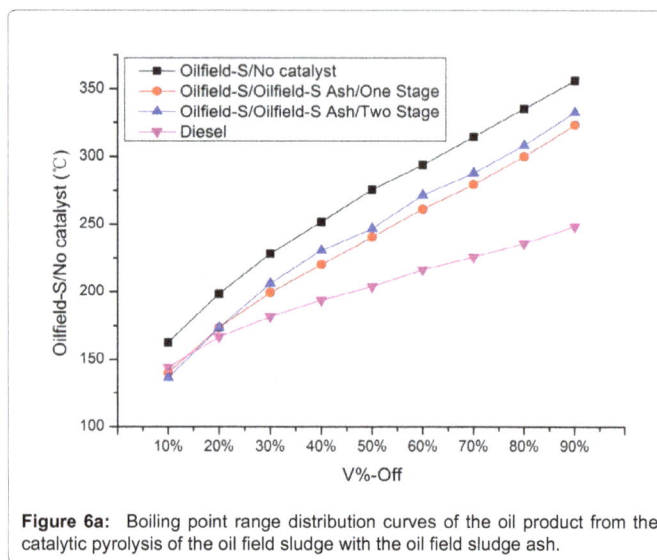

Figure 6a: Boiling point range distribution curves of the oil product from the catalytic pyrolysis of the oil field sludge with the oil field sludge ash.

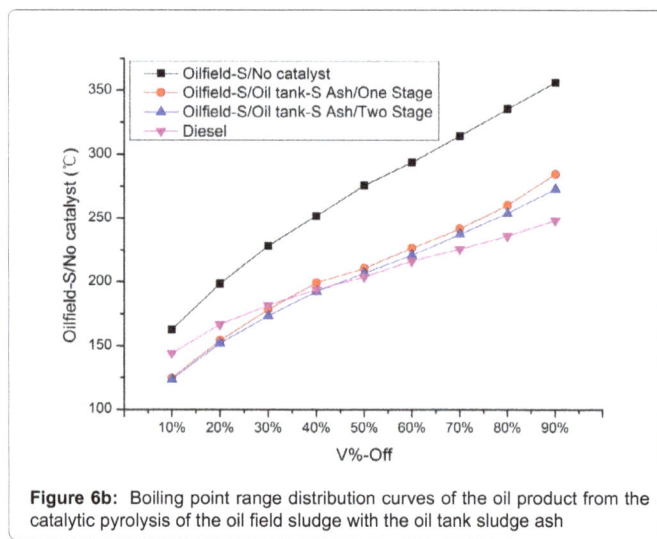

Figure 6b: Boiling point range distribution curves of the oil product from the catalytic pyrolysis of the oil field sludge with the oil tank sludge ash

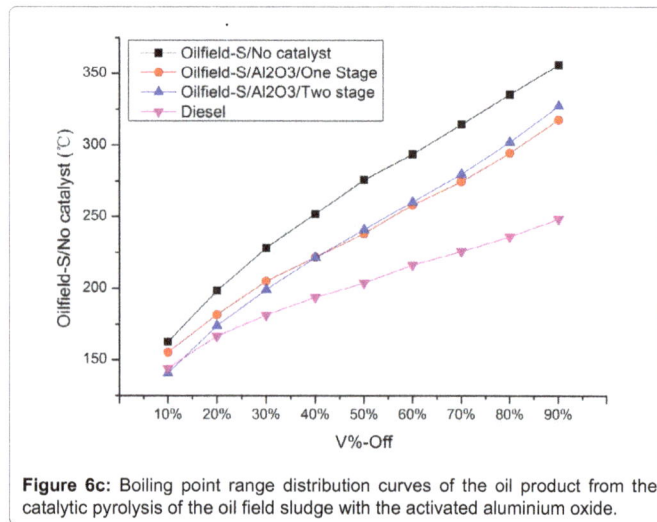

Figure 6c: Boiling point range distribution curves of the oil product from the catalytic pyrolysis of the oil field sludge with the activated aluminium oxide.

of remaining carbon at 450°C, which show the yield of coke in the oil product, are listed in Tables 3 and 4. The ratio of remaining carbon of remaining carbon are the important evaluation parameters of oil quality.

From Figures 6 and 7, we can see that the distillation characteristic

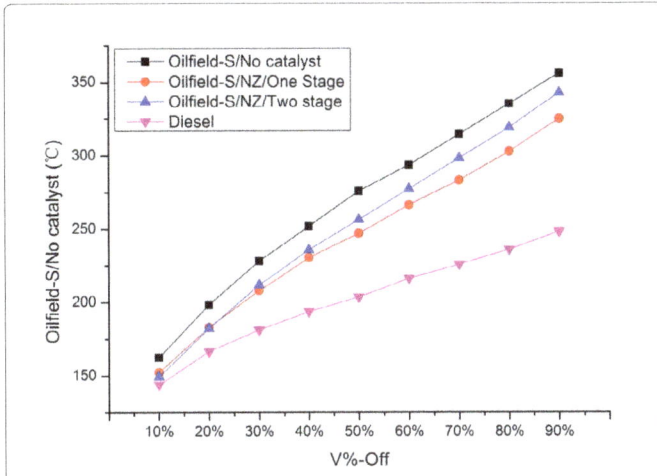

Figure 6d: Boiling point range distribution curves of the oil product from the catalytic pyrolysis of the oil field sludge with natural zeolite.

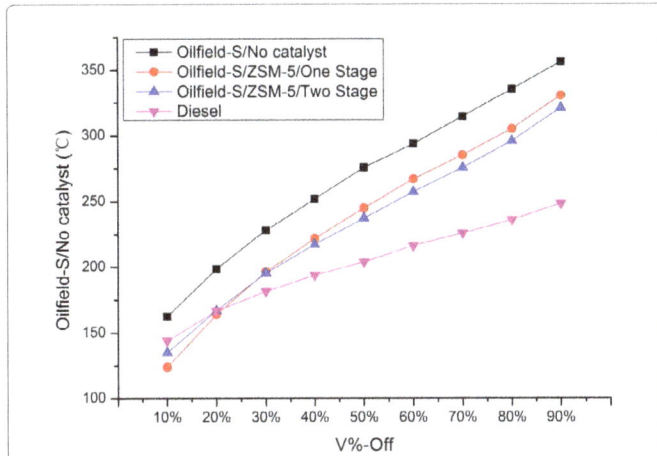

Figure 7b: Boiling point range distribution curves of the oil product from the catalytic pyrolysis of the oil tank sludge with the oil tank sludge ash.

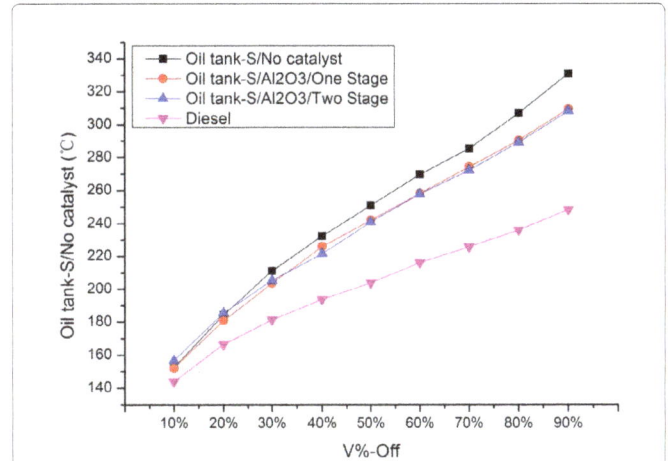

Figure 6e: Boiling point range distribution curves of the oil product from the catalytic pyrolysis of the oil field sludge with ZSM-5.

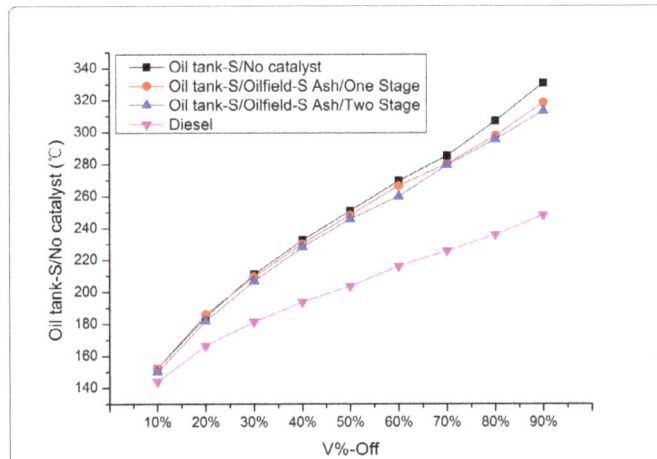

Figure 7c: Boiling point range distribution curves of the oil product from the catalytic pyrolysis of the oil tank sludge with activated aluminium oxide.

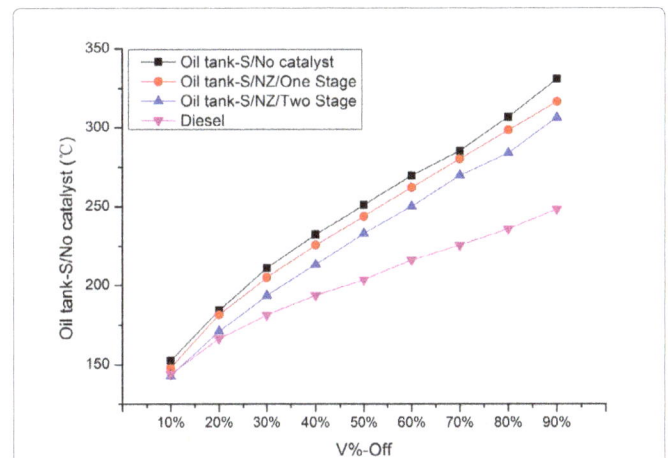

Figure 7a: Boiling point range distribution curves of the oil product from the catalytic pyrolysis of the oil tank sludge with the oil field sludge ash.

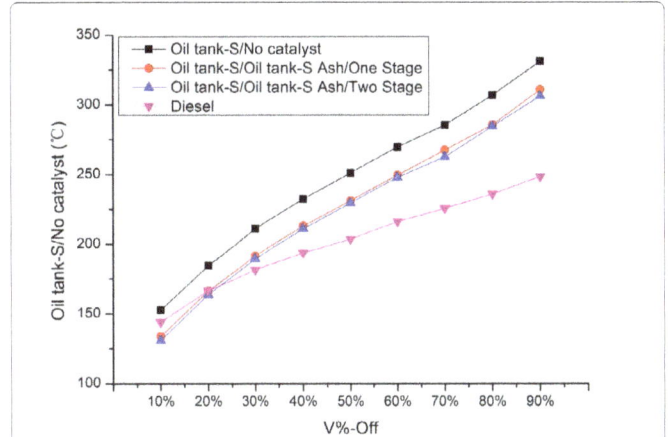

Figure 7d: Boiling point range distribution curves of the oil product from the catalytic pyrolysis of the oil tank sludge with natural zeolite.

can reflect the amount of carbon deposition when the oil products are used in an engine. Both the distillation characteristics and the ratio

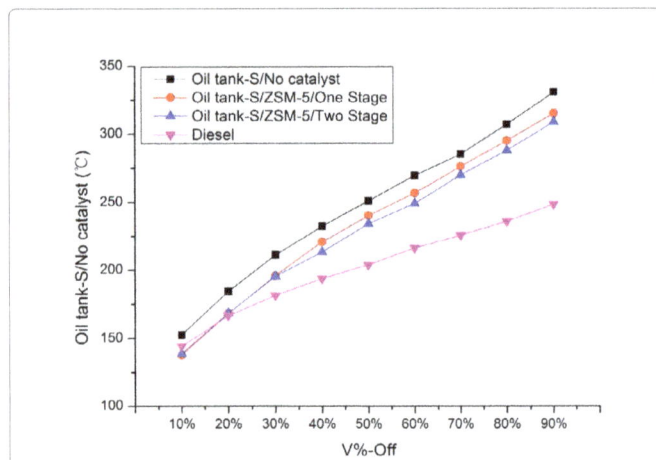

Figure 7e: Boiling point range distribution curves of the oil product from the catalytic pyrolysis of the oil tank sludge with ZSM-5.

	Oilfield-S Oilfield-S Ash One stage	Oilfield-S Oil field-S Ash Two stage	Oilfield-S No catalyst
Ratio of remaining carbon	6.25%	6.75%	8.25%

Table 3(a): Ratio of remaining carbon in the oil product from the catalytic pyrolysis of the oil field sludge with the oil tank sludge ash.

	Oilfield-S Oil tank-S Ash One stage	Oilfield-S Oil tank-S Ash Two stage	Oilfield-S No catalyst
Ratio of remaining carbon	5%	6.25%	8.25%

Table 3(b): Ratio of remaining carbon in the oil product from the catalytic pyrolysis of the oil field sludge with the oil tank sludge ash.

	Oilfield-S Al_2O_3 One stage	Oilfield-S Al_2O_3 Two stage	Oilfield-S No catalyst
Ratio of remaining carbon	3.75%	9.25%	8.25%

Table 3(c): Ratio of remaining carbon in the oil product from the catalytic pyrolysis of the oil field sludge with activated aluminium oxide.

	Oilfield-S NZ One stage	Oilfield-S NZ Two stage	Oilfield-S No catalyst
Ratio of remaining carbon	5.25%	10%	8.25%

Table 3(d): Ratio of remaining carbon in the oil product from the catalytic pyrolysis of the oil field sludge with natural zeolite.

	Oilfield-S ZSM-5 One stage	Oilfield-S ZSM-5 Two stage	Oilfield-S No catalyst
Ratio of remaining carbon	4.75%	7%	8.25%

Table 3(e): Ratio of remaining carbon in the oil product from the catalytic pyrolysis of the oil field sludge with ZSM-5.

curves of oil products, whether from the pyrolysis of the oil field sludge or the oil tank sludge, are located between the curve for no catalyst oil product and the curve for diesel standard. It means that the boiling

range of all the oil products produced by the catalytic pyrolysis are in between the boiling range of no catalyst oil product and diesel standard. The results indicate that in the catalytic pyrolysis process of the oil field sludge, all of the catalysts can improve the quality of oil products by decreasing their boiling point range. In Figure 6-b, we can see that the boiling range of the oil product with the use of the catalyst of the oil tank sludge ash (one stage case: 124.5~284.5°C; two stage case: 123.4~273°C) is the nearest to the boiling range of diesel standard (144~248.4°C).

Although there is no obvious differences between the one stage pyrolysis and the two stage pyrolysis in the boiling range, the ratio of remaining carbon in the oil product from the one stage pyrolysis process, shown in Table 3, are generally lower than those from the two stage pyrolysis process, and also lower than the oil product from the no catalyst pyrolysis process. Therefore, the oil products from the one stage pyrolysis process of the oil field sludge contained lower coke amount than the oil product produced by the two stage pyrolysis of the same feedstock.

For the distillation characteristic of the oil products from the catalytic pyrolysis of the oil tank sludge, it was found that neither the type of the catalyst nor the pyrolysis method affected the boiling range of the oil product significantly, except for the oil product with the catalyst of the oil tank sludge ash and ZSM-5. On the other hand, the ratio of remaining carbon in the no catalyst oil product (17.5 wt%) is

	Oil tank-S Oilfield-S Ash One stage	Oil tank-S Oilfield-S Ash Two stage	Oil tank-S No catalyst
Ratio of remaining carbon	7.25%	4.25%	17.5%

Table 4(a): Ratio of remaining carbon in the oil product from the catalytic pyrolysis of the oil tank sludge with the oil field sludge ash.

	Oil tank-S Oil tank-S Ash One stage	Oil tank-S Oil tank-S Ash Two stage	Oil tank-S No catalyst
Ratio of remaining carbon	7.5%	3%	17.5%

Table 4(b): Ratio of remaining carbon in the oil product from the catalytic pyrolysis of the oil tank sludge with the oil tank sludge ash.

	Oil tank-S Al_2O_3 One stage	Oil tank-S Al_2O_3 Two stage	Oil tank-S No catalyst
Ratio of remaining carbon	3.75%	2.75%	17.5%

Table 4(c): Ratio of remaining carbon in the oil product from the catalytic pyrolysis of the oil tank sludge with activated aluminium oxide.

	Oil tank-S NZ One stage	Oil tank-S NZ Two stage	Oil tank-S No catalyst
Ratio of remaining carbon	8.25%	5.25%	17.5%

Table 4(d): Ratio of remaining carbon in the oil product from the catalytic pyrolysis of the oil tank sludge with natural zeolite.

	Oil tank-S ZSM-5 One stage	Oil tank-S ZSM-5 Two stage	Oil tank-S No catalyst
Ratio of remaining carbon	7.25%	0.75%	17.5%

Table 4(e): Ratio of remaining carbon in the oil product from the catalytic pyrolysis of the oil tank sludge with ZSM-5.

(mole)	H/C	O/C	N/C	S/C
Oilfield-S No catalyst	1.67	0.0131	0.0044	0.0066
Oilfield-S/Oilfield-S Ash/One Stage	1.86	0.0038	0.0041	0.0059
Oilfield-S/Oil tank-S Ash/One Stage	1.81	0.0039	0.0031	0.0058
Oil tank-S/No catalyst	1.68	0.0998	0.0040	0.0071
Oil tank-S/Oilfield-S Ash/One Stage	1.89	0.0063	0.0050	0.0072
Oil tank-S/Oil tank-S Ash/One Stage	1.83	0.0051	0.0034	0.0069

Table 5: Mole H/C, O/C, S/C and N/C ratios of the oil products from the one stage catalytic pyrolysis of the oil sludge with the oil sludge ash.

far higher than those from the catalytic pyrolysis(0.75~8.25 wt%). It is shown that the catalytic pyrolysis prevented coke formation more during the oil recovery than no catalyst case.

Table 5 shows the mole H/C, O/C, N/C, and S/C ratios of a part of the oil products. In both oil field sludge and oil tank sludge cases, the mole H/C ratio of the oil product was increased from around 1.67 (no catalyst) to more than 1.8 (one stage pyrolysis with the oil field or oil tank sludge ash). The mole O/C ratio decreased due to the vaporization of the water and halogenated metal containing oxygen. Therefore, the presence of oil sludge ash plays an important role in increasing the mole H/C ratio and decreasing the mole O/C ratio of the oil product. However, the mole S/C and N/C ratios were not affected regularly.

FT-IR analysis results

The IR spectroscopy was used widely for evaluation of possible interaction and bonding mechanism of oil products. For this part of the study, the raw oil sludge and the oil products after the pyrolysis process of oil sludge with and without the catalyst of the oil sludge ash are analyzed by employing the FT-IR analysis. The liquid and solid samples were prepared by using the cast film and pressed-pellet method, respectively. The infrared spectra of the oil product with and without catalysts, and the raw oil sludge were compared in Figures 8 and 9. For brevity, only the oil products from the one stage pyrolysis with the catalyst of the oil sludge ash are provided.

The major vibration frequencies of the oil products are associated with the alkyl C-H modes of vibrations, such as –CH$_2$, CH$_3$ group. One would expect that the relative concentration of –CH$_2$ and CH$_3$ group are higher than others. In the case of oil sludge, whether the oil field sludge or the oil tank sludge, except for the dominant C-H frequency peaks, the O-H and Si-O group which are formed by the interaction of oil, water and soil in the oil sludge, can be observed easily. However, the O-H and Si-O frequency peaks disappeared completely from the infrared spectra of the oil products. This is because that the oil products were separated from water and soil due to the pyrolysis process.

Comparing with the oil product without catalysts, as shown in Figures 8 and 9, the intensity of C=C stretch in aromatics decreased in the infrared spectra of the oil product with the oil sludge ash catalyst. Oppositely, the C=O stretch in carbonyl/carboxylic became more prominent in the IR spectra of the oil product from the catalytic pyrolysis of the oil field sludge with the oil field sludge ash (Figure 8-a). These finding are in agreement with the result of [1], who pointed that this trend shows the effect of the dehydrogenation reaction with ketone and olefins formation during the pyrolysis of oil (Table 6).

NMR analysis and average structural parameter

In the NMR analysis, the oil products were dissolved in deuteriated-chloroform, and analyzed by the ^{1}H NMR and ^{13}C NMR spectroscopy. The average structural parameters of oil products were calculated by combining the results of the ^{1}H NMR spectra, ^{13}C NMR spectra with

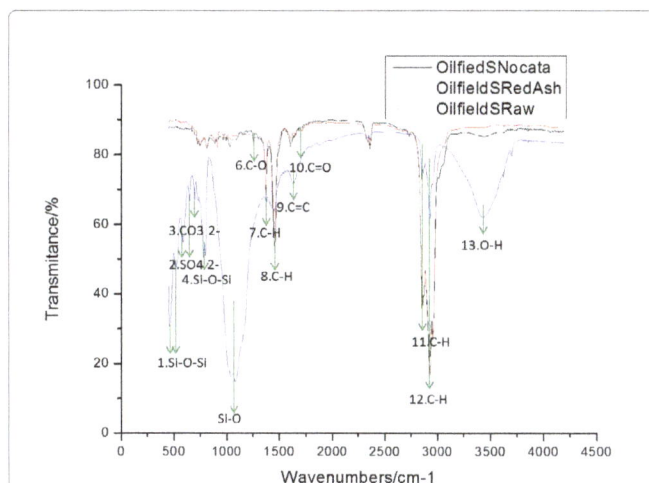

Figure 8: FT-IR spectra of the raw oil field sludge, the oil products from the one stage pyrolysis of the oil field sludge with and without the catalyst of the oil field sludge ash.

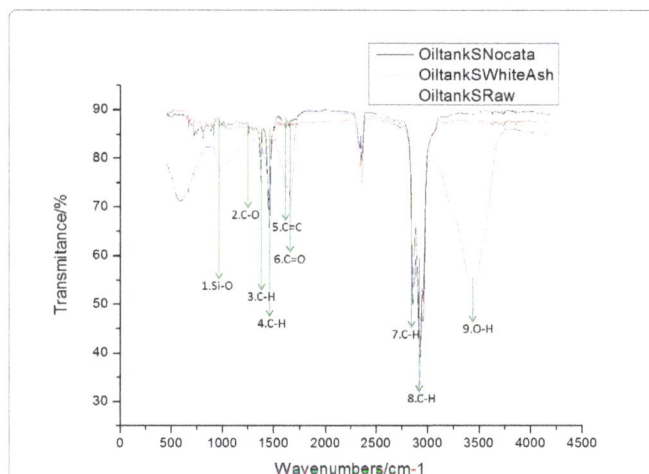

Figure 9: FT-IR spectra of the raw oil tank sludge, the oil products from the one stage pyrolysis of the oil tank sludge with and without the catalyst of the oil tank sludge ash.

Modes of vibration	Frequency range	Peak centered at(cm-1)
Si-O-Si bend	540-400	516
SO$_4$ 2- bend	680-580	647/586
CO$_3$ 2- bend	750-680	694
Si-O-Si stretch	800-600	796
Si-O stretch	1100-970	1037/981
C-O stretch	1260-1250	1259
C-H symmetric deform in CH$_3$	1385-1365	1376
C-H asymmetric deform in CH$_2$ and CH$_3$	1490-1430	1457
C=C stretch in aromatics	1642-1547	1604
C=O stretch in carbonyl/carboxylic	1752-1640	1716/1652
C-H out-of-phase stretch in CH$_2$	2863-2843	2854
C-H in-phase stretch in CH$_2$	2936-2916	2925
O-H stretch	3630-3330	3629/3428

Table 6: Major modes in the IR of the oil product and the raw oil sludge.

elemental analysis. The equations and parameters are listed in Table 7. It should be pointed out ,that the Brown-Ladner method is established under two assumed conditions: (1) Hs/Cs≈2, where Hs is the number of saturated hydrogen, Cs is the number of saturated carbon; (2) There is no other elements but only C and H exits in the oil product. According to the results of NMR analysis and the elemental analysis, the Hs/Cs of the tested oil products in this study are in the range of 2.12~1.87. And the weight percentages of C+H of the tested oil products are higher than 96.5%. Therefore, the actual situation of the oil products is generally in line with the two assumed conditions above.

Six groups of average structural parameter of oil products (A~F) were calculated, and presented in Figure 10 and Table 8. Figure 10 shows the ratios of the aliphatic carbon, the naphthenic carbon, and the aromatic carbon of the oil products. The main part of the carbon in all the oil samples is the aliphatic carbon. This result agrees with the result of FT-IR. Due to the hydrogenation of the aromatics and the presence of catalyst, the ratio of the aliphatic carbon of the oil sample increased from 65% (in A and D) to higher than 77% (in B, C, E and F). This increase is especially strong in the case with the oil field sludge ash, which might owe to the high Ni content of the oil field sludge and its ash. As known, the Ni/SiO_2 catalyst was widely used for hydrogenation of heavy oil and vacuum residuum. Although, in this experiment, no extra hydrogen was added to the reactor, and the presence of the paraffinic structure in the raw oil sludge provided a source of hydrogen in the hydrogenation. Moreover, the decarburization, which was presented by formation of coke, during the pyrolysis process, can also reduce the ratio of the naphthenic carbon and the aromatic carbon in the oil products.

The number of aromatic ring, naphthenic ring (RA and RN), and the average chain length of aliphatic chain or side chain (L) shown in Table 8 can reflect the average molecular structure of the oil product visually. Each molecule of sample A and B contain about three aromatic rings and two and a half naphthenic rings in average. After the catalytic pyrolysis with the oil field sludge ash and the oil tank sludge ash, the number went down to 0.8~1.6 and 1.5~1.9, respectively. This decrease can be explained by the hydrogenation and ring-open of naphthenic hydrocarbon during the catalytic pyrolysis reaction. As to the average chain length of the oil product, the reason of the decrease from the no catalyst cases (A and B) to the cases over the oil field sludge ash (B and E) might be attributed to the de-methylation of alkanes. However, this explanation is not suitable for the cases over the oil tank sludge ash (C and F). The L was calculated based on the value of Hγ and Cs, and this calculation method ignored the $-CH_3$ on α and β position. The simplification might lead to an unstable result or a mistake.

Fuel characteristics of the blended fuels of oil products with diesel

Several previous studies have shown that the characteristics of pyrolytic oil products of oil sludge are similar to commercial diesel. However, it still contains amount of impurities and vacuum residues. In this part, the oil products from the one stage pyrolysis process of the oil field sludge with the oil field sludge ash (B), and from the one stage pyrolysis process of the oil tank sludge with the oil tank sludge ash(F) were mixed with commercial diesel (the diesel standard) by one-to-five and one-to-ten volume ratio. We chose these two oil products in the

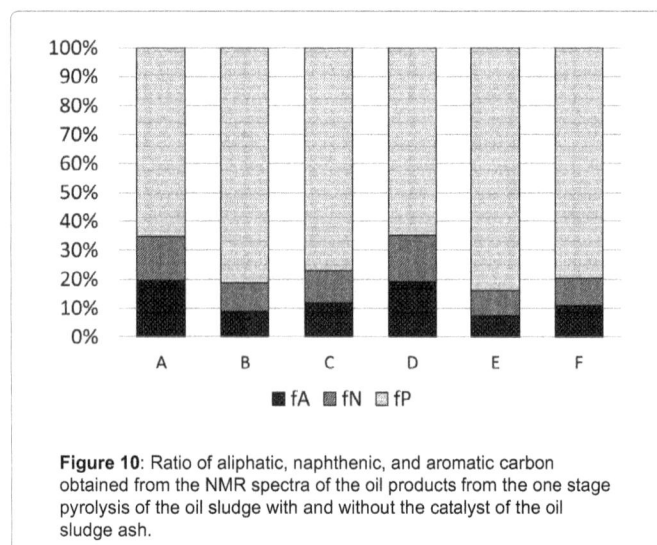

Figure 10: Ratio of aliphatic, naphthenic, and aromatic carbon obtained from the NMR spectra of the oil products from the one stage pyrolysis of the oil sludge with and without the catalyst of the oil sludge ash.

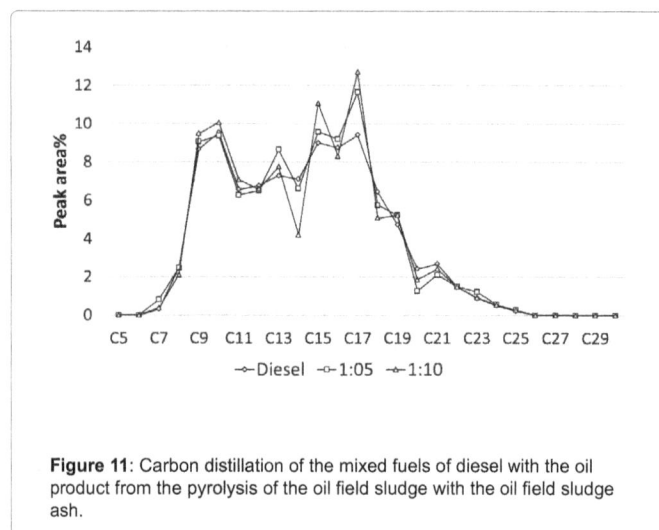

Figure 11: Carbon distillation of the mixed fuels of diesel with the oil product from the pyrolysis of the oil field sludge with the oil field sludge ash.

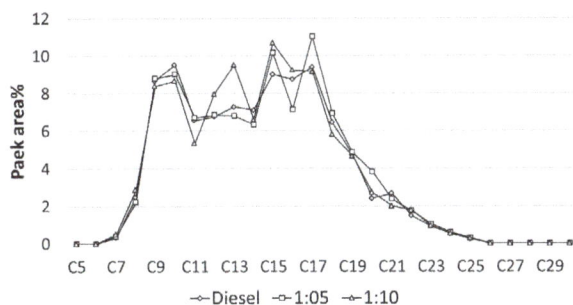

Figure 12: Carbon distillation of the mixed fuels of diesel with the oil product from the pyrolysis of the oil tank sludge with the oil tank sludge ash.

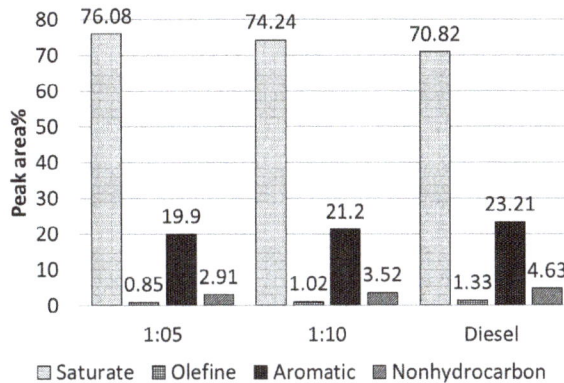

Figure 14: Chemical composition of the mixed fuels of diesel with the oil product from the pyrolysis of the oil tank sludge with the oil tank sludge ash.

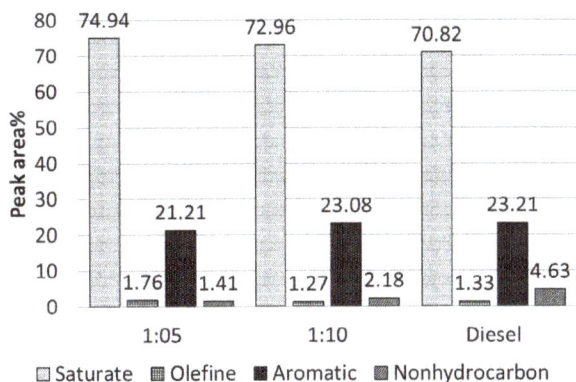

Figure 13: Chemical composition of the mixed fuels of diesel with the oil product from the pyrolysis of the oil field sludge with the oil field sludge ash.

	A*	B*	C*	D*	E*	F*
C (Wt%)	84.69	84.7	84.0	84.9	83.9	84.51
H (Wt%)	11.81	13.11	12.67	11.87	13.21	12.91
S (Wt%)	1.48	1.35	1.33	1.6	1.63	1.56
N (Wt%)	0.44	0.41	0.31	0.4	0.49	0.34
O (a) (Wt%)	1.48	0.43	0.44	1.13	0.7	0.58
RT	5.6	2.8	3.5	5.7	2.3	3.1
RA	3.0	1.1	1.6	2.9	0.8	1.4
RN	2.6	1.7	1.9	2.8	1.5	1.7
L	6.0	5.8	6.1	5.6	5.5	5.8

A: Oil product from the no catalyst pyrolysis of the oil field sludge;
B: Oil product from the one stage pyrolysis of the oil field sludge with the oil field sludge ash;
C: Oil product from the one stage pyrolysis of the oil field sludge with the oil tank sludge ash;
D: Oil product from the no catalyst pyrolysis of the oil tank sludge;
E: Oil product from the one stage pyrolysis of the oil tank sludge with the oil field sludge ash;
F: Oil product from the one stage pyrolysis of the oil tank sludge with the oil tank sludge ash.

Table 8: Elemental analysis results and distribution of ring number of the oil products from the one stage pyrolysis of the oil sludge with and without the catalyst of the oil sludge ash.

	Kinematic viscosity (40degC)	Density (20°c-g/cm3)	Heating value(HHV) (MJ/kg)	Cetane index
Oil product(oil field-S/oil field-S ash/one stage) : Diesel				
1 : 5	2.53	0.819	45.7	38.22
1 : 10	2.52	0.812	46.1	38.51
Oil product(oil tank-S/oil tank-S ash/one stage) : Diesel				
1 : 5	2.54	0.813	45.3	37.53
1 : 10	2.50	0.810	45.8	37.58
Diesel standard	2.49	0.809	45.2	35.44

Table 9: Kinematic viscosity, density, heating value and cetane index of the mixed fuels and diesel standard.

view of commercial angle for two reasons: (1) The presence of oil sludge ash in the one stage pyrolysis did improve the yield and quality of the oil product; (2) The source of oil sludge ash is the pyrolysis residue of oil sludge, which means that the oil sludge ash catalyst can be produced as long as the pyrolysis process of oil sludge will be continued.

The mixtures were filtered using 0.8 μm (47 mm) pore size filter film to remove the insoluble impurities and vacuum residues. The kinematic viscosity, the density, the higher heating value (HHV) and the catene index of the soluble material are summarized in Table 9. All of the mixed liquids have higher HHV than the commercial diesel. For the same oil product, HHV of the fuel with the mixing ratio of 1:10 is higher than that of 1:5. We can expect that the mixing ratio of 1:10 can support the burning and heat release of the mixed fuel better. Moreover, the catena index, which reflects the ignition characteristics of a diesel range fuel, of these mixed fuels also shows the same tendency with HHV. For further analysis of the chemical components of these mixed fuels, the fuel samples were measured by using the GC-MS. The chromatograms of the samples from GC-MS analysis were compared with the results of GC-FID, and the main identified components are

No.	R.T.(min)	Compounds	Area (%)		B*:diesel		F*:diesel	
			Diesel	1:5(v/v)	1:10(v/v)	1:5(v/v)	1:10(v/v)	
Saturate								
1	35.98	Hexadecane	4.62	4.33	4.6	4.8	4.36
2	39.66	Heptadecane	4.42	4.32	4.32	4.6	4.36
3	32.09	Pentadecane	4.41	4.46	4.44	4.66	4.49
4	43.13	Octadecane	3.75	3.99	3.89	3.16	3.77
5	46.43	Nonadecane	2.99	2.95	2.97	2.76	2.97
6	27.97	Tetradecane	2.94	2.86	2.85	2.92	3.03
7	49.57	Eicosane	2.85	2.44	2.56	2.1	2.51
8	52.57	Heneicosane	1.96	1.97	1.95	1.6	1.99
9	23.65	Tridecane	1.8	2	1.74	1.82	1.85
10	14.57	Undecane	1.73	1.73	1.74	1.77	1.74
11	19.14	Dodecane	1.61	1.82	1.67	1.6	1.8
12	10.11	Decane	1.32	1.36	1.38	1.37	1.35
13	6.04	Nonane	1.22	1.15	1.2	1.22	1.21
14	37.66	Hexadecane, 3-methyl-	1.2	1.22	1.2	1.24	1.21
15	30.47	Tridecane, 3-methyl-	1.29	1.25	1.27	1.31	1.27
16	43.36	Hexadecane, 2,6,10,14-tetramethyl-	1.31	1.58	1.52	1.4	1.31
17	-	Docosane	1.45	1.49	1.46	-	-
18	-	Pentadecane, 2,6,10,14-tetramethyl	-	-	-	1.06	1.2
Aromatic								
19	8.49	Benzene, 1,2,3-trimethyl-	1.53	1.32	1.43	1.32	1.31
20	5.36	p-Xylene	1.07	1.08	1.1	1.04	1.07
Nonhydrocarbon								
21	-	Octadecane, 1-(ethenyloxy)-	0.27	-	0.12	-	0.12
22	-	Oxirane, [(dodecyloxy) methyl]-	-	0.49	-	-	-
23	-	Ethanol, 2-(dodecyloxy)-	0.33	-	-	0.28	

Table 10: Main components identified by GC-MS for the mixed fuels and diesel standard.

From the carbon distillation of all the mixed fuels and the commercial diesel (Figures 11 and 12), it can be seen that the range of carbon distillation of these mixed fuels (C_7~C_{26}) is identical with diesel. However, the distillation curve of both oil product-diesel mixtures with the mixing ratio of 1:10 present a strong fluctuation in the range of C_{12}~C_{17}. Thus it can be seen that the mixture of the oil product and diesel maybe not a simple mixing of chemical components. Some unknown interactions between the chemical components in the oil product and the diesel may happen, which would affect the detection results of GC-MS.

In Figures 13 and 14, the mixed fuels and diesel are divided into saturates: alkane and cycloalkane; olefin; Aromatics; and non-hydrocarbon: resin and heterocyclic. From both Figures 13 and 14, we can see that the saturate content of the mixed fuel increased with the increase of the oil product proportion. On the other hand, the compositions of aromatic and non-hydrocarbon decreased with the increase of the oil product proportion. Combining with HHV and cetane index results above, one can expect that the adding of pyrolytic oil product improved the quality of diesel oil to same extent.

Conclusions

First, this experimental study indicates that the oil fraction can be separated from water and soil by the pyrolysis process of oil sludge with and without catalysts. On the whole, the yield of the oil products of the one stage pyrolysis is higher than that of the two stage case, whether the feedstock is the oil field sludge or the oil tank sludge. The highest oil product yield is obtained in the one stage pyrolysis of the oil field sludge and also the oil tank sludge with the oil field sludge ash. This can be partially attributed to the nature of the presence of Fe element and S element in the oil field sludge ash.

Second, the distillation characteristic curves of the oil products show that the boiling range of all the oil products produced from the catalytic pyrolysis are between the boiling range of the no catalyst oil product and diesel standard, though there is no obvious difference between the one stage pyrolysis and the two stage pyrolysis processes in the boiling range. In the catalytic pyrolysis process of the oil field sludge, all of the catalysts can improve the quality of the oil products by decreasing their boiling point range. However, for the distillation characteristics of the oil products from the catalytic pyrolysis of the oil tank sludge, it was found that neither the type of catalysts nor the pyrolysis method affected the boiling range of the oil product significantly, except for the oil product with the catalyst of the oil tank sludge ash and ZSM-5.

Third, according to the results of NMR and FT-IR analysis, the main part of the carbon in the oil product from the pyrolysis of the oil sludge with the oil sludge ash is aliphatic carbon. The total ring number of the oil products decreased owing to the hydrogenation and ring-open of naphthenic hydrocarbon during the catalytic pyrolysis process.

Last but not least, the highest HHV and cetane index can be obtained, when the oil product from the one pyrolysis process of the oil field sludge with the oil field sludge ash was mixed with diesel with the mixing ratio of 1: 10. The range of carbon distillation of all the mixed fuels (C7~C26) was identical with diesel standard. The saturate contents of mixed fuel increased with the increase of the oil product proportion. Combining with HHV and cetane index results, we can expect that addition of the pyrolytic oil product will improve the quality of diesel oil to some extent.

References

1. Ma Z, Gao N, Xie L, Li A (2014) Study of the fast pyrolysis of oilfield sludge with solid heat carrier in a rotary kiln for pyrolytic oil production. Journal of Analytical and Applied Pyrolysis 105: 183-190.

2. Wenying Z, Jingchun T, Fei W, Daming L (2012) Pollution of oil sludge and its disposal techniques in china and abroad. Petrochemical Industry Application.

3. Jin Y, Zheng X, Chu X, Chi Y, Yan J, et al. (2012) Oil recovery from oil sludge through combined ultrasound and thermochemical cleaning treatment. Ind Eng Chem Res 51: 9213-9217.

4. Yeung P, Johnson R, Acharya S (1993) An improved procedure for determining oil content in wet soil samples. ASTM Special Technical Publication 1221: 11.

5. Ilias AM (1993) Fuel isolation, identification and quantitation from soils.ASTM Special Technical Publication 1221: 15.

6. Wang Z, Guo Q, Liu X, Cao C (2007) Low temperature pyrolysis characteristics of oil sludge under various heating conditions. Energy Fuels 21: 957-962.

7. Sankaran S, Pandey S, Sumathy K (2008) Experimental investigation on waste heat recovery by refinery oil sludge incineration using fluidised-bed technique. Journal of Environmental Science and Health, Part A: Toxic/Hazardous Substances and Environmental Engineering 33: 829-845.

8. Yu Ling Wei, Chin Hua Wu (2012) PAH emissions from the fluidized-bed incineration of an industrial sludge. Journal of the Air & Waste Management Association 47: 953-960.

9. Schmidt H, Kaminsky W (2001) Pyrolysis of oil sludge in a fluidized bed reactor. Chemosphere 45: 285-290.

10. Shie J, Chang C, Lin J, Wu C, Lee D (200) Resources recovery of oil sludge by pyrolysis: kinetics study J Chem Technol Biotechnol 75: 443-450.

11. Chang C, Shie J, Lin J, Wu C, Lee D, et al. (2000) Major products obtained from the pyrolysis of oil sludge. Energy Fuels 14: 1176-1183.

12. Shie J, Chang C, Lin J, Lee D, Wu C (2001) Use of inexpensive additives in pyrolysis of oil sludge. Energy Fuels 16: 102-108.

13. Shie J, Lin J, Lee D, Wu C (2003) Pyrolysis of oil sludge with additives of sodium and potassium compounds. Resources, Conservation and Recycling 39: 51-64.

14. Shie JL, Chang CY, Lin JP (2004) Oxidative thermal treatment of oil sludge at low heating rates. Energy & fuels: An American Chemical Society journal 18: 1272-1281.

15. Shie J, Lin J, Chang C, Shih S, Lee D, et al. (2004) Pyrolysis of oil sludge with additives of catalytic solid wastes. Journal of Analytical and Applied Pyrolysis 71: 695-707.

16. Chang C, Shie J, Lin J, Wu C, Lee D, et al. (2000) Major products obtained from the pyrolysis of oil sludge. Energy Fuels 14: 1176-1183.

17. Tomita A, Watanabe Y, Takarada T, Ohtsuka Y, Tamai Y (1985) Nickel-catalysed gasification of brown coal in a fluidized bed reactor at atmospheric pressure. Fuel 64: 795-800.

18. Siddiqui MN(2010) Catalytic pyrolysis of Arab heavy residue and effects on the chemistry of asphaltene. Journal of Analytical and Applied Pyrolysis 89: 278-285.

19. Brown JK, Landner WR (1960) A study of the hydrogen distribution in coal-like materials by high-resolution nuclear magnetic resonance spectroscopy 1. A comparison with infra-red measurement and the conversion to carbon structure. Fuel 39: 87-96

20. Williams RB (1958) Characterization of hydrocarbons in petroleum by nuclear magnetic resonance. Symposium on composition of Petroleum Oils. ASTM, Spec Tech Publ 224.

21. Gonçalves MLA, Ribeiro DA, Mota DAPD, Teixeira AMRF, Teixeira MAG (2006) Investigation of petroleum medium fractions and distillation residues from brazilian crude oils by thermogravimetry. Fuel 85: 1151-1155.

22. Mondragon F, Ouchi K (1984) New method for obtaining the distillation curves of petroleum products and coal-derived liquids using a small amount of sample. Fuel 63: 61-65.

23. Kotanigawa T, Yamamoto M, Sasaki M, Wang N, Nagaishi H, et al. (1997) Active site of iron-based catalyst in coal liquefaction. Energy Fuels 11: 190-193.

10

Reduction of Nitrate Nitrogen from Alkaline Soils Cultivated with Maize Crop Using Zeolite-Bentonite Soil Amendments

Molla A[1], Ioannou Z*, Dimirkou A[1], and Mollas S[1]

Soil Science Laboratory, Department of Agriculture, Crop Production & Rural Environment, School of Agricultural Sciences, University of Thessaly, Greece

Abstract

The present study examines the efficiency of soil amendments regarding the retention of nitrate ions, from maize (*ZEA MAYS*). The experiments were performed in May – June 2010 at the greenhouse of the University of Thessaly in Volos (Central Greece). The soil amendments that have been used for the experiments were zeolite, bentonite and zeolite – bentonite in a proportion of 3:1 w/w. Two doses of nitrogen were used (400 and 800 kg N ha^{-1}) in the form of NH_4NO_3. Nine treatments occurred; six of them contained soil amendments. Each treatment repeated three times. According to the statistical analysis of the greenhouse experiments' data, bentonite and zeolite – bentonite increased the height of the plants in the dose of 800 kg N ha^{-1}. Moreover, all the used soil amendments reduced the concentration of nitrate nitrogen in soil and plants. Consequently, such materials can be used for the remediation of polluted soils with nitrogen and the production of high quality products.

Keywords: Bentonite; Zeolite; Nitrate nitrogen; Maize

Introduction

Most of the nitrogen (97–98%) in the soil is found in the organic matter and unavailable to plants. The percentage of nitrogen in the inorganic form such as nitrate (NO_3^-) and ammonium (NH_4^+) ions, which are available to plants, is only 2–3%. During the appropriate conditions of moisture, temperature and oxygen content, microorganisms are breaking down the organic matter, which is released as inorganic nitrogen into the soil (mineralization). During the mineralization process, most of the organic matter is first converted to ammonium (NH_4^+) and then to nitrate (NO_3^-) through the nitrification process, leading to the availability of nitrate ions by crops and microorganisms. Nitrates are very mobile in the soil. Most plants absorb the majority of their nitrogen in the nitrate (NO_3^-) form and in the ammonium (NH_4^+) form to a lesser extent [1,2].

Maize crops respond strongly to nitrogen (N) supply. Fertilizers enriched with nitrogen lead to high crop yields but such process also increases the risk of nitrate leaching and groundwater contamination. To achieve financial profitability and environmental sustainability, except for resource management rules, the use of soil amendments is the necessary measure to diminish the excess quantity of nitrogen from soil and crops [3].

Zeolite and bentonite are naturally occurring structured and silicate minerals, respectively, with high cation exchange and ion adsorption capacity. Zeolites are hydrated aluminosilicates of alkaline and alkaline-earth minerals [4]. Their structure is made up of a framework of AlO_4^{5-} and SiO_4^{4-} tetrahedra linked to each other by sharing oxygen atoms. The substitution of Si^{4+} by Al^{3+} in tetrahedral sites results in more negative charges and a high cation exchange capacity [5]. Bentonite is a 2:1 mineral with one aluminum octahedral sheet and two silica tetrahedral sheets, which form a layer. Layers are held together by Van der Waals forces. Because of these weak forces and some charge deficiencies in the structure, water can easily penetrate between layers and cations balance the deficiencies [6].

The objective of this study was to evaluate the removal of nitrate nitrogen ions from soils cultivated with maize crop using soil amendments such as zeolite, bentonite and zeolite-bentonite.

Materials and Methods

Greenhouse experiments were conducted for the determination of zeolite (Z), bentonite (B) and zeolite – bentonite (Z-B) in a proportion of 3:1 Z/B w/w. as soil amendments. Zeolite and bentonite was bought from S&B Company (Greece). Specifically, 1 kg of soil, 1.0 g of each soil amendment and 100 mg N kg^{-1} (400 kg N ha^{-1}) or 200 mg N kg^{-1} (800 kg N ha^{-1}) in the form of NH_4NO_3 were added to plant pots. Then, plant seeds of maize were cultivated.

Four treatments were realized for each dose of nitrogen (4x2) but only three of them contained soil amendment (Table 1). Each treatment was repeated three times. As a consequence twenty seven treatments were realized (eighteen with soil amendments, six with no soil amendment but with nitrogen and three with no soil amendment and nitrogen). Forty five days after germination the plants were collected, their morphological characteristics were identified, NO_3^--N were determined in plant and soil by copperized cadmium method [7]. The experiments took place from May to June 2010. The physicochemical characteristics of soil, which was selected from Velestino area in Volos (Central Greece), show soil pH equal to 8.82, low concentrations of organic matter (<2.3%) and nitrogen (<0.22%). The water storage capacity of soil remained stable at 65% and the temperature ranged from 25-35°C. The statistical results of the greenhouse experiments were analyzed by LSD test with significance 95% (p<0.05) using Strat graphics Plus 8.1.

The titration method was used for the determination of the Zero Point of Charge (z.p.c.) of Bentonite (B) and zeolite-bentonite (Z-B). This method is based on curves of surface charge versus pH at

*Corresponding author: Ioannou Z , Soil Science Laboratory, Department of Agriculture, Crop Production & Rural Environment, School of Agricultural Sciences, University of Thessaly, Fytoko St., N. Ionia, Volos, 38446, Greece
E-mail: zioan@teemail.gr

different ionic strength of the supporting electrolyte (KNO_3). The point of intersection of the above curves corresponds to the p.z.c. Specific Surface Area (SSA) of Z-B material was determined by the BET method, using N_2 as adsorbate in Sorptomatic 1900 Carlo Erba Surface area analyzer [8].

The experimental values of each parameter, e.g. crop height, nitrogen in soil or crop, from C or F treatment, which were measured, compared to the respective theoretical values, which come from mixtures law, follows [9]:

$$C_{theor.} = x_1 \cdot A_{exp.} + x_2 \cdot B_{exp.} \text{ or } F_{theor.} = x_1 \cdot D_{exp.} + x_2 \cdot E_{exp.} \quad (1)$$

where $C_{theor.}$ or $F_{theor.}$ is the theoretical value of the treatment according to the applied nitrogen dose, i.e. 400 or 800 kg N ha^{-1}, $A_{exp.}$ or $D_{exp.}$ is the experimental value of A or D treatment which contains zeolite as soil amendment at different nitrogen doses, $B_{exp.}$ or $E_{exp.}$ is the experimental value of B or E treatment which contains bentonite as soil amendment at different nitrogen doses, x_n is the weight percentage of the n soil amendment of the specific treatment.

Results and Discussion

Figure 1a shows that the use of soil amendments, i.e zeolite (Z), bentonite (B), zeolite-bentonite (Z-B) of A, B and C treatments, respectively, led to no significant statistical increase in the height of maize in correlation with the unamended control (G) for nitrogen dose equal to 400 kg N ha^{-1}. Comparing the experimental height of maize using Z-B as soil amendment ($C_{exp.}$ treatment) with the theoretical height of maize ($C_{theor.}$ treatment) using the laws of mixtures, the results are similar. On the contrary, the use of bentonite (E treatment) and zeolite-bentonite (F treatment) increased significantly the height of maize compared to the unamended control (G) for nitrogen doses around 800 kg N ha^{-1} (Figure 1b). Moreover, a significant decrease in the height of crops presented to soils enriched with zeolite (D treatment) as soil amendment.

Figures 2a and 2b shows that the use of bentonite (B treatment) and zeolite (D treatment) as soil amendments led to a statistical increase in the dry weight of crops compared to un-amended control (G) for nitrogen doses equal to 400 and 800 kg N ha^{-1}, respectively. Comparing the treatment with nitrogen but without soil amendment (H,I) with all the others which had soil amendments no statistical significant difference in the dry weight of plants was presented. Comparing treatments G (without soil amendment and nitrogen) and H, I (without soil amendment but with nitrogen), it seems that no statistical difference was presented to both nitrogen doses (Figure 2).

Comparing the experimental dry weight of maize using Z-B as soil amendment ($C_{exp.}$ or $F_{exp.}$ treatment) with the theoretical dry weight of maize using the law of mixtures ($C_{theor.}$ or $F_{theor.}$ treatment), the results showed small differences for the two doses of nitrogen.

The Zero Point of Charge (z.p.c.) of bentonite and zeolite was equal to 8.0 and 6.8 respectively (*Table 2*), while the soil solution pH was equal to 8.82. Since pH is higher than the z.p.c. of bentonite, negative charges are predominant and nitrates repelled from the region near the root system to deeper soil. Moreover, bentonite and zeolite appear also high specific surface area equal to 45.7 and 30.7 $m^2 g^{-1}$. Z-B amendment appears intermediate values of z.p.c. and specific surface area.

Comparing treatments H and I with all the others, it seems a statistical decrease in nitrate nitrogen to soils with soil amendments (Figure 3). For nitrogen dose equal to 400 kg N ha^{-1}, soil presented low concentrations of nitrate nitrogen using bentonite (B treatment) as soil

amendment compared to all the other soil amendments (Figure 3a). Soil with zeolite as amendment (A treatment) presented the highest concentration of nitrate nitrogen while soil with Z-B (C treatment) presented intermediate values of nitrate nitrogen. Higher doses than 400 kg N ha^{-1} increased the toxicity of soil to nitrate nitrogen and the use of Z-B as soil amendment (F treatment) presented a significant decrease of its amount to soil compared to the other soil amendments (Figure 3b). The increased amount of nitrate nitrogen in soil, which contained bentonite (E treatment), was caused due to the repelled nitrate ions from the root system through soil water, as it is illustrated in Figure 3b. Similar behavior presented also with the use of zeolite as soil amendment (D treatment) due to its negative surface charge.

A small amount of nitrate nitrogen in soils with Z, B and Z-B as amendments may be absorbed by soil amendments due to their extended porosity and high specific surface area. Comparing the experimental values of nitrate nitrogen in soil using Z-B as soil amendment ($C_{exp.}$ or $F_{exp.}$ treatment) with the theoretical values of nitrate nitrogen in soil using the law of mixtures ($C_{theor.}$ or $F_{theor.}$ treatment), the results showed no significant differences between them for the two doses.

As far as the uptake of NO_3^--N ions from crops is concerned, it seems that all soil amendments reduced significantly their uptake from crops but especially the addition of bentonite (B and E treatments) (Figure 4). The addition of bentonite to soil led to crops with the same NO_3^--N content as the unamended control (G). Moreover, Z-B soil amendment (F treatment) can also reduce significantly the amount of nitrate nitrogen in maize to the level of the unamended control (G treatment) (Figure 4b) in high nitrogen dose equal to 800 kg N ha^{-1}.

According to Figures 4a and 4b the addition of soil amendment to soil polluted with 400 and 800 kg N ha^{-1}, reduced significantly the amount of nitrate nitrogen to maize following the order: Bentonite (B) > Zeolite (Z) > Zeolite-Bentonite (Z-B) and Bentonite (B) > Zeolite-Bentonite (Z-B) > Zeolite (Z), respectively.

The repelled nitrate nitrogen ions from the root system of maize to the deep soil as a result of the negative charge of bentonite led to the lowest uptake of nitrate nitrogen from crops. Comparing the experimental nitrate nitrogen in maize using Z-B as soil amendment ($C_{exp.}$ or $F_{exp.}$ treatment) with the theoretical nitrate nitrogen in maize using the law of mixtures ($C_{theor.}$ or $F_{theor.}$ treatment), the results showed significant differences between them especially for the dose of 800 kg

Treatment	Maize cultivation	Amendment			Nitrogen dose	
		Z	B	Z-B	I	II
A	x	x	-	-	x	-
B	x	-	x	-	x	-
C	x	-	-	x	x	-
D	x	x	-	-	-	x
E	x	-	x	-	-	x
F	x	-	-	x	-	x
G	x	-	-	-	-	-
H	x	-	-	-	x	-
I	x	-	-	-	-	x

Table 1: Schematic representation of greenhouse experiments.

Specific surface area (SSA)			Zero point of charge (z.p.c.)		
	$m^2 g^{-1}$				
Z^a	B^b	Z-B	Z^a	B	Z-B
30.7	45.7	34.5	6.8	8.0	8.7

[a,b] The characteristics of Z and B were calculated elsewhere [8,10]

Table 2: Physical and chemical characteristics of different soil amendments.

Soil amendment	400 kg N ha^{-1}			800 kg N ha^{-1}		
	Treatments	% NO$_3^-$ - N reduction in soil	% NO$_3^-$ - N reduction in maize crop	Treatments	% NO$_3^-$ - N reduction in soil	% NO$_3^-$ - N reduction in maize crop
Zeolite (Z)	A-H	39.44	87.32	D-I	36.88	33.88
Bentonite (B)	B-H	57.40	98.05	E-I	17.89	98.27
Zeolite-Bentonite (Z-B)	C-H	48.36	73.87	F-I	46.33	91.57

Table 3: Percentage of NO$_3^-$ - N reduction in maize crop cultivation of each soil amendment compared to treatments H (treatment with maize cultivation and nitrogen dose I) and I (treatment with maize cultivation and nitrogen dose II)

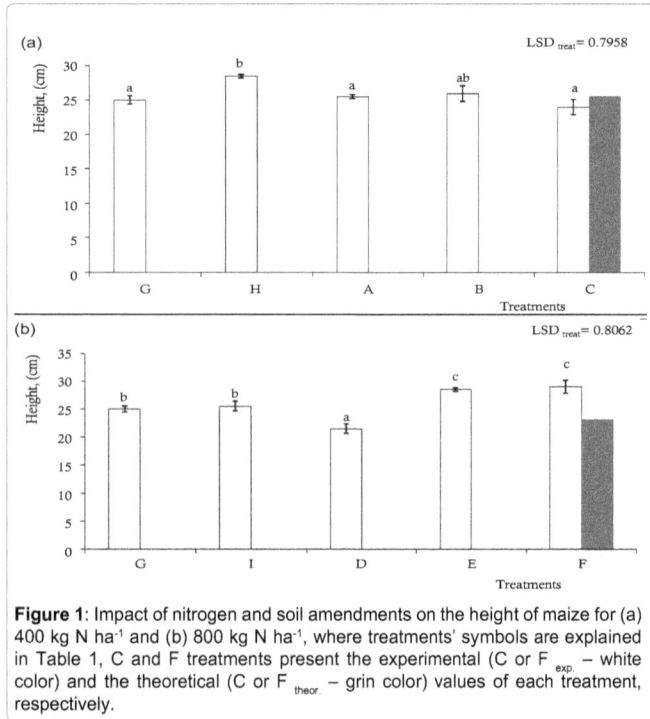

Figure 1: Impact of nitrogen and soil amendments on the height of maize for (a) 400 kg N ha^{-1} and (b) 800 kg N ha^{-1}, where treatments' symbols are explained in Table 1, C and F treatments present the experimental (C or F$_{exp.}$ – white color) and the theoretical (C or F$_{theor.}$ –grin color) values of each treatment, respectively.

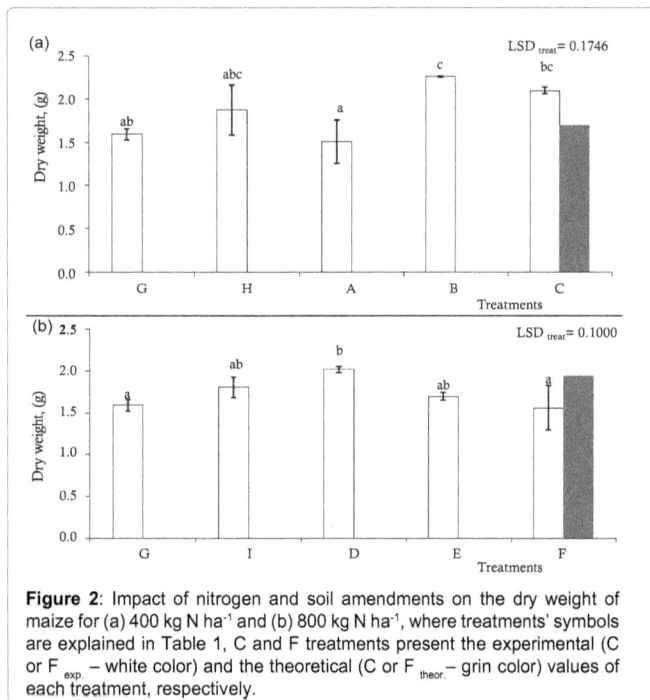

Figure 2: Impact of nitrogen and soil amendments on the dry weight of maize for (a) 400 kg N ha^{-1} and (b) 800 kg N ha^{-1}, where treatments' symbols are explained in Table 1, C and F treatments present the experimental (C or F$_{exp.}$ – white color) and the theoretical (C or F$_{theor.}$ – grin color) values of each treatment, respectively.

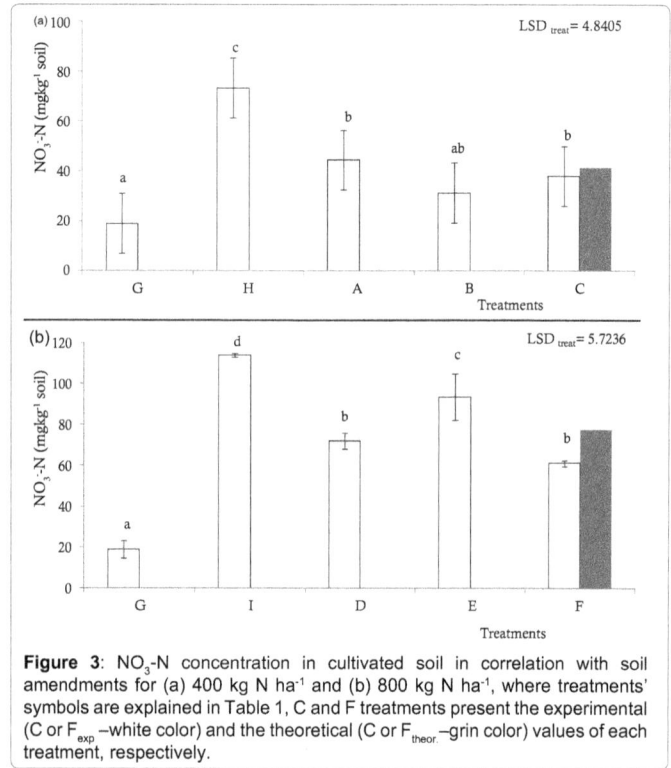

Figure 3: NO$_3$-N concentration in cultivated soil in correlation with soil amendments for (a) 400 kg N ha^{-1} and (b) 800 kg N ha^{-1}, where treatments' symbols are explained in Table 1, C and F treatments present the experimental (C or F$_{exp.}$ –white color) and the theoretical (C or F$_{theor.}$ –grin color) values of each treatment, respectively.

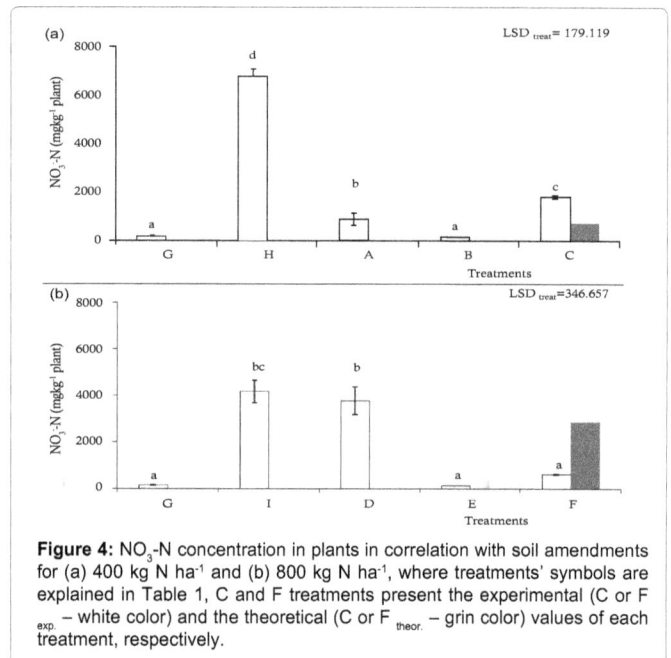

Figure 4: NO$_3$-N concentration in plants in correlation with soil amendments for (a) 400 kg N ha^{-1} and (b) 800 kg N ha^{-1}, where treatments' symbols are explained in Table 1, C and F treatments present the experimental (C or F$_{exp.}$ – white color) and the theoretical (C or F$_{theor.}$ – grin color) values of each treatment, respectively.

N ha^{-1} (Figure 4). Probably the high toxicity of the soil with nitrate nitrogen, led Z-B amendment (F$_{exp.}$ treatment) to have a similar behavior as bentonite (E treatment) although the amount of bentonite in Z-B is one third of the initial amount of bentonite.

The reduction percentage of NO$_3^-$ - N in soil and maize crop cultivation appear in Table 3. From the above table, it seems that all materials presented high decrease of NO$_3^-$ - N either to soil or maize crop, respectively. Zeolite and bentonite appeared the highest decrease of NO$_3^-$ - N in maize crop for the addition of 400 kg N ha^{-1}. Moreover, bentonite and zeolite-bentonite presented the highest decrease of NO$_3^-$ - N in maize crop for the addition of 800 kg N ha^{-1}. The comparison of treatments with soil amendment and different nitrogen doses (A,B,C,D,E,F) with the treatment G where only maize was cultivated is not acceptable due to the lack of additional nitrogen to treatment G.

Conclusions

- According to the results, soil amendments did not manage to increase the agronomic characteristics of maize, except for bentonite and Z-B, which managed to increase the height of the crop in the dose of 800 kg N ha^{-1}.

- No statistical significant difference in the dry weight of crops was presented with the use of soil amendments.

- All soil amendments decreased statistical significantly the nitrate nitrogen in soil and plant in both doses.

- Considering all soil amendments that were used, the most suitable for the removal of nitrate nitrogen from maize crop is bentonite and zeolite-bentonite for doses equal to 400 kg N ha^{-1} and 800 kg N ha^{-1}.

- The low concentration of nitrate nitrogen in maize crops cultivated in soil with bentonite as soil amendment was caused by the negative charges of bentonite which led to the repelled nitrate nitrogen ions from the root system of maize to the deep soil through soil water.

References

1. Nitrogen: An Essential Element In Crop Production (2014) Nachurs Alpine Solutions.

2. Nitrogen sources and transformations (2014) Colorado State University Extension. USA.

3. Li FY, Johnstone PR, Pearson A, Fletcher A, Jamieson PD, et al. (2009) A decision support system for optimizing nitrogen management of maize, NJAS -Wageningen Journal of Life Sciences 57: 93-100.

4. Mitchell JK (1993) Fundamentals of soil behavior (2ndedn). Wiley and Sons Inc., New York, USA 392-397.

5. Mumpton FP (1999) Uses of natural zeolites in agriculture and industry, Proceedings of the National Academy of Science 96: 3463-3470.

6. Akbar S, Khatoon S, Shehnaz R, Hussain T (1999) Natural zeolites: Structures, and importance. Science International (Lahore) 11: 73-78.

7. Wood ED, Armstrong FAJ, Richards FA (1967) Journal of the Marine Biological Association of the United Kingdom 47: 23-31.

8. Ioannou Z, Dimirkou A, Ioannou A (2013) Phosphate adsorption from aqueous solutions onto goethite, bentonite, and bentonite–goethite System. Water Air Soil Pollution 224: 1374-1388.

9. Ioannou Z, Simitzis J (2013) Production of carbonaceous adsorbents from agricultural by-products and novolac resin under a continuous countercurrent flow type pyrolysis operation. Bio resource Technology 129: 191-199.

10. Molla A, Ioannou Z, Dimirkou A, Skordas K (2014) Surfactant modified zeolites with iron oxide for the removal of ammonium and nitrate ions from waters and soils. Topics in Chemistry and Materials Science, Innoslab Ltd 7: 38-49.

11

Reducing the Indoor Odorous Charge in Waste Treatment Facilities

E Gallego[1]*, FJ Roca[1], JF Perales[1] and G Sánchez[2]

[1]Laboratory Centre for Environment, Polytechnic University of Catalonia (LCMA-UPC), Spain
[2]Department of Prevention and Waste Management of Greater Barcelona, Spain

Abstract

Characterising and determining the odorous charge of indoor air through Odour Units (OU) is an advantageous approach to evaluate indoor air quality and discomfort inside municipal solid waste facilities. The assessment of the OU can be done through the determination of Volatile Organic Compounds (VOC) concentrations and their odour thresholds. The aim of the study was to evaluate the differences in the odorous charge in the organic matter pit of a mechanical-biological waste treatment plant with a processing capacity of 287,500 tons year^{-1}. The sampling was carried out during the months of September 2012 (original situation) and October 2012 (after emptying the organic matter pit drain pipe). 150 chemical compounds were determined qualitatively in the studied location, from which 102 were quantified due to their odorous characteristics as well as their potentiality of having negative health effects. The results obtained demonstrated that after a maintenance cleaning operation such as draining the organic matter pit pipe, the odorous charge inside the facility can be diminished in a great way, up to a 95%.

Keywords: Indoor air quality; Maintenance cleaning; Municipal solid waste (MSW); Odour units; TD GC/MS; Volatile organic compounds (VOC)

Introduction

Perceptible malodours in the indoor air of waste treatment plants have a considerable impact on occupational comfort, hygiene, health and safety [1,2]. Volatile organic compounds (VOC) are the main causers of odorous nuisances [3], being formed and released to the indoor environment of waste treatment facilities either from degradation processes of the organic matter or by degradation and volatilization of other materials treated. Hence, determining the odorous contribution of each VOC or family to the total odorous charge in the indoor air is a helpful method to identify, characterize and evaluate the most annoying chemicals in order to prevent their generation during the waste treatment processes, as well as to find solutions to suppress them [3]. The working thesis assumes the superposition of the individual odorous concentrations calculated through VOC concentrations and their concrete odour thresholds [3,4]. It has to be taken into account that possible effects derived from masking or synergies between the evaluated compounds are not considered, and that the total odour units (OU) determined by olfactometry could differ in a certain way [5-8]. However, it has also to be considered that OU calculated using the presented methodology in a previous study were in the range of 1200-28,000 OU [3], in accordance with OU calculated using dynamic olfactometry (5000-30,000 OU m^{-3}) in similar facilities [9,10]. Additionally, several studies have demonstrated good correlations between olfactometrically determined OU and VOC concentrations [5,10,11]. The use of the presented procedure is advisable to be used when comparing differences in the odorous charge when changes in the processes developed into the facility are implemented [3]. The aim of this short report is to exemplify the effects of maintenance cleaning operations, such as draining the organic matter pit pipe, in the odorous charge of indoor air in a waste treatment facility.

Materials and Methods

Sampling strategy

The evaluation of the odorous charge in two scenarios (before and after a cleaning maintenance operation, i.e. draining the organic matter pit pipe) was done in the organic matter pit building of a mechanical-biological waste treatment (MBT) plant located in the metropolitan area of Barcelona, which has a processing capacity of 287,500 tons year^{-1} of municipal residues: selected organic fraction (100,000 tons year^{-1}), waste fraction (160,000 tons year^{-1}) and light packaging (27,500 tons year^{-1}). The selected organic matter fraction is discharged from the garbage trucks in a waste reception pit in a closed building. Organic matter is disposed in a conveyor belt by a bridge crane, led through a pre-treatment section, and eventually anaerobically digested to obtain biogas.

The organic matter pit building platform is cleaned twice a week according to a maintenance program consisting in the application of pressurized water. However, the organic matter pit pipeline is cleaned when lixiviates do not drain, without following a regular planification. Two samples from the organic matter pit indoor air were taken between 17th and 25th of September 2012 in the original conditions of the facility, without having emptied the lixiviate pipe of the pit for 2-3 weeks, respectively. Additionally, on the 1st of October 2012 a sample was taken after 4 days of having purged the lixiviate pipeline. VOC and VSC were dynamically sampled by connecting custom packed glass multi-sorbent cartridge tubes (Carbotrap 20/40, 70 mg; Carbopack X 40/60, 100 mg and Carboxen 569 20/45, 90 mg) and Tenax TA (60/80, 200 mg) tubes, respectively, to AirChek 2000 SKC pumps [12,13].

Analytical instrumentation

VOC and VSC were analysed by Automatic Thermal Desorption and capillary Gas Chromatography/Mass Spectrometry Detector using a Perkin Elmer ATD 400 (Perkin Elmer, Boston, Massachusetts, USA) and a Thermo Quest Trace 2000 GC (Thermo Quest, San Jose,

*Corresponding author: Eva Gallego, Laboratory Centre for Environment, Polytechnic University of Catalonia (LCMA-UPC), Avda Diagonal 647, 08028 Barcelona, Spain
E-mail: lcma.info@upc.edu

California, USA) fitted with a Thermo Quest Trace Finnigan MSD. Mass spectral data were acquired over a mass range of 20-300 amu. Samples were quantified by the external standard method. The methodology is described elsewhere [12].

Limits of detection, determined with a signal-to-noise ratio of 3, ranged from 0.001 to 10 ng. Compounds showed repeatabilities (% relative standard deviation values) ≤ 25%.

Results and Discussion

Indoor air VOC concentrations

150 chemical compounds were determined qualitatively in the studied location, as it had been observed in a previous study [3], from which 102 were quantified (those compounds with a low odour threshold as well as those with toxicity component or potential negative health effects). Table 1 shows the chemical familial concentrations for each sampling day. Concentrations obtained were of the same order of magnitude than the observed in previous studies regarding organic matter waste treatment, being terpenoids, alcohols, carboxylic acids and esters the mainly emitted compounds [3,6,14]. Generally, familial concentrations increase from 23-282%, with a global value of 45% between the two first samplings. Four days after cleaning, concentrations decreased a global value of 70% in respect to the sample taken 3 weeks after the last pipeline drainage (25th September). Terpenoid and aldehyde concentrations did not vary in a substantial way after the pipeline drainage process.

In the original scenario, alcohols, terpenoids, carboxylic acids, esters and ketones showed higher concentrations in respect to the other families evaluated, as observed in a previous study [3]. However, when the pipeline was drained, the most concentrated VOC were alcohols, terpenoids, ketones and aldehydes.

VOC concentrations did not exceed the VLA-ED (Table 2), the Spanish correspondence for Threshold Limit Value (TLV)-Time Weighted Average, as it had been observed in previous studies conducted in similar facilities [3,10,15]. However, as a great number

Family	17th September	25th September		1st October	
	μg m-3	μg m-3	Increase (%)[a]	μg m-3	Decrease (%)[b]
Alkanes	218	380	74	9.4	98
Aromatic hydrocarbons	551	761	38	47	94
Alcohols	91,161	124,464	37	21,129	83
Ketones	2854	4657	63	1064	77
Halocarbons	137	45	-67[c]	11	76
Aldehydes	290	795	174	802	-1[d]
Esters	2984	3664	23	178	95
Acids	3250	9884	204	163	98
Terpenoids	15,896	24,953	57	27,339	-10[d]
Organosulfurs	165	219	33	10	95
Ethers	1.7	6.5	282	5.7	12
Furans	4.5	6.2	38	0.2	97
Glycols	123	155	26	7.0	95
Organonitrogenated	5.5	3.3	-40[c]	1.0	70
Total VOC (mg m-3)	118	170	45	51	70

[a]Increase in concentrations (%) from 17th September to 25th September
[b]Decrease in concentrations (%) from 25th September to 1st October
[c]Decrease in concentrations.
[d]Increase in concentrations.

Table 1: Indoor familial concentrations (μg m-3) in the organic matter pit.

Compound	17th September	25th September	1st October	Odour threshold[a]	VLA-ED[b]	Vapour pressure[c]
Aromatic hydrocarbons						
m+p-Ethyltoluene	28.1	72.0	1.1	42	-[d]	0.4
Styrene	14.0	32.2	0.3	12	86,000	0.6
Alcohols						
1-Butanol	382	2521	31.6	480	61,000	0.9
1-Hexanol	28.7	56.0	1.1	40	-	0.1
1-Propanol	1733	4042	69.1	2000	246,000	3.5
2-Butanol	5161	7619	183	400	308,000	2.4
Ethanol	82,397	106,833	20,547	2000	1,910,000[e]	7.9
Ketones						
Biacetyl	32.8	162	1.7	0.7	-	7.6
Aldehydes						
Acetaldehyde	221	651	786	2.7	46,000[e]	120
Benzaldehyde	9.9	26.1	4.2	10	-	0.1
Isovaleraldehyde	2.6	3.3	0.3	1.6	-	6.6
Propanal	24.3	58.9	4.5	3.6	48,300[f]	39.9
Esters						
Ethyl butyrate	38.8	113	0.6	0.017	-	1.7

Ethyl hexanoate	74.2	255	1.7	10	-	0.2
Ethyl isovalerate	0.8	2.2	0.04	0.1	-	1.1
Ethyl octanoate	2.8	13.4	0.2	6	-	0.03
Methyl butyrate	5.5	9.6	0.1	7.7	-	4.2
Acids						
Acetic acid	572	4070	95	90	25,000	2.1
Butanoic acid	16.0	365	1.8	0.35	-	0.2
Hexanoic acid	2653	5245	59.5	20	-	0.02
Propanoic acid	7.2	194	6.8	5.1	31,000	0.6
Terpenoids						
D-Limonene	14,600	22,851	15,798	1700	110,000[f]	0.3
p-Cymene	640	726	11,457	200	-	0.2
α-Pinene	191	277	14.6	230	113,000	0.6
β-Myrcene	78.9	164	10.2	130	-	0.3
Organosulfurs						
Dimethyl disulfide	151	187	5.1	7	-	3.8
Dimethyl sulfide	13.3	24.2	0.3	1	25,800	86.3

[a]Source: "Compilations of odourthres hold values in air and water", L.J. van Gemert (TNO Nutrition and Food Research Institute). Boelens Aroma Chemicals InformationService (BACIS). The Netherlands (2003); "OdorThresholds for Chemicals with Established Occupational Health Standards" American Industrial Hygiene Association.USA (2009); "Reference Guide to Odor Thresholds for Hazardous Air Pollutants Listed in the Clean Air Act Amendments of 1990". EPA/600/R-92/047 (2009); and "Measurement of odor threshold by triangle odor bag method", Y. Nagata.Odor Measurement Review, 118-127, Japan Ministry of Environment (2003).

[b]Valor Límite Ambiental-Exposición Diaria: the Spanish correspondence for Threshold Limit Value-Time Weighted Average (TLV-TWA).

[c]Vapour pressure at 25ºC (kPa)

[d]Not established value

[e]As VLA-EC: Valor Límite Ambiental-Exposición de cortaduración (maximum of 15 min during the daily exposure)

[f]Proposed value

Table 2: Concentrations (μg m⁻³) of selected relevant odorous VOC in the organic matter pit. Concentrations with grey shading exceed the odour threshold of the compound.

Compound	17th September	25th September		1st October	
	OU	OU	Increase (%)[a]	OU	Decrease (%)[b]
Aromatic hydrocarbons					
m+p-Ethyltoluene	<1[c]	1.7	-[d]	<1	100
Styrene	1.2	2.7	125	<1	100
Total OU Aromatic hydrocarbons	**1.2**	**4.4**	**267**	**<1**	**100**
Alcohols					
1-Butanol	<1	5.3	-	<1	100
1-Hexanol	<1	1.4	-	<1	100
1-Propanol	<1	2.0	-	<1	100
2-Butanol	13	19	46	<1	100
Ethanol	41	53	29	10	81
Total OU Alcohols	**54**	**81**	**50**	**10**	**88**
Ketones					
Biacetyl	47	232	394	2.4	99
Total OU Ketones	**47**	**232**	**394**	**2**	**99**
Aldehydes					
Acetaldehyde	82	241	194	291	-21[e]
Benzaldehyde	<1	2.6	-	<1	100
Isovaleraldehyde	1.6	2.1	31	<1	100
Propanal	6.7	16	139	1.3	91
Total OU Aldehydes	**90**	**262**	**191**	**292**	**-11[e]**
Esters					
Ethyl butyrate	2284	6654	191	34	99
Ethyl hexanoate	7.4	25	238	<1	100
Ethyl isovalerate	7.6	22	189	<1	100
Ethyl octanoate	<1	2.2	-	<1	100
Methyl butyrate	<1	1.2	-	<1	100
Total OU Esters	**2299**	**6704**	**192**	**34**	**99**
Acids					
Acetic acid	6.4	45	603	1.1	98
Butanoic acid	46	1041	2163	5.1	99
Hexanoic acid	133	262	97	3.0	99
Propanoic acid	1.4	38	2614	1.3	97

Toal OU Acids	**186**	**1387**	**646**	**10**	**99**
Terpenoids					
D-Limonene	8.6	13	51	9.3	28
p-Cymene	3.2	3.6	13	57	-1483[e]
α-Pinene	<1	1.2	-	<1	100
β-Myrcene	<1	1.3	-	<1	100
Total OU Terpenoids	**12**	**20**	**67**	**66**	**-230[e]**
Organosulfurs					
Dimethyl disulfide	22	27	23	<1	100
Dimethyl sulfide	13	24	85	<1	100
Total OU Organosulfurs	**35**	**51**	**46**	**<1**	**100**
Total OU	**2724**	**8741**	**221**	**414**	**95**

[a]Increase in OU (%) from 17th September to 25th September
[b]Decrease in OU (%) from 25th September to 1st October
[c]Concentration of the compound below the odour threshold, hence, odour units below the unity.
[d]Increase not calculated due to the absence of OU from this compound the 17th September.
[e]Increase in OU.

Table 3: Odour units (OU) in the organic matter pit.

of VOC exceed their odour thresholds, they can lead to a lower employee satisfaction and productivity in the workplace, as well as to an increase of discomfort and personnel health hazards [15-17]. High VOC concentrations, even presenting lower values than TLV, can cause direct reactions such as sensory irritation of mucous membranes (eyes, nose and throat), and other individual's subjective symptoms like weakness, confusion, difficulty in making decisions, headache and memory loss [2,18]. In a previous study conducted in the same facility evaluated in the present paper, the total carcinogenic and non-carcinogenic risks (sum of selected VOC) were obtained in the ranges of 10^{-5}-10^{-4} and 10^{-2}-6, respectively [15]. Even though, long term epidemiological occupational health studies in municipal waste management plants are scarce. Additionally, major differences exist among developed and developing countries in relation to health and safety management in this kind of facilities. More research in this field has to be promoted before long, and the use of biomarkers can be a crucial step in order to detect biological effects in the exposed workers before the illnesses are diagnosed [1].

Odorous charge

The OU, calculated by dividing the concentration of a specific compound by its odour threshold limit, indicate how many times the threshold limit has been exceeded [3]. The OU in the organic matter pit are presented in Table 3. Familial OU increase from 46 to 646%, with a global value of 221% between the two first samplings, a higher increase than in concentrations, mainly due to the low odour thresholds that present several compounds. Additionally, some compounds only generate OU on the 25th of September, when their concentrations are relatively high (e.g. certain alcohols, aldehydes, esters and terpenes). These compounds present odour thresholds between 6 and 2000 μgm^{-3}. The above mentioned compounds present the lower vapour pressures, between 0.03-4.2 kPa at 25°C, being less volatile than the other evaluated compounds, and only present in concentrations that generate OU when their accumulation due to lack of cleaning is produced. Carboxylic acids, p-cymene and α-pinene also present similar vapour pressures, yet they are the main released compounds from organic matter degradation processes [3,19]. Once the pipeline was drained, OU in the organic matter pit decreased a global value of 95%. Previously to the draining process, esters were the family that contributed most to total OU (Figure 1), as already observed in a preceding study [3]. After the pipe was drained, ketones, esters, carboxylic acids and organosulfurs decreased in a relevant way their

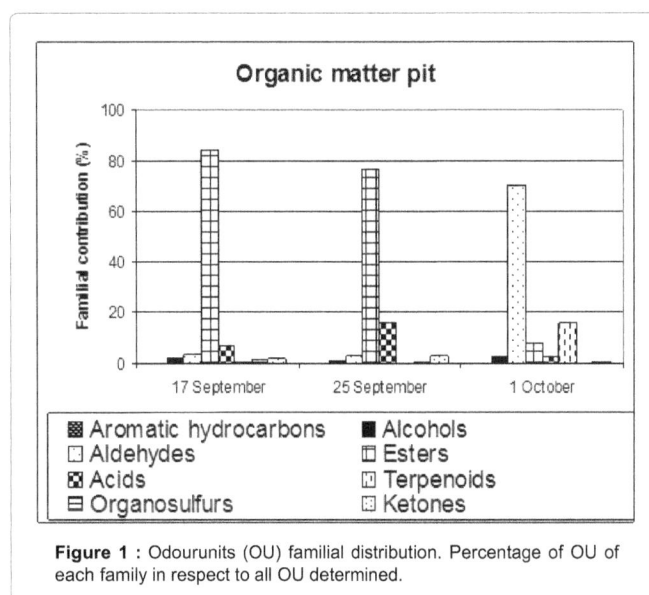

Figure 1 : Odourunits (OU) familial distribution. Percentage of OU of each family in respect to all OU determined.

contribution to the total OU. Hence, the main contributors to total OU in this second scenario were aldehydes and terpenoids.

Conclusions

The best way to avoid the nuisance produced by odours in a waste treatment plant is not generating them, reducing at a minimum the presence of VOC in the indoor air of the facility. The present study has demonstrated that with a simple maintenance operation odours can be reduced up to 95%. Programming a purging procedure instead of purging when the pipeline lixiviates do not drain would decrease indoor discomfort related to odours. This work is a preliminary approach to the effects of cleaning maintenance programs in respect to ambient VOC. Deeper evaluations and more research are needed in this field.

Acknowledgements

The authors acknowledge the support given through the EHMAN project (DPI2009-09386) financed both for the Spanish Ministry of Science and Innovation and the ERDF from the EU. E.G. acknowledges with thanks a Juan de la Cierva grant from the Spanish Ministry of Science and Innovation.

References

1. Giusti L (2009) A review of waste management practices and their impact on human health. Waste Manag 29: 2227-2239.

2. Lehtinen J, Tolvanen O, Nivukoski U, Veijanen A, Hanninen K (2013) Occupational hygiene in terms of volatile organic compounds (VOCs) and bioaerosols at two solid waste management plants in Finland. Waste Manag 33: 964-973.

3. Gallego E, Roca FJ, Perales JF, Sanchez G, Esplugas P (2012) Characterization and determination of the odorous charge in the indoor air of a waste treatment facility through the evaluation of volatile organic compounds (VOCs) using TD-GC/MS. Waste Manag 32: 2469-2481.

4. Schauberger G, Piringer M, Knauder W, Petz E (2011) Odour emissions from a waste treatment plant using an inverse dispersion technique. Atmospheric Environment 45: 1639-1647.

5. Tsai CJ, Chen ML, YeAD, Chou MS, Shen SH, et al. (2008) The relationship of odor concentration and the critical components emitted from food waste composting plants. Atmospheric Environment 42: 8246-8251.

6. Orzi V, Cadena E, D'Imporzano G, Artola A, Davoli E, et al. (2010) Potential odour emission measurement in organic fraction of municipal solid waste during anaerobic digestion: Relationship with process and biological stability parameters. Bioresource Technology 101: 7330-7337.

7. Almarcha D, Almarcha M, Nadal S, Caixach J (2012) Comparison of the depuration efficiency for VOC and other odoriferous compounds in conventional and advanced biofilters in the abatement of odour emissions from municipal waste treatment plants. Chemical Engineering Transactions 30: 259-264.

8. Dorado AD, Husni S, Pascual G, Puigdellivol C, Gabriel D (2014) Inventory and treatment of compost maturation emissions in a municipal solid waste treatment facility. Waste Manag 34: 344-351.

9. Schlegelmilch M, Streese J, Biedermann W, Herold T, Stegmann R (2005) Odour control at biowaste composting facilities. Waste Manag 25: 917-927.

10. Scaglia B, Orzi V, Artola A, Font X, Davoli E, et al. (2011) Odours and volatile organic compounds emitted from municipal solid waste at different stage of decomposition and relationship with biological stability. Bioresource Technology 102: 4638-4645.

11. Pierucci P, Porazzi E, Pardo MM., Adani F, Carati C, et al. (2005) Volatile organic compounds produced during the aerobic biological processing of municipal solid waste in a pilot plant. Chemosphere 59: 423-430.

12. Ribes A, Carrera G, Gallego E, Roca X, Berenguer MJX (2007) Development and validation of a method for air-quality and nuisance odors monitoring of volatile organic compounds using multi-sorbent adsorption and gas chromatography/mass spectrometry thermal desorption system. J Chromatogr A 1140: 44-55.

13. Gallego E, Roca FX, Perales JF, Rosell MG, Guardino X (2008) Development and validation of a method for air-quality and nuisance odours monitoring of volatile sulphur compounds (VSCs) using ATD-GC/MS. Proceedings of the 12th Day of Instrumental Analysis, 21-23 October, Barcelona.

14. Leguizamón DS (2003) The Air Qualityclosed a plant for the treatment and recovery of the organic fraction of municipal solidwaste. Spain.

15. Nadal M, Inza I, Schumacher M, Figueras MJ, Domingo JL (2009) Health risks of the occupational exposure to microbiological and chemical pollutants in a municipal waste organic fraction treatment plant. IntJ Hyg Environ Health 212: 661-669.

16. Harrison EZ (2007) Health impacts of composting air emissions. Biocycle 48: 44-50.

17. Domingo JL, Nadal M (2009) Domestic waste composting facilities: A review of human health risks. Environ Int 35: 382-389.

18. Smeets MAM, Dalton PH (2005) Evaluating the human response to chemicals: odor, irritation and non-sensory factors. Environ Toxicol Pharmaco I19: 581-588.

19. Staley BF, Xu F, Cowie SJ, Barlaz MA, Hater GR (2006) Release of trace organic compounds during the decomposition of municipal solid waste components. Environ Sci Technol 40: 5984-5991.

12

Principle Component Analysis of Longitudinal Dispersion Coefficient Parameters

Abbas Parsaie and **Amir Hamzeh Haghiabi**

Department of water Engineering, Lorestan University, Khorram Abad, Iran

Abstract

Study on the river water quality is a main part of environmental engineering. Longitudinal dispersion coefficient (D_L) is one of the main important parameters in the river water quality studies. Several parameters such as hydraulically and morphological rivers are affective on the D_L, whereas the mount effectiveness of some of them such as riverbed form cannot be measured. So, researchers proposed that the D_L is proportional to the flow velocity, channel width, river flow depth, and shear velocity. Defining the most influence parameters on the D_L leads to develop an optimal structure for empirical formulas and soft computing techniques which will be propose for estimation of D_L. In this paper the principle component analysis (PCA) was used to define the most affective parameters on the D_L. The PCA results indicated that the width of the river, flow depth, and flow velocity are the most important parameters on the D_L. Evaluating the performance of empirical formulas with considering the PCA results showed that the Tavakollizadeh and Kashefipur formula is accurate among the empirical formulas.

Keywords: Longitudinal dispersion coefficient; Water quality; River pollution; Soft computing; Principle component analysis

Introduction

River pollution has become one the main problem in the environmental health monitoring. Study on the river water quality is one of the main parameters in the part of the environmental engineering. Longitudinal dispersion coefficient definition is the important parameter in the river pollution studies. As shown in the Figure 1 when a contamination poured in the river flow rapidly emissions and moved to the downstream through the river. Figure 1 shows that at the first the dispersion mechanism is 3D dimensionality whereas by passing the time and moving through the river the dispersion mechanism is one dimensional which named in general longitudinal dispersion coefficient [1-3]. Study on the longitudinal dispersion coefficient in rivers usually conducted by field studies. Flied studies are usually conducted by injection a tracer which has not interaction with the water in the river and some station along the river is considered for sampling from the river water. The main notes that are more important in the field studies included the material of the tracer which should not interaction with water and has any destructive effect on the environment and another note is related to the place of the sampling station, the location of the first station should be considered after complete mixing the trace in the cross section of the river flow [4]. Several field and laboratory studies have been conducted on the mechanism of the D_{Lc} in the rivers. In this regard, the studies which conducted by [5-12] can be mentioned. Due to high cost of the field studies and laboratory equipment recently researcher welcomed to use the numerical approaches. In the field of the numerical modelling the government equation which usually is differential equation solves by numerical method such as finite difference, finite volume and finite element another numerical method which are widely uses in the environmental studies are artificial neural networks (ANN) such as Multilayer perceptron neural network (MLP), adaptive neuro fuzzy inference system (ANFIS) ,genetic programing(GP) and support vector machine. [13-23] The soft computing technique can be uses with numerical methods to increase the accuracy of the numerical simulation [24,25]. Developing the soft computing models are based on the data set, it means that the important parameters which are influence on the phenomena should be measured in the past. In this paper the principle component analysis (PCA) are used to derived most important parameters on the D_L. Developing the empirical formulas and soft computing models based on the PCA leads to prepare an optimal structure of the model by more reliability.

Methods and Material

Longitudinal dispersion coefficient is a function of the river geometries, fluid properties and hydraulic condition. The main parameters that are influence on the D_c are given on the Equation (1).

$$D_c = f_1\left(\rho, \mu, u, u_*, h, w, s_f, s_n\right) \quad \dots\dots\dots\dots (1)$$

Where ρ is fluid density; μ is dynamic viscosity; w is the width

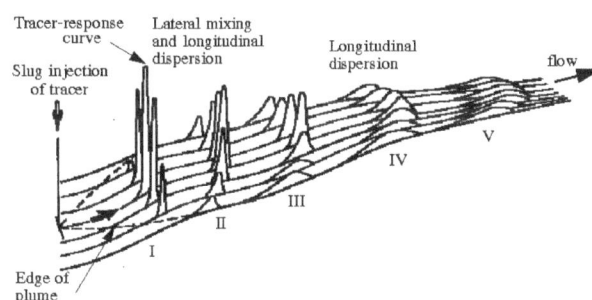

Figure 1: Schematic shape of the dispersion mechanism in rivers

*Corresponding author: Abbas Parsaie, Ph.D. Candidate of Hydro Structures, Department of Water Engineering, Lorestan University, Khorram Abad, Iran
E-mail: Abbas_Parsaie@yahoo.com

of cross-section; h is flow depth; u_* is share velocity, S_f is longitudinal bed shape and S_n is sinuosity of the river. To derive the dimensionless parameter on the D_L, the Buckingham theory was considered and dimensionless parameter will be derived as shown below

$$\frac{D_L}{hu_*u} = f_2\left(\frac{uh}{h}, \frac{u}{u_*}, \frac{w}{s}, s_f, n\right) \dots\dots (2)$$

The flow in the nature is always turbulent especially in the river. Therefore, the Reynolds number $\tilde{n}\frac{uh}{i}$ can be ignored and the bed form and sinuitis path parameters cannot be measured clearly, as well. Therefore, the effect of them can be considered as flow resistant, which is seen in the flow depth. The dimensionless parameters that can be clearly measured are given as below. Table 1 gives some of the famous empirical formula which proposed by researchers. As mentioned in past developing the ANN models is based on the data set so about 150 data set related to the D_L was collected and range of them given in the Table 2.

$$\frac{D_L}{hu_*} = f_2\left(\frac{u}{u_*}, \frac{w}{h}\right) \dots\dots\dots\dots (3)$$

Principle Component Analysis (PCA)

The Principle Component Analysis (PCA) is an advanced categorized method in the factor analysis approaches and usually uses for data reduction in the field of engineering. The main application of the PCA is in the compression and classification of data; in the other words the main usage of this approach is to reduce the dimensionality of a data set (sample) by finding a new set of variables, smaller than the original set of variables that nonetheless retains most of the sample's information. During the PCA process because of using the all initial variable, the new variable involved all the initial variable information. The process of PCA continued two steps which explained in the follow [26].

Results and Conclusion

The accuracy of the empirical formulas was conducted by compression with measurement data and results of these are shown in the Figure 2. Figure 2 shows the results of the empirical formulas versus the measured data and also the standard error indices such as correlation coefficient (R^2) and root mean square of error (RMSE) was added to this figure. As seems in the Figure 2 the Tavakollizadeh and Kashefipur formula is the accurate among the empirical formulas by ($R^2 = 0.45$ and RMSE=5861). In overall, assessing the performance of the empirical formulas shows that these formulas have unacceptable accuracy to use in management problems. To define the most influence parameters on the D_L, the PCA technique was carried out on the collected data which these range was given in the Table 2. The result of the PCA is given in the Figure 3. As shown in the Figure 3, the channel width, flow depth, flow velocity are the most important parameters for D_L prediction. Another scenario as similar to the equation (2 &3) was considered, the result of the scenario number (2) shows the ratio of the u/u* is more important than the W/H.

By attention to the PCA results and reviewing the structure of empirical formulas which was given in the table 1 and evaluating the results of the empirical formula which was given in the Figure 2 it could be found that the empirical formula such as Tavakollizadeh and Kashefipur which considered more weight for the parameters such as W

Equation	Author	Row
$D_L = 5.93hu_*$	Elder (1959)	1
$D_L = 0.58\left(\frac{h}{u_*}\right)^2 uw$	McQuivey and Keefer (1974)	2
$D_L = 0.011\dfrac{u^2 w^2}{hu_*}$	Fisher (1967)	3
$D_L = 0.55\dfrac{wu_*}{h^2}$	Li et al. (1998)	4
$D_L = 0.18\left(\frac{u}{u_*}\right)^{0.5}\left(\frac{w}{h}\right)^2 hu_*$	Liu (1977)	5
$D_L = 2\left(\frac{w}{h}\right)^{1.5} hu_*$	Iwasa and Aya (1991)	6
$D_L = 5.92\left(\frac{u}{u_*}\right)^{1.43}\left(\frac{w}{h}\right)^{0.62} hu_*$	Seo and Cheong (1998)	7
$D_L = 0.6\left(\frac{w}{h}\right)^2 hu_*$	Koussis and Rodriguez-Mirasol (1998)	8
$D_L = 5.92\left(\frac{u}{u_*}\right)^{1.2}\left(\frac{w}{h}\right)^{1.3} hu_*$	Li et al. (1998)	9
$D_L = 2\left(\frac{u}{u_*}\right)^{0.96}\left(\frac{w}{h}\right)^{1.25} hu_*$	Kashefipur and Falconer (2002)	10
$D_L = 7.428 + 1.775\left(\frac{u}{u_*}\right)^{1.752}\left(\frac{w}{h}\right)^{0.62} hu$	Tavakollizadeh and Kashefipur (2007)	11
$D_L = 10.612\left(\frac{u}{u_*}\right) hu$	Rajeev and Dutta (2009)	12

Table 1: Empirical equations for calculating the longitudinal dispersion coefficient

Data Range	W (m)	H(m)	U(m/s)	u*(m/s)	W/H	U/u*
min	1.40	0.14	0.03	0.00	2.20	1.06
max	711.20	19.94	1.74	0.55	156.54	20.77
AVG	68.23	1.72	0.53	0.09	44.57	7.54
STDEV	99.28	2.20	0.35	0.07	29.76	4.79

Table 2: Range of collected data.

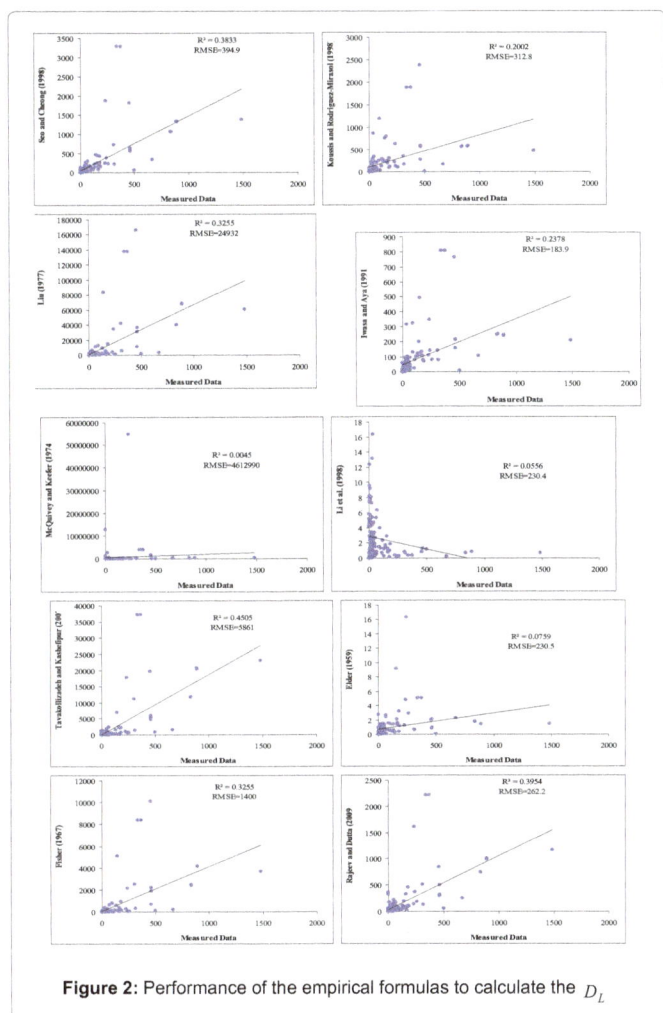

Figure 2: Performance of the empirical formulas to calculate the D_L

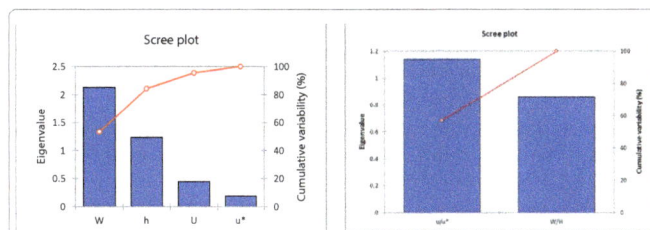

Figure 3: Results of the PCA techique, left: Scenario 1, right: Scenario 2.

and H or u/u_* are more accurate in the compare to the other formulas.

Conclusion

In this study, some famous analytical approaches for calculating the longitudinal dispersion coefficient (D_L) were assessed. To this purpose, 150 experimental data which published in the credible journal was collected. Calculation the standard error indices for analytical approaches results show that the Tavakollizadeh and Kashefipur formula by correlation coefficient about 0.45 is accurate among the empirical formulas. To define the most influence parameters on the D_L, the PCA technique was used. The PCA result indicated that the channel width, flow depth, flow velocity are the most important parameters on the D_L. Using the results of principle component analysis (PCA) techniques leads to develop an optimal model structure for empirical formula or soft computing models.

References

1. Chanson H (2004) Environmental Hydraulics for Open Channel Flows.

2. Holzbecher E (2012) Environmental Modeling: Using MATLAB.

3. Sahin S (2014) An Empirical Approach for Determining Longitudinal Dispersion Coefficients in Rivers. Environ Process 1: 277-285.

4. Atkinson TC, Davis PM (1999) Longitudinal dispersion in natural channels: I. Experimental results from the River Severn, U.K. Hydrol Earth Syst Sci 4: 345-353.

5. Baek KO, Seo IW (2010) Routing procedures for observed dispersion coefficients in two-dimensional river mixing. Advances in Water Resources 33: 1551-1559.

6. Davis PM, Atkinson TC (1999) Longitudinal dispersion in natural channels: 3. An aggregated dead zone model applied to the River Severn, U.K. Hydrol Earth Syst Sci 4: 373-381.

7. Kashefipour SM, Falconer RA (2002) Longitudinal dispersion coefficients in natural channels. Water Res 36: 1596-1608.

8. Nordin CF, Troutman BM (1980) Longitudinal dispersion in rivers: The persistence of skewness in observed data. Water Resources Research 16: 123-128.

9. Papadimitrakis I, Orphanos I (2004) Longitudinal Dispersion Characteristics of Rivers and Natural Streams in Greece. Water Air & Soil Pollution: Focus 4: 289-305.

10. Seo I, Park S, Choi H (2009) A Study of Pollutant Mixing and Evaluating of Dispersion Coefficients in Laboratory Meandering Channel. Advances in Water Resources and Hydraulic Engineering 485-490.

11. Azamathulla H, Ghani A (2011) Genetic Programming for Predicting Longitudinal Dispersion Coefficients in Streams. Water Resources Management 25: 1537-1544.

12. Azamathulla HM, Wu FC (2011) Support vector machine approach for longitudinal dispersion coefficients in natural streams. Applied Soft Computing 11: 2902-2905.

13. Benedini M, Tsakiris G (2013) Water Quality Modelling for Rivers and Streams.

14. Etemad-Shahidi A, Taghipour M (2012) Predicting Longitudinal Dispersion Coefficient in Natural Streams Using M5' Model Tree. Journal of Hydraulic Engineering 138: 542-554.

15. Fuat Toprak Z, Savci ME (2007) Longitudinal Dispersion Coefficient Modeling in Natural Channels using Fuzzy Logic." CLEAN – Soil, Air, Water 35: 626-637.

16. Li X, Liu H, Yin M (2013) Differential Evolution for Prediction of Longitudinal Dispersion Coefficients in Natural Streams. Water Resour Manage 27: 5245-5260.

17. Mirbagheri S, Abaspour M, Zamani K (2009) Mathematical modeling of water quality in river systems.

18. Riahi-Madvar H, Ayyoubzadeh SA, Khadangi E, Ebadzadeh MM (2009) An expert system for predicting longitudinal dispersion coefficient in natural streams by using ANFIS. Expert Systems with Applications 36: 8589-8596.

19. Sahay R (2011) Prediction of longitudinal dispersion coefficients in natural rivers using artificial neural network. Environ Fluid Mech 11: 247-261.

20. Szymkiewicz R (2010) Numerical Modeling in Open Channel Hydraulics.

21. Tayfur G, Singh V (2005) Predicting Longitudinal Dispersion Coefficient in Natural Streams by Artificial Neural Network. Journal of Hydraulic Engineering

131: 991-1000.

22. Toprak ZF, Hamidi N, Kisi O, Gerger R (2014) Modeling dimensionless longitudinal dispersion coefficient in natural streams using artificial intelligence methods. KSCE J Civ Eng 18: 718-730.

23. Parsaie A, Haghiabi A (2015) The Effect of Predicting Discharge Coefficient by Neural Network on Increasing the Numerical Modeling Accuracy of Flow Over Side Weir. Water Resour Manage 29 973-985.

24. Parsaie A, Yonesi H, Najafian S (2015) Predictive modeling of discharge in compound open channel by support vector machine technique. Model Earth Syst Environ 1: 1-6.

25. Alvarez PA (2011) Exploratory Data Analysis with MATLAB, (2nd Edn). International Statistical Review 79: 492-492.

26. Camacho J, Pérez-Villegas A, Rodríguez-Gómez RA, Jiménez-Mañas E (2015) Multivariate Exploratory Data Analysis (MEDA) Toolbox for Matlab. Chemometrics and Intelligent Laboratory Systems 143: 49-57.

13

Review of Clean Development Mechanism and use of Bundled Projects in Small and Medium Scale Enterprises

Mayuri Naik*, Anju Singh, Seema Unnikrishnan, Neelima Naik and Indrayani Nimkar

National Institute of Industrial Engineering (NITIE), Mumbai-400087, India

Abstract

Carbon finance through the Clean Development Mechanism (CDM) offers significant opportunity to a developing country like India for an array of greenhouse gas (GHG) emission reduction projects. However, the transaction cost associated with the development of CDM project is a serious barrier to many small scale CDM (SSC) projects due to which these proponents face many difficulties in attracting international investors. To reduce this transaction cost, individual small projects with similar project context can be bundled together to form a single CDM project. These SSC bundled projects that reduce GHG emissions can claim Certified Emission Reductions (CERs) under the concept of bundling. This paper presents 98 bundled CDM projects registered and issued worldwide till October 2014, out of which India has 29 projects, along with a case study on small scale hydro-electric power generation project. The visited project is a good example of clean technology that helps to reduce stress on conventional energy sources and is an improvement of social and economic life of local people. Energy efficiency, grid connected electricity generation, fossil fuels switching, thermal energy production and methane recovery are some of the methodologies in these types of projects. These methodologies reduce GHG emissions without harming the environment.

Keywords: Clean development mechanism; Small scale projects; Bundled; Transaction cost

Introduction

CDM is an agreement under Kyoto protocol, which is a multilateral effort taken by United Nation Framework Convention on Climate Change (UNFCCC) to tackle climate change. It is meant to promote changes in the pattern of GHG emission intensive activities in developing countries as discussed in Kyoto, Japan on 11th December 1997 and entered into force on 16th February 2005. 192 countries have ratified the treaty to date. It aims to reduce the GHG emissions by 5.2 % against the 1990 levels over the first commitment period of five years 2008-2012 [1,2]. Now it has been extended for a second commitment period which is from January 2013 to December 2020. During the second commitment period, parties are committed to reducing their GHG emissions by at least 18 percent below 1990 levels in the coming eight year period from 2013 to 2020 [3]. As a part of CDM, the developed (Annex-I) countries have a target of GHG emission reduction [4]. In order to achieve this target developed countries invest in or finance projects that reduce emissions in developing countries (Non Annex-I) using clean technologies. After implementing these projects in developing countries, for every tone of CO2 that does not enter into the atmosphere, a developing country earns one CER. These CERs can be further traded in international carbon market. Developed (Annex-I) countries exchange these CERs for money and technology transfer with Developing (Non-Annex-I) countries. Most of the demand for CERs from the CDM comes from the European Union Emissions Trading Scheme (EU-ETS) which is the largest carbon market. 93% of all the issued CERs come from five developing countries with maximum number of CDM projects. These are China, India, South Korea, Brazil and Mexico [5]. The Conference of Parties (COP), Executive Board (EB), Designated Operational Entity (DOE), Designated National Authority (DNA) are the important regulators in CDM. The COP is the supreme decision making body of the convention. EB supervises the CDM under the authority and guidance of the COP and reviews the regional/sub-regional distribution of the CDM project activities. It also maintains the registry of CDM projects and approves the methodologies for baseline and for monitoring project boundaries.

DOE is an independent auditor accredited by the EB to validate project proposals or verify whether implemented projects have achieved planned GHG emission reductions or CERs. DNA is the organization which has granted the responsibility to authorize and approve CDM projects in respected country. To attain CERs, project developer has to create first a Project Design Document (PDD) which requires the validation by the DOE, preferably by the DNA. Then it is sent to the EB for registration. After the registration, project can run in the developing (host) country. Monitoring is required for measuring CERs. The DNA then verifies and certifies the CERs which are finally issued by the EB. The project developer receives the certificate of CERs at this stage, which further can be sold out in the carbon market. CDM transaction costs are generally fixed. There are substantial diseconomies of scale that make it difficult for small projects to prove their financial viability [6-8]. The impact of such heavy cost warrants careful assessment of risks from potential buyers. Many investors, as well as administrators of carbon funds, would thus consider engaging into small scale CDM ventures only if the estimated potential credits exceed a minimum quota, depending upon their financial assessments. To conquer this, multiple small scale projects of the same type could be bundled together to form a single CDM project. Bundling of small scale projects is better option as it reduces overall costs for the administration of the CDM.

CDM Bundled Projects

Project activities with a maximum output capacity equivalent of up

*Corresponding author: Mayuri Naik, National Institute of Industrial Engineering (NITIE), Mumbai-400087, India
E-mail: mayuri.mrn02@gmail. com

to 15 megawatts are termed as small scale CDM projects (SSC). SSC projects can be bundled at various stages as PDD preparation, validation, registration, monitoring, and verification [9,10]. The requirements for bundling of SSC projects include: projects should be of the same type, within the same geographical area, at the same stage of development and be bundled by one organization [11]. Further, project activities in a bundle can also be divided in to a series of sub-bundles, provided that each sub-bundle should belong to same type of project activity [12]. Having the same crediting period and composition of bundles shall not change over a period of time. At registration, all project participants have to submit an agreement about their individual project activities to be bundled and their proposed bundling agency should represent all project participants in order to communicate with EB [28]. After implementation of a project, an overall monitoring plan that monitors performance of the constituent project activities further verifies or certifies the emission reductions achieved [13]. A single DOE certifies achieved emission reductions in terms of CERs. So the advantage is that a single validation and certification report for the entire bundle can be obtained, which streamlines these processes for project participants [14]. Bundling of projects could help in achieving a critical mass that would attract financing from lenders. Another advantage in this approach is the sharing of CDM and financing transaction costs among individual projects which could be prohibitive or even create a negative leverage if the transaction is made for a single project [15]. Sometimes it is important to distinguish between a large-sized CDM project and a bundle of small CDM projects because few large sized CDM projects with intension to use the simplified modalities and procedures, and reduce transaction costs may split large projects into two or more small projects. Therefore, the de-bundling of a large project into smaller projects in order to be able to use the simplified procedures is not allowed [16]. As a result of this EB approved a set of criteria to screen out these types of projects to check that a small project truly is a small-scale CDM project by showing that it is not closely related to or integrated with another small-size project. In a bundle, single entity acts on everyone's behalf with lower transaction cost and single monitoring report and allows small scale projects access to carbon market and allows it to benefit from carbon finance. These smaller bundled activities are effectively promoting local sustainable development in impoverished areas of developing countries. This paper provides an insight into these bundled projects which are registered and issued by CDM EB in order to achieve emission reduction in small scale projects. The paper also represents a case study related to small scale CDM bundled project which involves hydroelectric power generation from the water reservoir made available for irrigation. The project was analyzed as it is efficient and environmentally safe technology for power generation.

Results and Discussion

World

Figure 1 and Table 1 describe the country wise distribution of the registered CDM bundled projects. A total of 98 CDM bundled projects are registered and issued till July 2014. These projects are categorized on the basis of renewable energy sources used. India is having maximum number of CDM bundled projects (29 numbers), out of which 24 projects deal with wind power, 2 projects are biomass based, 2 projects relate to hydro power and 1project is on energy efficiency. Next to India, China has 26 projects, out of which 23 are power related and 3 are based on other renewable energy sources. 18 and 13 projects in Brazil and Mexico respectively are based on biomass energy. Republic of Korea has

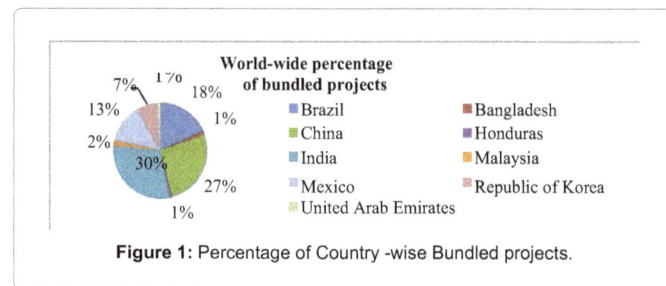

Figure 1: Percentage of Country -wise Bundled projects.

7 projects out of which 3 are based on hydro power, 1 on wind power, 1 is related to fuels switching and 2 projects are other renewable energy sources. Malaysia has 2 projects; out of which 1 is based on hydropower and other is biogas based project. Honduras has only 1 project based on biomass energy. Attracting investment to developing countries is a major goal of CDM. United Arab Emirates has 1 project on Energy efficiency. China, India, Brazil, Mexico and Malaysia are the top five countries with maximum number of registered CDM projects and number of issued CERs. These few developing countries are projected to emit more at a faster rate than the industrialized world and the rest of developing countries as these are at the stage of rapid industrialization [17,18]. Climate change mitigation projects can reinforce these countries pursuit to a low carbon path of development. Actual emission reduction is the achieved CERs which is the ratio of total CDM credits over a country's actual carbon emissions. It is the expected emission reduction achieved through CDM projects compared to a country's actual carbon emissions (Table 2). It gives a rough idea of the domestic emission reduction efforts via CDM and how much the CDM projects can contribute to national emission reduction efforts in a given country [19]. Up to July 2014, India has issued maximum number of CERs (1,296,695 metric tons) followed by China issuing 1,175,550 metric tons of CERs. Brazil and Mexico have issued 601,089 and 317,835 metric tons of CERs respectively. Republic of Korea and Malaysia have so far issued 137,485 and 109,485 metric tons of CERs respectively; United Arab Emirates has issued 2,004 metric tons of CERs. Honduras has so far issued 89,443 metric tons of CERs with maximum CER issuance rate (104.9%).

Renewable bundled projects

The CDM can also interface with renewable energy development to build up the capacity in the developing countries for renewable energy including wind, solar, biomass and hydro. The growth in CER issuance will be driven by capacity additions in the renewable energy sector and by the eligibility of more renewable energy projects to get CERs. Table 3 presents CERs issued under such renewable energy source used in CDM bundled projects across the world and in India respectively. Figure 2 presents renewable energy sources used in worldwide bundled projects. Out of total 98 bundled projects, 33 projects are based on biogas with 940,352 metric tons of CERs issued followed by 29 hydro power projects with maximum issued CERs of 1,393,979 metric tons. 25 projects are based on wind power with 928,386 metric tons. 5 projects are based on other renewable energy sources with 39,999 metric tons of issued CERs. Energy efficiency has 3 projects with 60,494 metric tons of issued CERs. 2 projects are based on biomass with 242,911 metric tons of CERs. Fuel switch have 1 project with issued CERs 6,290 metric tons.

Methodologies

Methodologies used in CDM projects help to reduce GHG emissions into the atmosphere without harming the environment. The function of

Sr. no.	Country	No. projects	Type of projects (No. of Projects)	Methodology used	CER Issued	CER Issuance Rate
1	Brazil	18	Biogas (18)	AMS III.D	601,089	53.4%
2	Bangladesh	1	Energy efficiency (1)	AMS-II.D	17,403	21%
3	China	26	Hydropower (23)	AMS. I.D	1,175,550	68%
			Other renewable energy (3)	AMS. I.D		
4	Honduras	1	Biogas (1)	AMS-I.D.	170,654	117.6%
5	India	29	Wind power (24)	AMS-I.D.	1,296,695	71.3%
			Biomass (2)	AMS-I.C. AMS-I.D.		
			Hydro (2)	AMS-I.D		
			Energy efficiency (1)	AMS-II.D		
6	Malaysia	2	Hydro (1)	AMS-I.D	109,485	85.60%
			Biogas (1)	AMS-III.H.		
7	Mexico	13	Biogas (13)	AMS-III.D.	317,835	34.70%
9	Republic of Korea	7	Hydro power(3)	AMS-I.D.	137,485	78.5%
			Wind power (1)	AMS-I.D.		
			Fuel switch (1)	AMS-III.B.		
			Other renewable energy (2)	ACM0002, AMS-I.D.		
10	United Arab Emirates	1	Energy efficiency (1)	AMS-II.D	2,004	20.9%
		98			3,828,200	

Table 1: Country-wise CDM Bundled Projects.

Sr. No.	Name of CDM Project Activity	State	Other Parties Involved	Project Participants (Authorized by Host Party)	Project Participants after registration (Authorized by other Parties involved)	Type of Project	Methodology	CER Issuance Audits	Total Issued CERs	Total Verified Days	CER Issuance Rate
1	8.75 MW Wind Power Project	Gujarat	Japan; Switzerland	Rolex Rings Private Ltd.	Emergent Ventures India Pvt. Ltd.; Mitsubishi Corporation	Wind power	AMS-I.D.	3	44,159	1,601	64%
2	12 MW Bundled Wind Power Project	Tamilnadu	Switzerland	NEG Micon (I) Private Ltd.	Vestas Wind Technology India Private Limited	Wind power	AMS-I.D.	2	54,732	1,027	82%
3	Bundled 15 MW Wind Power Project in India	Karnataka, Tamilnadu and Maharashtra	Spain; Sweden	GNA (HUF), Ferromar Shipping Private Ltd.	Kingdom of Spain (withdrawn); Swedish Energy Agency (withdrawn)	Wind power	AMS-I.D.	3	127,864	1,984	67%
4	3.7 MW Bundled Wind Power Project at Priyadarshini Polysacks Ltd.	Maharashtra	UK	Priyadarshini Polysacks Ltd.	The CarbonNeutral Company Limited	Wind power	AMS-I.D.	6	28,404	2,193	61%
5	13.4 MW bundled wind power project	Karnataka	Norway; Switzerland	NEG Micon (I) Private Ltd.	The Norwegian Ministry of Finance; Vestas Wind Technology India Private Limited	Wind power	AMS-I.D.	3	125,015	2,033	102%
6	11.2 MW Wind Power project	Tamilnadu	Switzerland	Amarjothi Spinning mills Ltd.	Vitol S.A.	Wind power	AMS-I.D.	2	158,386	2,555	92%
7	7.85 MW Bundled Wind Power Project in Southern India	karnataka and Tamilnadu	Switzerland	Vandana Ispat Ltd.	Bunge Emissions Fund Limited	Wind power	AMS-I.D.	1	16,494	385	66%
8	Bundled 3.0 MW Wind Energy Project	Tamilnadu	Switzerland	M/s Bhagyodaya Agencies	Emergent Ventures India (Pvt) Ltd.	Wind power	AMS-I.D.	1	10,373	750	68%
9	8.75MW Bundle Wind Power Project	Maharashtra	--	M/s Shahi Export Pvt. Ltd	--	Wind power	AMS-I.D.	1	32,977	976	59%
10	4 MW Bundled Grid Connected Wind Power Project	Tamilnadu	Spain	Pushpit Steels Private Ltd. ; V. S. Steel and Power Private Ltd. ; Garg Iron & Energy Private Ltd.	Zero Emissions Technologies S.A.	Wind power	AMS-I.D.	2	19,180	1,102	76%

11	7.2 MW Wind Project at Chitradurga	Karnataka	Japan	Mysore Mercantile Company Ltd.	Mitsubishi UFJ Morgan Stanley Securities Co., Ltd.	Wind power	AMS-I.D.	5	62,478	1,796	88%
12	8.5 MW Wind Energy Project by KS Oils Limited, India	Madhya Pradesh and Gujarat	Switzerland	K.S. Oils Ltd.	Emergent Ventures India Pvt. Ltd.	Wind power	AMS-I.D.	3	35,827	962	81%
13	2.5 MW Bundled Wind Power Project	Maharashtra	--	Raj Infrastructure Development, (India) Pvt. Ltd.	--	Wind power	AMS-I.D.	2	6,319	750	77%
14	Bundled Wind Power Project	Tamil Nadu	Switzerland	Jocil Ltd The Andhra Sugars Ltd	Bunge Emissions Fund Limited	Wind power	AMS-I.D.	2	57,479	765	80%
15	5.5 MW Bundled Wind Power Project by WMI Cranes Ltd.	Maharashtra, Gujarat and Tamil Nadu	Switzerland	WMI Cranes Ltd.	RWE Supply & Trading Switzerland S.A.	Wind power	AMS-I.D.	1	7,991	565	45%
16	7.5 MW bundled small-scale wind projec	Maharashtra	Spain; Sweden	Modelama Exports Ltd.	Kingdom of Spain (withdrawn); Swedish Energy Agency (withdrawn)	Wind power	AMS-I.D.	3	26,900	981	76%
17	9.75 MW Bundled wind power project	Kerala	--	Zenith Energy Services (P) Ltd.	--	Wind power	AMS-I.D.	1	19,160	370	89%
18	Bundled Wind Power Project by M/s. D. J. Malpani	Karnataka and Gujarat	--	M/s. D. J. Malpani	--	Wind power	AMS-I.D.	1	22,285	631	114%
19	Wind Power Generation Project activity by Interocean Shipping India Private Limited	Maharashtra, Tamilnadu and Rajsthan	Switzerland	Interocean Shipping India Private Ltd.	EGL AG	Wind power	AMS-I.D.	1	10,292	370	46%
20	Cleaner Technology in Electricity Production	Tamil Nadu	--	Sheela Clinic	--	Wind power	AMS-I.D.	1	6,403	322	86%
21	Bundled Wind Power Project	Gujarat	--	Vish Wind Infrastructure LLP ; J. N. Investment & Trading Co. Private Ltd.	--	Wind power	AMS-I.D.	1	17,732	396	54%
22	4.75 MW Bundled Wind Power Project by Associated Stone Industries (Kotah) Ltd	Tamil Nadu, Karnataka and Maharashtra	--	Associated Stone Industries (Kotah) Ltd.	--	Wind power	AMS-I.D.	1	4,010	366	34%
23	9.9 MW Bundled Wind Power Project	Maharashtra	--	REI Agro Ltd.	--	Wind power	AMS-I.D.	1	14,017	385	58%
24	Bundled 9.00 MW Wind Power Generation Project by Gangadhar Narsingdas Agrawal Group	Maharashtra and Karnataka	--	Gangadhar Narsingdas Agrawal (HUF) ; Ferromar Shipping Private Ltd. (FSPL)	--	Wind power	AMS-I.D.	1	17,615	355	99%
25	2 X 5 MW Upper khauli & Drinidhar small hydroelectric project for a grid system	Himachal Pradesh	--	Vamshi Industrial Power Ltd.	--	Hydro power	AMS-I.D.	1	31,643	731	43%
26	10 MW Renewable Energy Project for a Grid	Himachal Pradesh	--	AT Hydro (P) Ltd. ; Cimaron Power Ltd.	--	Hydro power	AMS-I.D.	1	54,962	730	70%

27	7.5 MW biomass plants using agricultural waste Limited	Tamilnadu	Germany	Shriram City Union Finance Ltd.• Shriram Transport Finance Company Ltd. (Shriram nvestments Ltd.	KfW ('withdrawn)	Biomass	AMS-I.D.	2	190,303	1,362	63%
28	Renewable biomass based thermal energy generation at Mahalaxmi Group of Companies	Gujarat	--	Mahalaxmi Fabric Mills Private Ltd. (MFM) ; Mahalaxmi Rubtech Ltd. (MRT)		Biomass	AMS-I.C.	2	52,608	1,430	39%
29	Fuel efficiency improvement in glass melting	West Bengal, Union Territory of Pondicherry and Haryana	Switzerland	Hindusthan National Glass & Industries Ltd.	Mercuria Energy Trading SA	Energy efficiency	AMS-II.D.	1	41,087	1,096	90%
		29			17	29		55	1,296,695	28,969	71.3%

Table 2: The List of Bundled CDM Projects in India.

Renewable Energy Source	No. of Projects		No. of CERs issued	
	World	India	World	India
Wind power	25	24	928386	9,26,092
Biomass	2	2	2,42,911	2,42,911
Biogas	33	0	11,56,141	0
Hydro power	29	2	13,93,979	86,605
Fuel Switch	1	0	6,290	0
Energy efficiency	3	1	60,494	41,087
Other Renewable	5	0	39,999	0

Table 3: CERs issued with respect to Renewable energy sources in World (including India).

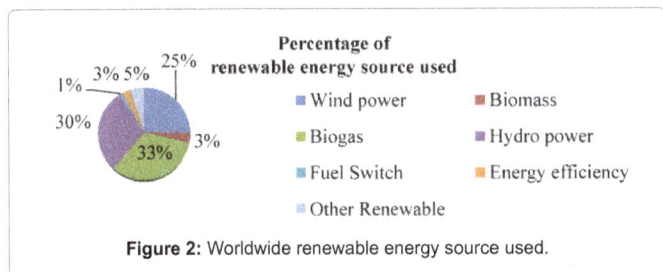

Figure 2: Worldwide renewable energy source used.

Sr. No.	Methodology	No. of Project	Name of the Methodology
1	AMS-I.D	58	Grid connected renewable electricity generation
2	AMS-II.D.	3	Energy efficiency and fuel switching measures for industrial facilities
3	AMS-III.D.	31	Methane recovery in animal manure management systems
4	AMS-I.C.	2	Thermal energy production with or without electricity
5	AMS-III.H.	1	Methane recovery in wastewater treatment
6	AMS-III.B.	1	Switching fossil fuels
7	ACM0002	2	Grid-connected electricity generation from renewable sources

Table 4: Methodologies used in worldwide Bundled CDM Projects.

these methodologies is easy to grasp, but the methodologies themselves can be quite complex. These are necessarily diverse in their composition and application in order to accommodate the wide range of activities and areas covered by the CDM [20]. The CDM Executive Board approves all baseline-monitoring methodologies. It takes around 200 days between the submission of a new methodology and final decision on rejection; about 300 days are required to get the final approval decision [21]. Energy efficiency, renewable energy, landfill gas recovery, biomass and methane recovery [22] are the types of methodologies mostly used in simplified small scale projects which helps reduce the transaction cost of the project [23,24]. Table 4 and Figure 3 provides the list of all the methodologies that are used in all 98 bundled projects. Only 7 different methodologies were used in 98 bundled projects. All are approved small scale methodologies. Methodology "Grid connected renewable electricity generation" (AMS-I.D) contributes to 58 projects followed by "Methane recovery in animal manure management systems" (AMS-III.D) used in 31 projects. Methodologies "Energy efficiency and fuel switching measures for industrial facilities" (AMS-II.D.) contributes to 3 projects followed by "Thermal energy production with or without electricity" (AMS-I.C.) in 2 projects, "Grid-connected electricity generation from renewable sources" (ACM0002) were executed in 2

projects. "Methane recovery in wastewater treatment" (AMS-III.H.) and "Switching fossil fuels" (AMS-III.B.) were implemented in 1 project each. Figure 4 shows methodology „Grid connected renewable electricity generation" (AMS.I.D) was used in maximum number of projects. The methodology involves construction and the operation of a power plant that uses renewable energy sources and supplies electricity to the grid or retrofit, replacement or capacity addition of an existing power plant that uses renewable energy sources and supplies electricity to the grid (CDM Rule Book, 2013). It is an approved methodology for small-scale grid-connected renewable energy projects that have less than 15 MW threshold capacity [25]. This methodology is applicable to all type of projects that use renewable energy sources like wind, hydro, solar and biomass. Energy generated through CDM projects is supplied to the national grid which increases energy availability. This energy can be supplied to areas having shortage of electricity and help in reducing load on national and local grids as well as in community welfare.

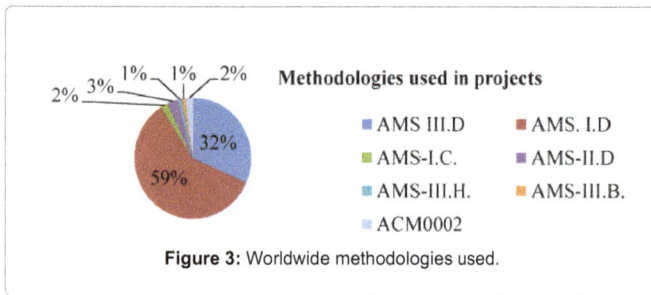

Figure 3: Worldwide methodologies used.

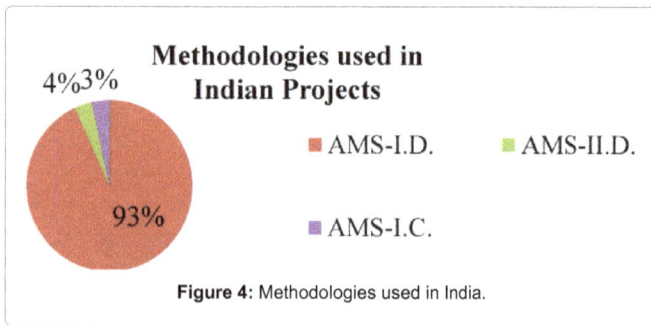

Figure 4: Methodologies used in India.

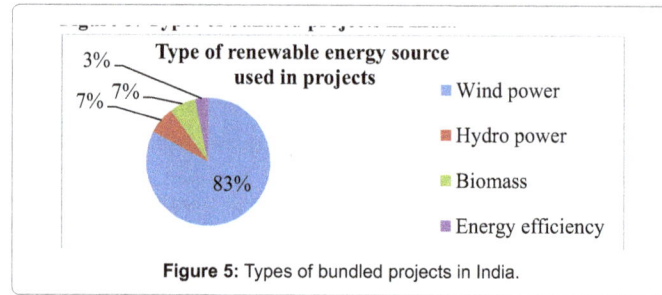

Figure 5: Types of bundled projects in India.

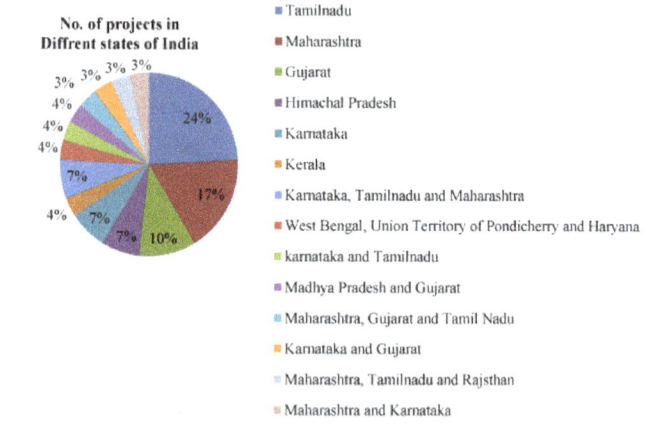

Figure 6: Bundled projects in different states of India.

India

India has a long dependency on traditional energy sources such as firewood, agricultural waste, animal dung etc., which are still continuing to meet the energy necessities especially in rural India. These traditional fuels are getting replaced by fossil fuels such as coal, petroleum and natural gas which are major cause of climatic change and air pollution. Thus, the focus of energy planners has shifted towards renewable resources like wind, solar, biogas for energy generation [26]. Table 2 presents the list of CDM bundled projects in India that use renewable resources and furnish the details of bundled project activities, such as project participants, use of renewable energy sources in the projects, methodologies used in these projects and estimated emission reduction per annum with respective number of issued CERs in the year 2014 for its total verified days and CERs issuance rate. India has 29 registered bundled projects till July 2014 having 1,296,695 metric tons of CERs issued with 71.3% issuance rate. In India bundled projects have been implemented in wind power, biomass based hydro power and energy efficiency sectors with the help of different methodologies. Figure 5 presents types of renewable energy source used in Indian bundled projects. Out of 29 bundled projects, 24 are based on wind power with 881,161 metric tons of issued CERs. 2 hydro power and biomass based project each with 54,962 and 242,911 metric tons of CERs respectively. 1 project is on energy efficiency with 41,057 metric tons of issued CERs. Wind power based projects have used AMS.I.D methodology with 926,092 CER issued with 73.5% CER issuance rate, biomass based projects implemented AMS.I.D and AMS.I.C methodologies with 242,911 CER issued with 50.8% CER issuance rate and energy efficiency used AMS.II.D methodology with 41,087 issued CER with 90% of CER issuance rate and hydropower based projects implemented AMS.I.D methodology with 86,605 CER issued with 56.5% of CERs issuance rate. Figure 6 represents number of bundled projects in different states of India. Karnataka, Tamil Nadu, Maharashtra, Gujarat, Madhya Pradesh, Kerala, Rajasthan, Himachal Pradesh, West Bengal, Haryana, Pondicherry are the states where bundled projects are being implemented. Out of 24 wind power

projects in India, maximum projects are carried out in Tamil Nadu (6) and Maharashtra (5) followed by Karnataka and Gujarat with 2 projects each and Kerala with only 1 project. Tamil Nadu and Gujarat each have 1 biomass based CDM bundled project. Himachal Pradesh has 1 hydropower based project. Within 29 projects in India, 5 projects have achieved below 50% CERs. 10 projects have achieved CERs between 50% - 70%. Other 10 projects have achieved CERs within 70-90%. Only 4 projects have achieved CERs more than 90%. In India maximum projects have used wind energy for power generation. These projects use wind energy for power generation thereby reducing usage of fossil fuel which leads to reduction of GHGs emission into the atmosphere. Wind turbine generators deployed in these projects ensure efficient and safe operation of the project activity. Thus the mechanism establishes technical expertise in countries with little prior experience, helping to incentivize increasing amounts of foreign investment (Hedger and Stokes, Undated). Grid-interactive power generated from renewable sources in the year 2013 was 4125 MW, out of which wind power generation was maximum 2500 MW, 350 MW power is obtained from Small Hydro Power and Bagasse Cogeneration each, 105 MW power is from Biomass & Gasification, Waste to power generation was 20 MW and power generated from solar energy was 800 MW [27].

Case Study

The visited project is the Dhom (Balkawadi) water reservoir which has capacity of 115.53 million cubic meters. It is built across flow of Krishna River in Balkwadi village, Wai Taluka, Satara district, Maharashtra state, India. The project is a 'bundled project activity' with a total installed capacity of 6.5 MW electricity generations. Projects are typical 'dam-toe' irrigation based river hydro projects with the installation of power house at the foot of an existing dam. The projects utilize irrigation discharges from the storage across river and surplus discharges through the spillway to generate electricity. The electricity

Figure 7: Total Catchment Area

Figure 8: Powerhouse

Figure 9: Penstock

is mainly supplied to co-operatives, industries, educational institutions and other sectors. The total catchment area (Figure 7) is around 42.77km^2. The length of the dam is 1.173 km. Dam serves the purpose of storing water for irrigation. Released water is stored downstream. The irrigation is from the submergence of Dhom backwaters. The water stored in the dam is released into the reservoir through tunnel. Project envisages the construction of a small powerhouse of (Figure 8) capacity 4 MW to be operated by the water released from the reservoir made available for irrigation. The annual net generation of the project is estimated at 9.37 MW [28]. A pipe between the surge tank and

prime mover and a butterfly wall, which is used to regulate the flow in a pipe, is called a penstock (Figure 9). Penstock is made of steel through reinforced concrete and usually equipped with the head gates at the inlet which can be closed during the repair of the penstocks. A sufficient water head should be provided above the penstock entrance in the fore bay or surge tank to avoid the formation of vortices. The butterfly valve consists of a disk built onto a rod, which when turned to the closed position completely stops the flow. When the butterfly valve is in the open position, the flow is nearly unrestricted. A turbine device with blades at one end and electromagnets at the other end, generates electricity as the blades moves. It consists of different parts viz. generator, shaft, and guard wheel. The online monitoring of electricity is done on continuously. The meters are capable of measuring the electricity parameters on real time basis. Data is archived in electronic form for two years after the end of the crediting period or of the last issuance of CERs. The generated electricity is connected to grid which supplies electricity to an identified consumer facility. On-site small scale renewable energy generation projects can produce significant clean energy, environmental, and economic benefits for a local government and the community. Hydropower is one of the highest renewable energy resources used after wind. Also, it contributes to grid stability by providing flexibility, as spinning turbines can be ramped up more rapidly than any other generation source [29]. Cited small scale hydro-power project is good example clean power generation. The project leads to conservation of resources as the power facility that would have come up in the absence of the conventional energy sources. It has the potential to shape the social and economic life of the local people by creating jobs on a regular and permanent basis and additional investment consistent with the needs of the local community. It is expected to accelerate the commercialization of grid connected renewable energy technologies such as small hydro in the market and thus, provide an alternative to the high-growth, coal-dominated business as-usual scenario.

Recommendation Suggested as per the Study

India is still depending on traditional energy sources to fulfill their energy needs. Whereas traditional energy sources are non-renewable and limited in supply also causes climatic change and air pollution. Thus, to minimize these effects energy-use pattern have changed towards renewable energy resources like wind, solar, hydro, biogas. Rapid deployment of renewable energy resources are resulting in significant energy security, climate change mitigation, and economic benefits. CDM can interface with renewable energy development projects to build up the capacity for renewable energy in developing countries. But small scale energy generation projects, which are typically found in less developed countries, face proportionally higher transaction costs. Therefore, approach in the concept of bundling of small scale CDM projects is to share CDM financing transaction cost between individual projects. Bundling of projects will facilitate the scaling-up of carbon finance business, while minimizing the transaction costs per unit emission reductions. Many small bundled and large scale renewable energy projects are serving as climate change mitigation projects. Also, these projects keep money circulating within the local economy and would reduce the need to spend money on importing coal and natural gas from other places. Thus, they projects have high potential to contribute local sustainable development, as these projects provide new opportunities for industries and economic activities to setup in the area thereby resulting in greater local employment, ultimately leading to overall development. With respect to renewable energy there are great opportunities for India as in emerging markets renewable energy potential is attracting foreign investment, generating new jobs and creating local supply chains. Government of India is also promoting use of renewable energy through its National Action Plan on Climate Change. The renewable energy goals require continues effort, strong implementation and improve utilization of capacity. Improvement in energy efficiency makes renewable energy source more affordable and attractive to finance source.

Conclusion

In recent years there has been increasing attention to the crucial issue of whether CDM has fulfilled its sustainable development objective. But the actual effects of CDM project activities on the host countries sustainable development at the national level are difficult to evaluate [30-33]. The global GHG emission reduction potential of such activities and empirical research is needed to identify and quantify actions that will yield the most emissions savings. Thus, the paper provides a probing insight into these small scale CDM project activities and their methodologies in order to achieve emission reduction. Small and medium scale industries occupy an important and strategic place in economic growth and equitable development in all the developing countries. Although CDM is devised to foster sustainable development, small scale CDM project activities which are known to be beneficial to the sustainable development of local communities, are often burdened with high costs and low returns. In order to leverage the development of small scale CDM projects, the UNFCCC introduced bundle concept. The concept is for the utilization of the CDM in simplified modalities and procedures for small-scale projects in order to secure carbon revenue for a community-scale. Small-scale project activities may be bundled up to the defined threshold level and eligible SSC project activities can be bundled to overcome transactions costs. By using bundle scheme, small projects can become cost effective and thus become sufficiently attractive with CER revenue.

References

1. Bharadwaj N, Parthan B, Coninck HC, Roos C, Van der Linden NH, et al. (2004) Realising the potential of small-scale CDM projects in India.

2. Bhardwaj B (2013) Future of Carbon Trading: A Business That Works For Global Environment. International Journal of Science, Environment & Technology 2: 115-121.

3. CDM rulebook (2013) What is bundling? Clean Development Mechanism rules, practice & procedures.

4. Central Electricity Authority (2013) CO2 Baseline Database for the Indian Power Sector User Guide Version 6.0, Ministry of Power, Government of India, New Delhi India.

5. Express hospitality (2013) Carbon credits: The Indian scenario.

6. Gonzales A (2001) Financing issues and options for small-scale industrial CDM projects in Asia.

7. Griffith-Jones S, Hedger M, Stokes L (2009) The role of private investment in increasing climate friendly technologies in developing countries. Columbia University and Institute of Development Studies.

8. Haites E, Yamin F (2004) Special Feature on the Kyoto Protocol Overview of the Kyoto Mechanisms, International Review for Environmental Strategies 5: 199-216.

9. Hinostroza M, Cheng CC, Zhu X, Fenhann J, Figueres C, et al. (2007) Potential

10. and barriers for end-use energy efficiency under programmatic CDM. CD4CDM Working Paper 3, United Nation Environment programme, UNEP Riso Centre.

10. Huang Y, He J, Tarp F (2012) Is the Clean Development Mechanism Promoting Sustainable Development? Working paper, 2012/72, United Nation University.

11. Humbad A, Kumar S, Babu B (2009) Carbon credits for energy self-sufficiency in rural India- A case study. Energy science and research 22: 187-197.

12. IGES (2014) CDM project database.

13. International Renewable Energy Agency (2012) SUMMARY FOR POLICY MAKERS: Renewable Power Generation Costs.

14. Kumar HV, Kulkarni SV, Thukral K (2004) Bundled Small-Scale CDM Projects, UNEP Risoe Centre, Denmark.

15. Mariyappan J, Bharadwaj N, Conick H, Linden N (2005) A guide to bundled small-scale CDM projects.

16. Miah D, Shin MY, Koike M (2011) Forest to climate change mitigation. Clean development mechanism in Bangladesh, Springer Heidelberg Dordrecht London New York.

17. Ministry of New and Renewable Energy (MNRE) (2013) Physical Progress (Achievements). Government of India.

18. Najam A, Rahman A, Huq S, Sokona Y (2003) Integrating sustainable development into the Fourth Assessment Report of the Intergovernmental Panel on Climate Change. Climate Policy Volume 3: S9-S17.

19. Novikova A, Urge-Vorsatz D, Liang C (2006) The "Magic" of the Kyoto Mechanisms: Will It Work for Buildings?

20. Pan J (2009) Meeting Human Development Goals with Low Emissions. IDS Bulletin on Climate Change and Development 35: 90-97.

21. Schneider L (2007) Is the CDM Fulfilling its Environmental and Sustainable Development Objectives? An Evaluation of the CDM and Options for Improvement. WWF Report.

22. Shishlov I, Bellassen V (2012) 10 Lessons from 10 Years of the CDM. Climate Report.

23. Shrestha RM, Abeygunawardana AMAK (2007) Small-scale CDM projects in a competitive electricity industry: How good is a simplified baseline methodology?. Energy Policy 35: 3717-3728.

24. Sreekanth KJ, Sudarsan N, Jayaraj S (2012) Achieving Certified Emission Reduction in Rural Domestic Energy Sector Through Alternative Fuel Replacement. International Journal of Renewable Energy Research 2: 38-43.

25. Subbarao S, Gadde B (2006) Analysis and Evaluation of CDM Potential of Biomethanation Sector in India. The 2nd Joint International Conference on "Sustainable Energy and Environment (SEE 2006)"F-024 (O) 21-23 November 2006, Bangkok, Thailand.

26. Sutter C (2001) Small-Scale CDM Projects: Opportunities and Obstacles.

27. The World Bank (2013) Mapping Carbon Pricing Initiatives report.

28. UNEP RISO CENRE (2013) Small Scale CDM & Bundled Criteria, Advantages and Status.

29. UNFCCC (2013) Introduction to Clean Development Mechanism (CDM).

30. UNFCCC (2013) Annexure III, Bundles of small-scale CDM projects vs. large-scale CDM projects.

31. UNFCCC (2013) CDM Methodology booklet. (6th edn) Information updated as of EB 79, United Nations.

32. UNFCCC (2013) Clean development mechanism project design document (CDM-PDD).

33. World Bank (2013) Carbon Funds & Facilities at the World Bank.

14

Characterization of Biofield Energy Treated 3-Chloronitrobenzene: Physical, Thermal, and Spectroscopic Studies

Mahendra Kumar Trivedi[1], Alice Branton[1], Dahryn Trivedi[1], Gopal Nayak[1], Ragini Singh[2] and Snehasis Jana[2*]

[1]Trivedi Global Inc., 10624 S Eastern Avenue Suite A-969, Henderson, NV 89052, USA
[2]Trivedi Science Research Laboratory Pvt. Ltd., Hall-A, Chinar Mega Mall, Chinar Fortune City, Hoshangabad Rd., Bhopal- 462026, Madhya Pradesh, India

Abstract

The chloronitrobenzenes are widely used as the intermediates in the production of pharmaceuticals, pesticides and rubber processing chemicals. However, due to their wide applications, they are frequently released into the environment thereby creating hazards. The objective of the study was to use an alternative strategy i.e. biofield energy treatment and analysed its impact on the physical, thermal and spectral properties of 3-chloronitrobenzene (3-CNB). For the study, the 3-CNB sample was taken and divided into two groups, named as control and treated. The analytical techniques used were X-ray diffraction (XRD), thermogravimetric analysis (TGA), differential scanning calorimetry (DSC), UV-Visible (UV-Vis), and Fourier transform infrared (FT-IR) spectroscopy. The treated group was subjected to the biofield energy treatment and analysed using these techniques against the control sample. The XRD data showed an alteration in relative intensity of the peak along with 30% decrease in the crystallite size of the treated sample as compared to the control. The TGA studies revealed the decrease in onset temperature of degradation from 140°C (control) to 120°C, while maximum thermal degradation temperature was changed from 157.61°C (control) to 150.37°C in the treated sample as compared to the control. Moreover, the DSC studies revealed the decrease in the melting temperature from 51°C (control) →47°C in the treated sample. Besides, the UV-Vis and FT-IR spectra of the treated sample did not show any significant alteration in terms of wavelength and frequencies of the peaks, respectively from the control sample. The overall study results showed the impact of biofield energy treatment on the physical and thermal properties of 3-CNB that can further affect its use as a chemical intermediate and its fate in the environment.

Keywords: Biofield energy treatment; 3-Chloronitrobenzene; X-ray diffraction study; Thermogravimetric analysis; Differential scanning calorimetry; UV-Visible spectroscopy; Fourier transform infrared spectroscopy

Abbreviations: 3-CNB: 3-Chloronitrobenzene; NCCAM: National Centre for Complementary and Alternative Medicine; NIH: National Institute of Health; XRD: X-ray diffraction; TGA: Thermogravimetric analysis; DTG: Derivative thermogravimetry; FT-IR: Fourier transform infrared

Introduction

Chlorobenzene is an aromatic, colourless, and flammable organic compound present in the form of liquid, that is widely used as intermediate for the manufacturing of other chemicals. The chlorination of benzene results in the production of monochlorobenzene that has been used for the synthesis of diphenyl oxide, chloronitrobenzenes (CNBs), and sulfone polymers. Apart from that, it is also used in the manufacturing of phenol, pigment intermediate, and dioctyl phenol [1,2]. CNBs that are an important end product of monochlorobenzenes possess three isomeric forms *i.e.* 2- CNB, 3- CNB, and 4- CNB. These isomers structurally differ from each other in terms of the position of the nitro group in the benzene ring with respect to the chloro group; however they possess similar chemical, pharmacological, and toxicological properties [3,4]. They are used as intermediates in the manufacturing of substitute phenyl carbamates, pharmaceuticals (*e.g.* acetaminophen), pesticides (*e.g.* parathion and carbofuran), and rubber-processing chemicals [5-7]. Moreover, 3-chloronitrobenzene (3-CNB), a yellow crystalline solid, plays a very important role as precursor due to the presence of two reactive sites. It can be chlorinated for producing pentachloronitrobenzene that is used as a fungicide and in the manufacturing of various agrochemicals. 3-chloroaniline (Orange GC Base), a dye intermediate, is produced from 3-CNB *via* hydrogenation process [8]. Due to their wide application in the chemical industry, the

CNBs are directly released into the environment. Their presence has been mainly found in water and fishes [9]. 3-CNB has the ability to enter in the environment through the chlorination of drinking water. Moreover, Rivera et al. found 3- CNB as a main pollutant during their research in Spain [10]. All these circumstances create a need for some alternative strategy which could be helpful for these chemicals to improve the yield efficiency and reducing the environmental hazards. Biofield energy treatment recently came in focus due to its ability to make alterations in various living organisms and non-living objects. It is a type of energy healing therapies that are also recommended by National Institute of Health (NIH)/National Centre for Complementary and Alternative Medicine (NCCAM) [11]. The term 'biofield' is related to the biological energy field central to the life and thought to be produced from the physical processes, emotions and thoughts of the human being [12]. It may interact with the environmental processes and the emissions of other individuals. The frequency of these radiations depends on the physiological, mental, emotional, and spiritual state of the person [13]. The non-living objects also possess the energy aura in the form of electromagnetic radiations due to their atomic and molecular

*Corresponding author: Snehasis Jana, Trivedi Science Research Laboratory Pvt. Ltd., Hall-A, Chinar Mega Mall, Chinar Fortune City, Hoshangabad Rd., Bhopal-462026, Madhya Pradesh, India
E-mail: publication@ trivedieffect.com

vibrations. The non-living objects cannot change this energy parameter by more than 2%, whereas, the human being can change it drastically by the natural exchange process from the environment [14,15]. Thus, the human has the ability to harness the energy from the environment or universe and can transmit it to any living or non-living object(s) around the Globe. The objects always receive the energy and responding to the useful way. This process is known as biofield energy treatment. Mr. Trivedi is well known to possess a unique biofield energy treatment (The Trivedi Effect') that has been reported for causing alterations in various research field *viz.* microbiology [16], agriculture [17], and biotechnology [18]. Besides that, the impact of biofield treatment was also reported on physical, thermal and spectral properties of various metals and organic compounds [19-21]. Hence, the current study was conceptualized to evaluate the impact of Mr. Trivedi's biofield energy treatment on the physical, thermal and spectral properties of 3-CNB using various analytical methods.

Materials and Methods

3-chloronitrobenzene (3-CNB) was procured from Loba Chemie Pvt. Ltd., India. The sample was divided into two parts; the first one was kept as a control while another was subjected to Mr. Trivedi's biofield energy treatment and coded as treated sample. The treated group was handed over to Mr. Trivedi in sealed pack for biofield treatment under standard laboratory condition. Mr. Trivedi provided the treatment to the treated group without touching the sample through his energy transmission process. The biofield treated sample was further characterized using the standard protocols of X-ray diffraction, thermogravimetric analysis, differential scanning calorimetry, UV-Vis, and FT-IR spectroscopic characterization.

Characterization

X-ray diffraction (XRD) study: X-ray powder diffractogram were obtained on Phillips, Holland PW 1710 X-ray diffractometer system. The X-ray generator was equipped with a copper anode with nickel filter operating at 35kV and 20 mA. The wavelength of radiation used by the XRD system was 1.54056 Å. The data were collected in the 2θ range of 10°-99.99°. The step size was 0.02° and the counting time was kept at 0.5 seconds per step. The data obtained from the XRD analysis was in the form of a chart of 2θ vs. intensity. It data showed a detailed table that contains peak intensity counts, d value (Å), peak width (θ'), and relative intensity (%).

The crystallite size (G) was calculated from the Scherrer equation with the method based on the width of the diffraction patterns obtained in the X-ray diffracted in the crystalline region.

$$G = k\lambda/(bCos\theta)$$

Where, k is the equipment constant (0.94), λ is the X-ray wavelength (0.154 nm), b in radians is the full-width at half of the peak and θ the corresponding Bragg angle. However, percent change in crystallite size was calculated using the following equation:

$$\text{Percent change in crystallite size} = [(G_t-G_c)/G_c] \times 100$$

Where, G_c and G_t are crystallite size of control and treated powder sample, respectively.

Thermogravimetric analysis/ Derivative Thermogravimetry (TGA/DTG): It showed the effect of temperature on the stability of the control and treated samples of 3-CNB. The samples were analysed using Mettler Toledo simultaneous thermogravimetric analyser (TGA/DTG) and heated from room temperature to 350ºC with a heating rate

of 5ºC/min under air atmosphere. From TGA/DTG curve, the onset temperature T_{onset} (temperature at which sample start losing weight) and T_{max} (maximum thermal degradation temperature) were recorded.

Differential scanning calorimetry (DSC) study: DSC analysis of control and treated sample was done to analyse the melting behaviour of sample, and it was carried out using Perkin Elmer/Pyris-1. Each sample was accurately weighed and hermetically sealed in aluminium pans and heated at a rate of 10ºC/min under air atmosphere (5 mL/min). The thermogram was collected over the temperature range of 45°C to 250ºC. An empty pan sealed with cover pan was used as a reference sample.

UV-Vis spectroscopic analysis: For UV-Vis spectroscopic analysis, the treated sample was divided into two groups, served as T1 and T2. The analysis was measured using Shimadzu UV-2400 PC series spectrophotometer. The spectrum was recorded with 1 cm quartz cell having a slit width of 2.0 nm over a wavelength range of 200-400 nm. In this method, the wavelength of light absorbed by the sample depends on the structure of the sample. With UV-Vis spectroscopy, it is possible to investigate electron transfer between orbitals or bands of atoms, ions and molecules from the ground state to the first excited state [22].

Fourier transform-infrared (FT-IR) spectroscopic characterization: For FT-IR characterization, the treated sample was divided into two groups named T1 and T2. The samples were crushed into fine powder for analysis and followed by mixing in the spectroscopic grade KBr in an agate mortar. Then the sample was pressing into pellets with a hydraulic press. FT-IR spectra were recorded on Shimadzu's Fourier transform infrared spectrometer (Japan). FT-IR spectra are generated by the absorption of electromagnetic radiation in the frequency range 4000-400 cm^{-1}. With the help of FT-IR analysis, the impact of biofield treatment on bond strength, rigidity and stability of 3-CNB compound can be analysed [23].

Results and Discussion

XRD study

The XRD diffractograms of control and treated 3-CNB are presented in Figure 1. The XRD diffractogram of control 3-NCB showed the occurrence of intense crystalline peaks at 2θ equal to 15.62º, 15.81º, 17.35º, 17.47º and 25.29º. However, the treated sample showed XRD peak at 2θ equal to 25.20º. The sharp peaks on the diffractograms of the control and treated samples confirm the crystalline nature of 3-CNB [24]. Moreover, a single sharp peak observed in the treated sample as compared to multiple peaks in the control. It is possible that the crystalline planes may reoriented in the same direction after biofield treatment and that might be the probable cause for the emergence of single diffraction peak in the treated sample. In addition, the relative intensity of the diffracted peak in the treated sample was higher as comparison to the control. It is reported that the alteration in relative intensities of the peaks may occur due to change in the crystal morphology [25]. Thus, it is assumed that the energy was transferred through the biofield treatment, and it probably altered the morphology of the 3-CNB molecules. Besides, the average crystallite size was found as 121.67 nm in the control and 84.8 nm in the treated sample of 3-CNB. The percentage reduction in crystallite size was 30.30%. It is assumed that there is presence of severe internal strain due to biofield energy treatment that might be a reason for fracturing the grains into sub grains that lead to decrease in crystallite size of the treated sample [19]. It was reported that decrease in crystallite size might fasten the rate kinetics in the chemical reactions [26]. 3-CNB is used as a precursor and intermediate in various chemical reactions. Hence, the decrease in crystallite size might enhance the percentage yield of end products by fastening the rate of chemical reaction. Besides, the smaller crystallite size exposed a higher crystallite edge surface that might help in increased rate of degradation [27]. Hence, the decreased crystallite size in biofield energy treated 3-CNB sample may fasten the rate of degradation of 3-CNB molecules from environment either through the process of volatilization or vaporization [28,29].

TGA studies

The TGA thermogram of the control and treated 3-CNB are shown in Figure 2, and data is reported in Table 1. The control 3-CNB showed the occurrence of one step thermal degradation pattern. The thermal degradation commenced at 140°C and completed at 182°C. During this process, the sample showed major weight loss (52.13%) that might be due to vaporization of the 3-CNB molecules. The temperature at which maximum vaporization occurred in the control sample (Tmax) was observed at 157.61°C; as shown by the DTG thermogram. The TGA thermogram of the treated 3-CNB showed single step thermal degradation between 120°C-179°C. During this thermal event, the treated sample showed rapid vaporization and weight loss (78.23%). The DTG thermogram of treated 3-CNB showed a decrease in Tmax and it was observed at 150.37°C. Moreover, the thermal stability was reported to be directly related to the crystallite size [30]. Hence, it was assumed that the decrease in crystallite size due to biofield energy treatment might be responsible for the decrease in thermal stability of the treated sample as compared to the control. Moreover, it was previously reported

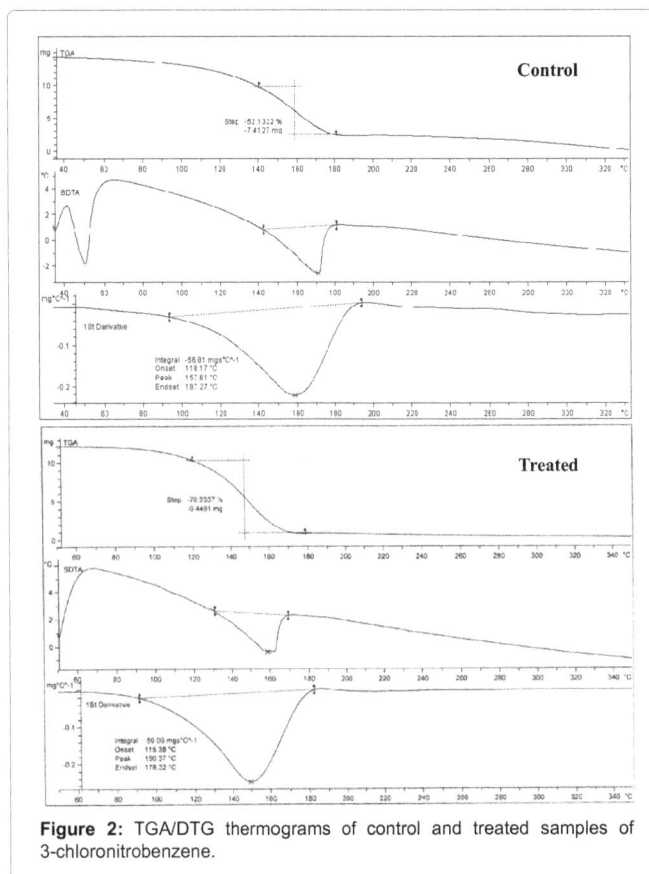

Figure 2: TGA/DTG thermograms of control and treated samples of 3-chloronitrobenzene.

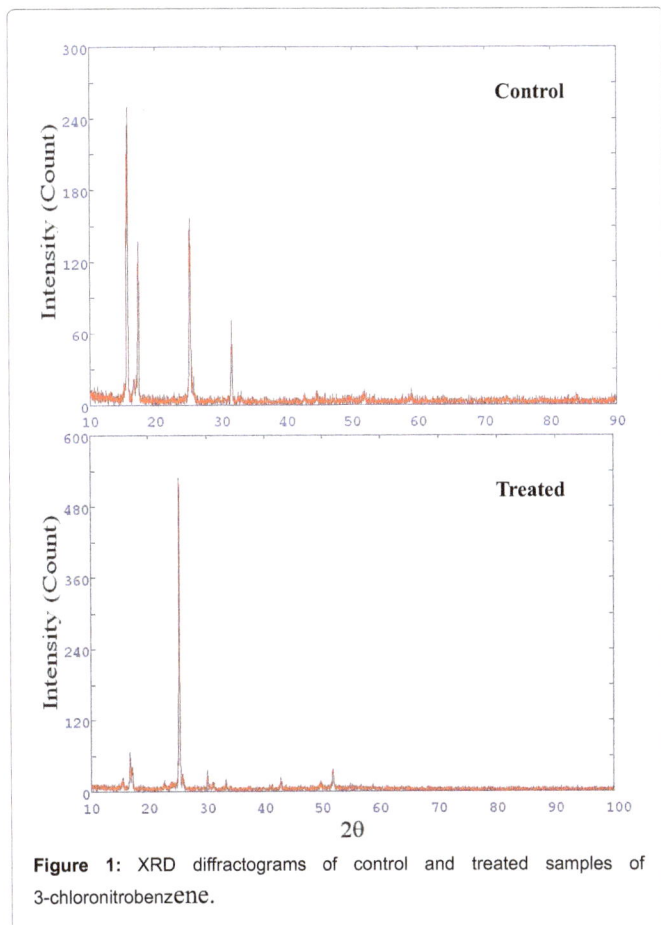

Figure 1: XRD diffractograms of control and treated samples of 3-chloronitrobenzene.

Parameter	Control	Treated	Percent change
Onset temperature (°C)	140	120	-14.29
Endset temperature (°C)	182	179	-1.65
T_{max} (°C)	157.61	150.37	-4.59
Percent weight loss (%)	52.13	78.23	46.23
Melting point (°C)	51	47	-7.84

Table 1: Thermal analysis of control and treated samples of 3-chloronitrobenzene. T_{max}: temperature at which maximum vaporization occur

that the rate of reaction was affected by the state of reactant, and gases reacts faster than solids and liquids. On the other hand, the decreased vaporization temperature indicates that the molecules of 3-CNB may change their phase at low temperature [31]. Also, the percent weight loss was more in the treated sample (78.23%) than the control sample (52.13%), which also supports the fast vaporization of the treated sample. Hence, it was assumed that the decrease in vaporization temperature and increased vaporization process in biofield treated sample might fasten the reaction kinetics. Besides, the environmental fate of 3-CNB from the aquatic surface and the moist soil surface is expected through the volatilization and vaporization process [28,29]. Through this process, the 3-CNB molecules reached in the atmosphere in vapour phase and degrade there by reacting with photochemically produced hydroxyl radicals [32]. Hence, due to the decrease in vaporization temperature, the biofield treated 3-CNB molecule may get easily vaporised from the water and soil surface. This process may help the fast degradation of 3-CNB from the environment by decreasing the volatilization and vaporization half-life.

DSC analysis

The result of DSC analysis was reported in Table 1. The control sample exhibited a sharp endothermic peak at 51°C, whereas the treated sample showed a sharp peak at 47°C. The peaks were due to the melting of control and treated samples, respectively of 3-CNB. The result showed about 8% decrease in melting temperature of the treated sample as compared to the control. It was reported that the melting points of the samples increased due to an increase in crystallite size and *vice versa* [33]. It might be a possible reason for the decrease in melting temperature of the treated sample as it was evident from XRD studies that crystallite size was reduced in treated sample as compared to the control. The decrease in melting temperature might be advantageous for 3-CNB to be used as a chemical intermediate as it helps in accelerating the reaction rate.

UV-Vis spectroscopic analysis

The UV spectra of control and treated samples of 3-CNB are shown in Figure 3. The UV spectrum of control sample showed characteristic absorption peaks at 209 and 257 nm. The spectrum was well supported by the literature [34,35]. The treated sample also showed absorption of light at the similar wavelength. The peaks were appeared at 208 and 256 nm in T1 while in T2 sample, at 209 and 257 nm. It suggested that biofield treatment may not cause any change in structure or position of the functional group as well as the energy that is responsible for electronic transitions between highest occupied molecular orbital and

lowest unoccupied molecular orbital.

FT-IR analysis

The FT-IR spectra of the control and treated samples are shown in Figure 4. The spectra showed characteristic vibrational frequencies as follows:

Nitrogen- oxygen vibrations: In the present study, the NO_2 asymmetric stretching vibration was observed at 1523 cm⁻¹ in all three samples, *i.e.* control, T1, and T2. Similarly, the NO_2 symmetric stretching vibration was observed at 1348 cm⁻¹ in the control and 1346 cm⁻¹ in the treated (T1 and T2) samples. The peak responsible for NO_2 deformation was observed at 538 cm⁻¹ in control and T1 sample and 540 cm⁻¹ in T2 sample. Moreover, the NO_2 rocking vibration peak was appeared at 499 cm⁻¹ in all three samples, *i.e.* control and T1, and T2.

Carbon- hydrogen vibrations: The peak of aromatic C-H stretching was observed at 3101 cm⁻¹ in the control sample; similarly, in T1 sample the peak was observed at 3101 cm⁻¹ and in T2, 3101 cm⁻¹. Moreover, the peaks due to C-H out of plane bending were appeared at 748 and 732 cm⁻¹ in the control sample. These peaks were observed at 743 and 732 cm⁻¹ in T1 sample and 748 and 731 cm⁻¹ in T2 sample.

Ring vibration: Several bands from overtone and combination were appeared in the range of 1992-1732 cm⁻¹ due to *meta* di- substituted benzene in the control sample. The similar bands were observed in T1 and T2 samples in the range of 1992-1782 cm⁻¹ and 1990-1782 cm⁻¹, respectively. Moreover, the peak due to C=C aromatic stretching was

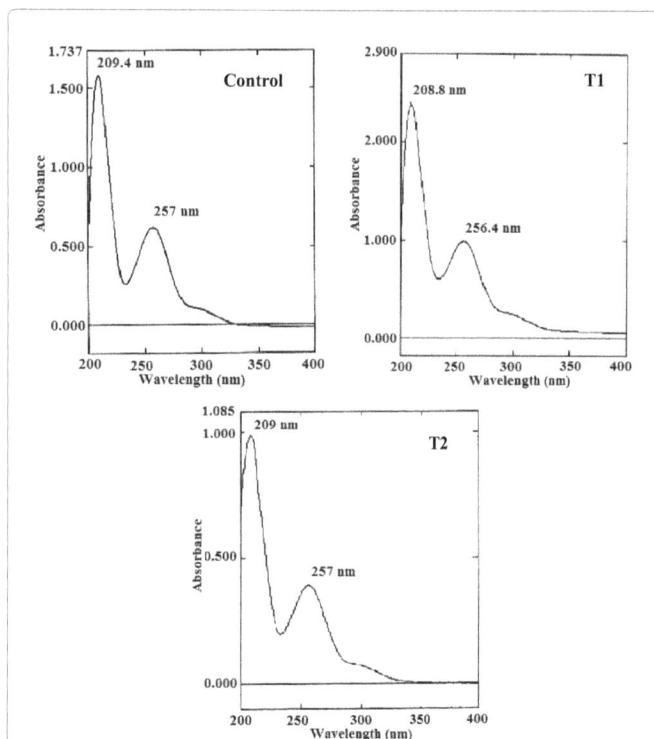

Figure 3: UV-Vis spectra of control and treated samples of 3-chloronitrobenzene.

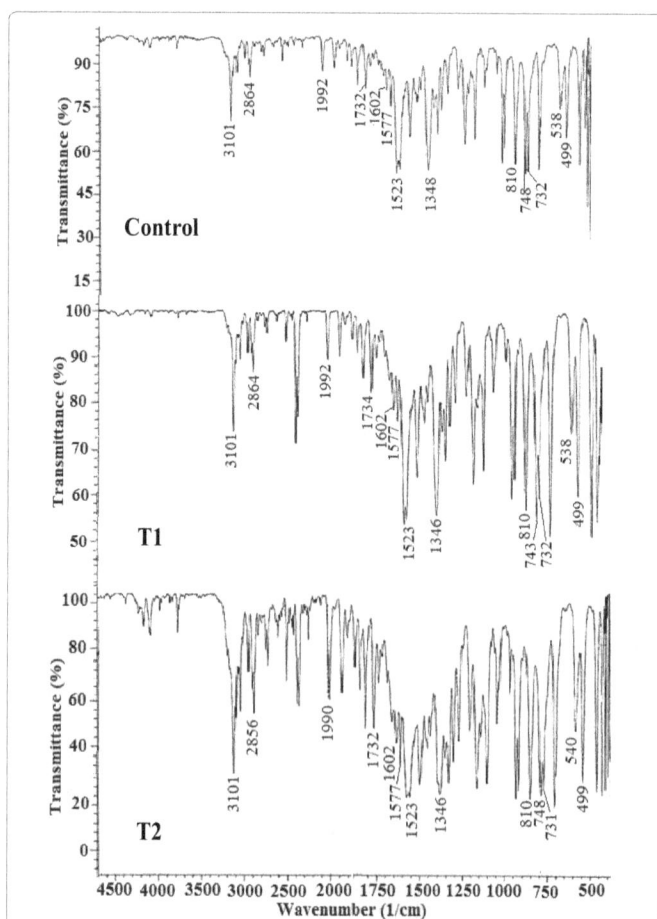

Figure 4: FT-IR spectra of control and treated samples of 3-chloronitrobenzene.

observed at 1602 cm^{-1} in all three samples, *i.e.* control and T1, and T2. Similarly, the peak due to ring stretching was observed at 1577 cm^{-1} in all three samples, *i.e.* control and T1, and T2.

C-Cl vibration: A prominent peak due to C-Cl stretching was observed at 810 cm^{-1} in all three samples, *i.e.* control, T1, and T2. The overall FT-IR analysis was supported by the literature data [36,37] and showed that there was no significant difference between observed frequencies of control and treated samples. Hence, it suggested that biofield energy treatment might not induce any significant change at bonding level.

Conclusion

From the overall study, it was observed that the crystallite size of the treated sample was reduced by 30% that suggests the probable increase in internal strain may be due to the impact of biofield energy treatment. The decreased crystallite size might help in fastening the reaction kinetics when used as intermediate as well as the enhanced rate of degradation of 3-CNB molecules. The XRD results were also supported by thermal analysis data. The TGA analysis revealed an increase in vaporization temperature and decrease in thermal stability of treated sample as compared to the control. It may occur due to the decrease in crystallite size of the treated sample, and it may help in the fast degradation of 3-CNB from the environment. The DSC analysis showed a decrease in melting temperature of the treated sample as compared to the control that might further relate with the decreased crystallite size of the treated sample. It also might advantageous for 3-CNB to be used as chemical intermediate as it helps in fastening the reaction rate. The study concluded the impact of Mr. Trivedi's biofield energy treatment on the physical and thermal properties of the 3-CNB sample that probably help in increasing the reaction kinetics of sample along with possible enhancement in its rate of degradation from the environment.

Acknowledgement

The authors would like to acknowledge the whole team from the Sophisticated Analytical Instrument Facility (SAIF), Nagpur and MGV Pharmacy College, Nashik for providing the instrumental facility. Authors are very grateful for the support of Trivedi Science, Trivedi Master Wellness and Trivedi Testimonials in this research work.

References

1. Popp P, Bruggemann L, Keil P, Thuss U, Weiss H (2000) Chlorobenzenes and hexachlorocyclohexanes (HCHs) in the atmosphere of Bitterfeld and Leipzig (Germany). Chemosphere 41: 849-855.

2. Kellersohn T (2003) Ullmann's encyclopedia of industrial chemistry. (6th edn), Wiley-VCH, Verlag, Weinheim, Germany.

3. Cralley LJ, Cralley LV, Bus JS (1982) Patty's industrial hygiene and toxicology. (3rd edn), John Wiley and Sons, New York.

4. Davydova SG (1967) A comparison of the properties of nitrochlorobenzene isomers for the determination of their permissible concentrations in water bodies. Hyg Sanit 32: 161-166.

5. Surrey AR, Hammer HF (1946) Some 7-substituted 4-aminoquinoline derivatives. J Am Chem Soc 68: 113-116.

6. Pilaniappan S (2000) Chemical copolymerization of aniline with *o*-chloroaniline: Thermal stability by spectral studies. Polym Int 49: 659-662.

7. Ding Y, Padias AB, Hall Jr. HK (1999) Chemical trapping experiments support a cation-radical mechanism for the oxidative polymerization of aniline. J Polym Sci A Polym Chem 37: 2569-2579.

8. Booth G (2000) Nitro compounds, aromatic. Ullmann's Encyclopedia of Industrial Chemistry. Wiley-VCH, Weinheim.

9. Dunnivant FM, Anders E (2006) A basic introduction to pollutant fate and transport: An integrated approach with chemistry, modeling, risk assessment and environmental legislation, John Wiley & Sons.

10. Rivera J, Ventura F, Caixach J, De Torres M, Figueras A, et al. (1987) GC/MS, HPLC and FAB mass spectrometric analysis of organic micropollutants in Barcelona's water supply. Int J Environ Anal Chern 29: 15-35.

11. NIH (2008) National Center for Complementary and Alternative Medicine. CAM Basics.

12. Tiller WA (1993) What are subtle energies? JSE 7: 293-304.

13. http://www.biofieldglobal.org/what-is-human-aura.html

14. Prakash S, Chowdhury AR, Gupta A (2015) Monitoring the human health by measuring the biofield "aura": An overview. Int J Appl Eng Res 10: 27654-27658.

15. Rubik B (2002) The biofield hypothesis: Its biophysical basis and role in medicine. J Altern Complement Med 8: 703-717.

16. Trivedi MK, Patil S, Shettigar H, Gangwar M, Jana S (2015) An effect of biofield treatment on Multidrug-resistant *Burkholderia cepacia*: A multihost pathogen. J Trop Dis 3: 167.

17. Lenssen AW (2013) Biofield and fungicide seed treatment influences on soybean productivity, seed quality and weed community. Agricultural Journal 8: 138-143.

18. Nayak G, Altekar N (2015) Effect of biofield treatment on plant growth and adaptation. J Environ Health Sci 1:1-9.

19. Trivedi MK, Nayak G, Patil S, Tallapragada RM, Latiyal O, et al. (2015) An evaluation of biofield treatment on thermal, physical and structural properties of cadmium powder. J Thermodyn Catal 6: 147.

20. Trivedi MK, Patil S, Shettigar H, Singh R, Jana S (2015) An impact of biofield treatment on spectroscopic characterization of pharmaceutical compounds. Mod Chem appl 3: 159.

21. Trivedi MK, Patil S, Mishra RK, Jana S (2015) Structural and physical properties of biofield treated thymol and menthol. J Mol Pharm Org Process Res 3: 127.

22. Hunger M, Weitkamp J (2001) *In situ* IR, NMR, EPR, and UV/Vis spectroscopy: Tools for new insight into the mechanisms of heterogeneous catalysis. Angew Chem Int Ed Engl 40: 2954-2971.

23. Coates J (2000) Interpretation of infrared spectra, a practical approach. Encyclopedia of analytical chemistry. John Wiley and Sons Ltd., Chichester.

24. Rudrangi SR, Bhomia R, Trivedi V, Vine GJ, Mitchell JC, et al. (2015) Influence of the preparation method on the physicochemical properties of indomethacin and methyl-β-cyclodextrin complexes. Int J Pharm 479: 381-390.

25. Inoue M, Hirasawa I (2013) The relationship between crystal morphology and XRD peak intensity on CaSO$_4$•2H$_2$O. J Cryst Growth 380: 169-175.

26. Chaudhary AL, Sheppard DA, Paskevicius M, Pistidda C, Dornheim M, et al. (2015) Reaction kinetic behaviour with relation to crystallite/grain size dependency in the Mg–Si–H system. Acta Mater 95: 244-253.

27. Scott G (2013) Degradable polymers: Principles and applications. (2nd edn), Springer Science & Business Media, Netherlands.

28. Lyman WJ, Reehl W, Rosenblatt DH (1990) Handbook of chemical property estimation methods: Environmental behaviour of organic compounds. American Chemical Society, Washington, DC.

29. Altschuh J, Bruggemann R, Santl H, Eichinger G, Piringer OG (1999) Henry's law constant for a diverse set of organic chemicals: Experimental determination and comparison of estimation methods. Chemosphere 39: 1871-1887.

30. Praserthdam P, Phungphadung J, Tanakulrungsank W (2003) Effect of crystallite size and calcination temperature on the thermal stability of single nanocrystalline chromium oxide: Expressed by novel correlation. Mater Res Innov 7: 118-123.

31. Espenson JH (1995) Chemical kinetics and reaction mechanisms. (2nd edn), Mcgraw-Hill, U.S.

32. Meylan WM, Howard PH (1993) Computer estimation of the atmospheric gas-phase reaction rate of organic compounds with hydroxyl radicals and ozone. Chemosphere 26: 2293-2299.

33. Farrow G (1963) Crystallinity, 'crystallite size' and melting point of polypropylene,

Polymer 4: 191-197.

34. Ungnade HE (1954) Near ultraviolet absorption spectra of halogenated nitrobenzenes. J Am Chem Soc 76: 1601-1603.

35. Weast RC (1979) Handbook of chemistry and physics. (60th edn), CRC Press Inc., Boca Raton, Florida.

36. Linstrom PJ, Mallard WG Evaluated infrared reference spectra. NIST chemistry webbook, National Institute of Standards and Technology, Gaithersburg MD.

37. Lambert JB (1987) Introduction to organic spectroscopy. Macmillan, New York, USA.

15

Cement-Based Solidification of Incinerated Sewage Sludge Ash by the Addition of a Novel Solidifying Aid

Xiaopeng Wang[1,3], Wenxiang Yuan[2] and Haiping Yuan[1*]

[1]School of Environmental Science and Engineering, Shanghai Jiao Tong University, 800 Dongchuan Road, Shanghai, 200240, China
[2]Shanghai Institute for Design & Research on Environmental Engineering Co., Ltd, 11 Shilong Road, Shanghai, 200232, China
[3]Shanghai Qingcaosha Investment Construction and Development Co., ltd. 700 Jinyu Road, Shanghai, China

Abstract

The effects of a novel solidifying aid on the solidification of incinerated sludge ash were investigated in this study. The compressive strength and the heavy metal leaching toxicity of the solidification block were measured, and the composition and the microstructure were also detected by XRD and SEM. The results showed that the optimal solidifying agent was as follows: incinerated sewage sludge ash (ISSA): Portland cement: Kaolin: solidifying aid= 100:40:10: 0.7. The compressive strength of 12.74 MPa was observed when ISSA was mixed with the best solidification condition after 28 days curing. TCLP test results showed that the concentrations of all the metals in the leachate of the solidified samples are below those set in the maximum solubility limit issued by the Environmental Protection Agency in China. The XRD and SEM analysis indicated that the structure of the solidification block was of many acicular crystals and very dense. Furthermore, quartz, $CaAl_2Si_2O_8$, $Ca_2Al_2SiO_7$ and other materials could be found in the solidification blocks, which were known as improving the compressive strength of the solidification blocks.

Keywords: Incinerated sewage sludge ash (ISSA); Solidification; Compressive strength; Solidifying aid

Introduction

The limited space and the high cost of land disposal led to the development of recycling technologies and the reuse of sewage sludge ash in structural and construction materials [1]. Incineration, as a method of solid waste management, could reduce the sludge volume by 90-95%, has become an important method for disposal sewage sludge following the EU wide ban on sea disposal in 1998 [2]. It could generate solid residues, such as bottom and fly ash as well as off-gas cleaning residues with high levels of heavy metals, inorganic salts and other organic compounds. It is estimated that 1.2 million tonnes of incinerated sewage sludge ash (ISSA) are currently produced annually in the EU and North America [3], and a further 0.5 million tonnes/yr in Japan alone [4]. For this reason it requires special management. In recent years, ISSA has been widely used in the production of construction material [5], concrete and cement [6,7], tile [8,9], pavement materials [10], and lightweight aggregate [11,12]. The alternative utilization of ISSA used as construction material according to its cementitious properties by the method of solidification was reported little. It basically involves waste containment within a solid matrix using different binder materials such as cement, pozzolans, clay and Kaolin [13]. It is a relatively new treatment process that has the potential to reduce leachability of hazardous constituents from the disposed waste and the research work on solidification of ISSA has not been extensive. In this study, an unburned solidification technology was applied to the treatment of sewage sludge ash with the use of different types of solidifying agent and aid. The objective of this study was to optimize the solidification agent based on the Portland cement and a new solidifying aid, and analyze solidification mechanism by using X-ray diffraction (XRD) and scanning electron microscope (SEM) techniques. Furthermore, leaching tests was examined to investigate the risk of hazardous heavy metals leaching following the use of solidified sewage sludge ash as construction materials.

Materials and Methods

Materials

The incinerated sewage sludge ash (ISSA) used in this research work was collected from TaoPu Wastewater Treatment Plant located in PuTuo District, Shanghai, China. Commercial Portland cement (PC) typed CEM I 42.5 was used in the experiments as the main binder. The Kaolin used in this research was typed SD-C6 with 300 meshes. Both of their chemical and physical properties were shown in Tables 1 and 2. The solidifying aid (SS-SA) used in this study was a mixture of H_2SO_4, H_3PO_4, and Na_2SiO_3, which mass ratio was $H_2SO_4:H_3PO_4:Na_2SiO_3=1:8:91$.

Sample preparation

Twenty-three treatments were shown in Table 3, the dosages of the components were determined according to the previous experiments. All the samples were prepared by mixing the incinerated sewage sludge ash and solidification agents together with distilled water. The water cement ratio was fixed at 0.35. The mortar mixtures were mixed following the standard test method ASTM C 305-94 to achieve a uniform distribution of the mixtures before transferred to the 40 mm×40 mm×40 mm mould [14]. The mortar mixtures were cast in moulds for the first 24 h at room temperature and then the blocks were demolded and set in a temperature- controlled curing box at 25±1°C for 7, 14 and 28 days to determine the optimum solidification agent formula.

*Corresponding author: Haiping Yuan, School of Environmental Science and Engineering, Shanghai Jiao Tong University, 800 Dongchuan Road, Shanghai, 200240, China
E-mail: hpy_002@163.com

'%wt	SiO$_2$	Fe$_2$O$_3$	Al$_2$O$_3$	TiO$_2$	CaO	MgO	SO$_3$	P$_2$O$_5$	K$_2$O	Na$_2$O	Total
ISSA	40.87	9.63	16.89	0.69	11.66	2.25	5.03	8.19	1.69	1.01	98.85
Cement	20.30	3.56	7.65	0.32	61.25	2.13	1.98	-	1.02	0.36	100.97
Kaolin	54.96	0.47	42.35	0.09	0.52	0.42	0.19	-	0.53	0.32	99.86

Table 1: Chemical compositions of materials used in experiment.

	Fineness (m^2/kg)	Density (kg/m^3)
ISSA	1980	1.56
Cement	347	3.15
Kaolin	524	2.43

Table 2: Physical properties of materials used in experiment.

			Solidified agent		
Group	No.	ISSA	Cement	kaolin	Solidifying Aid
G1 ISSA/PC	1	100	0	10	0
	2	100	10	10	0
	3	100	20	10	0
	4	100	30	10	0
	5	100	40	10	0
	6	100	50	10	0
	7	100	60	10	0
G2 ISSA/PC/ Kaoline	8	100	40	0	0
	9	100	40	10	0
	10	100	40	20	0
	11	100	40	30	0
	12	100	40	40	0
	13	100	40	50	0
	14	100	40	60	0
	15	100	40	70	0
G3 ISSA/PC/ Kaoline /SS-SA	16	100	40	10	0
	17	100	40	10	0.1
	18	100	40	10	0.3
	19	100	40	10	0.5
	20	100	40	10	0.7
	21	100	40	10	1.0
	22	100	40	10	1.5
	23	100	40	10	5.0

Table 3: Mixture proportions (wt%).

Compressive strength test

The physical strength of the solidified matrix is significant since it determines the suitability of the solids to be used as construction material and for secure landfill stacking. Samples curing for 7 days, 14 days and 28 days were tested by determination of compressive strength according to the methods described by the European Standard EN 196.1 [15]. For each mortar and curing age, three specimens were tested.

Toxicity characteristic leaching procedure (TCLP)

The leaching test for heavy metals from the solidified blocks was assessed using the toxicity characteristic leaching procedure (TCLP) as defined by the U.S. EPA on the samples cured under 25 ± 1°C for 28 days [16]. The TCLP test was carried out by crushing the sample into powder using a mortar and pestle to reduce the particle size to smaller than 9.0 μm with 0.1M acetic acid and 0.0643 M NaOH solutions (pH 2.8), at a liquid/solid ratio of 20:1. The extraction vessels were rotated at 120 rpm for 20 h (TDL-5-A, Anke Ltd,Shanghai). The leachate was filtered through a 0.45 μm membrane filter to remove suspended solids and then divided into two portions, one of which was used

for pH measurement and the other for the determination of metals present in the leachate by ICP-AES (Iris Advantage, Thermo Electron Corporation,US) according to the standard methods. Each extraction was done in triplicate, and the average value was reported to ensure the reproducibility of the data.

XRD and SEM analysis

In order to investigate the influences of different mix ratios on specimen properties, chemical compositions were analyzed using X-ray diffraction (D/max-2200/PC, Rigaku Corporation,Japan) and sample microstructures were studied using LV UHR FE-Scanning Electron Microscopy (Na- no SEM 230, NOVA FEI, US) to microscopically analyze the physicochemical variations of the ISSA before and after the solidification in terms of the strength-enhancing effects of the solidifying reaction. The XRD and SEM analyses were conducted for each of the solidified samples, which were prepared by mixing fixed proportions of ISSA/solidifying agent/solidifying aid and then curing them under 20± 1°C for 28 days. An X-ray diffraction analyzer was used to investigate the constituents of solidified sludge ash samples at the 30 kV of acceleration voltage condition. SEM imaging was performed to obtain the micrographs of the sludge samples before and after solidification at 20 kV of acceleration voltage condition.

Results and Discussion

Effect of solidifying agent on the compressive strength

The compressive strength of the solidification blocks treated by ISSA/PC and ISSA/PC/KL systems were shown in Figures 1 and 2. In ISSA/PC/KL system, incinerated sludge ash was mixed with fixed amount of PC (40 g/ 100g ISSA) and blended with different amounts of KL. The compressive strength of all solidified wastes incorporating the incinerated sludge ash was very low in the early age, and thereafter, the compressive strength increased to higher values gradually, as shown in Figure 1. This is because the calcium hydroxide crystal which was produced during the PC hydration process was consumed during the pozzolanic reaction of ISSA [17]. It also could be observed in Figure 1 that the more content of Portland cement in the solidified blocks, the higher compressive strength the solidified blocks show. The compressive strength of the solidification blocks increase little when the PC content of

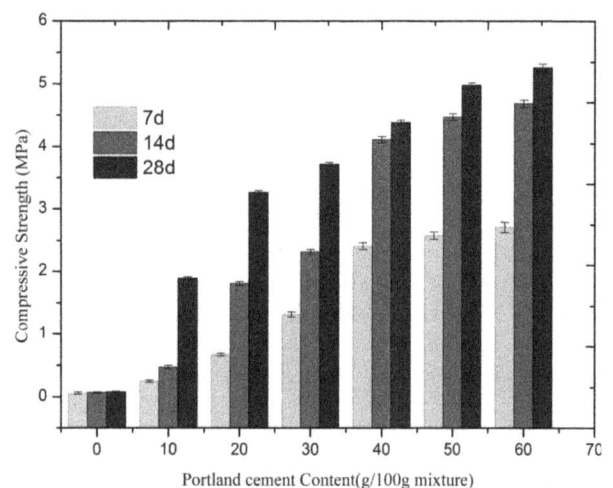

Figure 1: Effect of PC content on the compressive strength of solidification blocks under various curing time.

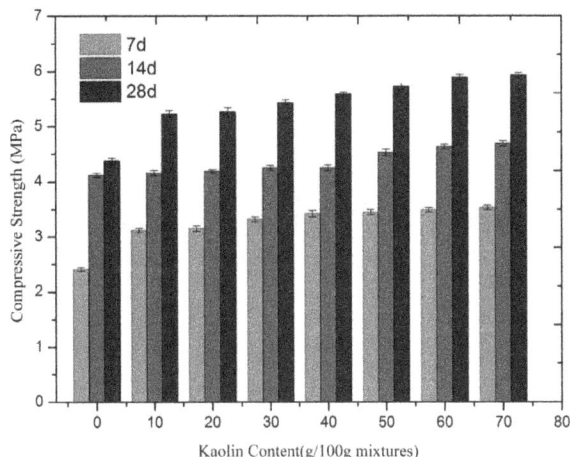

Figure 2: Effect of KLcontent on the compressive strength of solidification blocks under various curing time.

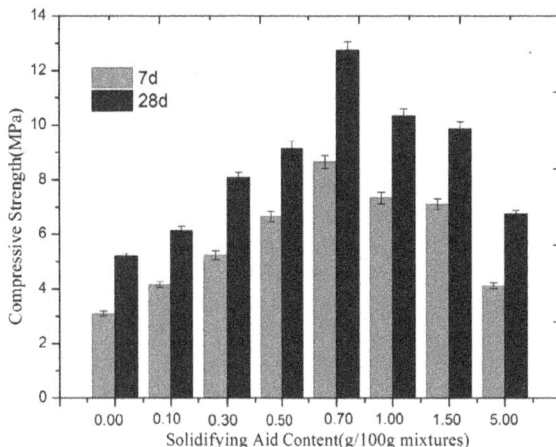

Figure 3: Effect of solidifying aid content on the compressive strength of solidification blocks.

the solidified agent exceeded 40g/100g mixture Figure 1. So, according to the cost of the solidification, 40g/100g mixture was chosen as the optimal content of PC. All the PC-solidified-samples met the minimum requirement for highway or the national highway subbase, but none met the minimum requirement for fired common bricks. It could be seen from Figure 2, when 10g/100g mixture of Kaolin was added to the system, the 28-day-compressive strength could develop from 4.38 MPa to 5.20 MPa. Thereafter, the compressive strength increased slowly with the increasing content of Kaolin. As a result, the 10g/100g mixture of Kaolin was chosen as the optimal content.

Effect of solidifying aid on the compressive strength

As shown in Figure 3, with different proportions of solidifying aid added to the solid mixtures including ISSA and solidifying agent with the fixed ratio mentioned in Table 3, the early-phase compressive strength of the samples was significantly improved compared to that of the samples without solidifying aid addition. It was probably because that the Na_2SiO_3-based solidifying aid accelerated the formation of ettringite, which has positive effect on the compressive strength of solidification block. The highest compressive strength of 12.74 MPa (shearing strength: 3.54 Mpa) was observed from solid mixtures blended with 0.70g/100g mixtures solidifying aid after 28 days curing, which meets the minimum requirement for fired common bricks. Subsequently, the compressive strength of the solidification blocks dropped when the content of the solidifying aid exceeded 0.70g/100g mixture. As a result, the 0.70 g/100g mixture was chosen as the optimal content of solidifying aid when the PC content was fixed at 40 g/100g mixture.

TCLP results

Leaching tests were conducted to examine the potential toxicity of heavy metal leaching from the solidified sludge ash, solidified block A (ISSA blended with PC and KL at the ratio of 100: 40: 10) and solidified block B (ISSA blended with PC, KL and SS-SA at the ratio of 100: 40: 10: 0.7). All the samples were cured under 20± 1°C for 28 days. The results of the metal concentrations in TCLP leachates were shown in Table 4. TCLP results showed that the concentrations of all the heavy metals in the leachate from solidified samples were all lower than the maximum

Component (mg/L)	ISSA	Block A	Block B	EPA hazardous Criteria
As	0.1217 ± 0.0002	0.0134 ± 0.0001	0.0073 ± 0.0001	<5.0
Ba	0.3352 ± 0.0003	0.2780 ± 0.0003	0.1592 ± 0.0002	<100
Cd	0.2930 ± 0.0003	0.2739 ± 0.0002	0.2672 ± 0.0002	<1.0
Cr	0.1367 ± 0.0002	0.0968 ± 0.0001	0.0887 ± 0.0001	<5.0
Ni	0.2170 ± 0.0002	0.0388 ± 0.0001	0.0331 ± 0.0001	<10.0
Pb	0.0122 ± 0.0001	ND	ND	<5.0

ND, not detected

Table 4: Toxic characteristics leaching procedure (TCLP) test results.

solubility limit set by the Environmental Protection Agency in the Republic of China, and thus the risk of hazardous heavy metals leaching following the use of solidified sewage sludge ash as construction materials may be safely concluded to be quite low. In addition, metal concentrations in the TCLP leachates extracted from the cement-based solidified ISSA were much lower than that extracted from the untreated ISSA. It could be concluded that the heavy metal could be immobilization by cement-based solidification, and the addition of solidifying aid could enhance the immobilization of heavy metals.

X-ray diffraction analysis

The crystalline phases presented in the solidified waste samples were characterized by X-ray diffraction (XRD) analysis. The samples were mixtures of ISSA, PC, KL and Solidifying aid at a mass proportion of 100:40:10:0.7, which was obtained from above research results. Samples cured at 20 ± 1°C for 28 days were crushed into powder using a mortar and pestle. Then the powder was sieved through a 150 μm mesh and scanned from 0° to 60°. The XRD analysis results were shown in Figure 4. As presented in Figure 4(a) and Figure 4(c), X-ray diffraction experiments provided variation information about the crystalline material among the different samples. The results of the XRD analysis revealed that the major crystalline phases presented in the Figures were quartz (SiO_2), CSAH ($CaO \cdot SiO_2 \cdot Al_2O_3 \cdot nH2O$), ettringite ($3CaO \cdot Al_2O_3 \cdot 3CaSO_4 \cdot 32H_2O$) and CSH ($CaO \cdot SiO_2 \cdot nH_2O$). These hydrated compounds are generally formed by a Pozzolanic reaction through the reaction of $Ca(OH)_2$ and SiO_2 [18]. It could be observed from Figure 4(a) that, quartz, one of the main chemical components of ISSA, is prominent in the diffraction patterns of the untreated ISSA blocks. No other crystalline phases matched sufficient peaks to be positively identified, although analysis using an XRD pattern data base (International Centre for Diffraction Data, ICDD) suggested minor peaks may correspond to aluminium phosphate ($AlPO_4$), iron silicite ($FeSi_2$) and calcium copper fluoride ($CaCuF_4$) [19]. When ISSA was blended with solidifying agent at a ratio of 5:2 Figure 4(b), ettringite and CSH were both detected, which complied with relevant research reporting that the hydrated products of the cement were known to be CSH, CAH, CSAH, ettringite, C_3AF ($3CaO \cdot Al_2O_3 \cdot Fe_2O_3$), $Ca(OH)_2$, $CaSO_4$ and $CaCO_3$ [20]. Similar XRD pattern was observed when solidifying aid was added to the ISSA/PC/KL system Figure 4(c).

Scanning electron microscopy analysis

Figures 5(a) and 5(b) showed the SEM observations of the incinerated sludge ash samples before and after solidification, respectively. Figure 5(a) was the SEM image of the unsolidified incinerated sludge ash and Figure 5(b) was the solidified block A, ISSA blended with PC and KL at the ratio of 100:40:10. As shown in the SEM micrographs, honeycomb-like hydrated products, known to be distinct CSH compounds [21] which have positive relationship with the compressive strength, were not observed in the incinerated sludge ash before solidification Figure 5(a). On the contrary, they were found to be widespread on the surface of the solidified sludge ash mixed with PC and KL.

Conclusion

The solidification of ISSA mixed with PC and KL was investigated

Figure 5: SEM Photographs of unsolidified ISSA and solidified block A. (a): unsolidified ISSA; (b): block A.

to examine the feasibility of its utilization as construction materials with the purpose of the economical recycling of waste materials. Compressive strength tests have showed that incinerated sludge ash solidification using PC, KL and SS-SA in appropriately mixed proportions enhanced the geotechnical characteristics of the sludge ash, with an ultimate strength meeting the minimum requirement for fired common bricks. TCLP results showed that the metal concentrations in leachates extracted from the cement- based solidified ISSA were much lower than that from the untreated ISSA, far below the EPA (China) toxicity characteristic criteria, suggesting that the solidification can minimize the hazard of heavy metals leaching from the solidified ISSA. Quartz, CSAH, ettringite and CSH were found in the XRD patterns of the 28-day solidified samples, which play an important role in enhancing the solidifying reaction of the sludge ash. SEM micrographs showed that the structure compaction was due to the formation of honeycomb-like C-S-H compounds caused by the addition of solidifying agent. It could be concluded that solidification and recycling for construction materials could be an inexpensive and effective treatment method for incinerated sewage sludge ash.

Acknowledgements

This work was supported by the Shanghai Science and Technology Commission (No. 09DZ1200108) and the Key project of Science and Technology Commission of Shanghai Municipality (No. 11DZ0510200).

References

1. Carbone LG, Gutenmann WH, Lisk DJ (1989) Element immobilization in refuse incinerator ashes by solidification in glass, ceramic or cement. Chemosphere 19: 1951-1958.

2. Donatello S, Tyrer M, Cheeseman CR (2010) EU landfill waste acceptance criteria and EU Hazardous Waste Directive compliance testing of incinerated sewage sludge ash. Waste Manage 30: 63-71.

3. Cyr M, Coutand M, Clastres P (2007) Technological and environmental behavior of sewage sludge ash (SSA) in cement-based materials. Cement Concrete Res 37: 1278-1289.

4. Murakami T, Suzuki Y, Nagasawa H, Yamamoto T, Koseki T, et al. (2009) Combustion characteristics of sewage sludge in an incineration plant for energy recovery. Fuel Process Technol 90: 778-783.

5. Kinuthia MJ, O'Farrell M, Sabir BB, Wild S (2001) A preliminary study of the cementitious properties of wastepaper sludge ash-ground granulated blastfurnace slag (WSA-GGBS) blends. Proceedings of the International Symposium, London. 93-104.

6. Tay J H (1991) Clay-blended sludge as lightweight aggregate concrete material. J Environ Eng 117: 834- 844.

7. Yagüe A, Valls S, Vàzquez E (2002) Use of cement Portland mortar of stabilized dry sewage sludge in construction applications. Waste Manage Environ 34: 527-536.

8. Lin DF, Luo HL, Sheen YN (2005) Glazed tiles manufactured from incinerated sewage sludge ash and clay. J Air Waste Manage 55: 163-172.

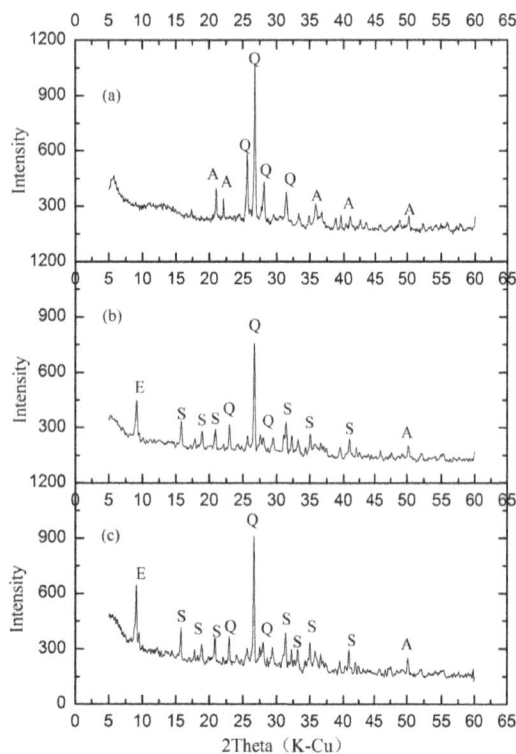

Figure 4: XRD patterns of three different samples. (a): untreated ISSA; (b): block A; (c): block B.

9. Lin DF, Lu HL, Zhang SW (2007) Effects of Nano-SiO2 on tiles manufactured with clay and incinerated sewage sludge ash. J Mater Civil Eng 19: 801-808.

10. Lin CF, Wu CH, Ho HM (2006) Recovery of municipal waste incineration bottom ash and water treatment sludge to water permeable pavement materials. Waste Manage 26: 970-978.

11. Chiou IJ, Wang KS, Chen CH, Lin YT (2006) Lightweight aggregate made from sewage sludge and incinerated ash. Waste Manage 26: 1453-1461.

12. Chen CH, Chiou IJ, Wang KS (2006) Sintering effect on cement bonded sewage sludge ash. Cement Concrete Comp 28: 26-32.

13. Monteiro SN, Alexandre J, Margern JI, Saanchez R, Vieira CMF (2008) Incorporation of sludge waste from water treatment plant into red ceramic. Constr. Build Mater 22: 1281-1287.

14. ASTM Standards. C494/C494M Specification for Chemical Admixtures for Concrete.

15. EN196-1 (2000) "Methods of testing cements-determination of compressive strength".

16. US EPA. EPA Method 1311 Toxicity Characteristic Leaching Procedure (TCLP).

17. Detwiler RJ, Bhatty JI, Bhattacharja S (1996) Supplementary cementing materials for use in blended cements. in: Research and Development Bulletin, Cement Association Publishers, USA, 87-89.

18. Qiao XC, Poon CS, Cheeseman CR (2007) Investigation into the stabilization/solidification performance of Portland cement through cement clinker phases. J Hazard Mater 139: 238-243.

19. Cheesemnan CR, Sollars C, McEntee S (2003) Properties, microstructure and leaching of sintered sewage sludge ash. Resour Conserv Recy 40: 13-25.

20. Kim EH, Cho JK, Yim S (2005) Digested sewage sludge solidification by converter slag for landfill cover. Chemosphere 59: 387-395.

21. Taylor HFW, Newbury DE (1984) An electron microprobe study of a mature cement paste. Cem Concr Res 14: 565-573.

16

Framework for Low Carbon Precinct Design from a Zero Waste Approach

Queena K Qian[1]*, Steffen Lehmann[2], Atiq Uz Zaman[3] and John Devlin[3]

[1]*Endeavour Australian Cheung Kong Post-Doc Fellow, sd+b Centre, University of South Australia, Australia*
[2]*Chair Professor of Sustainable Design; Director, sd+b Centre and CAC-SUD Centre, University of South Australia, Australia*
[3]*PhD Candidate, sd+b Centre, University of South Australia, Australia*

Abstract

The consumption-driven society today produces an enormous amount of waste, which puts pressures on land, pollutes the environment and creates economic burden. 'Zero waste' concept, a whole system approach aiming to achieve no waste along the materials flow through society, has become one of the most visionary concepts for tackling growing waste problems. System Dynamics (SD) approach is applied in the proposed framework for designing the waste management in a zero waste residential precinct. A cost-benefit analysis (CBA) is incorporated to supplement the SD framework to evaluate the total cost and benefit of waste and resources throughout the material flow chain. The authors proposed a list of parameters under the categories of process, technology and infrastructure, socio-economic and institutional, social- environment, to be tested in future case study of Bowden village, SA, Australia. The framework provides an inventory of leverage points that helps policy-makers design waste policies and allocate resources effectively, with minimum environmental impact and optimum social benefits. It also helps planning professionals and business stakeholders better understand the costs and benefits of different scenarios for achieving a zero waste residential precinct.

Keywords: Cost-benefit analysis; Low-carbon residential precinct; Solid waste management; System dynamics; Zero waste

Introduction

The development of science and technology as well as global levels of economic activity causes a dramatic increase in the production of urban solid waste [1]. The generation of waste over time has become a serious environmental problem for the world, and been affecting the balance of natural resources [2]. Solid Waste Management (SWM) has become crucial for protecting the environment and the human well-being. Various national and international initiatives for SWM are in place, which takes considerations of environmental, administrative, regulatory, scientific, market, technology, institutional and socio-economic factors [3].

The sustainable SWM is becoming essential at all phases of the waste chain from production, waste generation, collection, transportation, treatment, recycling till disposal. 'Zero waste' is, therefore, becoming a popular concept. It is a closed-loop concept aiming of optimum recycling or resource recovery, as well as elimination of unnecessary waste in the first place [4,5]. With a whole system approach, it seeks for an end–of-pipe solution for waste diversion along the materials flow through society. It encourages waste elimination through recycling and resource recovery, with a guiding design philosophy to reduce waste at source and at all points down the supply chain [6]. 'Zero waste' commitments have been made across the world, including US, Europe, Australia, New Zealand, etc. [7], and becomes trendy for the rest of the world.

A sustainable SWM approach is systematic, flexible and long term visionary. A sustainable society requires sophisticated ways to manage solid waste. A systems approach that reveals the relationships and explains its interactions among the parties in the system contributes to greater sustainable practice [8]. Based on reviewing and comparing different researchers' work on the waste management, our research aims to propose a research framework of zero-waste management and strategies for low carbon residential precincts. This approach selected needs to accommodate the fact that zero waste management can be achieved by identifying the leverage points during the entire zero waste chain and altering or redesigning the processes accordingly. Kytzia and Nathani believe a "combination between analyses of economic/ physical structures on the one hand and economic behaviour on the other hand is most promising" to achieve the zero waste concept [9]. The methodological framework presented will contributes to the understanding of the overall process of the zero waste management by combining system characteristics as well as the cost/ benefit impact with the attitudes and requirements of a specific stakeholder group (i.e., the city planner, government, and/or households). This paper highlights the dynamic interrelationships of the sustainable SWM practices, supplemented with the cost/benefit factors into the SD process. The system-oriented research framework serves the decision-makers to draw the forward-looking and preventative insights and reach a scientific understanding of the carbon and cost consequences relating to various sustainable SWM scenarios.

Literature Review

Similar researchin SWMand system approach

Research interests in addressing waste management issues have resulted in a large amount of publications during the last decade. Ossenbruggen and Ossenbruggen apply SWAP programs - a linear programming algorithm, to aid in the strategic plan and decision-makings of SWM, and weigh the cost associated and the benefits from various waste recovery alternatives [10]. Chung and Poon has applied multiple criteria analysis (MCA) in SWM to find out the preferred

*Corresponding author: Queena K Qian, Endeavour Australian Cheung Kong Post-Doc Fellow, sd+b Centre, University of South Australia, Australia
E-mail: kun.qian@fulbrightmail.org

waste management options. The merit of MCA is more objective and transparent and it accommodates quantitative and qualitative data [11]. Bovea et al. apply the Life Cycle Assessment technique to obtain parameters that quantifies the environmental impact of waste transportation and operating a transfer station in municipal SWM systems [12]. Beigl et al. review the modelling approaches for SWM and propose an implementation guideline with a compromise between information gain and cost-efficient model development [13]. Lu and Yuan develop a framework to understand the C and D WM research as archived in selected journals, and give useful references attempting the research of C and D WM research [14]. Chang and Davila simulate the predetermined scenarios with a minimax regret optimization, to achieve improved SWM strategies from different environmental, economic, legal, and social conditions [15].

Planning sustainable SWM has to address several interdependent issues including public health, the environmental impact, the treatment potential, the landfill capacity, and present and future economic and social costs, and financial expenditures, etc. It, therefore, becomes increasingly necessary to understand the dynamic nature of their interactions, and the complex, and multi-faceted system. How to combine all the correlated factors into the consideration when making the optimal sustainable SWM strategy among the alternatives? Pries et al. believes that system approach enriches the analytical framework of SWM, specially designed to understand the dynamics and intercalations among the factors, and develop better SWM strategies for both the SWM industry and the government [16]. It plays an important role to simulate and assess the integrated SWM systems, and inform the stakeholders with insightful strategies and rational decision-makings.

System dynamics (SD) approach in MSW management

SD is a well-established methodology that provides a theoretical framework and concepts for modelling complex social, economic and managerial systems [17]. It deals with the interrelationships and complex of the system, where the dynamic behaviour can be reflected and simulated by the feedback loops based on the control theory [18–21]. The SD approach is widely applied in the areas of environmental sustainability and regional sustainable development issues [22-26] environmental management and environmental systems [27,28], and waste management [29-31].

Thirumuthy applied SD approach to evaluate the investments required for various environmental services in Madras city [32]. Mashayekhi explored a dynamic analysis for analysing the transition from the landfill method of disposal to other forms of disposal for the city of New York [33]. Sudhir et al. proposed a system dynamics model to capture the dynamic nature of interactions among the various components in SWM for developing countries [28]. Karavezyris et al. studied the quantitative impact of different variables, such as voluntary recycling participation and regulation, on SWM [34]. Ulli-Beer presented a SD model for understanding local recycling systems [35]. Dyson and Chang applied a SD approach to predict solid waste generation in a fast growing urban area [36]. Duran et al. developed a model to assess the economic viability of creating markets for recycled construction and demolition (CandD) waste in scenarios using different economic instruments [37]. Rehan et al. proposes SD approach to develop a causal loop diagram for water and wastewater network management, as a complex system with multiple interconnections and feedback loops [38]. It demonstrates the significance of feedback loops for financial management of the complexity of the system by incorporating all feedback loops. Yuan et al. proposes a SD model to serve as a decision support tool for waste projection, and as a platform

for simulating effects of various waste reduction management strategies [39].

A sustainable SWM system incorporates feedback loops, focusing on processes, embodies adaptability and diverts wastes from disposal [8]. Due to the SWM hierarchy, the challenges lie in how to diversify the waste reduction options, increase the reliability of infrastructure systems, and leverage the redistribution of waste streams among production, transportation, compost, recycling, and other facilities. It depends on factors such as technology and infrastructure, socio-economic and institutional, social- environment, culture, as well as market considerations [40,41]. Transitioning to a sustainable SWM system requires identification and application of leverage points that stimulate positive change [8]. SD approach recognizes the sustainable SWM process and accommodates the zero waste achievement by identifying the leverage points during the entire zero waste chain and altering or redesigning the processes accordingly.

Cost-Benefit Analysis (CBA) supplemented in SD

Economic instruments to minimize waste play a crucial role in encouraging environmentally-friendly SWM practices [30]. The rising pressure in terms of cost efficiency on public services and facilities pushes governments to share those services with the industry. The partnership among the government, business and individual household collectively contributes to the overall effectiveness and efficiency of the MSW management. Therefore, CBA of waste management are essential to provide the evidence that will motivate stakeholders throughout waste management chains, and SD modelling helps examine the relationships between waste management activities in a holistic view. Yuan et al. analysed the CBA of the dynamics and interrelationships of CandD waste management practices using a SD approach [30]. Farel et al. proposes a SD approach to simulate the net economic balance of the recycling network under different future scenarios [42]. Therefore, a good balance between the cost and benefits is an important factor to select among the scenarios for the use of different stakeholders.

CBA has been widely acknowledged as a tool for policy and project analysis throughout the world. It helps the policy-makers as well as stakeholders to justify their decisions in a more systematic, rigorous and unambiguous way Gramlich [43]. It allows us to identify and assess positive and negative economic and physical effects independently. Particularly, it supports the simulation and optimization models for system analysis. Well-defined CBA parameters may translate environmental aspects into economic terms.

Methodological Framework

The life cycle of zero- waste in diagram (Figure 1)

The waste management chain consists of a series of potential lifecycle stages. Products pass through the manufacturing, production, and consumption stages before entering into the waste management system where various processes can minimize the impact of the waste. It is not a collection of independent waste management activities but rather a system of interdependent activities. It is to put together all the potential acts to reduce the original waste to be generated, until the waste can finally be disposed of with minimum costs involved. Some typical factors affecting waste management activities are based on an extensive literature review as well as works previously carried out at the zero waste research centre [4].

System dynamics (SD) approach Causal loop SD diagram for a zero-waste precinct

This causal loop diagram is designed based on the interaction of different components in a zero waste management system. It identifies the carbon emission loops and external factors as well as interrelationships in SWM from waste generation (manufacture, household consumption, separation) to collection, treatment, recycling and disposal. This study frames the scope of the zero-waste loop and establishes the initial step of a SD study framework. It is used to identify and explain the causal relationships between acts and stages of waste management in a residential precinct. Based on the above extensive literature reviews on SD and conceptual model of the zero-waste management chain in low carbon residential precinct (Figure 1), the casual loop diagram of zero-waste precincts can be developed (Figure 2).

Cost benefit analysis (CBA) approach

CBA is an analytical procedure to evaluate the desirability of a program or project by weighing the resulting benefits against the corresponding costs in order to see whether the benefits outweigh the costs [44-46]. In this research, CBA consider sand reflects the tangible factors (such as environmental costs) as well as invisible benefits, from the improved SWM scenarios. It attempts to evaluate effects on users (policy-makers as well as stakeholders), external effects, quantify values

and social benefits. In this research, the current value of a collective SWM scenario is considered on a net present value (NPV) basis. NPV reflects a stream of current and future benefits and costs, and results in a value in today's dollars that represents the present value of an investment's future financial benefits minus any initial investment. Typically, financial benefits for individual elements are calculated on a present value basis and then combined in the conclusion with net costs to arrive to NPV, as the function below:

$$NPV = \sum_{i=1}^{n} \frac{values_i}{(1+rate^i)}$$

If positive, the investment should be made, otherwise not. The value of the NPV of the proposed scenario gives a foundation for comparing alternative options. A bigger NPV indicates a better option.

Of particular interest to both waste industry stakeholders and policy-makers, this research looks into both private cost-benefit ratios (to establish how much benefit can be derived from every dollar is spent for improving waste management) and social cost benefit ratios (as the amount of return is perceived to have a direct impact on the degree of success of the waste management regulation or incentive scheme). The

Category	Parameters	-	+
Process	• Waste generation rate / Average waste generation per household /Waste generation after adopting waste reduction measures		x
	• Cost of waste collection		x
	• Recycling waste /cost of waste recycling / ratio of recycling		x
	• waste disposal cost / disposal cost saving / illegal disposal		x
	• transportation costs / transportation cost saving		x
	• Cost of waste sorting /ratio of reuse / cost of waste reuse / purchasing cost saving		x
	• Impact of landfill space limit on waste reduction /Promotion of waste reduction via landfill charge/ Unit land filling charge fee		x
	• Waste reduction rate/ Amount of reduced waste /Efforts to reduce waste /Impact of waste reduction costs		x
	• Management capacity for reducing waste		x
Technology & Infrastructure	• Actual investment in waste management (increasing rate)		x
	• Low-waste technologies application		x
	• Frequency of technology changes/updates		x
	• Training in waste management		x
	• Actual industry stakeholders' initiative to minimize the waste		
	• Changing of investment in waste management / Impacts of technology changes on waste reduction		x
	• Impact of measures taken for product design to reduce waste/ Decreasing of design changes		
Socio-economic, Institutional	• Regulation changes to SWM		x
	• Social performance value		
	• Regulating the illegal waste disposal to improve the society image		x
	• Public satisfaction of the waste performance		
	• Gaining experience of managing waste		
	• Impact of public's willingness to reduce waste / Changing of public's willingness		
	• Expected increasing rate of public's willingness to reduce waste		
by incentives		x	
	• Zero waste awareness and skills / Actual household's initiative to minimize waste		
	• Industry's willingness to reduce waste		
	• Improvements on SWM culture and public behaviour in the society		
	• Provision of job opportunities in waste management / Weight of provision of job opportunities		
Socio-environment	• Incentive to manage waste		x
	• Public appeal for zero waste to improve the environment		x
	• Environmental awareness / Environmental behaviour		
	• Accumulated environmental impact of zero waste to the environment		x
	• Accumulated impacts of physical living environment of SWM		
	• Impacts of SWM on participants' long-term health		
	• Accumulated performance value of waste emission		x
	• Weight of public's long-term health conditions		

Table 1: Selection of SD waste management Parameters for CBA ("-": cost; "+": benefit).

amount of investment return of implementing a particular measure is evaluated using private benefit cost ratios. The mathematical function of adopting incentives can be obtained as following:

$$private\ cost\text{-}benefit\ ratio_i^1 = \frac{\Delta B_i^{Econ}}{\Delta C_i^{Econ}}$$

Where ΔB_i^{Econ} and ΔC_i^{Econ} are the additional benefit derived, and additional life-cycle cost required for the private industry stakeholders and/or household by implementing the proposed waste management project or incentives, respectively. The private benefit–cost ratio can help identify which options are financially beneficial. The ratio can give an indication for selecting a particular measure if the ratio of the present value of benefits to the present value of costs is greater than 1.0.

$$social\ cost\text{-}benefit\ ratio_i^1 = \frac{\Delta B_i^{Sco}}{\Delta C_i^{Sco}}$$

Where ΔB_i^{Soc} is the net benefit in monetary value derived for society by implementing the proposed improvement measure; ΔC_i^{Soc}, which is the additional life-cycle cost required by implementing the proposed incentives. The social benefit–cost ratio can assist government officials or policy-makers in judging the environmental viability of the measure under consideration.

Discussions and Findings

Table 1 lists the selection of SD waste management parameters base on the zero-waste casual loop diagrams (Figures 1 and 2), to be considered in CBA. All the parameters selected are based on the literature review and methodological framework done in the earlier session. They are assigned in the categories of Process, Technology and Infrastructure, Socio-economic, Institutional, and Socio-environment. All the parameters are assigned to be either under the cost (-) or benefit (+) or both, depending on whether they are mainly involving the costs or benefits, or both to the stakeholders along the SWM process. Very often, the parameters selected involve both costs and benefits depending on the different stakeholders.

The benefits to get CBA supplemented to the SD approach, in that it quantifies the potential returns and expenses of a program or project and balances the pros and cons to arrive at a decision. This line of research has important implications both for assessing the cost of correcting market failures – such as environmental externalities – as well as clarifying the role of policies that are oriented to correct behavioural failures and market barriers. The benefits and costs are usually quantified in real monetary terms, i.e. converted to the present value, to enable assessment of different the benefits and the costs over time. The major CBA indicators include present value, net present

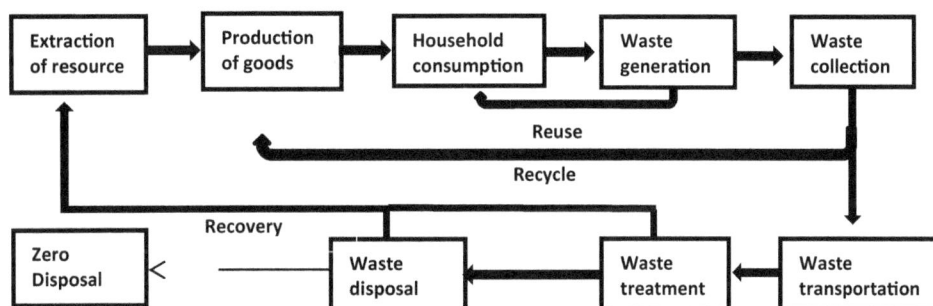

Figure 1: A conceptual model of zero-waste management chain for residential precinct.

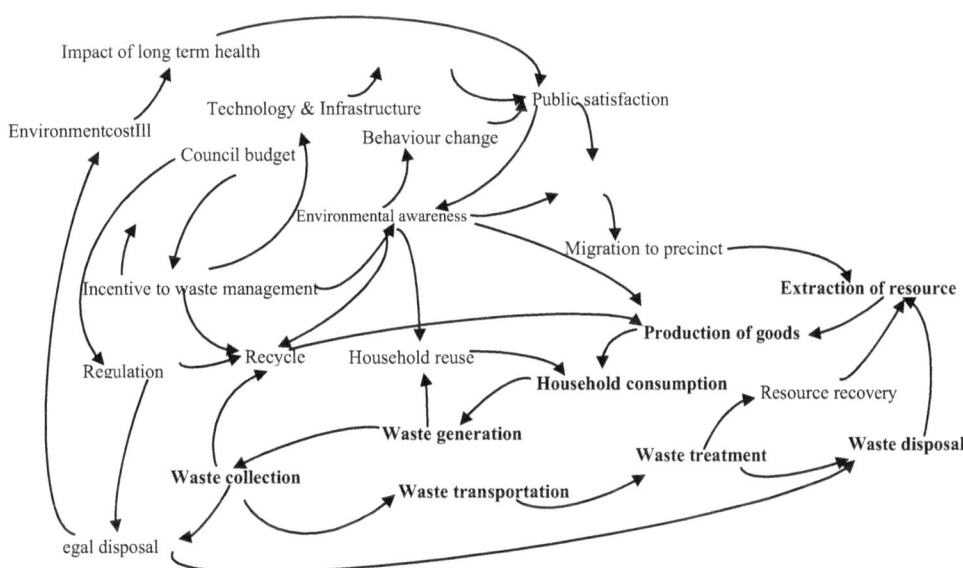

Figure 2: Causal loop SD diagram for a zero-waste precinct.

value and benefit cost ratio, which are applied in this study [47]. One of the aspects evaluated by CBA in this study is the hidden costs and benefits during the waste management loop, which has not been widely considered.

Future studies: scenarios setting and data collections- case of Bowden village, SA, Australia.

The framework proposed is to be tested in future study using real-life data, in three different residential density scenarios – mixed use high rise (10-storey), low rise (3 to 8-storey) and town house scale (3-storey). The data will be derived from Bowden Urban Village, a new residential precinct in Adelaide, the capital city of South Australia. Upon successful testing of the SD model using the Bowden village in future case study, it will be adjustable to a wider application to assist decision-making in different precincts. Most of the research on MSW management is examining systems at the city or national level. There are many benefits to this study's use of the precinct level: precincts provide more flexibility and precision, it also allows the model to be closer to reality as the dynamics and external factors at the city and/or national level are too complicated to be modelled.

The model described above is a theoretical framework for examining MSW loop and its management system at precinct level. For the future study case of Bowden urban village, three scenarios are set for different options and desire of the development plan- mixed use high rise (10-story), low rise (3 to 8-story) and town house scale (3-story). Using the SD simulation program, e.g. i think, it will examine the potential ways to reduce municipal solid waste (MSW) throughout generation, collection till disposal and concludes with overall carbon consequences and reduction options.

Following the general research framework developed in this paper, it is to collect and evaluate a broad spectrum of costs and benefits with the available data and the data from the survey, and to develop reasonable net present value estimation for comparison of each decision-making scenario. System dynamics approach is applied to capture and frame the scope of the waste life cycle in order to forecast the costs and benefits of waste management options from beginning to end. The overarching purpose is to answer the question: "does it make financial and economic sense for a given stakeholder, with particular incentives, to implement these waste management options?"; which serves the ultimate aim of this study: to balance the financial and economic interests of the private sector and those of the whole society (and environment) with minimum environmental impact and optimum social benefits, on behalf of the urban policy-maker.

Conclusions

Based on an extensive literature review on waste management of system approach, this paper proposes a holistic methodological framework for designing a SWM system for a zero-waste low carbon residential precinct. This study attempts to employ both a SD approach, incorporated with a cost-benefit analysis to simulate the changes in various MSW management scenarios for different low-carbon precincts. The causal loop diagram made it easier to understand and identify the critical activities throughout the waste management chain, the essential stakeholders and also external factors such as financial incentives. The methodological framework considers a list of parameters under the categories of socio-economic, institutional, socio-environmental, infrastructure and technology, and process. The framework is designed in such a way that it can be adapted to other local conditions by changing the local parameters and data for

whatever the regional case. In future studies, the proposed framework will be modelled using a computer program, i.e., i think, with a stock-flow diagram simulating different scenarios. The cost-benefit changes for each scenario are to provide rational options among the simulation plans for decision-makers, city planners and other stakeholders, and to help predict future waste management needs.

Acknowledgement

This article was supported by the Zero Waste Research Centre for Sustainable Design and Behaviour (sd+b) and the China–Australia Centre for Sustainable Urban Development (CAC_SUD) in the University of South Australia. The study is a part of an ongoing collaborative research project funded through the Co-operative Research Centre (CRC) for Low Carbon Living involving researchers from the University of South Australia, the University of New South Wales, the CSIRO and various government bodies. Special thanks to the Endeavour Research Fellowship Program for the support.

References

1. Su J, Xi BD, Liu HL, Jiang YH, Warith MA (2008) An inexact multi-objective dynamic model and its application in China for the management of municipal solid waste Management 28: 2532-2541.

2. Kollikkathara N, Feng H, Stern E (2009) A purview of waste management evolution: Special emphasis on USA. Waste Management 29: 974-985.

3. Kollikkathara N, Feng H, Yu DL (2010) A system dynamic modelling approach for evaluating municipal solid waste generation landfill capacity and related cost management issues. Waste Management 30: 2194-2203.

4. Zaman AU, Lehmann S (2013) The Zero Waste Index: A Performance Measurement Tool for Waste Management Systems in a Zero Waste City. Journal of Cleaner Production. 50: 123-132.

5. City of Austin (2008) Zero waste strategic plan the zero waste economy Prepared by Gary Liss & Associates, 4395 Gold Trail Way, Loomis, CA 95650-8929, USA.

6. ACT Government (1996) A Waste management strategy for Canberra- No waste Canberra.

7. Greyson J (2007) An economic instrument for zero waste, economic growth and sustainability, Journal of Cleaner Production 15: 1382-1390.

8. Seadon, JK (2010) Sustainable waste management systems. Journal of Cleaner Production 18: 1639-1651.

9. Kytzia S, Nathani C (2004) Bridging the gap to economic analysis: economic tools for industrial ecology. Prog Ind Ecol 1: 143–164.

10. Ossenbruggen PJ, Ossenbruggen PC (1992) Swap: A computer package for solid waste management Computers, Environment and Urban Systems, 16: 83-100.

11. Chung SS, Poon CS (1996) Evaluating waste management alternatives by the multiple criteria approach Resources, Conservation and Recycling 17: 189-210.

12. Bovea, MD, Powell JC, Gallardo A, Capuz-Rizo SF (2007) The role played by environmental factors in the integration of a transfer station in a municipal solid waste management system. Waste Management 27: 545-553.

13. Beigl P, Lebersorger S, Salhofer S (2008) Modelling municipal solid waste generation. A review Waste Management 28: 200-214.

14. Lu WS, Yuan HP (2011) A framework for understanding waste management studies in construction Waste Management 31: 1252-1260

15. Chang NB, Davila E (2007) Minimax regret optimization analysis for a regional solid waste management system. Waste Management 27: 820-832.

16. Pires A, Martinho G, Chang NB (2011) Solid waste management in European countries A review of systems analysis techniques. Journal of Environmental Management 92: 1033-1050.

17. Forrester JW (1958) Industrial dynamics a major breakthrough for decision makers. Harvard Business Review 36: 37-66.

18. Sufian MA, Bala BK (2007) Modelling of urban solid waste management system: The case of Dhaka city Waste Management 27: 858–868.

19. Forrester JW (1968) Principals of systems Cambridge USA Productivity Press.

20. Bala BK (1998) Energy and Environment: Modelling and Simulation Nova Science Publisher New York.

21. Bala BK (1999) Principles of System Dynamics. Agrotech Publishing Academy Udaipur India.

22. Forrester JW (1971) World Dynamics. Wright-Allen Press MIT Massachusetts.

23. Meadows DH, Meadows DL, Randers J (1992) Beyond Limits Chelsea Green Publishing Vermont.

24. Bach NL, Saeed K (1992) Food self-sufficiency in Vietnam: a search for a viable solution System Dynamics Review 8: 129–148.

25. Saeed K (1994) Development Planning and Policy Design: A System Dynamics Approach Chelsea Green Publishing Vermont.

26. Saysel AK, Barlas Y, Yenigun O (2002) Environmental sustainability in an agricultural development project a system dynamics approach. Journal of Environmental Management 64: 247–260.

27. Mashayekhi AN (1990) Rangelands destruction under population growth: The case of Iran. System Dynamics Review 6: 167–193.

28. Sudhir V, Srinivasan G, Muraleedharan VR (1997) Planning for sustainable solid waste management in urban India. System Dynamics Review 13: 223–246.

29. Talyan V, Dahiya RP, Anand S, Sreekrishnan TR (2007) Quantification of methane emission from municipal solid waste disposal in Delhi. Resource Conservation & Recycling 50: 240-259.

30. Yuan HP, Shen LY, Hao JJL, Lu WS (2011) A model for cost-benefit analysis of construction and demolition waste management throughout the waste chain. Resources Conservation and Recycling 55: 604-612.

31. Yuan HP (2012) A model for evaluating the social performance of construction waste management. Waste Management 32: 1218-1228.

32. Thirumurthy AM (1992) Environmental facilities and urban development: a system dynamics model for developing countries New Delhi India Academic Foundation.

33. Mashayekhi AN (1993) Transition in New York State solid waste system a dynamic analysis. Sys Dynam Rev 9: 23–48.

34. Karavezyris V, Timpe K, Marzi R (2002) Application of system dynamics and fuzzy logic to forecasting of municipal solid waste. Math Comput Simul 60: 149–158.

35. Ulli-Beer S (2003) Dynamic interactions between citizen choice and preference and public policy initiatives a System Dynamics model of recycling dynamics in a typical Swiss locality.

36. Dyson B, Chang NB (2004) Forecasting municipal solid waste generation in a fast-growing urban region with system dynamics modelling. Waste Manage 25: 669–679.

37. Duran X, Lenihan H, ORegan B (2006) A model for assessing the economic viability of construction and demolition waste recycling the case of Ireland Resources Conservation and Recycling 46: 302-320.

38. Rehan R, Knight MA, Haas CT, Unger AJA (2011) Application of system dynamics for developing financially self-sustaining management policies for water and wastewater systems. Water Research 45: 4737-4750.

39. Yuan HP, Chini AR, Lu YJ, Shen LY (2012) A dynamic model for assessing the effects of management strategies on the reduction of construction and demolition waste. Waste Management 32: 521-53.

40. Goddard HC (1995) The benefits and costs of alternative solid waste management policies. Resour Conserv 13: 183–213.

41. Nordhaus WD (1992) The ecology of markets. Proc Natl Acad Sci USA 89: 843–850.

42. Farel R, Yannou B, Ghaffari A, Leroy Y (2013) A cost and benefit analysis of future end-of-life vehicle glazing recycling in France. A systematic approach Resources Conservation and Recycling 74: 54-65.

43. Gramlich EM (1981) Benefit-cost analysis of government programs Englewood Cliffs N J. Prentice-Hall.

44. Cowen T (1998) Using cost-benefit analysis to review regulation, New Zealand Business Roundtable.

45. Posner RA (2000) Cost-benefit analysis definition justification and comment on conference paper. Journal of Legal Studies 29: 1153-1177.

46. EB (2008) Policy and consultation papers: a proposal on the mandatory implementation of the building energy codes.

47. Thomas JM (2007) Environmental economics applications policy and theory Mason Ohio: South-Western.

17

Characterization of Polyphenolic Phytochemicals in Red Grape Pomace

Dimitris P Makris[1]* and Panagiotis Kefalas[2,3]

[1]*School of Environment, University of the Aegean, Mitr Ioakim Str., Myrina-81400, Lemnos, Greece*
[2]*Food Quality and Chemistry of Natural Products, Mediterranean Agronomic Institute of China, China*
[3]*Centre International de Hautes Etudes Agronomiques Méditerranéennes, P.O. Box 85, Chania-73100, Greece*

Abstract

Polyphenolic phytochemicals are of particular importance to the food, pharmaceutical and cosmetics industry, because of their unique antioxidant properties. Red grape pomace is a solid waste of the wine manufacturing process and possesses a very high polyphenolic load, hence its significance as a rich and abundant residua source. However, in many instances there is a lack of analytical data regarding its polyphenolic composition. In this study, red grape pomace originating from the Greek native cultivar *Vitis vinifera var.* Agiorgitiko was efficiently extracted with 57% aqueous ethanol, which is a non-toxic and environmentally benign solvent, with the aim of obtaining a polyphenol-enriched extract. The extract was subsequently analyzed by liquid chromatography-diode array-mass spectrometry analysis, in order to tentatively characterize the major phytochemicals recovered. The compounds identified were a *p*-coumaric acid derivative, three flavonol conjugates (two glucosides and a glucuronide), along with three anthocyanin pigments that occur in grape berries. Three other major phenolics detected could not be assigned to a tentative formula and their structural elucidation merits further investigation. The data generated from this study could be used in assessing the overall polyphenolic profile of the pomace from this particular Greek, native variety, which could be of value in producing commercial formulations with high antioxidant activity.

Keywords: Antioxidants; Liquid chromatography-mass spectrometry; Polyphenols; Red grape pomace

Abbreviations: AAR, Antiradical Activity (mM TRE g^{-1} dpw); dpw: Dry Pomace Weight (g); GAE: Gallic Acid Equivalents; RGP: Red Grape Pomace; Rt: Retention Time; TRE: Trolox Equivalents; WISW: Wine Industry Solid Wastes; Y_{TP}: Total Polyphenol Yield (mg GAE g^{-1} dpw)

Introduction

Solid wastes in wine industry mainly consist of solid by-products, such as pomace and stems. The waste material may account on an average 30% (w/w) of the grapes used for wine production. Vinification wastes contain a relatively high content of polyphenolic phytochemicals [1,2], which depends on the type of grape (white or red), the part of the tissue (skins, seeds etc), as well as the processing conditions (e.g. pomace contact).

Solid wastes have attracted considerable attention as potential sources of bioactive phenolic compounds, which can be used in the pharmaceutical, cosmetics and food industry. Studies regarding WISW are mainly focused on the polyphenolic composition of seeds, which are very rich in flavanols [3-5], but red grape pomace (RGP), which is composed of seeds and skins, has also been evaluated as potential source of antioxidant polyphenols [6-9]. However, although several methods of extraction have been developed for the efficient recovery of pomace polyphenols [10], there is still a significant lack of analytical data on the polyphenolic profile of RGP originating from different cultivars. The composition of RGP is defined by the polyphenols occurring in both seeds and skins, which are mainly flavanols [11,12] anthocyanins and flavonols [13,14], although other minor constituents, such as stilbenes [15] and dihydroflavonols [16] have been reported. All these components are considered nutritionally important, since they may possess a variety of bioactivities [17]. Therefore, the investigation of the analytical polyphenolic composition of RGP is of undisputed significance in the development of tools and methodologies for extraction and final product formulation.

Because of the lack of analytical data regarding the polyphenolic composition, the scope of the present study was an examination on the polyphenolic composition of grape pomace from the native Greek variety Agiorgitiko (*Vitis vinifera* sp.). This variety is widely cultivated in the region of Peloponnese and it is the most important native species, in terms of quality wine production. The approach attempted was the examination of a polyphenol- rich pomace extract, obtained with a hydroalcoholic solution, employing liquid chromatography-diode array-mass spectrometry (LC-DAD-MS) analysis.

Materials and Methods

Chemicals

Folin-Ciocalteu reagent and gallic acid were from Fluka (Steinheim, Germany). Trolox™, gallic acid, and 2, 2-diphenyl-picrylhydrazyl (DPPH•) stable radical were from Merck (Darmstad, Germany). p-Coumaric acid was from Sigma (St. Louis, MO, U.S.A.).

Vinification solid waste

RGP was from Agiorgitiko cultivar (*Vitis vinifera* sp.), obtained from a winery located in Nemea (Peloponnese). The pomace was left in contact with the fermenting must for 7 days. The material was obtained immediately after pressing the pomace, transferred to the laboratory within a few hours and stored at -40°C until used.

Extraction procedure

A suitable quantity of RGP (approx 4.5 g) was chopped into small

---

*Corresponding author: Dimitris P Makris, School of Environment, University of the Aegean, Mitr. Ioakim Str., Myrina-81400, Lemnos, Greece
E-mail: dmakris@aegean.gr

pieces with a sharp, stainless steel cutter to facilitate extraction. The chopped material was ground with sea sand and a small portion of the extraction solvent, with a pestle and a mortar, and then left to macerate for 30 min in the dark. The paste formed was placed in a 100 mL conical flask with 25 mL of solvent (solvent-to-solid ratio 5.5) and extraction was performed under stirring at 700 rpm on a magnetic stirrer for 15 min. The extract was filtered through paper filter and this procedure was repeated twice more. The extracts were then combined in a 100 mL volumetric flask and made to the volume. All extracts were centrifuged at 4,500 rpm and filtered through 0.45 µm syringe filters prior to analyses.

Determination of total polyphenol yield (Y_{TP})

Analysis was carried out employing the Folin-Ciocalteu methodology [18]. In a 1.5-mL Eppendorf tube, 0.78 mL of distilled water, 0.02 mL of sample appropriately diluted, and 0.05 mL of Folin-Ciocalteu reagent were added and vortexed. After exactly 1 min, 0.15 mL of aqueous sodium carbonate 20% was added, and the mixture was vortexes and allowed to stand at room temperature in the dark, for 60 min. The absorbance was read at 750 nm (A_{750}), and the total polyphenol concentration was calculated from a calibration curve, using gallic acid as a standard. Yield in total polyphenols (Y_{TP}) was expressed as mg gallic acid equivalents (GAE) per g of dry weight, using the following equation:

$$Y_{Tp}(mgGAEg^{-1}) = \frac{(951 \times A_{750} - 1.49) \times V}{m} \quad (1)$$

Where, V is the volume of the extraction medium (mL) and m the dry weight of RGP (g).

Measurement of the antiradical activity (A_{AR})

Sample (0.025 mL), appropriately diluted, was added to 0.975 mL DPPH• solution (100 µM in methanol), and the absorbance at 515 nm was read at t=0 ($A_{515}{}^{t0}$) and t=30 min ($A_{515}{}^{t30}$). Results were expressed as Trolox equivalents (mM TRE) per g of dry weight, using the following equation:

$$A_{AR}\left(mMTREg^{-1}dw\right) = \left(\frac{0.018 \times \%\Delta A_{515} + 0.017}{m}\right) \times F_D \quad (2)$$

As determined from linear regression, after plotting % ΔA_{515} of known solutions of Trolox against concentration; where $\%\Delta A_{515} = \frac{A_{515}^{t=0} - A_{515}^{t=30}}{A_{515}^{t=0}} \times 100$, m the weight of dry material (g), and F_D the dilution factor.

Liquid chromatography-diode array-mass spectrometry (LC-DAD-MS)

A Finnigan MAT Spectra System P4000 pump was used coupled with a UV6000LP diode array detector and a Finnigan AQA mass spectrometer. Analyses were carried out on a Superspher RP-18, 125×2 mm, 4 µm, column (Macherey-Nagel, Germany), protected by a guard column packed with the same material, and maintained at 40°C. Analyses were carried out employing electrospray ionization (ESI) at the positive ion mode, with acquisition set at collision energies of 12 and 70 eV, capillary voltage 4 kV, source voltage 45 V, detector voltage 650 V and probe temperature 400°C. Eluent (A) a nd MeOH, respectively. The flow rate was 0.33 mL min^{-1}, and the elution programme used was as follows: 0-2 min, 0% B; 2-52 min, 100% B; 52-60 min, 100% B.

Statistical analysis

All determinations were carried out at least in triplicate and values

were averaged and given along the standard deviation (S. D.). For all statistics, SigmaPlot™ 12.0 and Microsoft Excel™ 2010 were used.

Results and Discussion

Effect of solvent composition

Ethanol percentage in the solvents used varied from 28.5 to 85.5, a range that has been previously shown to provide high yield for grape seed extraction [3,19], but also grape pomace [9] and other plant material, including olive leaves (59%) [20], black currants [21], onion peels [22] and white grape seeds, peels and stems (57%) [23]. A hydroalcoholic solution of 57% was found to be the most effective for high polyphenol recovery, as this was manifested by estimating both Y_{TP} and A_{AR} (Table 1). Thus the extract obtained with 57% ethanol was chosen for the examination of the analytical polyphenolic composition.

Tentative identification of major phytochemicals

The principal compounds detected in the extract analyzed (Figure 1) belonged to flavonol and anthocyanin classes. In particular, three flavonol and three anthocyanin conjugates were tentatively identified, along with a p-coumaric acid derivative (Figure 2). Three other substances could not be assigned to any known grape constituent and their identification merits further investigation.

Compound (2) (Table 2) showed an ion at m/z=327, which was assigned to its molecular ion, after considering a Na$^+$ adduct (m/z=349) and a dehydration ion at m/z=309. The UV-vis spectrum was identical to the original p-coumaric acid standard. These data concurred for the identification of this compound as p-coumaroyl glucoside. Compound (4) displayed a molecular ion at m/z=479. Since anthocyanins are positively ionised at acidic pH, this represents the actual molecular mass [24]. Characteristic fragment indicating the aglycone (m/z=301) were also observed. These findings were consistent with the anthocyanin petunidin 3-O-glucoside [25]. Likewise, peak 5 gave a molecular ion at m/z=493, a diagnostic fragment of the loss of two methyl units (m/z=463), the aglycon ion (m/z=315), and the demethylated aglycone (m/z=301). This peak was assigned to malvidin 3-O-glucoside. Similarly, peak 10 that displayed a molecular ion at m/z=639 and the ion corresponding to the aglycone (m/z=331), was assigned to malvidin 3-O-p-coumaroyl glucoside [26]. Compounds (6) and (7) with corresponding molecular ions at m/z=479 and 465 were found to yield the same daughter ion (m/z=303), and corresponding Na$^+$ adducts at m/z=501 and 487. Compound (6) gave also a characteristic fragment at m/z=561, indicating the formation of a double adduct with CH_3COOH and Na$^+$. These compounds were identified as quercetin 3-O-glucuronide and quercetin 3-O-glucoside, respectively [27]. In a similar fashion, compound (8) displayed a molecular ion at m/z=479 and adducts with both Na$^+$ and CH_3COOH at m/z=561. Ions at m/z=509 and 501 were also assigned to adducts with MeOH and Na$^+$, respectively. The m/z=317 also yielded a MeOH adduct at m/z=347. Based on these data, this compound was tentatively identified as isorhamnetin 3-O-glucoside.

Conclusions

Red grape pomace is an industrial by-product with a wide

Solvent (% v/v EtOH)	Y_{TP} (mg GAE g^{-1} dpw)	A_{AR} (mM TRE g^{-1} dpw)
28.5	25.46 ± 2.40	2.15 ± 0.09
57.0	72.59 ± 2.21	2.67 ± 0.10
85.5	48.17 ± 4.07	1.89 ± 0.04

Table 1: Values determined for Y_{TP} and A_{AR} of the extracts obtained using aqueous solvents with varying amounts of EtOH.

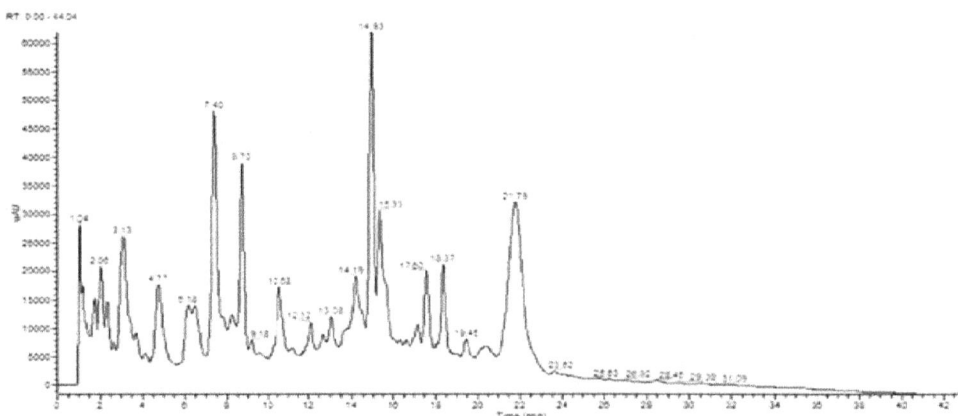

Figure 1: Chromatogram of the RGP extract monitored at 320 nm. Peak assignment according to retention time (Rt) is given in Table 2.

Peak No.	Rt (min)	λmax	[M]⁺ (m/z)	[M+H]⁺ (m/z)	Fragment ions (m/z)	Tentative identity
1	3.13	280, 296, 328(s)		-	385, 329, 203	Unknown
2	7.40	310		327	349, 309	p-Coumaroyl glycoside
3	8.70	277		675	409, 349, 327, 259, 231	Unknown
4	13.08	250, 342(s), 522	479		301	Petunidin 3-O-glucoside
5	14.19	250, 524	493		463, 331, 315, 301	Malvidin 3-O-glucoside
6	14.93	254, 354		479	561 [CH₃COOH+Na]⁺, 501 [Na]⁺, 303	Quercetin glucuronide
7	15.33	252, 356		465	487 [Na]⁺, 303	Quercetin glucoside
8	17.60	252, 354		479	531[MeOH+Na]⁺, 509 [479+MeOH]⁺, 501[Na]⁺, 347[317+MeOH]⁺, 317	Isorhamnetin glucoside
9	18.37	272		695	517, 387, 355, 195, 181, 163	Unknown
10	21.79	254, 534	639		331	Malvidin 3-O-p-coumaroyl glucoside

Table 2: Spectral characteristics and tentative identification of the major polyphenols detected in the RGP extract analysed.

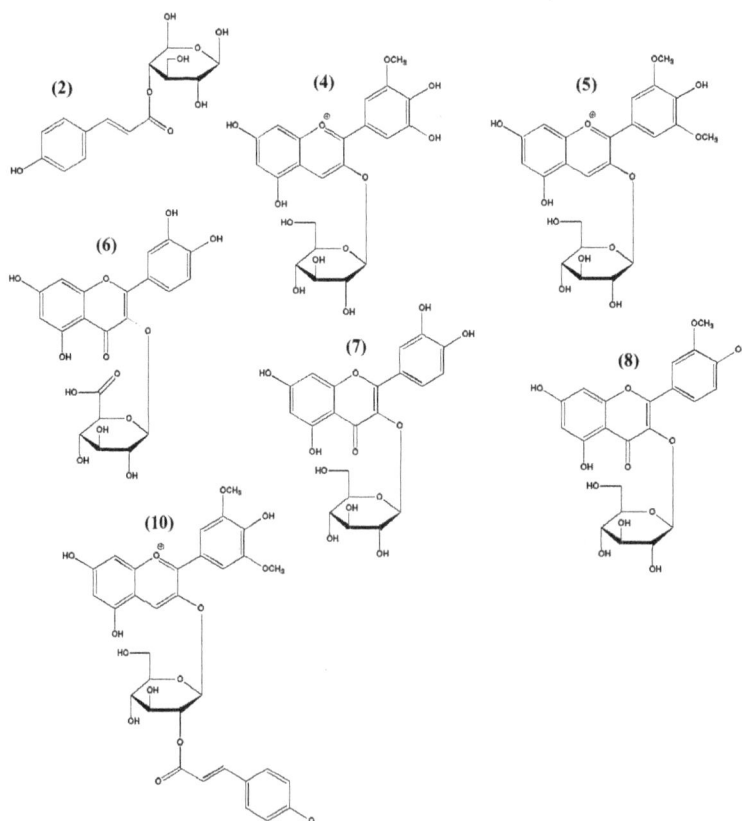

Figure 2: Chemical formulae of the major polyphenolic components tentatively identified in the RGP extract. Assignments are given in Table 2.

diversification, which depends on several factors, such as genetic (varietal) potential, treatment, post-disposal handling etc. Therefore, the examination of such waste materials from various sources might reveal the occurrence of a spectrum of substances. In the study presented herein, red grape pomace originating from the Greek native cultivar V. vinifera var. Agiorgitiko was efficiently extracted with 57% aqueous ethanol, to retrieve polyphenolic compounds. Ten principal polyphenols were detected and seven of them were tentatively identified on the basis of UV-*vis* and mass spectral data. The analyses revealed that the substances detected were a phenylpropanoid, derivative of *p*-coumaric acid, as well as anthocyanin pigments and flavonol glycoconjugates. Further research is needed to better illuminate the complex composition of this particular food industry by-product.

References

1. González-Paramás AM, Esteban-Ruano S, Santos-Buelga C, de Pascual-Teresa S, Rivas-Gonzalo JC (2004) Flavanol content and antioxidant activity in winery byproducts. Journal of Agricultural and Food Chemistry 52: 234-238.

2. Makris DP, Boskou G, Andrikopoulos NK (2007a) Polyphenolic content and in vitro antioxidant characteristics of wine industry and other agri-food solid waste extracts. Journal of Food Composition and Analysis 20: 125-132.

3. Yilmaz Y, Toledo RT (2006) Oxygen radical absorbance capacities of grape/wine industry byproducts and effect of solvent type on extraction of grape seed polyphenols. Journal of Food Composition and Analysis 19: 41-48.

4. Guendez R, Kallithraka S, Makris DP, Kefalas P (2005) An analytical survey of the polyphenols of seeds of varieties of grape (*Vitis vinifera* sp.) cultivated in Greece: implications for exploitation as a source of value-added phytochemicals. Phytochemical Analysis 16: 17-23.

5. Karvela E, Makris DP, Kalogeropoulos N, Karathanos VT, Kefalas P (2009) Factorial design optimization of grape (*Vitis vinifera*) seed polyphenol extraction. European Food Research and Technology 229: 731-743.

6. Louli V, Ragoussis N, Magoulas K (2004) Recovery of phenolic antioxidants from wine industry by-products. Bioresoure Technology 92: 201-208.

7. Kammerer D, Claus A, Schieber A, Carle A (2005) A novel process for the recovery of polyphenols from grape (*Vitis vinifera* L.) pomace. Journal of Food Science 70: C157-C163.

8. Pinelo M, Rubilar M, Jerez M, Sineiro J, Núñez JM (2005) Effect of solvent temperature, and solvent-to-solid ratio on the total phenolic content and antiradical activity of extracts from different components of grape pomace. Journal of Agricultural and Food Chemistry 53: 2111-2117.

9. Makris DP, Boskou G, Chiou A, Andrikopoulos NK (2008a) An investigation on factors affecting recovery of antioxidant phenolics and anthocyanins from red grape (*Vitis vinifera*) pomace employing water/ethanol-based solutions. American Journal of Food Technology 3: 164-173.

10. Fontana AR, Antoniolli A, Bottini R (2013) Grape pomace as a sustainable source of bioactive compounds: extraction, characterization, and biotechnological applications of phenolics. Journal of Agricultural and Food Chemistry 61: 8987-9003.

11. Mildner-Szkudlarz S, Zawirska-Wojtasiak R, Gośliński M (2010) Phenolic compounds from winemaking waste and its antioxidant activity towards oxidation of rapeseed oil. International Journal of Food Science and Technology 45: 2272-2280.

12. Rubilar M, Pinelo M, Shene C, Sineiro J, Nuñez MJ (2007) Separation and HPLC-MS identification of phenolic antioxidants from agricultural residues: almond hulls and grape pomace. Journal of Agricultural and Food Chemistry 55: 10101-10109.

13. Kammerer D, Claus A, Carle R, Schieber A (2004) Polyphenol screening of pomace from red and white grape varieties (*Vitis vinifera* L.) by HPLC-DAD-MS/MS. Journal of Agricultural and Food Chemistry, 52: 4360-4367.

14. Amico V, Chillemi R, Mangiafico S, Spatafora C, Tringali C (2008) Polyphenol

15. Anastasiadi M, Pratsinis H, Kletsas D, Skaltsounis AL, Haroutounian SA (2010) Bioactive non-coloured polyphenols content of grapes, wines and vinification by-products: evaluation of the antioxidant activities of their extracts. Food Research International 43: 805-813.

16. Jin ZM, Bi HQ, Liang NN, Duan CQ (2010) An extraction method for obtaining the maximum non-anthocyanin phenolics from grape berry skins. Analytical Letters 43: 776-785.

17. Pezzuto JM (2008) Grapes and human health: a perspective. Journal of Agricultural and Food Chemistry 58: 6777-6784.

18. Arnous A, Makris DP, Kefalas P (2002) Correlation of pigment and flavanol content with antioxidant properties in selected aged regional wines from Greece. Journal of Food Composition and Analysis 15: 655-665.

19. Shi J, Yu J, Pohorly J, Young JC, Bryan M, Wu Y (2003) Optimization of the extraction of polyphenols from grape seed meal by aqueous ethanol solution. Journal of Food, Agriculture and Environment 1: 42-47.

20. Japón-Luján R, Luque-Rodríguez JM, Luque de Castro MD (2006) Dynamic ultrasound-assisted extraction of oleuropein and related biophenols from olive leaves. Journal of Chromatography A 1108: 76-82.

21. Cacace JE, Mazza G (2003) Optimization of extraction of anthocyanins from black currants with aqueous ethanol. Journal of Food Science 68: 240-248.

22. Kiassos E, Mylonaki S, Makris DP, Kefalas P (2009) Implementation of response surface methodology to optimise extraction of onion (Allium cepa) solid waste phenolics. Innovative Food Science and Emerging Technologies 10: 246-252.

23. Makris DP, Boskou G, Andrikopoulos NK (2007b) Recovery of antioxidant phenolics from white vinification solid by-products employing water/ethanol mixtures. Bioresource Technology 98: 2963-2967.

24. Kefalas P, Makris DP (2006) Liquid chromatography-mass spectrometry techniques in flavonoid analysis: recent advances. In Antioxidant Plant Phenols: Sources, Structure- Activity Relationship, Current Trends in Analysis and Characterization. Research Signpost, Kerala, India.

25. Monrad JK, Howard LR, King JW, Srinivas K, Mauromoustakos A (2010) Subcritical solvent extraction of anthocyanins from dried red grape pomace. Journal of Agricultural and Food Chemistry 58: 2862-2868.

26. Sidani B, Makris DP (2011) Interactions of natural antioxidants with red grape pomace anthocyanins in a liquid model matrix: stability and copigmentation effects. Chemical Industry & Chemical Engineering Quarterly 17: 59-66.

27. Makris DP, Boskou G, Andrikopoulos NK, Kefalas P (2008b) Characterisation of certain major polyphenolic antioxidants in grape (*Vitis vinifera* cv. Roditis) stems by liquid chromatography-mass spectrometry. Food Research and Technology 226: 1075- 1079.

enriched fractions from Sicilian grape pomace: HPLC–DAD analysis and antioxidant activity. Bioresource Technology 99: 5960-5966.

18

Enhanced Separation Performance of Cellulose Acetate Membrane For Brackish Water Separation Using Modification of Additives and Thermal Annealing

TD Kusworo*, Budiyono, J Supriyadi and DC Hakika

Department of Chemical Engineering, Diponegoro University, Indonesia

Abstract

Membrane is an alternative technology of water treatment with filtration principle that is being widely developed and used for water treatment. The main objective of this study was to make an asymmetric membrane using cellulose acetate polymer and study the effect of additive and annealing treatment on the morphology structure and performance of cellulose acetate membranes in brackish water treatment. Asymmetric membranes for brackish water treatment were casted using a casting machine process from dope solutions containing cellulose acetates and acetone as a solvent. Membranes was prepared by phase inversion method with variation of polyethylene glycol (PEG) concentration of 1, 3, and 5wt% and the annealing temperature at 60oC and 70oC for 5, 10, and 15 seconds. Membrane characterization consists of calculation of membrane flux and rejection with brackish water as a feed, SEM and FTIR analysis. The research concluded that asymmetric cellulose acetate membrane can be made by dry/wet phase inversion method. The more added concentration of PEG will be resulted the larger pore of membrane. Meanwhile the higher temperature and the longer time of annealing treatment, the skin layer of membrane become denser. Membrane with the composition of 18 wt% cellulose acetate, 5 wt% PEG, 1 wt% distilled water, with heat treatment at temperature of 70oC for 15 seconds is obtained optimal performance: flux 6.52 $L.m^{-2}.h^{-1}.bar^{-1}$, 71% of total dissolved solid (TDS) rejection, 63.75% of turbidity rejection, 52.9% rejection of Ca^{2+}, and 41.9% rejection of Mg^{2+}.

Keywords: Asymmetric membrane; PEG; Cellulose acetate; Annealing; Brackish water

Introduction

Membrane is an alternative technology to the water treatment with filtration principle that is being widely developed [1,2]. Membrane technology is considered to have more advantageous to be applied water treatment because it does not require any necessary chemical additives such as the existing conventional technologies [1]. Additionally, membrane technology does not require a lot of equipment because the membrane component is portable so that the investment costs are lower than conventional systems. Research related to the condition of making membrane is still interesting. This is because of the many parameters that affect characteristics of the resulting membrane [3,4]. The combination of various parameters allows obtaining tailor-made membranes that are specific to a particular separation purposes [1].

A common type of polymer used in production of membranes are cellulose acetate. Manufacture of cellulose acetate membranes is performed using dry/wet phase inversion technique which polymer is converted from liquid phase to solid phase with the precipitation in a evaporation and immersion. For certain purposes, additives are often added to the dope solution. Kind of additive that is often used is polyethylene glycol (PEG).The manufacture of membranes with addition of PEG additive has the effect of increasing rate membrane permeation because PEG is known as a pore-forming organic material on the membrane [5-7]. Increasing amout of PEG can increase the porosity of the chitosan-cellulose composite membrane, which is shown through an increase in flux of the membrane [6-8]. PEG is a biocompatible compound, highly hydrophilic and anti-fouling [6].

In the membrane separation process, success of the separation process can be affected by the morphological structure of the membrane [7]. An optimal condition of the membrane performance is generally expressed by the magnitude of membrane permeability and selectivity of a particular chemical compound. The larger value of the permeability and selectivity of the membrane will have a better performance. But, in fact, on the membrane separation process will be found a common phenomenon that when the permeability of the membrane is high then selectivity will be low. Ismail et al. [5,8] stated that many factor affected the membrane performance such as: the type and concentration of polymer, solvent type, solvent evaporation time, additive concentration and shear rate. The membranes production with addition of PEG additive has the effect of increasing rate membrane permeation because PEG is known as a pore-forming organic material on the membrane [9].

The other important factor affected the membrane performance has heat treatment or annealing of membrane. A different of temperature of heat treatment can be resulted in different membrane performance. By the annealing processes, the resulting membrane will have a lower flux and higher selectivity than membrane without annealing [10]. Therefore, the main objective of the present study is to investigate the effect of PEG addition in dope solution and thermal annealing process on the performance of asymmetric cellulose acetate membranes for brackish water treatment. To our knowledge, there is no documentation on the use of PEG addition and thermal annealing

*Corresponding author: TD Kusworo, Department of Chemical Engineering, Diponegoro University, JlProf.Soedarto, SH.Semarang, 50239, Indonesia
E-mail: tutukjoko@yahoo.com

process to increase the performance of asymmetric cellulose acetate membrane for brackish water treatment.

Experimental

Materials

Materials used in the making of membranes are cellulose acetate from MKR Chemicals, 99.75% acetone from Mallinckrodt Chemicals, distilled water, PEG 4000 and brackish water from Demak.

Fabrication of asymmetric cellulose acetates

In this study, the dope solution consists of 18 wt % cellulose acetates, acetone, distilled water and PEG as additives with various concentration of 1% wt, 3% wt, 5% wt. The homogeneous dope solutions were prepared according to the following procedure; the cellulose acetate polymers were dispersed in to the solvent and stirred for 6 hours followed by the addition of a desired amount of PEG. The dope solution was agitated with a stirrer at least 6 hours to ensure complete dissolution of the polymer. A desired amount of distilled water was added to the homogenous solution. This dope solution was than agitated at high speed for at least 12 hours. After all the ingredients mix completely fit variable, then the dope solution allowed to stand for 1 hour to remove bubbles. Casting membrane using the method of phase inversion that is scored on a glass plate using a casting knife and allowed to correspond with the time variation of evaporation and then dipped into a coagulation bath containing distilled water in place for 1 day at room temperature. Defect on the membrane surface were repaired by a heat treatment method. Asymmetric membranes module after the air drying were dried in an oven at different temperatures (60 and 70°C). Furthermore, heat treatment for different durations (5, 10, and 15 seconds) at 60 and 70°C was also carried out. After the treatment the membranes were cooled down slowly to room temperature. The treated membrane after being subjected to different heat treatment methods were tested using a dead-end nanofiltration testing system. Subsequently membrane filtration cell is cut to size for the characterization of the flux and rejection.

Characterization of cellulose acetates membranes

In this study, the cellulose acetate membranes were tested in a brackish water from Demak using a dead-end filtration system. Figure 1 illustrated the schematic diagram of a testing apparatus to measure the flux and rejection values. Permeation experiments were carried out at room temperature. Before the permeability test, membrane first did compaction using distilled water for 30-45 minutes, hence the polymer

chains could arrange themselves. After the compaction process, distilled water was replaced with brackish water. Brackish water flux values was measured by measuring the volume of brackish water every 5 minutes. Determination of membrane rejection was performed by determining the concentration of total disolved solid (TDS), Ca^{2+}, Mg^{2+}, and brackish water turbidity before and after passing through the membrane. Determination of brackish water TDS was performed using a TDS meter, the analysis of brackish water turbidity was determined by turbidimeter, while the determination of Ca^{2+} and Mg^{2+} ion is using substitution and hardness titration. The flux was calculated using the equation as stated by Dasilva [11]:

$$J=V/(A.t.p) \qquad (1)$$

J=Flux, $L.m^{-2}.h^{-1}\ bar^{-1}$

V=Volume of Permeate, litre

A=Membrane Surface Area, m^2

t=Time, hour

p=Pressure, bar

A Scanning Electron Microscopy (SEM) was used to determine the asymmetric structure and the dimension of the asymmetric membranes. Membrane samples were fractured in liquid nitrogen. The membranes were mounted on an aluminium disk with double surface tape and then the sample holder was placed and evacuated in a sputter-coater with gold. Through this analysis, it can be seen the cross-sectional and the surface morphology of the membrane with a certain magnification. The changes in the chemical structure during the blending were identified using Fourier transform infrared spectroscopy (FTIR). The IR absorption spectra were measured at room temperature from 4000 to 500 cm^{-1} with a spectral resolution of 8 cm^{-1} and averaged over 16 scans. This test is done to ensure the presence of cellulose acetate and PEG on the membrane.

Results and Discussions

Effect of PEG additive concentration on the membrane performance

The permeation flux of the membranes was measured by dead-end filtration cell for measuring the membrane permeate flow rate per unit area per unit time. The effect of PEG additive is presented in Figures 2 and 3. Figure 2 shows that the more PEG additive added to the dope solution, the greater the flux membranes obtained. This increase in flux due to membrane pore formed by the addition of greater concentrations

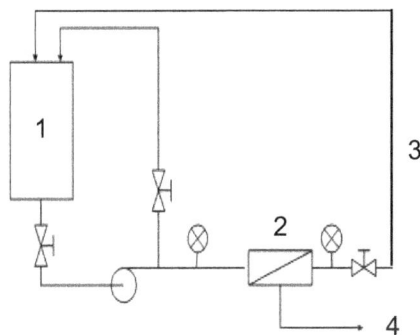

(1) Feed; (2) Membrane; (3) Retentate; (4) Permeate

Figure 1: Dead-end nanofiltration cell.

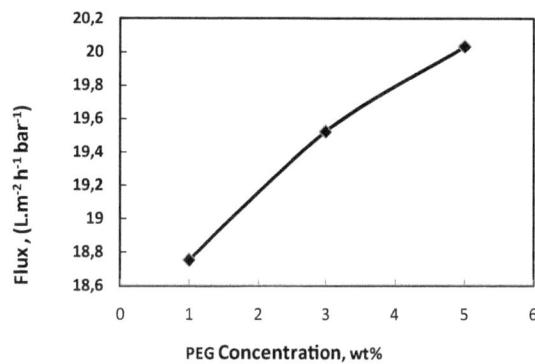

Figure 2: Effect of PEG additive concentration on membrane performance.

of PEG. PEG is an additive that has hydrophilic characteristic, thus increasing PEG concentration will lead to the formation of larger macrovoidon the pore structure of the membrane [8]. The larger the pore size of the membrane will increasethe flux value.

Performance of the membrane can also be determined by the value of membrane rejection. Membrane rejection was performed using a dead-end filtration cell, together with measurements of membrane flux. On rejection measurement, the indicators are based on the values of total dissolved solid (TDS), turbidity degree and level of Ca^{2+} and Mg^{2+} concentration before and after passing through the membrane. The effect of PEG additive on the performance of membrane rejection is shown in Figure 3. Based on Figure 3, it can be concluded that increasing the concentration of PEG additive will decrease membrane rejection values for TDS, turbidity (NTU), Ca^{2+} and Mg^{2+}. This phenomenon was because the more concentration of PEG is added to dope solution increases the number of pores or pore size that formed in membranes. PEG serves as porogen or pore-forming that are soluble in water, moreover, the PEG molecules will be diffused into the coagulation bath containing water and leaving pores in the matrix of cellulose acetate [12]. As a result, the increasing of the concentration of PEG will be increased the macrovoid in the membrane and will be caused a lot of species qualify and are not filtered as it passes through the membrane. Therefore, the membrane rejection will be decreased.

Effect of annealing treatment onthe membrane performance

Before applied in brackish water treatment, cellulose asetat membrane were given annealing treatment first at temperature of 60°C and 70°C with annealing time for 5,10, and 15 seconds. Then, the membrane flux measurements were performed using dead end cell filtration. The results of membrane flux measurements for the brackish water treatment with annealing treatment attemperature of 60°C and 70°C during 5,10, and 15 seconds are presented in Table 1.

Table 1 shows that the longer the heating time the membrane

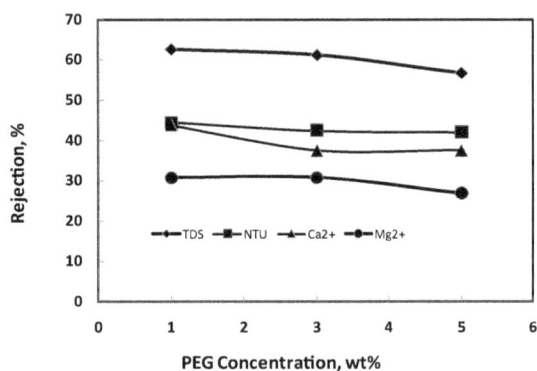

Figure 3: Effect of PEG additive concentration on the membrane rejection.

Temperature (°C)	Time (s)	Average flux (L.m⁻².h⁻¹.bar⁻¹)
60	5	18.17
	10	16.13
	15	12.05
70	5	14.15
	10	7.45
	15	6.52

Table 1: Effect of temperature and time of annealing treatment on the membrane performance.

flux decreased. This is because the heat treatment on the membrane will cause some membrane molecules that are more stable and more meetings and will be the narrowing of the pore membrane [13,14]. Therefore, the flux is decreased. As shown in Table 1, it appears that the increasing of the heating time and heating temperature was also affected the membrane flux, i.e the higher the heating temperature will decrease the value of the membrane flux. The higher the heating temperature of the membrane that forms the lining will be tight and smooth. Formed pore size also will be smaller, hence the flux value has decreased.

Rejection value of cellulose acetate membrane modified with annealing treatment attemperature 60°C and 70°C for annealing time of 5,10, and 15 seconds is determined by comparing the value of TDS brackish water, the degree of turbidity, and concentration of Ca^{2+} and Mg^{2+} ion before and after passing through the membrane. The results of cellulose acetate membrane rejection are presented in Figure 4. Figure 4 indicated that the higher temperature of annealing treatment given to membrane, the rejection value for TDS, turbidity, Ca^{2+} and Mg^{2+} ions increased. Increasing the value of this rejection also applies along with the length time of annealing treatment given. The annealing treatment given to the membrane will be improved the membrane performance because the arrangement of molecules on the membrane are more stable. The higher annealing temperature and the longer annealing treatment time on membrane, the membrane pores will shrinkage moreover the membrane surface layer becomes denser [13-15]. As a result, the membrane pore is narrowed which then increases the value of the membrane rejection.

Effect of PEG additive concentration on the characterization of membrane by FTIR analysis

Figures 5 and 6 were presented the FTIR spectra of asymmetric cellulose acetate membranes. Analysis of FTIR (Fourier Transform Infrared) on the membrane is used to determine the functional groups present in membrane. The effect of PEG concentration on the spectra of FTIR is depicted in Figure 5. Figure 5 illustrates that cellulose acetate membranes having functional groups of chemical compounds with wavelengths shown in Table 2.

Figure 5 showed that the cellulose acetate membrane has a group of -OH,-C=O,-CH₃,-COOH, C - C, and-CH. In Table 2 it can be seen the shift of the wavelength from chemical compounds in cellulose acetate membranes with PEG concentration of 1% and 5 %wt. The wavelength shift indicated that the PEG concentration in dope solution affects the membrane morphology structure. PEG as additive in cellulose acetate membrane is indicated by the presence of functional groups C=O and the re-unit $-CH_2-CH_2O-$. In cellulose acetate membrane with PEG concentration of 5 wt%, for wave numbers 1381.03 and 1435.04 cm⁻¹ showed greater extents than cellulose acetate membrane with PEG concentration of 1 wt% for the same functional group, in this case for the wave number 1381.03 cm⁻¹. The wider absorption area showed that increasingaddition of PEG concentration, the more the re-unit of $-CH_2-CH_2O-$ in membrane.

Effect of annealing treatment on the characterization of membrane byFTIR analysis

Figure 6 shows the FTIR characterization results on the cellulose acetate membrane with annealing treatment attemperature of 60°C and 70°C and annealing time for 5 seconds and 15 seconds.The functional group of chemical compounds with wavelength is presented Table 3.

Figure 6 indicated that the cellulose acetate membrane has a group

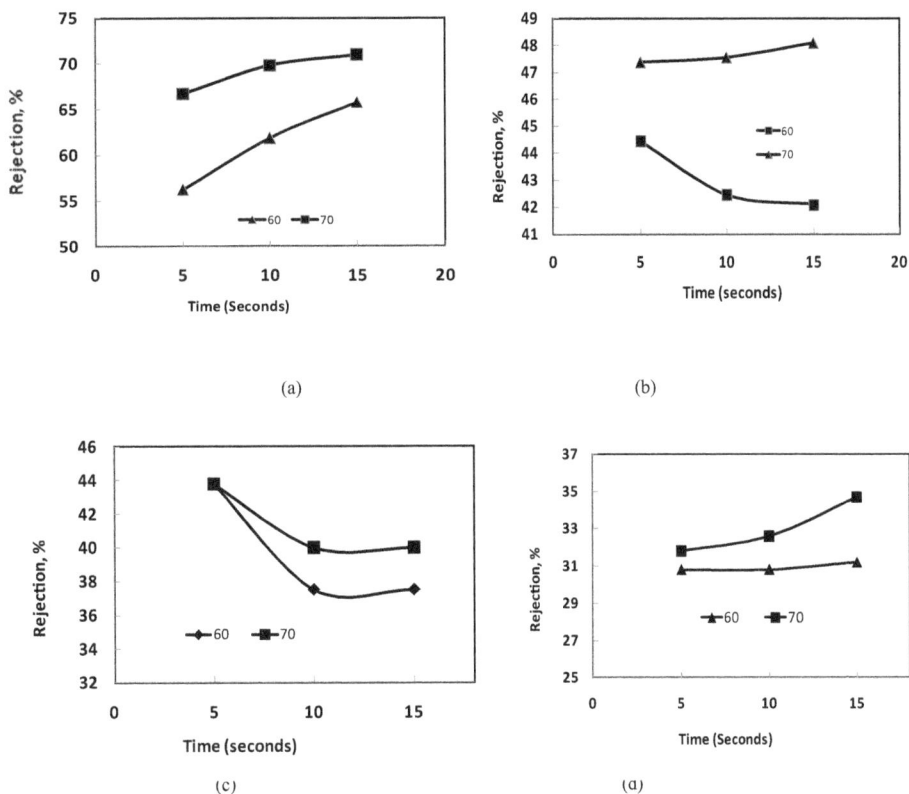

Figure 4: Effect of annealing time and temperature on the membrane rejection for: (a) TDS (b) turbidity (c) Ca²⁺ (d) Mg²⁺.

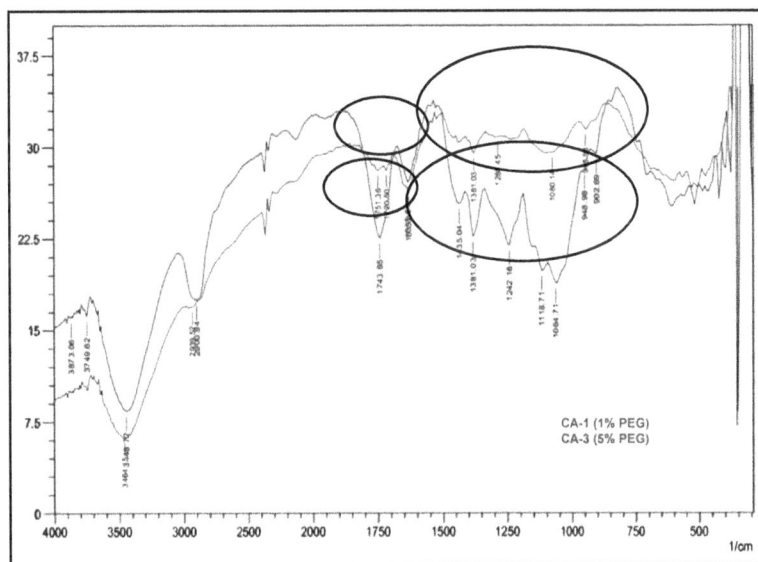

Figure 5: FTIR result for cellulose acetate membrane with PEG concentration of 1% and 5%wt.

-OH,-C=O,-CH₃, -COOH, C-C, and-CH. Meanwhile, in Table 3 was shown the shift of the wavelength from the chemical compounds contained in the cellulose acetate membrane with annealing treatment at temperature of 60°C and 70°C and annealing time for 5 seconds and 15 seconds. Post treatment was given on cellulose acetate membrane after the membrane is formed. Based on the functional groups contained in the cellulose acetate membrane, it can be observed that the functional group of -OH can be represented the water content in membrane. On cellulose acetate membrane with annealing treatment at temperature of 70°C, for wave numbers 3464.15 cm⁻¹ has smaller area than cellulose acetate membranes with annealing treatment at temperature of 60°C, in wave number 3464.15 cm⁻¹, respectively.

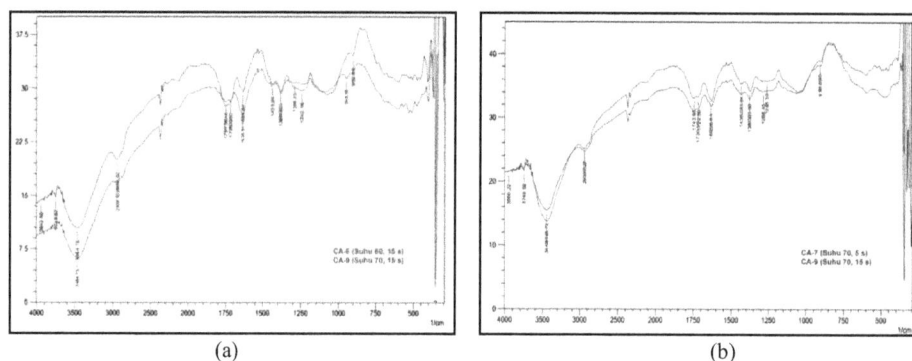

(a)   (b)

Figure 6: FTIR result for cellulose acetate membrane with: (a) annealing treatment attemperature 60 °C and 70 °C, and (b) annealing treatment for 5 seconds and 15 seconds.

No	Chemical Compound	Wavelength (cm⁻¹)	
		PEG 1 wt%	PEG 5 wt%
1	-OH	3464.15	3448.72
2	-CH	2939.52	2900.94
3	C=O	1635.64; 1720.5; 1751.36	1635.64; 1743.65
4	-CH₃	1381.03	1381.03; 1435.04
5	-COOH	1288.45	1242.16
6	C-C	948.98	902.69; 948.98

Table 2: Functional groups in cellulose acetate membranes with PEG concentration of 1% and 5% wt.

No.	Chemical compound	Wavelength (cm⁻¹)			
		60°C	70°C	5 detik	15 detik
1.	- OH	3464.15; 3749.62	3448.72	3448.72	3448.72
2.	- CH	2939.52	2900.94	2939.52	2939.52
3.	C=O	1635.64; 1720.5	1635.64; 1720.5	1635.64; 1743.65	1635.64; 1720.5
4.	CH3	1381.03	1381.03; 1435.04	1381.03; 1435.04	1381.03; 1435.04
5.	- COOH	1288.45	1242.16	1265.3	1288.45
6.	C – C	948.98	902.69	910.4	910.4

Table 3: Functional groups in cellulose acetate membranes with effect of temperature and time of annealing treatment.

While on cellulose acetate membrane with annealing time for 15 seconds, to wave number 3448.72 cm⁻¹ also has smaller extents than cellulose acetate membrane with annealing time for 5 seconds, which is the wave number 3448.72 cm⁻¹, respectively. The absorbtion area and extent showed indicate the water content in membrane and the intensity of the interaction between water molecules. Therefore, the smaller the absorption area showed that the interaction between water molecules in the membrane become smaller, which means the lower water content in the membrane [15].

Effect of PEG additive concentration on the characterization of membrane by SEM analysis

The performance of asymmetric cellulose acetate membranes depends on many factors.The structure and geometrical characteristics of the produced asymmetric cellulose acetate membranes were studied by scanning electron microscopy (SEM). Figures 7 and 8 shown the cross section near outer surface of the membranes at different PEG. As shown in Figure 7, all the structures of membrane consisted of a dense skin layer supported by a spongy porous substructure with small macrovoids. Generally, production membrane by coagulation process typically generates microporous structure containing macrovoids structure [16].

Figures 7 and 8 was also showed that the cellulose acetate membrane

is formed asymmetric structure membrane, where increasing PEG concentration also increasing the number and uniformity of pore membrane. In this case, PEG as an additive initially filling the matrix of prepared cellulose acetate membranes. Furthermore, in the process of dissolution with non solvent, additives together with solvent will dissolveinto non-solvent, leaving a cavity or pore in the membrane. Consequently, the pore become larger and uniformly [8,9]. Based on cross-sectional morphology of membrane it appears that the pores of membrane with PEG concentration of 5 wt% are more evenly than in membrane with PEG concentration of 1% wt. This shows the function of PEG as a pore-forming and increase the porosity of the cellulose acetate membrane.

Effect of annealing treatment on the characterization of membrane by SEM analysis

The effect of annealing on the membrane morphology is depicted in Figures 9-12. The membranes subjected to annealingpost-treatmentat 60°C and 70°C, with annealing time for 5 seconds and 15 seconds. As can be seen from the SEM pictures of all treatment temperature of membranes contained the porous structure in the cross-section of asymmetric membrane. The porous structures became increasingly denser with increasing the heat-treatment temperatures. Moreover, the

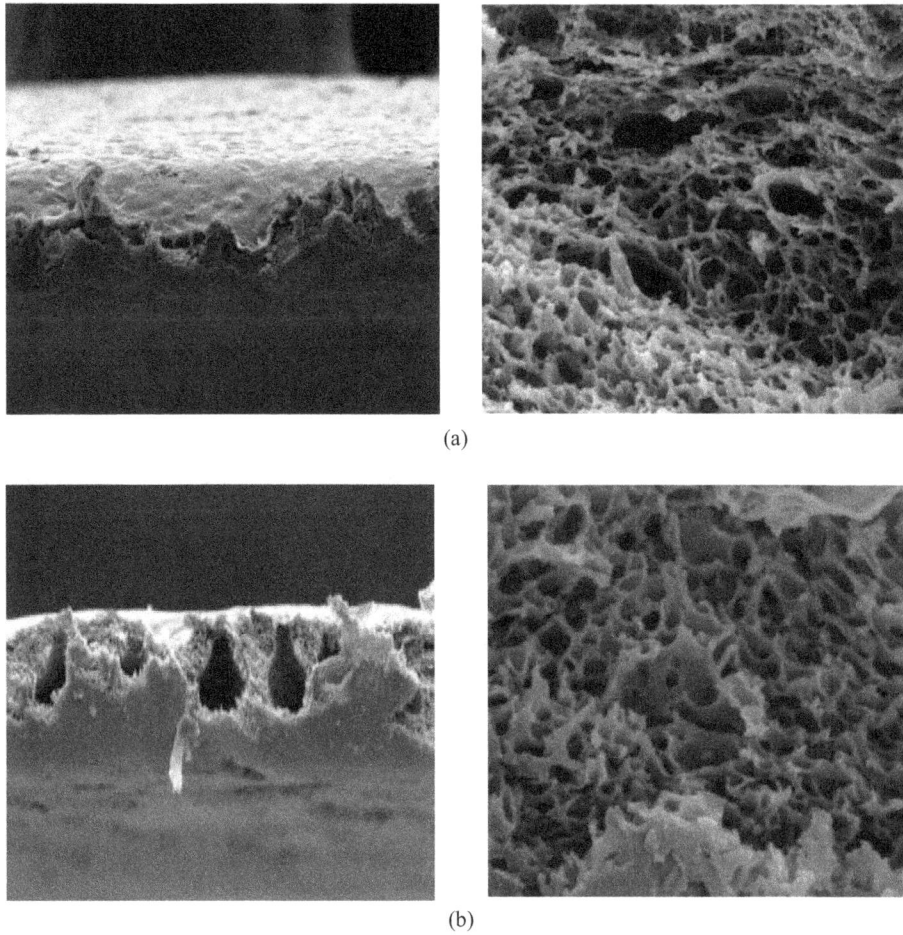

(a)

(b)

Figure 7: The cross-section morphology of cellulose acetate membrane with PEG concentration: (a) 1% wtand (b) 5%wt.

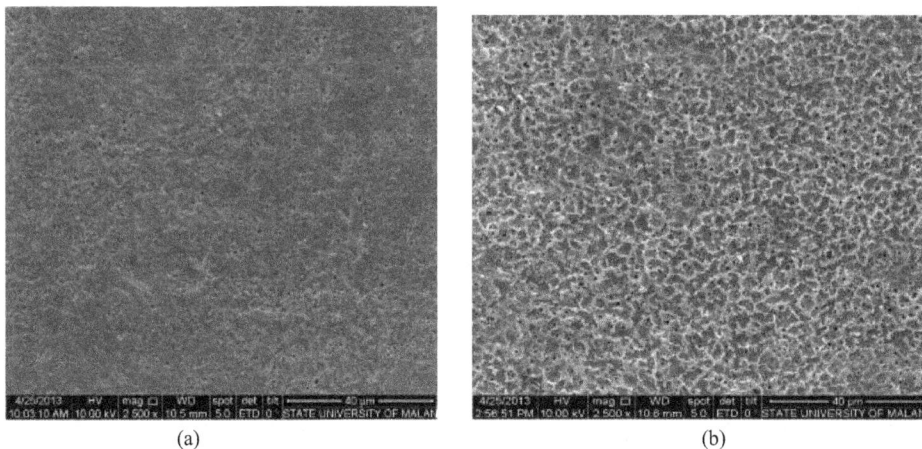

(a)   (b)

Figure 8: The surface morphology of cellulose acetate membrane with PEG concentration: (a) 1% wt and (b) 5%wt.

SEM pictures also reveal that the dense skin layer formed when the heat-treatment temperature was higher than 70°C. Furthermore because of the thermal effect the packed chains in the polymer membrane denser and the packed structure in the skin layer provide a high degree of size and shape discrimination between the particles. Therefore, after heat treatment asymmetric membranes became more compact in the outer skin and substructure as compared to the untreated membranes. As result the flux for treated membrane was lower than untreated membrane.This is consistent with the reduction of flux and increasing the rejection as shown in Table 1 and Figure 4.

(a)

(b)

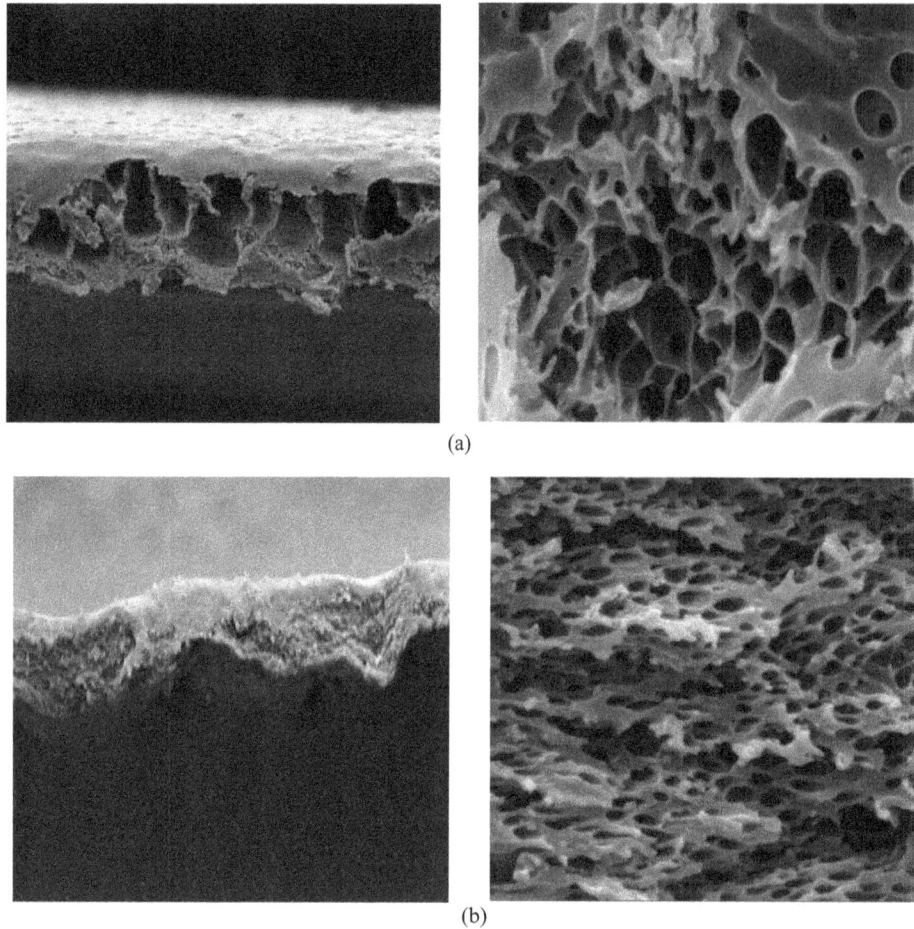

Figure 9: The cross-section morphology of cellulose acetate membrane with annealing treatment attemperature: (a) 60°C dan (b) 70°C

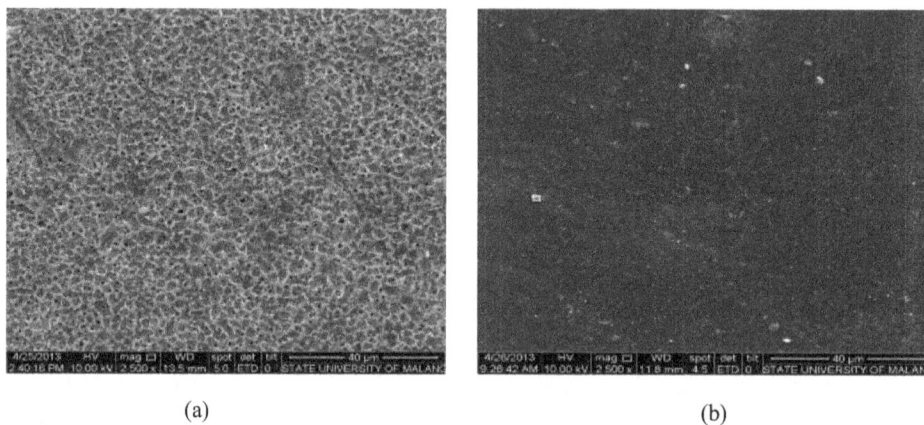

(a)   (b)

Figure 10: The surface morphology of cellulose acetate membrane with annealing treatment attemperature: (a) 60°C dan (b) 70°C.

Furthermore the effect of annealing time at 5-15 seconds on the morphology of cellulose acetate membranes is shown in Figures 11-12. Figure 11 and 12 show that the surface morphology of the membrane with annealing temperature of 70 °C and annealing time for 15 seconds has a smoother surface than the membrane by annealing at 60°C and the annealing time for 5 seconds. Moreover, the pores or cavities formed in membrane are smaller and denser.Annealing treatment on membrane causes an adjustment of the movement of the polymer chains. When cellulose acetate membrane were annealed, the movement of the molecules from polymer chains become easier and affectthe morphology structure of membrane. In addition, annealing treatment also decreases the free volume formed in membrane, due to

(a)

(b)

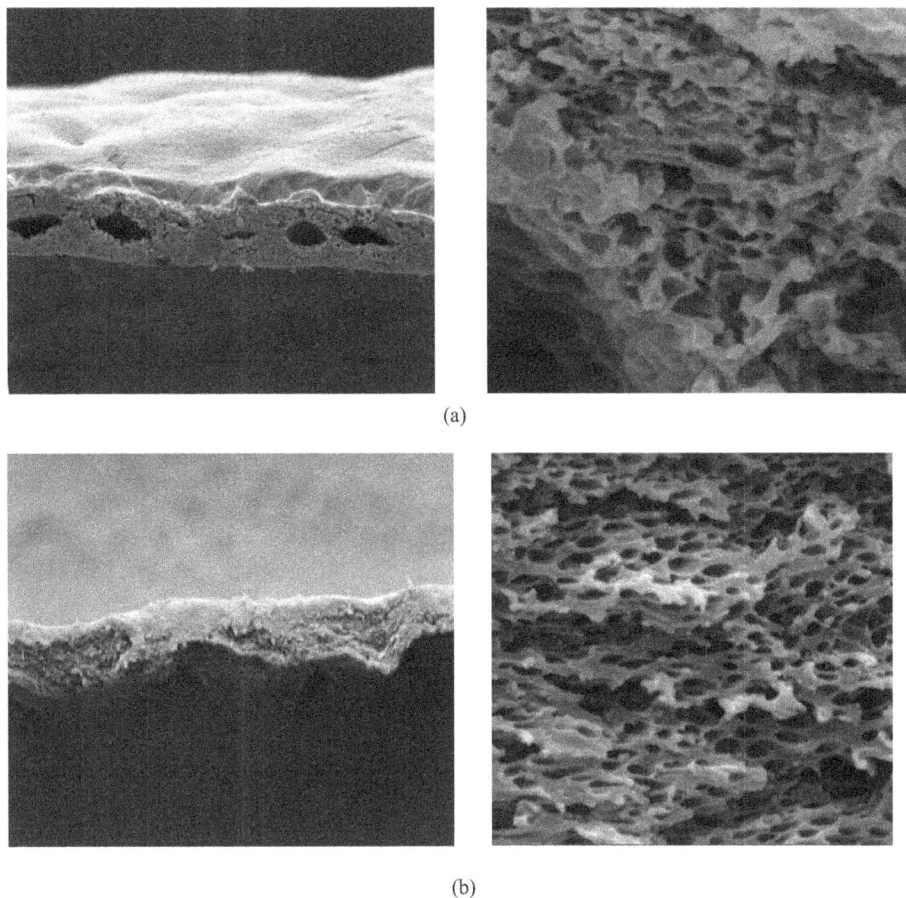

Figure 11: The cross-section morphology of cellulose acetate membrane with annealing time for: (a) 5 seconds and (b) 15 seconds.

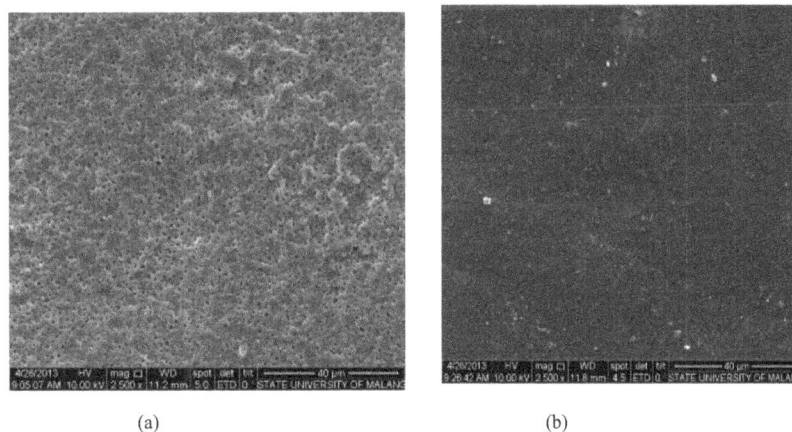

(a)   (b)

Figure 12: The surface morphology of cellulose acetate membrane with annealing time for: (a) 5 seconds and (b) 15 seconds.

the increasing of molecular movement in membrane [13-14,17]. The fewer free volume in membrane results in the smaller pores or voids are formed, so that the membrane becomes denser.

Conclusions

Asymmetric cellulose acetate membranes were fabricated at different PEG concentration in the dope solution. The effects of PEG concentration on the membrane morphology and filtration performance were analyzed. An attempt to improve the performance of the membranes also has been done by thermal annealing. Based on the experimental results and analysis, the following conclusions can be made.

i. Asymmetric cellulose acetate membranes can be produced using dry-wet phase inversion and the PEG can be used as additive to repair the membrane pore formation

ii. Both SEM and filtration test suggested that the annealing treatment can be used as a method to modify surface of membrane. The membrane surface layers become denser and smoother with increasing the time and temperature annealing.

iii. It is found that the combinations of PEG in dope solution and heat treatment can effectively improve the performance of cellulose acetate membrane with the flux 6.52 $L.m^{-2}.h^{-1}.bar^{-1}$, TDS rejection 71%, turbidity rejection 63.75%, Ca^{2+} ion rejection at 52.9%, and Mg^{2+} ion rejection was 41.9%, respectively.

Acknowledgements

The authors would like to the staff of Laboratory of Waste Treatment and Laboratory of Mer-C for its contributions as the research site

References

1. Baker RW (2004) Membrane Technology and Applications. Chichester: John Wiley & Sons, Ltd (2nd ed).

2. Mulder M (1996) Basic Principle of Membrane Technology. London: Kluwer Academic Publ.

3. Kusworo TD, Ismail AF, Mustafa A, Matsuura T (2008) Dependence of membrane morphology and performance on preparation conditions: The shear rate effect in membrane casting. Separation and Purification Technology 61: 249-257.

4. Ismail AF, Kusworo TD, Mustafa A (2008) Enhanced gas permeation performance of polyethersulfone mixed matrix hollow fiber membranes using novel Dynasylan Ameo silane agent. Journal of Membrane Science 319: 306-312.

5. Ismail AF, Hassan AR (2006) Formation and Characterizayion of Asymetric Nanofiltration Membrane : Effect of Shear Rate and Polymer Concentration. Journal of Membrane Science 270: 57-72.

6. Yang L, Hsiao WW, Chen P (2002) Chitosan-cellulose composite membrane for affinity purification of biopolymers and immunoadsorption. Journal of Membrane Science 197: 185-197.

7. Tsai HA, Ma LC, Yuan F, Lee KR, Lai JY (2008) Investigation of Post-Treatment Effect on Morphology and Pervaporation Performance of PEG Added PAN Hollow Fiber Membranes. Desalination 234: 232–243.

8. Ismail AF, Hassan AR, Ng BC (2002) Effect of Shear Rate on the Performance of Nanofiltration Membrane for Water Desalination. Songklanakarin Journal Science Technology Membrane Sci. & Tech. 24: 879-889.

9. Saljoughi E, Sadrzadeh M, Mohammadi T (2009) Effect of Preparation Variables on Morphology and Pure Water Permeation Flux Through Asymmetric Cellulose Acetate Membranes. Journal of Membrane Science 326: 627 - 634.

10. Kim IC, Yun HG, Lee KH (2002) Preparation of Asymmetric Polyacrylonitrile Membrane with Small Pore Size by Phase Inversion and Post-Treatment Process. Journal of Membrane Science 199: 75-84.

11. Dasilva MSF (2007) Polyamide and Polyetherimide Organic Solvent Nanofiltration Membranes. Dissertation. University Of Nova De Lisboa.

12. Chou WL, Yu DG, Yang MC, Jou CH (2007) Effect of Molecular Weight and Concentration of PEG Additives on Morphology and Permeation Performance of Cellulose Acetate, Separation and Purification Technology 57: 209-219.

13. Kawakami H, Mikawa M, Nagaoka S (1996) Gas transport properties in thermally cured aromatic polyimide membranes, Journal of Membrane Science 118: 223-230.

14. Ismail AF, Lorna W (2003) Suppression of Plasticization in Polysulfone Membranes for Gas Separation by Heat-treatment Technique. Separation and Purification Technology 30: 37-46.

15. Murphy D, De Pinho MN (1995) An ATR-FTIR Study of Water in Cellulose Acetate Membranes Prepared by Phase Inversion. Journal of Membrane Science 106: 245 - 257.

16. Chung TS, Teoh SK, Hu X (1997) Formation of Ultrathin High-Performance Polyethersulfone Hollow Fiber Membranes. Journal of Membrane Science 133: 161-175

17. Han MJ, Bhattacharyya D (1995) Thermal Annealing Effect on Cellulose Acetate Reverse Osmosis Membrane Structure. Desalination 101: 195-200.

19

Improving Anaerobic Methane Production from Ammonium-rich Piggery Waste in a Zeolite-fixed Bioreactor and Evaluation of Ammonium Adsorbed on Zeolite A-3 as Fertilizer

Cang Yu[1], Dawei Li[1], Qinghong Wang[2], Zhenya Zhang[1,*] and Yingnan Yang[1,*]

[1]Graduate School of Life and Environmental Science, University of Tsukuba, 1-1-1 Tennodai, Tsukuba, Ibaraki 305-8572, Japan
[2]State Key Laboratory of Heavy Oil Processing, China University of Petroleum, Changping, Beijing 102249, China

Abstract

To mitigate ammonia inhibition and enhance methane production, a zeolite-fixed bioreactor was developed for anaerobic digestion of ammonium-rich piggery wastes. Ammonium adsorption on zeolite A-3 fitted with the pseudo-second-order kinetic model and can be described by both Langmuir and Freundlich isotherms. Desorption of ammonium from saturated zeolite fits the first-order reversible reaction kinetic. The zeolite-fixed bioreactor with zeolite loading rate of 10 g l-1 showed the shortest startup period of 13 days and achieved the highest methane yield of 354.2 ml g-1-VS and the largest COD removal rate of 75.37%. Due to the effective mitigation of ammonia inhibition and enhancement of methane production, zeolite-fixed bioreactor is a good option for practical anaerobic digestion of ammonium-rich piggery wastes. Ammonium saturated zeolite can be directly used as fertilizer to decrease annual production of nitrogen fertilizer. Besides, regeneration of zeolite using Na2SO4 solution also obtained a (NH4)2SO4 by-product, which is nice nitrogenous fertilizer.

Keywords: Anaerobic digestion; Piggery wastes; Ammonium inhibition; Zeolite-fixed bioreactor

Introduction

In the past decades, anaerobic digestion of piggery wastes has attracted considerable attention because of the bioenergy recovery in the form of methane and mitigation of environment pollution [1]. However, digestion of pure piggery wastes has been observed to be unsuccessful, due to the inhibition of ammonia produced during biodegradation of nitrogenous compounds such as proteins and amino acids [2,3]. Although ammonia is an essential nutrient for growth of microorganisms [4], its undissociated form at high concentration has potential toxicity to methanogens [2]. Hobson and Shaw [5] reported that ammonium concentration of 2500 mg NH_4^+-N l-1 resulted to partial inhibition of methane production, while a complete failure of methanogenesis occurred whe n the concentration up to 3300 mg NH_4^+-N l-1. Consequently, to improve methane production from ammonium-rich piggery wastes, it is necessary to mitigate ammonia inhibition in the anaerobic digestion process using effective techniques.

Many physicochemical and biological methods have been employed for alleviating ammonia inhibition, such as air stripping [6], adsorption [7], chemical precipitation [8], microorganisms acclimation [9] and co-digestion [10]. Among these methods, adsorption has drawn more attention because of its in-situ ammonia removal, easy operation, high safety and low cost. Comparing with activated carbon [11], fly ash [12] and activated alumina [13], zeolite is the most promising adsorbent for ammonia removal [14] owing to its porous structure, biochemical stability and abundance on the earth. On the other hand, zeolite seems to be a potential support material for the immobilization of microorganisms as a porous surface. These characteristics make zeolite a promising option for counteracting ammonia inhibition in the anaerobic digestion of ammonium-rich piggery wastes.

In recent years, effects of variety, particle size, doses and dosage procedure of zeolite addition on anaerobic digestion of piggery wastes have been investigated [15-17]. Kotsopoulos et al. [18] showed that adding natural zeolite increased methane production from piggery wastes by reducing the toxicity of ammonia and regulating the C/N

(carbon/nitrogen) ratio through ammonia adsorption. Zeolite addition in anaerobic digestion of piggery wastes achieved the maximum ammonia removal at a dosage of 0.10 g-zeolite g-1-VSS, regardless of particles sized [16]. Milán et al. [15] found that addition of natural zeolite (doses: 2-4 g l-1) contributed to enhance the anaerobic digestion of piggery wastes by reducing the inhibitory effect of ammonia, but inhibition could not be overcome at doses higher than 6 g l-1. Continuous anaerobic digestion of piggery wastes in terms of chemical oxygen demand (COD) removal efficiency and methane production was effectively promoted by addition of natural zeolite on a daily basis [19]. According to these previous studies, addition of zeolite at an appropriate dosage could effectively mitigate ammonia inhibition thereby enhance the methane production from piggery wastes. However, the enhanced methane production was only attributed to ammonium removal by zeolite, neither the immobilization of microorganism nor the fixed mode of zeolite for mitigating ammonia inhibition were investigated in all of these studies. On the other hand, ammonium desorption by using brine solution [20] is of great significance for nitrogen recovery and sustainable utilization of zeolite in the anaerobic digestion of ammonium-rich piggery wastes. Nevertheless, when using zeolite as an additive to migrate ammonia inhibition in the anaerobic digestion of

*Corresponding author: Yingnan Yang, Graduate School of Life and Environmental Science, University of Tsukuba, 1-1-1 Tennodai, Tsukuba, Ibaraki 305-8572, Japan
E-mail: yo.innan.fu@u.tsukuba.ac.jp

Zhenya Zhang, Graduate School of Life and Environmental Science, University of Tsukuba, 1-1-1 Tennodai, Tsukuba, Ibaraki 305-8572, Japan
E-mail: zhang.zhenya.fu@u.tsukuba.ac.jp

piggery wastes, the ammonium desorption from zeolite had never been concerned by the previous researchers.

In this work, a zeolite-fixed bioreactor with advantages of ammonia adsorption and desorption of the adsorbed ammonium as fertilizer for future using and microorganism immobilization was developed for the anaerobic digestion of ammonium-rich piggery wastes. The purposes of this study are to investigate (1) the adsorption mechanism of ammonium on zeolite A-3 by kinetic and isotherm analyses; (2) the desorption efficiency of ammonium from zeolite A-3 in sodium sulfate solution; (3) the performance of zeolite-fixed bioreactor for ammonia removal and microorganism immobilization during anaerobic digestion of ammonium-rich piggery wastes.

Materials and Methods

Piggery wastes and seed sludge

Ammonium-rich piggery wastes used in the experiment was stale manure that had been kept at room temperature for almost two years after it had been obtained from a pig farm located in Tokyo. The stale manure compared with fresh piggery waste has a higher concentration of ammonium which can reach levels of up to 22,310 mg l^{-1}. The piggery waste was inoculated with 25% sludge (w/w) after diluted with tap water and pH adjustment with HCl. General characteristics of the diluted substrate were: COD: 76700 mg l^{-1}, total nitrogen (TN): 9400 mg l^{-1}, total solid (TS): 35000 mg l^{-1}, volatile solid (VS): 27725 mg l^{-1}, NH_4^+-N: 3770 mg l^{-1} and pH: 7.2.

The digested sludge collected from a municipal wastewater treatment plant in Ibaraki, Japan was used as seed sludge. After it was collected, the digested sludge was storage under 4°C in a refrigerator. Before used as inoculums, 900 ml digested sludge was cultured by putting into a fermenter bottle (1000 ml). After two days, 2 g raw piggery wastes was added to this reactor every day until the methane concentration reached 80% approximately. The cultivation of methanogens was carried out at 35°C for 7 days. The characteristics of seed sludge were: COD: 6500 mg l^{-1}, TN: 5489 mg l^{-1}, TS: 9850 mg l^{-1}, VS: 7415 mg l^{-1}, NH_4^+-N: 1547 mg l^{-1}, pH: 7.1.

Zeolite

The artificial zeolite A-3 used for ammonium adsorption and microorganism immobilization in the experiments was provided by Wako Pure Chemical Industries, Ltd. It has the following characteristics: pore diameter (Å): 3, Particle size (mm): 2.36-4.75, absorbable molecule: H_2O, NH_3, He, unabsorbable molecule: CH_4, CO_2, C_2H_2, O_2, H_2S, C_2H_5OH, Water absorbing capacity (wt %): 20, General formula: $(0.4 K + 0.6Na)_2O \cdot Al_2O_3 \cdot 2SiO_2$.

Ammonium adsorption experiment

The experiments of ammonium adsorption on zeolite A-3 were carried out in batch mode. For the ammonium nitrogen adsorption experiments and analysis, ammonium solution with a certain concentration ranging from 1000 to 5000 mg l^{-1} was prepared immediately by dissolving NH_4Cl in deionized water. Zeolite A-3 was added into 50 ml NH_4Cl solution at a loading rate of 10 g l^{-1} in a triangular flask (100 ml). Then, continuously shaking (100 rpm) of the triangular flasks were conducted in a constant temperature shaker with water bath at 35°C for 24 h.

Ammonium desorption experiment

For nitrogen recovery in the form of ammonium sulfate $((NH_4)_2SO_4)$

which is a nice nitrogenous fertilizer, desorption of ammonium from saturated zeolite A-3 were performed in sodium sulfate solution. According to the ion equivalent exchange principle: $2NH_4^+$-zeolite + Na_2SO_4 = $2Na^+$-zeolite + $(NH_4)_2SO_4$, the calculated concentration of Na_2SO_4 solution was 7.1 mol l^{-1} for the ammonium desorption from 0.5 g saturated zeolite A-3 (adsorbed NH_4^+-N: 20 mg). In the ammonium desorption experiment, 0.5 g of saturated zeolite was added into 100 ml as prepared Na_2SO_4 aqueous solution in a 200 ml triangular flask. Then, continuously shaking (100 rpm) of the triangular flasks were carried out in a shaker with water bath at 25°C for 24 h.

Anaerobic digestion experiment

In the previous experiment, we found that both the methane yield and COD removal in the zeolite-fixed bioreactor are much higher than those in the bioreactor with sunken zeolite at the bottom. When zeolite was dropped into the bottom of the reactor, it is difficult to contact with the ammonium in the liquid, because the swine waste would cover on the surface of the adsorbents. In zeolite-fixed bioreactor, the zeolite was suspended in the upper layer of the digested liquid, where ammonium could easily be trapped by the adsorbent. Therefore, in this present study, the zeolite was contained in a porous bag and suspended in the digested liquid, rather than directly dispersed in the bioreactor.

A number of Duran bottles (300 ml, SIBATA) with silicon rubbers were used as bioreactors in this study. The methane fermentation experiments were performed in two groups of bioreactors: zeolite-fixed bioreactors and bioreactors without zeolite as the control. The zeolite-fixed bioreactor was developed by hanging zeolite A-3 fixed in a porous nylon bag (pore diameter: 3 mm) in the Duran bottle. The schematic of zeolite-fixed bioreactor was shown in Figure 1. In the fermentation experiments, 200 ml of diluted swine waste including 25% (w/w) digested sludge was added into each bioreactor. After that, nitrogen flush was used to keep an anaerobic condition in the bioreactors. Then, the methane fermentation of piggery wastes was carried out in a batch mode at 35°C for 33 days. The biogas was collected using 50 ml plastic syringes, and the volume was read directly using the scale on the syringe. Each group of experiments was performed in duplicate.

Analytical methods

The gas composition was detected by a gas chromatography (GC-8A, SHIMAZU, Japan) using a machine equipped with a thermal conductivity detector (80°C) and a Porapak-Q column (60°C). Nitrogen was used as the carrier gas. COD, TS, VS, and TN were detected according to standard methods [21], and pH was determined using a pH meter (TES 1380). The amount of ammonium nitrogen was measured by an ion meter (Ti 9001, Toyo Chemical Laboratories Co., Ltd.). The activity of microorganisms was indicated by ATP analysis using a BacTiter-Glo™ Microbial Cell Viability Assay (Promega, USA). Morphological features of microorganisms immobilized on the zeolite after anaerobic digestion was observed using a scanning electron microscope (SEM) (JSM-6330F, JEOL, Japan).

Results and Discussion

Ammonium adsorption on zeolite A-3

Adsorption kinetic analyses: Prior to batch adsorption equilibrium studies, it is essential to confirm the equilibrium contact time required for the ammonium adsorption. Adsorption kinetic model is required for surveying the mechanism of adsorption. Several models have been utilized for the adsorption kinetic analyses. The most well-known models are Lagergren's pseudo-first-order and Ho's pseudo-second-

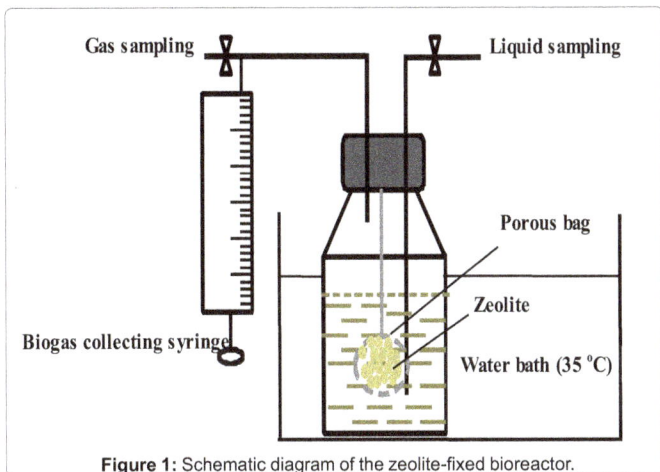

Figure 1: Schematic diagram of the zeolite-fixed bioreactor.

order. In order to assess the adsorption process of ammonium on the zeolite A-3, the above two models were applied to analyze the obtained experimental data under initial ammonium concentrations of 5000 mg l^{-1}, adsorbent loading rate of 10 g l^{-1} and contact time from 0 to 24 h. The integration of the pseudo-first-order kinetic equation is expressed as [22]:

$$\log(q_e - q_t) = \log q_e - \frac{k_1}{2.303}t \tag{1}$$

The integration of the pseudo-second-order model can be described by the following equation:

$$\frac{t}{q_t} = \frac{1}{k_2 q_e^2} + \frac{1}{q_e} \tag{2}$$

where k_1 is the pseudo-first-order rate constant (min^{-1}), k_2 is the pseudo-second-order rate constant (g min^{-1} mg^{-1}), q_t is the amount of ammonium nitrogen adsorbed at time t (mg g^{-1}), q_e is the adsorption capacity at equilibrium (mg g^{-1}), and t is the contact time (min).

The regressed curves and the correlation coefficients for the pseudo-first-order and the pseudo-second-order were shown in Figures 2A and 2B and Table 1, respectively. With regard to the pseudo-first-order model, the correlation coefficient was relatively low (R^2 = 0.9047), and the experimental adsorbed masses (78.83 mg g^{-1}) was much higher than the theoretical value (34.04 mg g^{-1}) at the equilibrium time. These results indicated a bad fit between the model and the experimental data; therefore, the adsorption of ammonium on zeolite A-3 was not

compliant with the pseudo-first-order reaction.

For the pseudo-second-order model, the correlation coefficient (R^2 = 0.987) was much higher than that of the pseudo-first-order model (R^2 = 0.905), and no obvious distinct occurred between the experimental (78.83 mg g^{-1}) and the theoretical adsorption capacity (77.52 mg g^{-1}) at equilibrium. The good accordance between the experimental data and the pseudo-second-order kinetic model showed that the adsorption of ammonium on the zeolite A-3 was well described by the pseudo-second-order kinetic model. As a result, this adsorption could be dominated by a chemical process, mainly ion exchange, which was in accordance with the results obtained by many other researches [12,23].

Adsorption isotherms: Adsorption isotherms are essential to describe how adsorbate masses will interact with adsorbent media and are useful to optimize the use of media as adsorbents. Therefore, empirical equations such as Langmuir and Freundlich isotherm models are important for investigating the adsorption mechanism. The linearized forms of Langmuir and Freundlich isotherms were applied to analyze the adsorption process under initial ammonium concentrations ranging from 1000 to 5000 mg l^{-1}, adsorbent loading rate of 10 g l^{-1} and contact time of 24 h.

The Langmuir model assumes only one solute molecule per site, and also assumes a fixed number of sites. The linear form of the Langmuir isotherm equation can be expressed as followings [22]:

$$\frac{C_e}{q_e} = \frac{1}{b\ q_m} + \frac{C_e}{q_m} \tag{3}$$

Freundlich isotherm assumes that the uptakes of adsorbate occur on a heterogeneous surface by multilayer adsorption and the amount of adsorbate adsorbed increases infinitely with an increase in concentration. The linear forms of the Freundlich isotherm equation is given as:

$$\ln q_e = \ln k_f + \frac{1}{n}\ln C_e \tag{4}$$

where C_e is the liquid phase concentration of the ammonium nitrogen at equilibrium (mg l^{-1}), q_e is the amount of ammonium nitrogen adsorbed on the ceramic adsorbent at equilibrium (mg g^{-1}), q_m is the maximum adsorption capacity (mg g^{-1}), b is the Langmuir constant related to the adsorption energy (l mg^{-1}), K_f (mg$^{1-1/n}$ l$^{1/n}$ g^{-1}) is the Freundlich isotherm model constant indicating the adsorption capacity of the adsorbent, and 1/n is an empirical parameter related to the intensity of adsorption, which varies with the heterogeneity of the material [14]. The plot of $\ln q_e$ versus $\ln C_e$ for the adsorption of ammonium nitrogen onto the zeolite A-3 was employed to generate

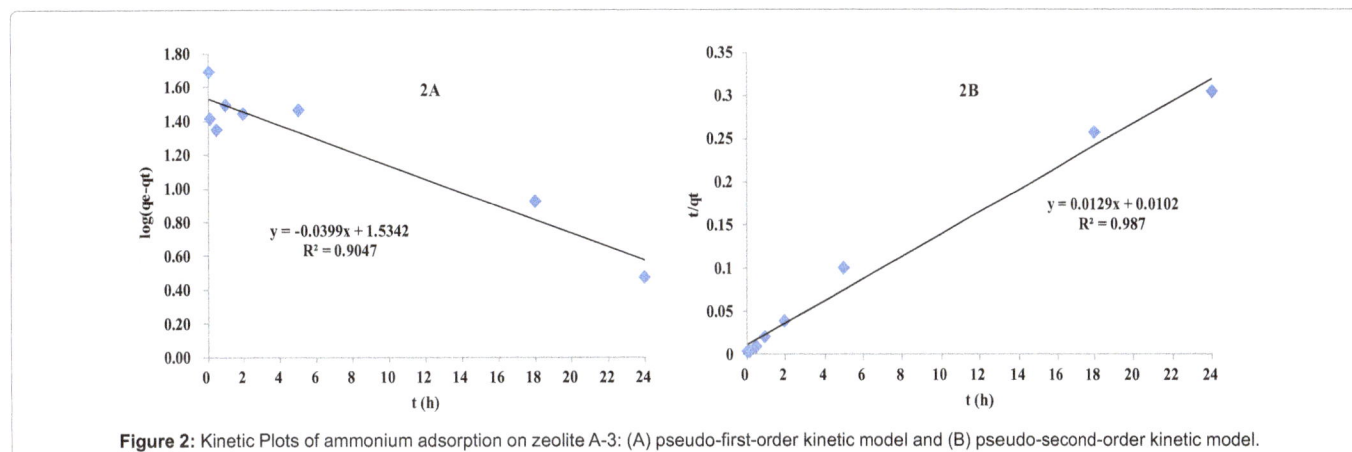

Figure 2: Kinetic Plots of ammonium adsorption on zeolite A-3: (A) pseudo-first-order kinetic model and (B) pseudo-second-order kinetic model.

Pseudo-first-order model			Pseudo-second-order model		
K_1(min^{-1})	q_e(mg g^{-1})	R^2	K_2(g mg^{-1} min^{-1})	q_e(mg g^{-1})	R^2
0.09189	34.04	0.9047	0.016317	77.52	0.987

Table 1: Pseudo-first-order model and Pseudo-second-order model constants for the ammonium adsorption on the zeolite A-3 adsorbent.

Langmuir isotherm			Freundlich isotherm		
b (L mg^{-1})	q_m(mg g^{-1})	R^2	K_f(mg$^{1-1/n}$ L$^{1/n}$ g^{-1})	1/n	R^2
0.000348	84.03	0.9863	0.243781	0.6464	0.9846

Table 2: Langmuir isotherm and Freundlich isotherm constants for the ammonium adsorption on the zeolite A-3 adsorbent.

the intercept value of K_f and the slope of 1/n.

The fitted curves for the Langmuir and Freundlich isotherms were shown in Figures 3A and 3B, and the isotherm parameters for the adsorption of ammonium nitrogen onto the zeolite A-3 were listed in Table 2. It can be seen that both Langmuir and Freundlich model were applicable for the adsorption of ammonium on the zeolite A-3, according to the high values of the regression correlation coefficients (R^2>0.98). The similar result was reported by Halim et al., 2010 [24], who compared the ammonia adsorption on zeolite, activated carbon and composite materials in the treatment of landfill leachate. The good compliance to Langmuir and Freundlich isotherms showed that the ammonium removal by zeolite A-3 via both the cation exchange and physical adsorption mechanism. The q_m of 84.03 mg g^{-1} calculated by the Langmuir model was higher than the measured value (78.83 mg g^{-1}). The values of the empirical parameter 1/n lying between 0.1 < 1/n < 1 indicated favorable adsorption for ammonium [22]. The 1/n value (0.6464) in the present study was lower than 1, which represented favorable removal conditions.

Ammonium desorption from saturated zeolite A-3

Sodium sulfate (Na$_2$SO$_4$) solution was used for ammonium desorption from saturated zeolite A-3, due to the advantages of nitrogen recovery in the form of ammonium sulfate ((NH$_4$)$_2$SO$_4$) which is a nice nitrogenous fertilizer and zeolite regeneration. Figure 4A shows the efficiency of ammonium desorption from zeolite A-3 and effluent NH$_4^+$-N concentration in the bulk solution. Both the desorption efficiency of ammonium and the effluent NH$_4^+$-N concentration in the bulk solution increased with reaction time and gradually reached equilibrium after 20 hours. The maximum desorption efficiency (38.2%) and highest effluent NH$_4^+$-N concentration (76.4 mg l^{-1}) were obtained under the equilibrium state.

Desorption kinetic of NH$_4^+$ can be described by a first-order reversible mechanism [25], which is expressed as:

$$C_t = C_e\left(1 - e^{-(k_1+k_{-1})*t}\right) \qquad (5)$$

where, C_e and C_t (mg l^{-1}) are the time-dependent concentration of the dissolved NH$_4^+$ at equilibrium and time t (h); k_1 and k_{-1} (h^{-1}) are the adsorption and desorption rate constants, respectively. Its logarithm form can be given as Eq. (6).

$$\ln\frac{C_e - C_t}{C_e} = -\left(k_1 + k_{-1}\right)*t \qquad (6)$$

The chemical response time (τ_{resp}) for a first-order reversible

reaction is:

$$\tau_{resp} = \frac{1}{\left(k_1 + k_{-1}\right)} \qquad (7)$$

The kinetic plot of $\ln((C_e-C_t)/C_e)$ versus t of ammonium desorption was illustrated in Figure 4B. The high linear regression coefficient (R^2 = 0.982) indicated that desorption of ammonium from saturated zeolite A-3 well fits the first-order reversible reaction kinetic. Value for (k_1 + k_{-1}) obtained from the regression line was 0.179 h^{-1}. The reaction constant ($k_1 + k_{-1}$) was used in Eq. (5) to predict desorption as a function of time. The calculated τ_{resp} was 5.59 h.

Performance of anaerobic digestion

In a previous study [15], it was found that addition of natural zeolite (doses: 2-4 g l^{-1}) contributed to enhance the anaerobic digestion of piggery wastes with NH$_4^+$-N concentration of 410 mg l^{-1} by reducing the ammonium inhibitory. In this present study, the NH$_4^+$-N concentration of ammonium-rich piggery waste was as high as 3770 mg l^{-1}, which is approximately 9-fold of that in the previous study. Thus, to obtain an optimum addition of zeolite A-3 for methane fermentation of ammonium-rich piggery wastes, the dosages loading rates of 10 g l^{-1} and 30 g l^{-1} were used in the zeolite-fixed bioreactors. The adjusted piggery wastes with an initial ammonium nitrogen concentration of 3770 mg l^{-1} fed to each bioreactor. Ammonium inhibition has occurred above pH 7.4 within the ammonium nitrogen concentration range of 1500-3000 mg l^{-1} during the anaerobic digestion process [3].

Figure 5A shows that the startup period for anaerobic digestion was 13 days and 20 days in the zeolite-fixed bioreactors and the control bioreactor, respectively. Beginning from the 13th day, methane production in 10 g l^{-1} and 30 g l^{-1} zeolite-fixed bioreactors increased gradually to the daily maximum of 583.5 ml l^{-1} and 543.3ml l^{-1} on 21th day, respectively. The corresponding methane concentration increased respectively from 72.5% to 87.3% and from 78.0% to 85.5% in these two bioreactors (Figure 5B). After that, the daily methane yield decreased gradually, whereas the methane concentration maintained at approximately 80% until the end of the digestion process. According to methane production and concentration during the first 20 days (Figures 5A and 5B), the zeolite-fixed bioreactors showed better performance than the control.

Above results indicates that the zeolite-fixed bioreactor is effective for improving methane production in anaerobic digestion of ammonium-rich piggery wastes. Because inhibitory ammonium in the anaerobic digestion of piggery wastes was partially removed by the zeolite, the NH$_4^+$-N levels in 10 g l^{-1} and 30 g l^{-1} zeolite-fixed bioreactors decreased respectively from 3770 to 3050 and 2958 mg l^{-1} during the first 4 day (Figure 5C). However, the NH$_4^+$-N level in the control bioreactor increased to 3896 mg l^{-1} on the 4th day. The zeolite suspended in the upper layer of the digested liquid, where ammonium could easily be removed by the adsorbent. The ammonium concentration increased during the anaerobic digestion process, because ammonia is produced by the biological degradation of the nitrogenous matter [26]. After the 4th day, the ammonium concentration in the zeolite-fixed bioreactors increased gradually. At the end of methane fermentation experiment which lasted for 33 days, the total NH$_4^+$-N concentration in the zeolite-fixed bioreactors (10 g l^{-1}, 30 g l^{-1}) and the control bioreactor increased to 3904, 3757 and 4940 mg/l, respectively (Figure 5C).

Overall, the pH value in the three bioreactors was between 7.0 and 8.1 and thus fulfilled the favorable pH level for methane fermentation (Figure 5D). In the zeolite-fixed bioreactors, the pH level decreased

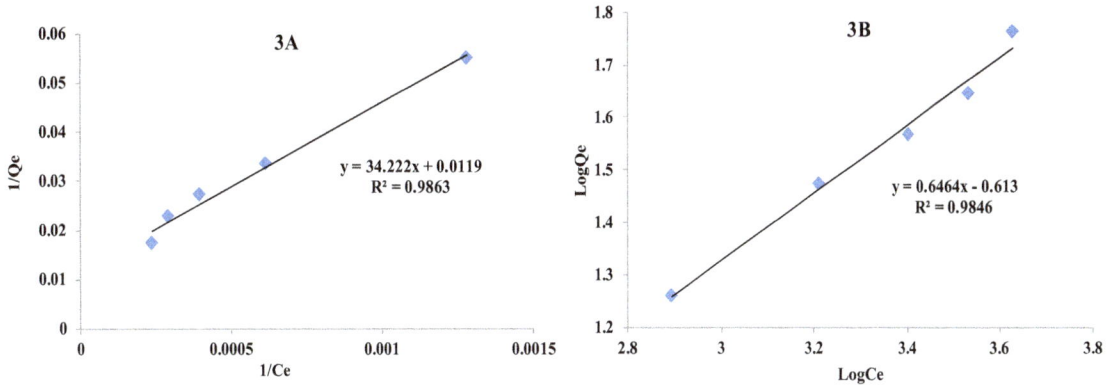

Figure 3: Ammonium adsorption isotherms on zeolite A-3: (A) Langmuir isotherm model and (B) Freundlich isotherm model.

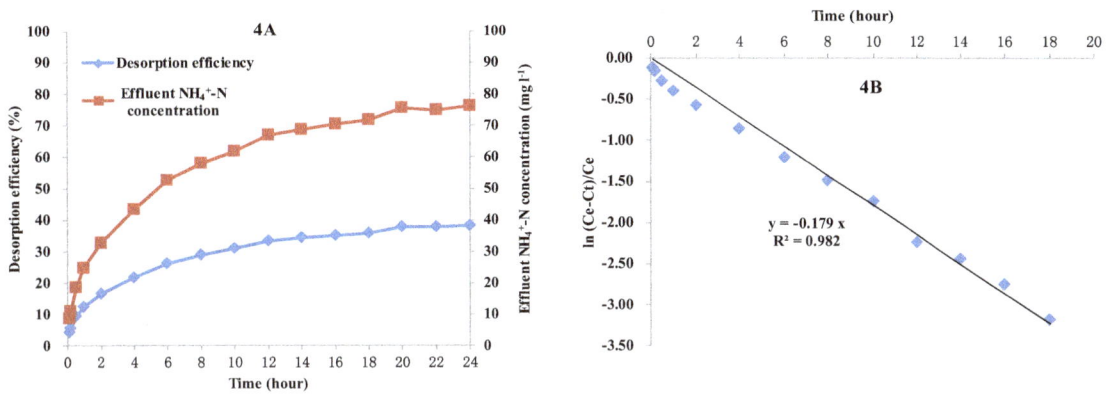

Figure 4: (A) The efficiency of ammonium desorption from zeolite A-3 and effluent NH_4^+-N concentration in the Na_2SO_4 solution, and (B) kinetic plots of ammonium desorption from the saturated zeolite A-3.

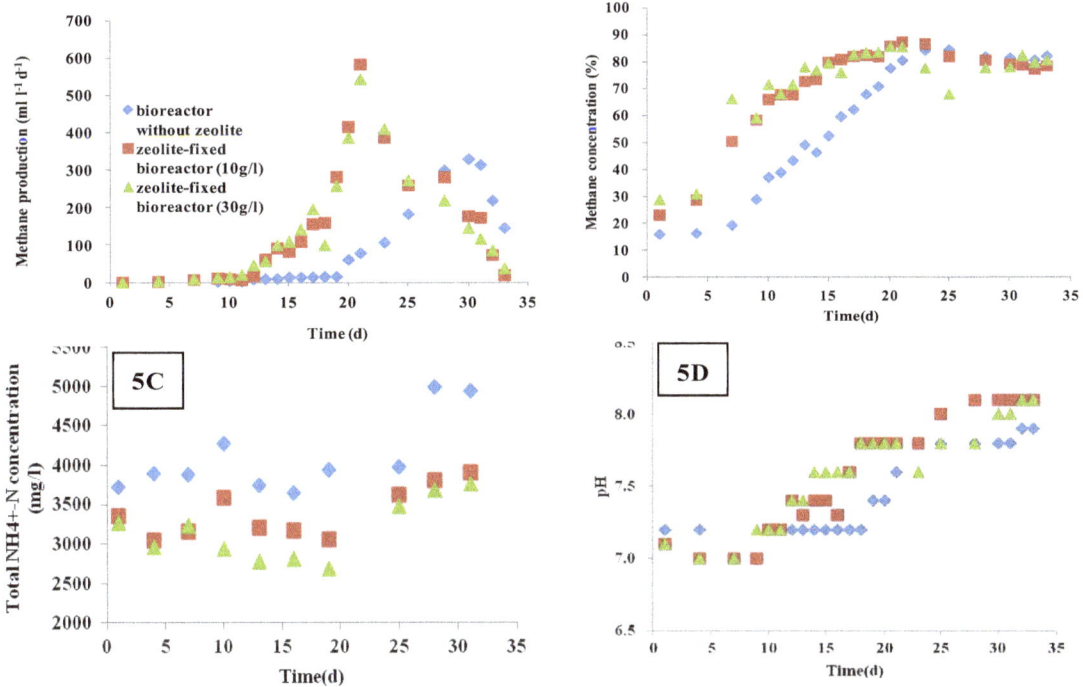

Figure 5: The performance of the zeolite-fixed bioreactors and bioreactor without zeolite as control for the anaerobic digestion of piggery wastes during the experiment: (A) methane production; (B) methane concentration; (C) ammonium nitrogen concentration and (D) pH value.

slightly (from 7.1 to 7.0) during the startup period, because of the production of volatile fatty acid (VFA). Beginning on the 15th day, the pH level gradually increased to 8.1. Then it remained at 8.1 until the end of the digestion process. The increase in pH can be explained by the increasing of ammonium concentration and the biodegradation of VFA into methane. The pH level in the control bioreactor remained constant until day 18 and then increased to 7.9 on day 33. This pH variation trend is consistent with that of methane production.

The 10 g l^{-1} and 30 g l^{-1} zeolite-fixed bioreactors showed similar trend of methane concentration and methane yield. Nevertheless, from the viewpoint of reducing mass transfer resistance and economic cost in the zeolite-fixed bioreactor, the optimum addition loading rate of zeolite was 10 g l^{-1} in this study.

Microorganism activity

Generally, the quantity and activity of the microorganisms in a bioreactor are two conclusive parameters [27]. ATP is an indicator of metabolically active cells and an index of microbial density, which has been shown to reflect the microorganism activity in the anaerobic digestion [28]. In this study, ATP concentration was examined on the surface of the zeolite in the zeolite-fixed bioreactor, and in the liquid from all the bioreactors at the end of the digestion experiment. The similar ATP values obtained in the liquid phase of zeolite-fixed bioreactor and control bioreactor were 0.026 and 0.023 μ mol l^{-1}, respectively. However, the ATP concentration (0.25μ mol l^{-1}) on the surface of the zeolite is much higher than that in the liquid phase in the zeolite-fixed bioreactor. This indicated that the high activity levels of the immobilized microorganisms on the zeolite surface, which could be understood as pointing that the fixed zeolite is a stable and suitable carrier for microorganisms, and most of them propagated on the surface of fixed zeolite. A number of microbes assembled on the surface of fixed zeolite (Figure 6B) resulted to the high concentration of ATP. The distribution of microbes in the liquid phase and on the surface of the support materials were about 5% and 95%, respectively [29].

On the other hand, the surface morphology of the zeolite A-3 before and after the anaerobic digestion process were observed by SEM at a magnification of 6000×. As illustrated in Figure 6A, the zeolite A-3 shows a porous structure and is covered with fractures. After anaerobic digestion the porous surface of zeolite was colonized subsequently by a great deal of methanogens (Figure 6B). This phenomenon confirmed the immobilization of microorganisms on the zeolite surface in the zeolite-fixed bioreactor.

In the anaerobic digestion of ammonium-rich piggery wastes, it has been found that free ammonia (NH$_3$) is the active form causing ammonia inhibition. The high concentration of free ammonia is the major causes of digester upset or failure. However, the adsorption of ammonium on zeolite surface mainly via the approach of cation exchange. Ammonium ion (NH$_4^+$) rather than NH$_3$ was adsorbed on the surface of zeolite. The toxicity of ammonia on the zeolite surface is much lower than that in the digested liquid. Therefore, the microbes tend to grow on the surface of the fixed zeolite in order to avoid the potential toxicity of free ammonia in the liquid and utilize the nitrogen source on the zeolite surface. Integrating the results of ATP analysis and SEM observation, it can be concluded that immobilization of microorganisms can be well performed using zeolite A-3 as carrier material in the zeolite-fixed bioreactor for effectively mitigating ammonia inhibition, thereby enhance the microorganism activity.

Effectiveness of zeolite-fixed bioreactor for the anaerobic digestion of ammonium-rich piggery wastes

As shown in Figure 7, the total methane yield (354.2ml g^{-1}-VS)

and COD removal rate (75.37%) in zeolite-fixed bioreactor are both higher than those in the control bioreactor (146.4 ml g^{-1}-VS and 35.10%). The methane yields are lower than the theoretical value (516 ml g^{-1}-VS) for piggery wastes [30]. However, the theoretical value is based on the assumption that all of the carbon substrate transformed into methane, a fraction of the substrate is in fact used to synthesize bacterial mass [31]. In addition, the quite high initial concentrations of NH$_4^+$-N and COD are another factor that contributed to the lower actual methane yield for piggery wastes. Sánchez et al. [32] investigated piggery waste treatment using an upflow anaerobic sludge bed reactor (UASB) and an anaerobic fixed bed reactor (AFBR) at initial COD and NH$_4^+$-N concentrations less than 12600 and 650 mg l^{-1}, respectively. Their study obtained 60% COD removal in the AFBR and 40% of that in UASB, respectively. Here, at much higher initial concentrations of COD (76700 mg l^{-1}) and NH$_4^+$-N (3770 mg l^{-1}), the COD removal rate reached as high as 75.37% in the zeolite-fixed bioreactor. This result indicated that the zeolite-fixed bioreactor developed in this study is effective for improving the methane production from ammonium-rich piggery wastes. In further research, the practical effectiveness of developed zeolite-fixed bioreactor should be determined by carrying out the continuous anaerobic digestion of ammonium-rich piggery wastes. Due to the easy replacement and regeneration of ammonium saturated zeolite, it can be expected that the zeolite-fixed bioreactor would be stable and sustainable in continuous anaerobic digestion.

Evaluation of ammonium adsorbed on zeolite A-3 as fertilizer

In general, the utilization efficiency of nitrogenous fertilizer in soil in developed countries could reach 50-70% (data is from the Food and Agricultural Organization), whereas that in most of the developing countries like China is only about 30% [33]. That means most of the nitrogenous nutrient was lost when using for crop growth. Fortunately, the ammonium saturated zeolite can be directly utilized as fertilizer to effectively avoid this lost, because of its slow-release of nitrogenous nutrient into the soil [23]. Besides, the zeolite itself was considered as a soil enhancer due to its nutrient retention capacity for potassium and phosphorus [34]. In this study, the ammonium adsorption capacity of zeolite A-3 is 78.83 mg NH$_4^+$-N g^{-1}-zeolite, which means that approximately 7.9% of the ammonium nitrogen was adsorbed on zeolite. Direct utilization of ammonium saturated zeolite as fertilizer shows great potential to decrease annual production of nitrogen fertilizer, thereby save fertilizer cost $ 0.20 per acre-inch [35] and mitigate the environmental pollution.

On the other hand, from the viewpoint of regeneration and reuse of zeolite, desorption of ammonium from saturated zeolite is of great interest. In this present study, the regeneration of saturated zeolite A-3 was successfully achieved by desorption of ammonium into Na$_2$SO$_4$ aqueous solution. In addition, a nice nitrogenous fertilizer ((NH$_4$)$_2$SO$_4$) was obtained as a by-product during the ammonium desorption process.

Conclusions

A zeolite-fixed bioreactor was developed to mitigate ammonia inhibition and enhance methane production in the anaerobic digestion of ammonium-rich piggery wastes. Adsorption of ammonium on zeolite A-3 is compliant with the pseudo-second-order kinetic model and fits well to both Langmuir and Freundlich isotherms. Using zeolite-fixed bioreactor could decrease the startup period, enhanced methane yield and COD removal. Direct utilization of ammonium saturated zeolite as fertilizer could be great potential to increase the utilization

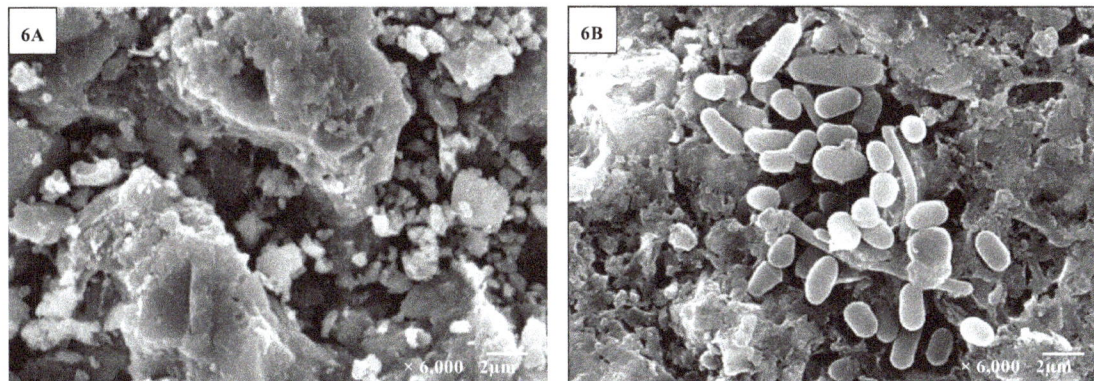

Figure 6: SEM images of (A) artificial zeolite A-3 and (B) microorganism immobilized zeolite A-3.

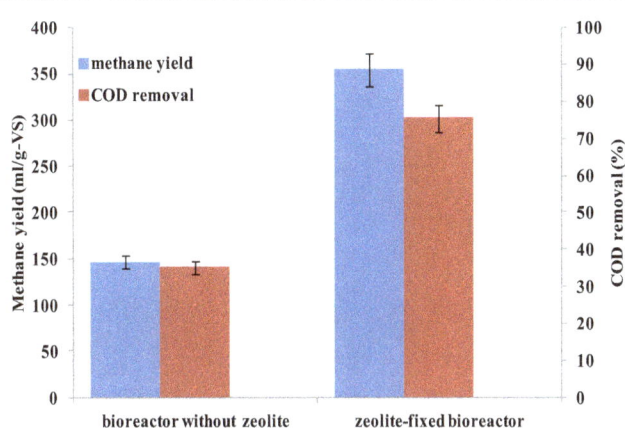

Figure 7: SMethane yield and COD removal in the zeolite-fixed bioreactor and bioreactor without zeolite as control.

efficiency of nitrogen fertilizer and decrease the environmental impact. Moreover, regeneration of zeolite A-3 using Na_2SO_4 solution also obtained a $(NH_4)_2SO_4$ by-product which is nice nitrogenous fertilizer.

Acknowledgement

The authors would like to thank the financial support from the Japan Society for the Promotion of Science (JSPS), Grants-in-Aid for Scientific Research (B) No. 25281048.

References

1. Wang Q, Yang Y, Yu C, Huang H, Kim M, et al. (2011) Study on a fixed zeolite bioreactor for anaerobic digestion of ammonium-rich swine wastes. Bioresour Technol 102: 7064-7068.

2. Hansen KH, Angelidaki I, Ahring BK (1998) Anaerobic digestion of swine manure: inhibition by ammonia. Water Res 32: 5–12.

3. Van Velsen AFM (1979) Adaptation of methanogenic sludge to high ammonia–nitrogen concentrations. Water Res 13: 995–999.

4. Gallert C, Bauer S, Winter J (1998) Effect of ammonia on the anaerobic degradation of protein by a mesophilic and thermophilic biowaste population. Appl Microbiol Biotechnol 50: 495-501.

5. Hobson PN, Shaw BG (1976) Inhibition of methane production by Methanobacterium formicicum. Water Res 10: 849–852.

6. Yabu H, Sakai C, Fujiwara T, Nishio N, Nakashimada Y (2011) Thermophilic two-stage dry anaerobic digestion of model garbage with ammonia stripping. J Biosci Bioeng 111: 312-319.

7. Wirthensohn T, Waeger F, Jelinek L, Fuchs W (2009) Ammonium removal from anaerobic digester effluent by ion exchange. Water Sci Technol 60: 201-210.

8. Uludag-Demirer S, Demirer GN, Chen S (2005) Ammonia removal from anaerobically digested dairy manure by struvite precipitation. Process Biochem 40: 3667-3674.

9. Calli B, Mertoglu B, Inanc B, Yenigun O (2005) Methanogenic diversity in anaerobic bioreactors under extremely high ammonia levels. Enzyme Microb Technol 37: 448-455.

10. Resch C, Wörl A, Waltenberger R, Braun R, Kirchmayr R (2011) Enhancement options for the utilisation of nitrogen rich animal by-products in anaerobic digestion. Bioresour Technol 102: 2503-2510.

11. Santhosh G, Venkatachalam S, Ninan KN, Sadhana R, Alwan S, et al. (2003) Adsorption of ammonium dinitramide (ADN) from aqueous solutions. 1. Adsorption on powdered activated charcoal. J Hazard Mater 98: 117-126.

12. Ugurlu M, Karaoglu MH (2011) Adsorption of ammonium from an aqueous solution by fly ash and sepiolite: Isotherm, kinetic and thermodynamic analysis. Microporous Mesoporous Mater 139: 173-178.

13. Baker HM, Massadeh AM, Younes HA (2009) Natural Jordanian zeolite: removal of heavy metal ions from water samples using column and batch methods. Environ Monit Assess 157: 319-330.

14. Huang H, Xiao X, Yan B, Yang L (2010) Ammonium removal from aqueous solutions by using natural Chinese (Chende) zeolite as adsorbent. J Hazard Mater 175: 247-252.

15. Milán Z, Sánchez E, Weiland P, Borja R, Martín A, et al. (2001) Influence of different natural zeolite concentrations on the anaerobic digestion of piggery waste. Bioresour Technol 80: 37-43.

16. Montalvo S, Díaz F, Guerrero L, Sánchez E, Borja R (2005) Effect of particle size and doses of zeolite addition on anaerobic digestion processes of synthetic and piggery wastes. Process biochem 40: 1475-1481.

17. Nikolaeva S, Sánchez E, Borja R, Travieso L, Weiland P, et al. (2002) Treatment of piggery waste by anaerobic fixed bed reactor and zeolite bed filter in a tropical climate: a pilot scale study. Process biochem 38: 405-409.

18. Kotsopoulos TA, Karamanlis X, Dotas D, Martzopoulos GG (2008) The impact of different natural zeolite concentrations on the methane production in hermophilic anaerobic digestion of pig waste. Biosyst Eng 99: 105-111.

19. Montalvo S, Guerrero L, Borja R, Travieso L, Sánchez E, et al. (2006) Use of natural zeolite at different doses and dosage procedures in batch and continuous anaerobic digestion of synthetic and swine wastes. Resour Conserv Recy 47: 26–41.

20. Demir A, Günay A, Debik E (2002) Ammonium removal from aqueous solution by ion-exchange using packed bed natural zeolite. Water SA 28: 329–335.

21. Eaton AD, Clesceri LS, Rice EW, Arnold E (2005) Standard methods for the examination of water and wastewater. American Public Health Association, USA.

22. Yusof AM, Keat LK, Ibrahim Z, Majid ZA, Nizam NA (2010) Kinetic and equilibrium studies of the removal of ammonium ions from aqueous solution by rice husk ash-synthesized zeolite Y and powdered and granulated forms of mordenite. J Hazard Mater 174: 380-385.

23. Zhao Y, Yang Y, Yang S, Wang Q, Feng C, et al. (2013) Adsorption of high ammonium nitrogen from wastewater using a novel ceramic adsorbent and

the evaluation of the ammonium-adsorbed-ceramic as fertilizer. J Colloid Interface Sci 393: 264-270.

24. Halim AA, Aziz HA, Johari MAM, Ariffin KS (2010) Comparison study of ammonia and COD adsorption on zeolite, activated carbon and composite materials in landfill leachate treatment. Desalination 262: 31–35.

25. Millward GE, Liu YP (2003) Modelling metal desorption kinetics in estuaries. Sci Total Environ 314-316: 613-23.

26. Kayhanian M (1999) Ammonia inhibition in high-solids biogasification: an overview and practical. Environ Technol 20: 355–365.

27. Chen H (2004) ATP content and biomass activity in sequential anaerobic/ aerobic reactors. J Zhejiang Univ Sci 5: 727-732.

28. Chu CP, Lee DJ, Chang BV, You CH, Liao CS, et al. (2003) Anaerobic digestion of polyelectrolyte flocculated waste activated sludge. Chemosphere 53: 757-764.

29. Yang Y, Tsukahara K, Sawayama S (2008) Biodegradation and methane production from glycerol-containing synthetic wastes with fixed-bed bioreactor

under mesophilic and thermophilic anaerobic conditions. Process Biochem 43: 362–367.

30. Moller HB, Sommer SG, Ahring BK (2004) Methane productivity of manure, straw and solid fractions of manure. Biomass Bioenergy 26: 485–495.

31. Angelidaki I, Ahring BK (2000) Methods for increasing the biogas potential from the recalcitrant organic matter contained in manure. Water Sci Technol 41: 189-194.

32. Sánchez E, Milán Z, Borja R, Weiland P, Rodriguez X (1995) Piggery waste treatment by anaerobic digestion and nutrient removal by ionic exchange. Resour Conserv Recy 15: 235–244.

33. Wu J, Pan X (2008) Study progress on nitrogen losses in paddy field. Sci technol 6: 117-124.

34. Xie Z, Wang X, Niu S, Tong Z, Sun B, Zhao F (2006) Zeolite and modified application as soil amendment. Crops 26: 142-144.

35. Lambert DK (1990) Risk considerations in the reduction of nitrogen fertilizer use in agricultural production. Western J Agric econ 15: 234–244.

20

Coagulation Processes for Treatment of Waste Water from Meat Industry

Zueva SB[1,2]*, Ostrikov AN[2], Ilyina NM[2], De Michelis I[1] and Veglìò F[1]

[1]*Laboratory of Integrated Treatment of Industrial Wastes and Wastewaters, Department of Industrial and Information Engineering and Economics of the University of L'Aquila, Italy*
[2]*Department of Environment and Chemical Engineering, Voronezh State University of Engineering Technology, Russia*

Abstract

Application of the coagulants in treatment of wastewater from food industry is one of the most promising techniques to establish environment-friendly industries. This article describes results of experiment and the chemistry of wastewater treatment with coagulants and so called "coagulant aids". The coagulation efficiency was determined combining different techniques (photocolorimetric method, the microscopying of the samples, zeta potential measurements). The results showed that aluminum sulfate was nearly twice more effective in presents of alumina powder. The reason for this behavior can be explained by the negative value of Zeta potential on its surface. It was found that the aggregation and sedimentation speed can be greatly enhanced.

Keywords: Alumina; Coagulation; Sorption; Zeta potential

Introduction

Coagulation has been subject of many research, most of which has been related to wastewater treatment [1-5], however, it may differ from depending on chemical and physical parameters of contaminants. The particles to be removed include organic material, which can react differently to a coagulant.

Wastewater of a meat plant is full of contaminants, such as fats, carbohydrates, proteins, organic acids. Macromolecules of protein roll up into compact globules, which are negatively charged and have hydrated coating. Their aggregative stability is provided by a series of factors: a decrease in surface energy of dispersed phase (i.e. reduction of driving force in coagulation) as a result of formation of an electric double-layer by the presence of kinetic barriers for coagulation in the form of electrostatic repulsion between colloidal particles and counter-ions of the similar charges. One more reason for stability of colloids is connected with the process of hydration (solvation) of ions. The coat of the diffusion layer composed of solvated counter-ions also prevents adhesion of the particles. However, if the charge on colloidal particles is destroyed, they are free to come nearer and grow in size. The following forces play an important role in the interaction of colloid particles: included volume repulsion, electrostatic interaction, Van der Waals force, entropic force, steric interactions.

In wastewater treatment coagulation process is particles adhesion process with formation of large flocs, as a result of addition of a chemical reagent (coagulating agent) for the purpose of destabilization of suspended colloidal particles and their subsequent coagulation (aggregation).

Inorganic coagulating agents (low-molecular inorganic and organic electrolyte), are neutralize negatively charged colloidal particles in water with following particles aggregation (coagulation) and precipitation. For example, aluminium sulfate ($Al_2(SO_4)_3 \cdot 18H_2O$), consists in hydrolysis of sulfates with formation at low pH positively charged sol of aluminum hydroxide $\left[mAl(OH)_3 \, 2nAl(OH)_2^+ (2n-x)SO_4^{2-} \right]^{2x+}$.

Together with coagulants can be used various clay, pastes, a metal dust, ashes, small quartz sand [4]. They can play a role of the additional centers of condensation of products of hydrolysis, promotes acceleration of coagulation, speed of their sedimentation. Some of them can be an adsorbents and increase effectiveness of purification of waste water from dissolved impurities.

During the precipitation of metal hydroxides organic macromolecules are enclosed in the flocks [6] and removed in the subsequent filtration process.

High-molecular organic or inorganic compound give simultaneous adsorption on two or several particles of disperse phase connects them in aggregates by polymeric bridges and as a result reduces stability of disperse system. Such bridges can form both macromolecules of water-soluble polymers, and insoluble substances in water, for example Aluminum Oxide powder (γ-Al_2O_3) [7].

One of the major uses of activated metal oxides is for adsorbing metal or nonmetal ions from aqueous wastewater streams. For instance, activated alumina is suggested by the Environmental Protection Agency (EPA) as a Best Available Treatment (BAT) for removing selenium and arsenic ions from water [8].

The free energy of adsorption for charged ions onto highly charged metal oxides is due to three main contributions: specific chemical interaction, columbic attraction or repulactivated and desolvation of the ion [9].

As it was shown in the metal ions are chemically bound to the $-Al-O^-$ group by an ionic exchange with H^+ group which is favored by the basic conditions of the medium. The metal ions retain a positive charge, then coagulation can occur by electrostatic interaction between the positive sites $=Al-O-M^+$ on one particle and the negative groups $=Al-O^-$ on another; in this process, the metal ions are acting as bridging units between the particles. This is the mechanism to flocculate and to adsorb the metal ions in solution on the particles surface [10-13].

*Corresponding author: Svetlana Zueva, Laboratory of Integrated Treatment of Industrial Wastes and Wastewaters, Department of Industrial and Information Engineering and Economics of the University of L'Aquila, Italy
E-mail: svetlana.zueva@graduate.univaq.it

Knowledge of the zeta potential provides an important parameter for the explanation of the adsorption mechanism of an adsorbate at the metal oxide–water interface [14]. The isoelectric point (i.e.p.) or point of zero charge (pzc) of the any metal oxide surfaces provides information on the cleanness of a metal oxide surface.

In this study the zeta-potential method was used to determine coagulations and sorption processes in meat industry wastewater treatment. This rate information was then analyzed to establish the mechanisms of coagulation and adsorption.

Materials and Methods

Activated γ-alumina (γ-Al_2O_3) and aluminum sulfate ($Al_2(SO_4)_3$ 20 % w/v) were used as the reagents for all phases of studies.

The subject of the research is wastewater of a dairy plant with the following pollution characteristics: Suspended Substance (SS)=1350 mg/L, Fats=64.3 mg/L, Iron salts=2.59 mg/L, pH=6.5-7.0, t=20°C.

The mass fraction of fat in the experimental samples was estimated using the Gerber method. The method is based on the allocation of fat under the action of strong sulphuric acid and isoamyl alcohol in the form of a continuous layer, the amount of which is measured in the graded part of the butyrometer.

The content of suspended substances in the samples was estimated using the turbidimetric method. The optical density (D) was measured at a fixed wavelength of 500 nm using a UV–visible spectrophotometer.

The content of Iron salts was monitored by photometric method at the wavelength 420 nm. This method based on the fact that sulfosalicylic acid in weakly alkaline medium form a colored complex salts with iron (III) and (II) (yellow color).

The zeta potentials and the average size of the particles were measured with Zetasizer (Malvern Instruments) based dynamic light scattering and electrophoretic light scattering method using M3-PALS technology.

The microscopying of the samples has been conducted with Biomed 4 trinocular microscope with a lens increased by x10 and additional increase of the camera x4. Biomed 4 is designed to monitor substance in transmitted light in the bright and dark field, phase contrast.

Results and Discussions

Wastewater from meat industry can be considered as a dispersed system (suspension). It is known that above pH 7 nano-particles in a sol-gel system are charged negatively. This negative charge stabilizes the dispersion by electrostatic repulsion, preventing the aggregation of the particles.

According results of measuring the initial particle size was of 3.54 μm with the zeta-potential of -15.9 mV at pH=6.8 (Figures 1 and 2). Indeed, particles with zeta potentials less than -30 mV are normally considered stable, the process of coagulation occurs rather slowly. Multi component structure of wastewater acts as a stabilizing factor here. To stimulate the process (of rapid coagulation) it was necessary to add special reagents – coagulants or flocculants – which should band together small particles of dispersed system into bigger ones under the influence of cohesive forces.

For the purpose of studying the process of coagulation purification of waste water of the meat plant series of experiments to identify the

Z-Avarage		Diam.(nm)	% Volume	Width (nm)
(d.nm): 6754	Peak 1:	3	100,0	588,2
PdI: 0,297	Peak 2:	0,000	0,0	0,0
Intercept: 0,983	Peak 2:	0,000	0,0	0,0

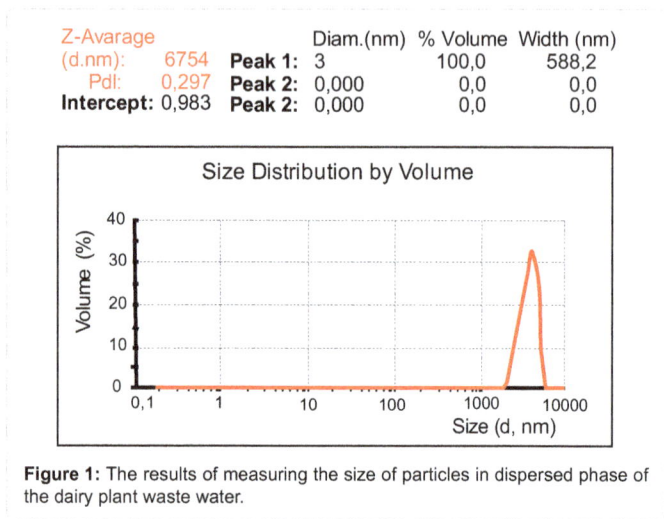

Figure 1: The results of measuring the size of particles in dispersed phase of the dairy plant waste water.

Zeta Potenzial (mV): -15,9		Mean (mV)	Area (%)	Width (mV
Zeta Deviation (mV): 3,96				
Conductivity	Peak 1:	-15,9	100,0	3,96
(mS/cm): 1,16	Peak 2:	0,00	0,0	0,00
	Peak 2:	0,00	0,0	0,00

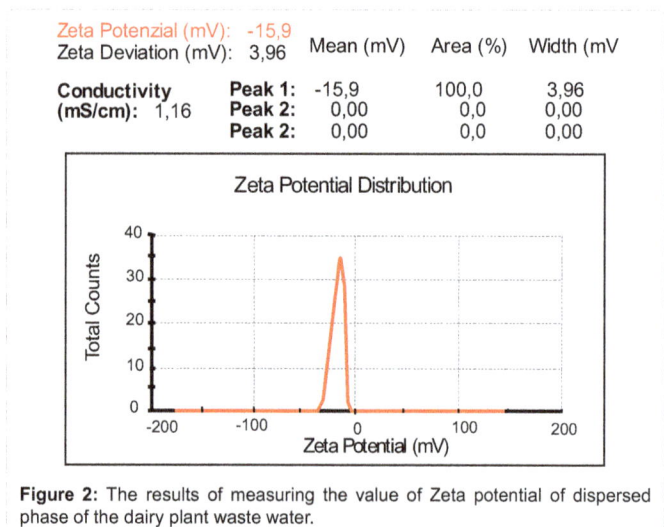

Figure 2: The results of measuring the value of Zeta potential of dispersed phase of the dairy plant waste water.

particles size and value of zeta-potential of resulting agglomerations against the dose and the type of the reagent were conducted.

Aluminum sulfate at low pH (<7, 8) form positively charged sol of aluminum hydroxide, which can neutralize negatively charged particles in wastewater with following aggregation (coagulation). Flock growth rate as a function of time for different dosage of aluminum sulfate presented on the Figure 3.

When the sizes of the flocs were large enough, the sedimentation effects become possible. The maximum of the flocs size were in correlation with their maximum sedimentation speed and with isoelectric point (i.e.p.) (Figure 4).

Effectiveness of wastewater treatment as a function time is shown on the Figure 5. Simultaneously with coagulation and precipitation adsorption effects took place. However, according with result of experiments the fats removal was much more effective than adsorption of iron salts at any dosage of aluminum sulfate. This phenomenon seemed to indicate that aluminum sulfate effective to enhance solids and fats removal and not so effective in removing of soluble components.

The properties of any specific fat molecule depend on the particular fatty acids that constitute it. Long chains of fatty acids are susceptible to intermolecular forces of attraction (in our case, van der Waals forces).

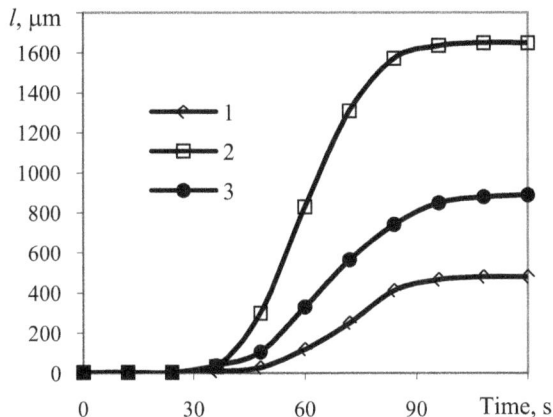

Figure 3: Flocs size as a function of time for different doses of aluminum sulfate: 1 – 10 µL/L; 2 – 30 µL/L; 3 – 50 µL/L.

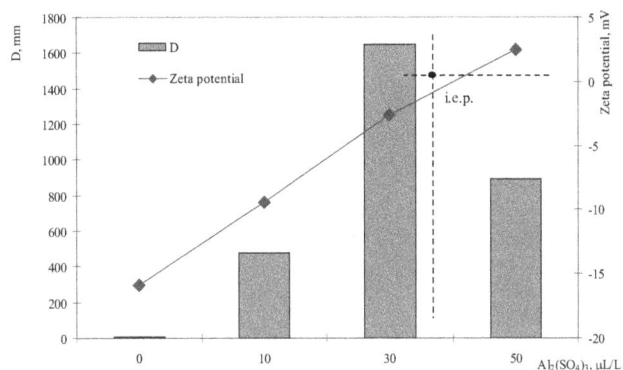

Figure 4: The influence of the dose of aluminum sulfate (10, 30 and 50 µL/L) on Zeta potential of the formed flakes (after 4 min).

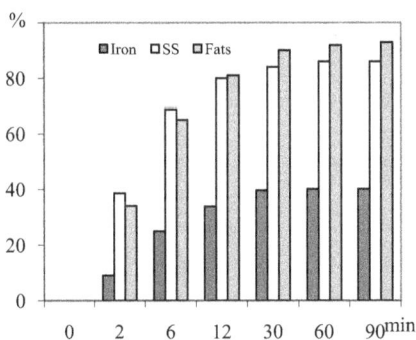

Figure 5: Effectiveness of wastewater treatment as a function time at the dosage of aluminum sulfate 30 µL/L.

Obviously the mechanism of coagulation both SS and fats include double layer compression and charge neutralization.

In order to increase effectiveness of purification activated γ-alumina (γ-Al$_2$O$_3$) obtained from the waste of aluminum alloys etching process [15] was studied as a "coagulant aid". The mean particle diameter of this sample was determined to be 0.6 µm and the specific surface area measured by BET method was 2.95 m^2/g.

According to Charge Distribution Multisite Complexation model

[8], on a surface of particles of γ-Al$_2$O$_3$ (hydro gels in water solutions), there are few versions of adsorption sites: on plane sides of γ-Al$_2$O$_3$ there is one version active adsorption sites at pH> 9; on costal are two versions, active in all interval pH.

Depending on pH suspensions it is formed positively or negatively charged surface, which charge is compensated by oppositely charged ions from a solution. At pH 3.5 the zeta potential of alumina particles is appropriately positive.

The results of the research have shown that the average diameter of the particles of powder of the sorbent was 5.22 µm, and the initial value of Zeta potential on the surface of the sorbent in wastewater was -15.6 mV at pH=6.7.

The kinetic and isotherm of the adsorption process of Fe^{3+} ions on the γ-Al$_2$O$_3$ are presented on the Figures 6 and 7.

The results of experiment showed expediency joint usage of sorbents and coagulants (Table 1).

The value of zeta potential decreased due to positive charge ions sorption. The metal ions are chemically bound to the $= Al-O^-$ group by an ionic exchange with H^+ group which is favored by the basic conditions of the medium [16]. We can suppose, that Fe^{2+} ions retain a positive charge, then coagulation can occur by electrostatic interaction between the positive sites $= Al-O-Fe^+$ on one particle and the

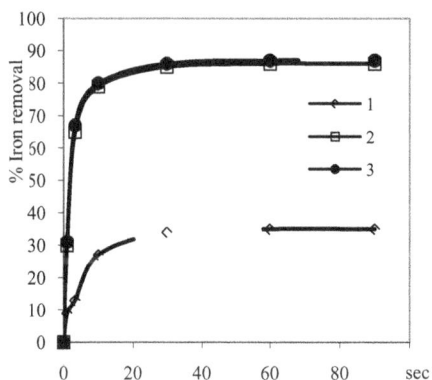

Figure 6: Kinetic plot of Fe^{3+} ions adsorption on γ-Al$_2$O^3 at three different dosages 1 – 10 µL/L; 2 – 30 µL/L; 3 – 50 µL/L.

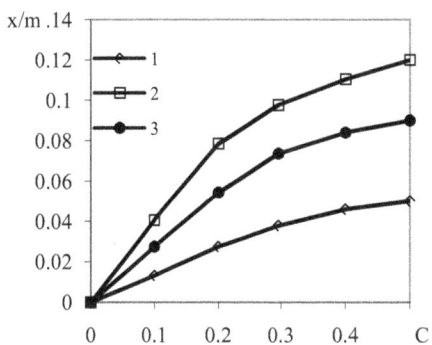

Figure 7: Isotherm plot of Fe^{3+} ions adsorption on γ-Al$_2$O$_3$ at three different dosages, were x=mass of adsorbate; m=mass of adsorbent; c=Equilibrium concentration of adsorbate in solution.

Way of treatment of waste water	Zeta potential, mV	Size of particles, μm	Iron removing, %	SS removing, %	Sedimentation speed, cm/min
Wastewater	-15.9	3.54	-	-	-
Al_2O_3	-42.8	0.61	-	-	-
Wastewater + Al_2O_3	-15.6	0.61	26	12	-
Wastewater + $Al_2(SO_4)_3$	-2.6	1650	40	86	15
Wastewater + $Al_2(SO_4)_3$ + Al_2O_3	-1,07	1890	85	95	75

Table 1: The composition of the wastewater before and after treatment.

negative groups $= Al - O^-$ on another; in this process, the Fe ions are acting as bridging units between the particles.

This is the mechanism to flocculate and to adsorb the metal ions in solution on the adsorbent's particles surface. As a result we had a very high effect of iron salts removing.

Indeed the value of zeta potential was still too high for flocculation of particles (in waste water there are particles of sorbent with zeta potential -15.6 mV and suspended particles -15.9 mV) and beginning of sedimentation process.

The residual concentration of iron salts in wastewater (effect of removing 90 %) can be explained by following. At low concentrations only the most thermodynamically favorable active surface groups participate in the adsorption process. The ion potential value for Fe^{2+} is 44.8 and for Fe^{3+} is 21.4; the energy of hydration for Fe^{2+} is -87.9 kJ/mol, and for Fe^{3+} is -135.6 kJ/mol [17]. So Fe^{2+} has more value of ion potential and less hydration degree and can be more adsorbed by negative charged surface of aluminum oxide. In the end of sorption we have residual concentration of Fe^{3+}.

Conclusions

Double layer compression and electric charge neutralization appeared to be the major factor in the coagulation process, as suggested by plotting zeta potential measurements and particle size during the experiments.

Joint usage of an inorganic sorbent and aluminum sulfate allows us to increase the value of Zeta potential and the effectiveness of purification almost twice. The sorbent particles act as additional condensation centers for hydrolysis products, which contributes to the increase of coagulation, the flakes of suspended solid particles become heavier, their hydraulic size increases. Besides, dissolved impurities are adsorbed from the wastewater. Use of dual aluminum sulfate – alumina systems can markedly enhance coagulation at a much lower dosage.

Acknowledgement

Authors are very grateful to Dr. Gudkova E.A., the Center of High Technologies of the Belgorod State National Research University for helpful collaboration during the zeta potential analyses.

References

1. Dikarevsky VS (2006) Disposal and purification of waste waters: study manual for university students majoring in "Water supply, waste-water disposal systems, efficient use and conservation of natural resources" Stroyizdat, Moscow.

2. Wang YF, Gao BY, Yue QY, Wang Y, Yang ZL (2012) Removal of acid and direct dye by epichlorohydrin–dimethylamine: Flocculation performance and floc aggregation properties. J Bioresour Technol 113: 265–271.

3. Wang YF, Gao BY, Yue QY, Wang Y (2011) Effect of viscosity, basicity and organic content of composite flocculant on the decolorization performance and mechanism for reactive dyeing wastewater. J Environ Sci 23: 1626–1633.

4. Nazarov VN (2007) A laboratory manual and a problem book on colloidal chemistry. Surface phenomena and dispersed systems. ICC, Moscow.

5. Lurie M, Rebhun M (1997) Effect of properties of polyelectrolytes on their interaction with particulates and soluble organics. Water Sci Technol 36: 93–101.

6. Haberkamp J, Ruhl AS, Ernst M, Jekel M (2007) Impact of coagulation and adsorption on DOC fractions of secondary effluent and resulting fouling behaviour in ultrafiltration. Water Research 41: 3794–3802.

7. Hiemstra T, van Rjemsdijk WH (1996) A surface structural approach to ion adsorption: the charge distribution (CD) model. J Colloid Interface Sci 179: 488-508.

8. Goldberg S, Davis JA, Hem JD (1995) The Surface Chemistry of Aluminum Oxides and Hydroxides. University of California, Berkeley, California.

9. Hiemstra T, Venema P, Van Rjemsdijk WH (1996) Intrinsic proton affinity of reactive surface groups of metal(hydr)oxides: the bond valence principle. J Colloid Interface Sci 184: 680-692.

10. Venema P, Hiemstra T, Weidler PG, van Riemsdijk WH (1998) Intrinsic proton affinity of reactive surface groups of metal (hydr)oxides: application to iron (hydr)oxides. J Colloid Interface Sci 198: 282-295.

11. Hiemstra T, van Rjemsdijk WH (1999) Surface structural ion adsorption modeling of competitive binding of oxyanions by metal (hydr)oxides. J Colloid and Interface Sci 210: 182-193.

12. Rietra RP, Hiemstra T, van Rjemsdijk WH (1999) The relationship between molecular structure and ion adsorption on variable charge minerals. Geochim et Cosmochim Acta. 63: 3009-3015.

13. Shevchenko TV, Mandziy UV, Tarasova MR (2003) Purification of waste waters with non-conventional sorbents. Ecology and Industry of Russia 1: 35–37.

14. Kosmulski M (2002) The pH-dependant surface charging and points of zero charge. J Colloid Interface Sci 253: 77-87.

15. Zueva SB (2011) Production aluminum oxide Russian Federation patent.

16. Pacheco S, Rodrigues R (2001) Adsorption properties of metal ions using alumina nano-particles in aqueous and alcoholic solutions. Sol-Gel Science and Technology 20: 263–273.

17. Kalukova EN, Pismenko VT, Ivanskaya NN (2010) Adsorption of Fe and Mn cations by inorganic sorbents. Sorption and Chromatographic process 10: 194-200.

21

Colour and COD Removal from Palm Oil Mill Effluent (POME) Using Pseudomonas Aeruginosa Strain NCIM 5223 in Microbial Fuel Cell

Hassan Sh Abdirahman Elmi*, Muhamad Hanif Md Nor and Zaharah Ibrahim

Environmental Biotechnology, Amoud University, Borama, Somalia

Abstract

Palm oil industries are the largest agricultural based industries in Malaysia and in processing palm oil, high pollutant liquid waste known as palm oil mill effluent (POME) is being generated. Currently, treatment of POME to meet the standard discharge has become an important issue. Therefore, this study was conducted to treat final discharge POME in microbial fuel cell (MFC). Double chamber MFC fabricated using polyacrylic sheets with a working volume of 1 L, proton exchange membrane (Nafion 115) were used. The anodic solution consisted of final discharge pond POME, overnight *Pseudomonas aeruginosa strain NCIM 5223* inoculum (10% v/v) and phosphate buffer (pH 7) while the cathodic solution consisted of phosphate buffer (pH 7) and potassium hexacyanoferrate (III). The results showed 58% of COD removal and 60% of colour removal in 8 days. In conclusion *Pseudomonas aeruginosa strain NCIM 5223* was able to remove colour and COD from final pond POME.

Keywords: Microbial fuel cell; COD; Wastewater treatment; POME; ADMI

Introduction

Palm oil tree (*Elaeis gunineensis*) is an equatorial plant rich in edible oils belonging to the family of Palmae. In the last decade, cultivation of palm oil increased rapidly in Malaysia and as reported it covered more than 11% of the land in 2003 [1]. Palm oil tree is planted in nursery and transferred to the plantation where it reaches the first harvesting time of 3 years [2]. Currently, Malaysia and Indonesia are the largest palm oil producing countries in the worlds palm oil export [3]. Palm oil industries are the largest agro based industries in Malaysia and the production of crude palm oil increased from 10.6 million tonnes in 1999, to more than 17.7 million tonnes in 2008, then from more than 45.9 million tonnes in 2010, to more than 50.2 million in 2011 tonnes [4,5]. In extracting crude palm oil, more water is used and liquid waste known as Palm oil mill effluent (POME) is being generated [6,7]. Wet palm oil milling process is the extraction method of crude palm oil from the Fresh Fruit Bunches (FBB) adapted in Malaysia [8]. Wet milling process consists of several stages including sterilisation, stripping, digesting, and oil extraction generating huge quantities of POME wastewater. Raw POME has many characteristics including that it is acidic in pH, brownish in colour, and contain environmental pollutant elements such as; COD, BOD, total solids, suspended solids, oil and grease [3]. Discharging POME without proper treatments can cause problems to the environment [9]. For this reason, Malaysian government has set Environmental Quality Act 1917 which defines the standard discharge limit of effluent. Biological treatment is the common treatment method of POME adopted in Malaysia though other treatments such as; physicochemical and membrane filtration is considered. Improving treatment methods of POME can contribute to minimize environmental pollution. In this study a double chamber microbial fuel cell separated by proton exchange membrane was fabricated from polyacrlylic sheets for the removal of chemical oxygen demand (COD) and colour from palm oil mill effluent (POME) in facultative anaerobic condition. POME and phosphate buffer was filled in the anode chamber while potassium hexacyanoferrate (III) and phosphate puffer was filled in the cathode chamber.

Methods and Materials

POME sampling and preparation

POME sample was collected from palm oil milling in Sedinak, Johor and stored at 4°C. Then POME was centrifuged at 4000 rpm for 15 minutes to remove the suspended solid. Then the supernatant was autoclaved at 121°C, 15 psi for 15 minutes to sterilize before it was used for treatment in the MFC. The process of sterilization was to kill indigenous microorganisms in the POME. Nine biological replicates were done during this study each with its control.

Preparation of the inoculum

Many colonies were isolated form POME sludge using bacterial isolating techniques. *Pseudomonas aeruginosa strain NCIM 5223* was among the best once in treating POME during the pre-testing process.

Designing MFC and analysis

As Figure 1 shows double chamber MFC was fabricated using polyacrylic sheets with a working volume of 1 L, proton exchange membrane (Nafion 115) were used. The anodic solution consisted of autoclaved final discharge pond POME, overnight *Pseudomonas aeruginosa strain NCIM 5223* inoculum (10% v/v) and phosphate buffer (pH 7) while the cathodic solution consisted of phosphate buffer (pH 7) and potassium hexacyanoferrate (III). After that, 10 mL of POME was removed in time interval from the MFC to analyze the removal of colour and COD from POME using APHA method on HACH DR 5000 [10].

*Corresponding author: Hassan Sh Abdirahman Elmi, Environmental Biotechnology, Amoud University, Borama, Somalia
E-mail: rabiic23@hotmail.com

DNA extraction

Isolation of genomic DNA was carried out according to Promega extraction kit (Wizard® Genomic DNA Purification Kit). Extracted DNA and 1 kb DNA ladder was loaded into 1% (w/v) polyacrylamide gel prepared from 0.50 g of agarose powder dissolved into 50 mL of 1 x TAE buffer stained with Ethidium Bromide (EtBr). Electrophoresis was run in 1 x TAE buffer at 90 V for 60 minutes.

Polymerase Chain Reaction (PCR)

Bacterial 16S rRNA gen was amplified by PCR using primers Fd1 (5'-AGA GTT TGA TCC TGGCTC AG-3') with 50% CG clamp and rp1 (5'-ACG GCT ACC TTG TTA CGA CTT-3') the PCR reaction and the thermal cycles were carried out as described by [11]. PCR products were again loaded in 1% (w/v) polyacrylamide gel then sent for sequences. Obtained sequences were compared with NCBI genomic database through BLAST and the phylogenetic tree were constructed using MEGA version 5 [12].

Result and Discussion

COD removal

The chemical oxygen demand (COD) indicates organic pollutants in the wastewater. Bacteria oxidises organic compounds in the wastewater for their growth and metabolism. The initial concentration of COD in the final discharge pond was 991 mg/L before the treatment and the final concentration of COD after the treatment was 441 mg/L. So *Pseudomonas aeruginosa strain NCIM 5223* was able to remove 550 mg/L of the COD (58%) in 8 days. Figure 2 shows, the percentage of COD removal in the MFC. According to [13] development of biofilm increase COD removal performances compared to the suspended system. *P. aeruginosa strain NCIM 5223* was developed biofilm on the anode and as the bacterial growth increased removal of the COD also increased. Significant COD removal was observed in the first three days of the treatment comparing to other days of the treatment. Similarly *P. aeruginosa strain NCIM 5223* reached the maximum growth in the first 3 days of the experiment. From the glucose concentration result more than 90 % of the glucose content was used up in the first 3 days. Therefore the relationship between the COD removal, *P. aeruginosa strain NCIM 5223* growth and the glucose removal was directly related.

Colour removal

The American dye manufacture intensity (ADMI) is the method of measuring colour intensity in the wastewater. The initial colour concentration in the final pond POME was 4440 ADMI and the final colour concentration of POME was 1740 which means 2700 ADMI (60 %) of the colour was removed. As show in Figure 3, there was no significant ADMI removal in the first days of the treatment. However; after the 4th day of the treatment the removal of ADMI was very high. This indicates that when almost all the glucose in the system was used bacteria started to remove the colour of the effluent. From the literature colour of the effluent is mainly contributed by the organic compounds, using these organic compounds can remove the colour of the effluent [14].

Bacterial analysis

Bacterium isolated was gram positive having morphological appearances of coccus when stained. Then genomic DNA was extracted following Promega extraction kit and run the PCR. After that the PCR product was sequenced and the obtained results were BLAST and found that the bacterium was *Pseudomonas aeruginosa strain NCIM*

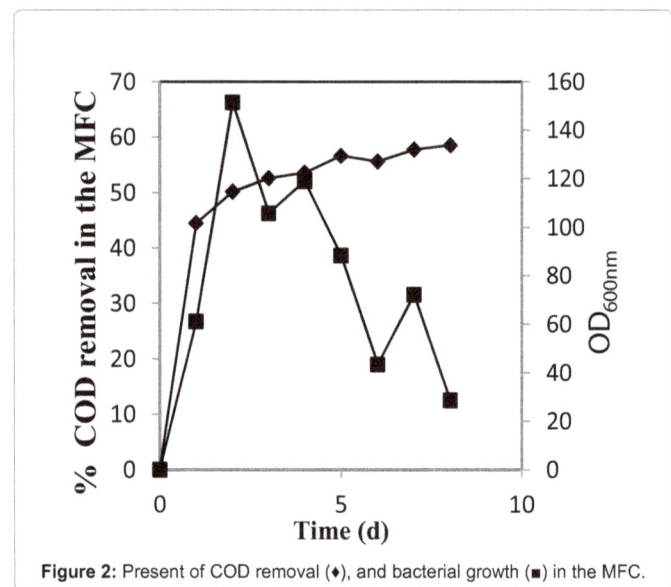

Figure 2: Present of COD removal (♦), and bacterial growth (■) in the MFC.

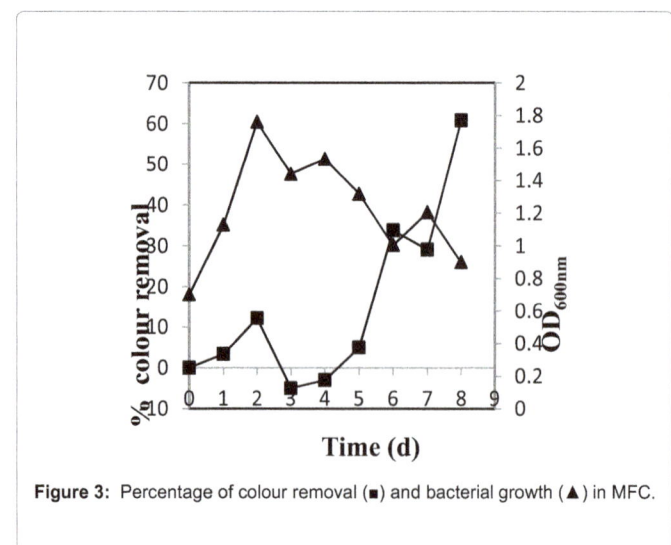

Figure 1: schematic diagram of the MFC used for the removal of COD and colour from POME.

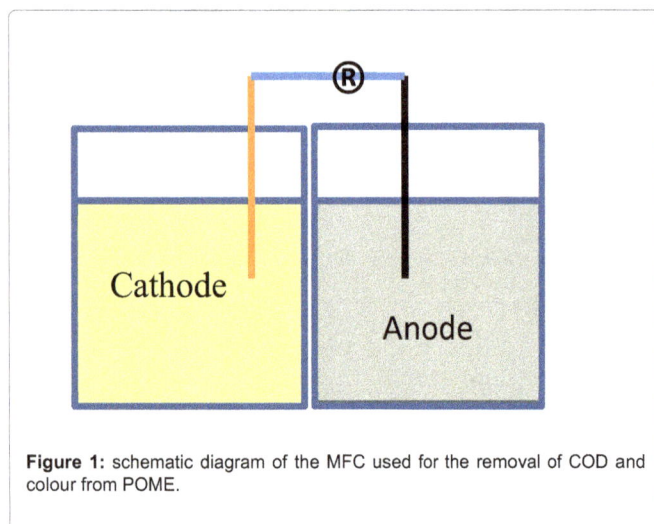

Figure 3: Percentage of colour removal (■) and bacterial growth (▲) in MFC.

5223. Figure 4, shows the phylogenetic tree of 15 different *Pseudomonas* groups. During the experiment it was observed that the bacterium was developed biofilm on the electrode and as reported by [3] biofilm formation improves the COD removal from the effluent [7,9,12].

Conclusion

Treatment of the final discharge POME was successfully conducted focusing on colour and COD removal using APHA method. The maximum COD removal achieved was 58% from the initial reading of 991 mg/L while the colour reduction was 60% from the initial reading of 4440 ADMI in 8 days.

References

1. Ahmad AL, Ismail S, Ibrahim N, Bhatia S (2003) Removal of suspended solids and residual oil from palm oil mill effluent. Journal of Chemical Technology and Biotechnology 78: 971-978.

2. APHA, AWWA, WEF (2005) Standard methods for the examination of water and wastewater. (21st ed.) Washington D.C: American Public Health Association.

3. Assas N, Ayed L, Marouani L, Hamdi M (2002) Decolorization of fresh and stored-black olive mill wastewaters by Geotrichum candidum. Process Biochemistry 38: 361-365.

4. Cappuccino J, Sherman N (2002) Microbiol. a laboratory manual. Pearson Education, Inc. San Francisco, CA.

5. Ibrahim AH, Dahlan I, Adlan MN, Dasti AF (2012) Comparative Study on Characterization of Malaysian Palm Oil Mill Effluent Research Journal of Chemical Sciences 2: 1-5.

6. Jia C, Xiuping Z, Jinren N, Alistair B (2010) Palm oil mill effluent treatment using a two-stage microbial fuel cells system integrated with immobilized biological aerated filters. Bioresource technology 101: 2729: 2734.

7. Li XN, Song HL, Li W, Lu XW, Nishimura O (2010) An integrated ecological floating-bed employing plant, freshwater clam and biofilm carrier for purification of eutrophic water. Ecological Engineering 36: 382-390.

8. Ma A (2000) Environmental management for the palm oil industry. Palm Oil Dev 30: 1-10.

9. Neoh CH, Lam CY, Lim CK, Yahya A, Ibrahim Z (2013) Decolorization of palm oil mill effluent using growing cultures of Curvularia clavata. Environmental Science and Pollution Research 21: 4397- 4408.

10. Neoh CH, Yahya A, Adnan R, Majid ZA, Ibrahim Z (2012) Optimization of decolorization of palm oil mill effluent (POME) by growing cultures of Aspergillus fumigatus using response surface methodology. Environmental Science and Pollution Research 20: 2912- 2923.

11. Rupani P, Singh R, Ibrahim M (2010) Review of current palm oil mill effluent (POME) treatment methods: Vermicomposting as a sustainable practice. World Applied Sciences 11: 70:81.

12. Tamura K, Peterson D, Peterson N, Stecher G, Nei M, et al. (2011) MEGA5: molecular evolutionary genetics analysis using maximum likelihood, evolutionary distance, and maximum parsimony methods. Molecular biology and evolution 28: 2731-2739.

13. Wood BJ, Pillai KR, Rajaratnam JA (1979) Palm oil mill effluent disposal on land. Agricultural Wastes 1: 103-127.

14. Wu TY, Mohammad AW, Jahim JM, Anuar N (2010) Pollution control technologies for the treatment of palm oil mill effluent (POME) through end-of-pipe processes. Journal of Environmental Management 91: 1467-1490.

22

Recommendations through a Complete Study on Healthcare Solid Waste Management Practices of Government Hospitals in Colombo, Sri Lanka

Liyanage Bundunee Chanpika[1]*, Athapattu Prathapage Priyantha[1] and Tateda Masafumi[2]

[1]Department of Civil Engineering, Faculty of Engineering Technology, The Open University of Sri Lanka
[2]Department of Environmental Engineering, Toyama Prefectural University, Toyama, Japan

Abstract

Eighteen hospitals in Colombo, Sri Lanka were investigated in terms of understanding the current situation of healthcare management. Out of these, central government and the rest are administrating ten hospitals by the provincial government. The focus points of this study were the following: 1) general information (i.e., names and types of hospital selected for investigation, amount of waste generated, numbers of beds, and so on); 2) waste types; 3) sources of waste generated; 4) segregation of healthcare waste; 5) waste storage, transportation, and disposal; and 6) adherence to regulations. Questionnaires and investigations involving direct visit were done to obtain more concrete data. The results of our study suggested that most of the hospitals investigated were neither satisfactory in terms of World Health Organization (WHO) guidelines for waste disposal, nor on environmental grounds. Several recommendations were made based on information obtained through this study.

Keywords: Healthcare solid waste; Hospital; Storage; Segregation; Treatment; Disposal; Regulation

Introduction

Healthcare wastes are defined to include all types of wastes produced by health facilities such as general hospitals, medical centers, and dispensaries. Blenkharn [1] introduced comprehensive literature on healthcare waste. Healthcare wastes represent a small amount of total residues generated in a community. However, such residues can potentially transmit diseases and present an additional risk to the staff of the healthcare facilities, patients, and the community when they are not managed properly [2,3]. Healthcare wastes may be classified into different types according to the source, type, and risk factors associated with their handling, storage, and ultimate disposal. Komilis et al. [4] investigated healthcare waste generation from different categories of healthcare facilities. Throughout the world, the health sector is developing and improving more rapidly compared to other economic sectors. However, it seems that the fraction of waste generated at healthcare institutions has not attracted the same level of attention as other types of wastes, particularly in developing countries, despite the fact that healthcare wastes are labeled as hazardous waste because they pose serious and direct threats to human health [5-10]. Healthcare waste poses serious threats not only to human health but also to economy of a country. An illicit economy based on healthcare waste has also been reported [11].

The healthcare service system in Sri Lanka can be divided into government and private hospitals. Governmental hospitals serve around 95% of the Sri Lankan population, and the private sector serves the rest 5%. The governmental service system in the country is divided into curative services and preventive services. Health care is provided free of cost to all people including the super specialty services. For preventive services, there is one Medical Officer of Health (MOH) in each Assistant Government Division (AGD), per 60,000 people. Peripheral Health Mid Wives (PHMW) and Public Health Inspectors support each MOH. They are responsible for the mother and child health programme and for the food hygiene and sanitation respectively. In each district, there is a Deputy Director Provincial Health officer (DDPHS); for each province, there is a Provincial Health Officer (PHO); and at the top of this hierarchy is the Director General of Health Services (DGHS). The number of private medical services is relatively small in Sri Lanka, and these services are predominantly located in Colombo and the other larger cities. Management of healthcare wastes is a major environmental issue in the country. Though healthcare services are responsible to manage healthcare wastes they generate, most of them fail to do this efficiently, which results in environmental pollution through such wastes.

This study aims at examining healthcare waste, especially solid waste management practices by surveying the current practices followed by the government hospitals in the Colombo District. Colombo district was selected as the survey area because all the hospitals in the district are government administered.

Methods of Research

Hospitals for investigation

Eighteen out of 26 government hospitals in the Colombo District were selected in order to characterize the healthcare wastes in the national context. Colombo District was selected as the survey area because it is the most important district in the country and has the most comprehensive healthcare centers in the country. Colombo is also the most populated, industrialized, urbanized, and developed district in the country based on the social and economic sectors. Additionally, the population in the district represents every ethnic group and every religious group in the country. The social framework of the district consists of municipal, urban, suburban, and rural areas. Every type of government hospital is located within this district. These include

*Corresponding author: Liyanage Bundunee Chanpika, Department of Civil Engineering, Faculty of Engineering Technology, The Open University of Sri Lanka, Nawala, Nugegoda, Sri Lanka
E-mail: bcliy@ou.ac.lk

categories like national hospitals, general hospitals, teaching hospitals, base hospitals, district hospitals, peripheral, rural and others hospitals. The central government while base, district, peripheral administrate national, general, base and other hospitals, and the provincial government administrates rural hospitals. There is a national hospital in Colombo, which has all the specialty and super specialty services, and is the apex referral center in the health system. In each district, there is one general hospital with specialties like ENT/ophthalmology, dermatology, and radiology apart from medical, surgical/pediatrics, and obstetrics and gynecology (OBG). Teaching hospitals are attached to the medical colleges. Base hospitals act as referral units with medical, surgical, pediatrics, and obstetrics and gynecology specialties. Further, there are five to eight district hospitals in each district, depending on district size and population. For curative services, the government has established peripheral units (PU) in rural villages. Rural hospitals offer basic treatments.

Research design

Research design for conducting this study consisted of four major tasks: 1) developing a questionnaire, 2) identifying the hospitals in the Colombo District to be visited for data collection, 3) conducting site visits to selected hospitals and collecting data and information through interviews and observations, and 4) analyzing survey results to make recommendations regarding sound healthcare waste management in the government hospitals in the Colombo District.

Questionnaire development

The questionnaire was developed based on the recommendations of the World Health Organization (WHO) for evaluation of hospital waste management in developing countries [6]. After taking into consideration specific differences which may exist in hospitals in Sri Lanka and the views of environmental specialists regarding the present problems in the management of medical waste and the expected results from the questionnaire, some modifications were made to the questionnaire suggested by the WHO. The questionnaires were divided into seven sections. They were 1) general information, such as names and types of hospitals selected for investigation, amount of waste generated, numbers of beds, and so on; 2) waste types; 3) sources of waste generated; 4) segregation of healthcare waste; 5) waste storage, transportation, and disposal; and 6) regulation adherence.

Site visits

Authors made two to five visits to each of the selected hospitals. Five visits to the national and teaching hospitals, three visits to base hospitals and two visits to the rest were made during this study. Each visit consisted of spending time in the different departments of the hospital, recording notes, and making observations about the healthcare waste management practices followed by staff responsible for the task. Regular visits were conducted to general medical wards, maternity wards, surgical and intensive care wards, operation theaters, and orthopedic sections, as well as waste collecting, treatment, and disposal areas. During those visits, the authors collected information examining the rules, procedures, and regulations to be followed by the personnel regarding the management of medical waste generated at the hospital. During the visits, one or two members of hospital staff, the head of the hospital, who is in charge of the infectious control unit, and a public health officer, accompanied the authors; in addition, two personnel engaged in waste management were interviewed for collecting data and information included in the questionnaire. Site visits were helpful in obtaining firsthand knowledge of handling and disposal

practices of healthcare wastes. The authors also made use of both primary and secondary data. Secondary data were obtained from the hospital documentation. Information was also obtained from published and unpublished books, journals, newsletters, periodicals, articles, and internet. Primary data were collected through questionnaires and interviews during the site visits.

Data analysis

The analysis is essentially descriptive. Data for the analysis were extracted from questionnaires and personal interviews carried out by the authors.

Results and Discussion

General information

The collected data was analyzed explaining the present healthcare waste management practices adopted by the government healthcare establishments in the Colombo District. To analyze the data, the investigated hospitals were divided into Groups A and B, based on the following facts. The hospitals in Group A are administrated by the central government and available for specific treatment facilities of operation theaters, radiotherapy, testing laboratory, intensive care, blood bank, wards, and clinics. The hospitals in Group B are administrated by the provincial government and include an outpatient department, dental unit, isolation ward, wards, and clinics. General information on the selected hospitals is summarized in Table 1(ANNEX).

A total of 14 hospitals in Group A were located in Colombo and consisted of one national and general, nine teaching, one base, and three other hospitals. Eight of the hospitals in Group A were investigated. On the other hand, a total of 12 hospitals from Group B existed in Colombo i.e., three base, three district, five peripheral unit, and one rural, out of which 10 hospitals were investigated in this study. Waste generation and bed occupancy represent averaged values at the time the authors received the data from the hospitals. As has been reported by Komilis et al. [4], our results also showed that waste generation increases with a higher number of beds. As shown in Table 1, the eight hospitals investigated from Group A and 10 from Group B generated 7,920 kg and 1,624 kg of healthcare waste per day, respectively.

Waste types identified

Table 2(ANNEX) shows the result of waste types identified by hospitals. In this study, it was revealed that, two hospitals out of the eight investigated (25%) in Group A separate all healthcare wastes into seven categories called as general, pathogenic, infectious, sharps, radioactive, recyclable, and other wastes, and only these two hospitals (25%) generate radioactive wastes. Another six hospitals out of the eight investigated (75%) in Group A separate their healthcare wastes into five categories. According to Pruss et al. [6], wastes are generated from the various activities carried out in the hospital. Types of wastes and their amount may vary from hospital to hospital and may depend on climatic season, location of the hospital, and many other factors. General wastes produced at hospitals are related to food preparation in the hospital kitchen or canteens, administrative activities, and land clearing. This type of waste is similar to domestic and municipal wastes. With regard to healthcare wastes in hospitals, different kinds of therapeutic activities such as cobalt therapy, chemotherapy, dialysis, surgery, delivery, resection of gangrenous organs, autopsy, biopsy, paraclinical testing, and injections, among other treatments, are carried out and result in the production of infectious and pathogenic wastes, contaminated sharps with patients' blood and secretions, radioactive wastes, and chemical materials, which are considered to be hazardous wastes.

Except for Kalubowila (34.0%) in Group A and Angoda Fever (27.0%) and Aturugiriya (52.6%) in Group B, it was found that general waste occupied a large portion of the total waste generated and was more than 60% in both the groups.

Waste source generated

Sources of waste generation may vary from hospital to hospital; the main categories of such sources being patients' services, theaters, laboratories, kitchens, canteens, staff hostels, pharmacies, and gardens.

Table 3 (ANNEX) shows the presence of different categories of sources of waste generation in the hospitals investigated. Our study revealed that all hospitals investigated in Group A have patients services, laboratories, kitchens, canteens, staff hostels, pharmacies, and gardens and only one out of the eight hospitals investigated in Group A did not have theaters as a waste generation source. All hospitals investigated in Group B have patients' services, pharmacies, and gardens as their sources of waste generation. Meanwhile four out of ten hospitals have theaters, eight out of ten have laboratories, two out of ten have kitchens, four out of ten have canteens, and three out of ten have staff hostels. The number of waste generation sources of the hospitals in Group B as a whole is less than that of Group A.

Segregation of healthcare waste

Rao et al. [12] have specified identification and segregation of wastes as the main steps in healthcare waste management. The data collected regarding segregation status by hospitals (Table 4 (ANNEX)) in this study shows that 100% of the hospitals in Group A separate healthcare by waste category and also adopt a color code system. Meanwhile, three hospitals out of ten in Group B separate their healthcare wastes into four categories: infectious, sharps, recyclable, and general wastes. Four hospitals out of ten separated only sharps as their healthcare wastes and the remaining healthcare wastes were classified as general wastes. The remaining hospitals investigated in Group B did not practice any segregation of healthcare wastes and collected and disposed of all wastes as garbage. This also demonstrated that none of the hospitals investigated in Group B practice a color code system. Figure 1 shows

Figure 1: Segregation as per guidelines of color code system at Castle Hospital.

segregation of healthcare waste by Castle Hospital as per the color code system.

According to the WHO guidelines [6], healthcare wastes are categorized as infectious, pathogenic, sharp, chemical, pharmaceutical, radioactive, wastes with high content of heavy metals, and pressurized containers. However, the healthcare waste categorization used by the hospitals investigated in our study is different from that of WHO. For instance, wastes with heavy metals, chemical wastes, and pressurized containers are not categorized or collected as separate wastes and are disposed of with other wastes in those hospitals. Pharmaceutical wastes in those hospitals are not categorized as wastes, but also do not enter the waste stream since outdated pharmaceuticals are returned to the manufacturing companies to be destroyed. The hospitals basically separate medical wastes from the general waste stream at the waste generation points. Thereafter, they are stored and disposed of separately. However, in terms of qualitatively considerations, the segregation of healthcare wastes differed from hospital to hospital or even from ward to ward within a given hospital. Generally, in the wards, doctors and nurses are required to drop the used sharps into different containers, but this is not diligently followed. Users of sharps sometimes leave them on hospital beds, which could be very dangerous to patients. The mixing of different categories of waste is common in these hospitals. According to recommendations by WHO, hospitals have to provide plastic bags and strong plastic containers, such as empty containers of antiseptics used in the hospital, for infectious waste. Bags and containers for infectious waste should be marked with the biohazard symbol [6]; however, some hospitals do not label infectious waste in this way. According to Franka et al. [13], maintaining a clean environment and disposal of medical waste are social obligations of hospitals. Meanwhile, Johannessen et al. [14] stated that proper management of medical waste could minimize the risk both within and outside healthcare facilities. The first priority is to segregate wastes, preferably at the point of generation, into reusable and non-reusable, hazardous and non-hazardous components. However, considering these recommendations or WHO guidelines, existing practices followed by the Health-care workers for segregation in most of these hospitals are not satisfactory. The importance of training and education with regards to reduction and proper segregation of healthcare waste has been stated by several researchers [15,16] and might prove as a key factor in bringing a positive change in this direction.

Waste collection, storage, transportation, and disposal

Table 5 (ANNEX) shows frequency of waste collection by hospitals investigated in Groups A and B. Six of eight hospitals in Group A transport wastes collected in wards or clinics to waste storage areas daily, and the other two hospitals transport these wastes to storage areas once every three days. In Group B, out of ten hospitals, transport of wastes collected in wards or clinics to storage areas occur daily in three hospitals, once every two days in two hospitals, once every three days in four hospitals, and weekly in one hospital.

The place where the hospital waste is kept before it is transported to the final disposal site is termed a "temporary waste storage area." This area must be well sanitized and secured in such a way that it should be accessible only to authorized persons [6]. As per recommendations of WHO for healthcare waste storage, storage areas are to be free of odor and must discourage the harborage of vermin. Healthcare facilities must provide an enclosed structure such as a shed, garage, cage, fenced area, or separate loading bay to store waste. The holding area should be located away from food and clean storage areas; it must not be accessible to the public, have a lockable door and rigid impervious flooring. Clean-up facilities, spill kits, appropriate drainage, and so on should be provided where wastes are stored in locked bins. A specific area with adequate drainage for washing equipment should be designated. All hospitals investigated in this study have temporary storage areas. The wastes are kept in these temporary storage areas until they are disposed of or transported off-site.

Table 6 (ANNEX) shows conditions of temporary wastes storage areas of hospitals investigated in Groups A and B. None of the hospitals investigated in this study carry out compliance checks against the recommendations of WHO for healthcare wastes storage. Only one out of eight hospitals investigated in Group A has a waste storage area with a cemented floor and roof. Meanwhile, five out eight hospitals have provided open containers for temporary waste storage and two other hospitals in this group store their healthcare wastes on the open ground, at a place designated for the purpose, in the hospital premises. Three out of ten hospitals investigated in Group B store their wastes on the open ground, but these areas have been separated with fixed fences and lockable doors. However, the rest of hospitals in Group B store their healthcare wastes at the designated areas on the open ground. Figures 2 and 3 show current healthcare waste storage conditions at the hospitals investigated.

Table 7 (ANNEX) shows the ways of disposal for each category of wastes by the investigated hospitals in Groups A and B. The upper and lower values in the table (Upper/Lower) represent the values of Groups A and B, respectively. In Group A, all hospitals dispose of their general wastes through municipal councils following WHO recommendations. Six hospitals dispose of their pathogenic wastes through funeral parlors, but the process is not monitored by these hospitals, and is also not recommended as per the WHO guidelines. One out of eight hospitals disposes the pathogenic waste through outsourced incineration. In terms of disposing of their infectious wastes, five hospitals dispose them of through municipal councils, two hospitals burn them openly on the hospital premises, and the other one incinerates them. As for the disposal of sharps, two hospitals incinerate them on the premises, four incinerate them through outsourcing, and the other two burn them openly on the hospital premises. Two hospitals that generate radioactive wastes dispose of their radioactive wastes by outsourcing. Seven hospitals outsource recyclable wastes for recycling and the other one disposes them of openly on the hospital premises. Two hospitals that generate radioactive wastes segregate them from the other types of wastes, and incinerate them on the hospital premises.

Figure 4 shows open burning and dumping conditions in some of the hospitals investigated. As is evident, the methods for healthcare waste management followed by these hospitals are quite basic. Soares et al. [17] discussed the importance of Life Cycle Assessment (LCA) and cost analysis for healthcare waste management. These methods should be applied for healthcare waste management in developing countries.

As per the Gazette No: 1534/18 dated 02/01/2008, published under the National Environmental Act, all healthcare centers including hospitals should obtain an Environmental Protection License and a Scheduled Waste Management License from the Central Environmental Authority.

Table 8 (ANNEX) shows the status of investigated hospitals in terms of regulation adherence. However, our investigation revealed that eight hospitals from Group A have applied for licenses and have been taking corrective actions to meet the requirements. Meanwhile, in Group B, only one hospital out of ten, i.e., Thalangama hospital, has applied for the license and is working towards fulfilling the conditions. While five hospitals, i.e., Angoda Fever, Avissawella, Homagama, Moratuwa, and Premadasa Memorial Maligawatta hospitals, are planning to apply for these licenses and another four out of ten have taken no action in this regard. Table 9 shows number of hospitals in Groups A and B where workers wear safety equipment. According to the guidelines of WHO, workers who engage in healthcare wastes handling activities should wear safety equipment such as overalls, boots, hand gloves, face masks. But this investigation reveals that in eight hospitals in Group A, healthcare workers wear only hand gloves, and in eight hospitals out of ten in Group B wear only hand gloves as safety equipment. Caniato et al. [18] mention that not only is regulation effective, but involvement of stakeholders is also vital to the improvement of healthcare management.

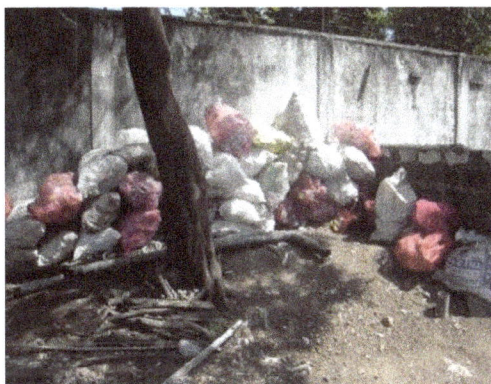

Figure 3: Temporary storage on open land at the hospital investigated in this study.

Figure 2: Uncovered storage of segregated recyclable healthcare waste for outsource at the hospital investigated in this study.

Figure 4: Waste treatment at some of the hospitals investigated in this study; (a) open burning of sharps; (b) open burning in metal barrels; (c) dumping in a shallow pit.

Conclusions and Recommendations

The following are summarized as conclusions of this study:

- Most of the investigated hospitals are aware of the risks or health impacts of healthcare wastes, but are not aware of the environmental damage or pollution that may result from the poor management of healthcare wastes.

- Policies and regulations of healthcare waste management do not address the lack of physical and financial facilities, or the absence of adequate and responsible staff in hospitals.

- In considering sound and environmentally friendly healthcare waste management or the guidelines of WHO in this regard, some healthcare waste practices in the hospitals investigated were satisfactory, but most of them were unsatisfactory.

Using the information obtained through this study, especially observations and literature, the following recommendations should be made for healthcare waste management in government hospitals in Sri Lanka, particularly in terms of the Environmental Protection License and Schedule Wastes License.

- Formal regulations on healthcare waste segregation, treatment, and final disposal must be established.

- There should be a healthcare waste management plan, training, and education in each hospital with respect to waste minimization, efficiency from waste segregation up to disposal, and minimization of health risks and environmental pollution.

- Adequate and necessary physical or financial facilities, especially waste containers, waste storage, transport, treatment, disinfection, disposal facilities, etc., should be provided for hospitals.

- A separate department with necessary staff including a healthcare waste expert, at least for general hospitals, teaching hospitals, and base hospitals should be employed with respect to healthcare waste management.

- It is recommended that the Central Environmental Authority in coordination with the Ministry of Health formulate a manual of guidelines with respect to healthcare waste management.

- Needle cutters should be provided for hospitals to separate needle sharps from the waste stream at their generation points.

- The hospitals should provide necessary training and education for all staff including doctors and other workers with respect to safety, health risks, and environmental issues in healthcare waste handling.

- Healthcare waste should be transported using dedicated, wheeled, leak-proof containers or vehicles.

Acknowledgement

The authors deeply appreciated the selected hospitals and their staffs who were willing to support the authors' investigation through questionnaires or direct visits. The study would not have been possible without the understanding and cooperation extended by the stakeholders at the hospitals.

References

1. Blenkharn JI (2011) Clinical waste management. Encyclopedia of Environmental Health. Elsevier Science, Philadelphia, USA.

2. Silva CED, Hoppe AE, Ravanello, MM, Mello N (2005) Medical waste management in the south of Brazil. Waste Management 25: 600-605.

3. Hossain MS, Santhanam A, Norulaini NAN, Omar AKM (2011) Clinical solid waste management practices and its impact on human health and environment – a review. Waste Management 31: 754-766.

4. Komilis D, Fouki A, Papadopoulos D (2012) Hazardous medical waste generation rates of different categories of health-care facilities. Waste Management 32: 1434-1441.

5. Coad A (1992) Inter-Regional Consultation on Hospital/Infectious Wastes Management in Developing Countries. World Health Organization, Geneva, Switzerland.

6. Prüss A, Girout E, Rushbrook P (1999) Safe management of wastes from healthcare activities'. World Health Organization, Geneva, Switzerland.

7. Oweis R, Mohamad AW, Ohood AL (2005) Medical waste management in Jordan: a study at the King Hussien Medical Centre. Waste Management 25: 622-625.

8. Blenkharn JI (2006) Standards of clinical waste management in UK hospitals. Journal of Hospital Infection 62: 300-303.

9. Patwary MA, O'Hare WT, Sarker (2011) Assessment of occupational and environmental safety associated with medical waste disposal in developing countries: a qualitative approach. Safety Science 49: 1200-1207.

10. Patwary MA, O'Hare WT, Sarker (2012) Occupational accidents: an example of fatalistic beliefs among medical waste workers in Bangladesh. Safety Science 50: 76-82.

11. Patwary MA, O'Hare WT, Sarker (2011) An illicit economy: scavenging and recycling of medical waste. Journal of Environmental Management 92: 2900-2906.

12. Rao SKM, Ranyal RK, Bhatia SS, Sharma VR (2004) Biomedical waste management' an infrastructural survey of hospitals. Medical Journal Armed Force India 60: 379-382.

13. Franka E, Zoka AHE, Hussein AH, Elbakosh MM, Arafa AK, Ghenghesh KS (2009) Hepatisis B virus and hepatitis C virus in medical waste handlers in Tripoli, Libya. J Hosp Infect 72: 258-261.

14. Johannessen, JM, Dijkman M, Bartone C, Hanrahan , Boyer G, et al. (2000) Health Nutrition and Population Discussion Paper: Healthcare Waste Management Guidance Note. World Bank, Washington DC, USA.

15. Mosquera M, André-Prado MJ, Rodríguez-Caravaca G, Latasa P (2014) Evaluation of an education and training intervention to reduce health care waste in a tertiary hospital in Spain. Am J Infect Control 42: 894-897.

16. Oroei M, Momeni M, Palenik CJ, Danaei M, Askarian M (2014) A qualitative study of the causes of improper segregation of infectious waste at Nemazee Hospital, Shiraz, Iran. J Infect Public Health 7: 192-198.

17. Soares SR, Finotti AR, Silva VPd, Alvarenga RAF. (2013) Applications of life cycle assessment and cost analysis in health care waste management. Waste Management 33: 175-183.

18. Caniato M, Tudor T, Vaccari M (2015) Understanding the perceptions, roles and interactions of stakeholder networks managing health-care waste: a case study of Gaza Strip. Waste Management 35: 255-264.

ANNEX:

Group	Hospital Name	Type of Hospital	Waste Generation (tons/day)	Number of Beds	Bed Occupancy (%)
A (Central Government)	Colombo General	National and General	4,444	2,996	84.0
	Angoda Mental	Teaching	329	900	33.3
	Cancer Institute	Teaching	495	879	96.7
	Castle	Teaching	723	485	90.7
	De-Soysa	Teaching	232	227	113
	Eye	Teaching	524	463	64.8
	Kalubowila	Teaching	441	1,094	91.4
	Lady Ridgeway (LR)	Teaching	732	873	26.7
B (Provincial Government)	Angoda Fever	Base	22.2	144	53.3
	Avissawella	Base	654	477	73.8
	Homagama	Base	591	333	78.5
	Moratuwa	District	48.6	64	37.5
	Premadasa Memorial-Maligawatta	District	29.4	53	34.0
	Wetara	District	43.8	107	35.7
	Nawagamuwa	Peripheral Unit	78.5	60	25.0
	Piliyandala	Peripheral Unit	89.6	106	45.5
	Thalangama	Peripheral Unit	48.0	48	54.6
	Aturugiriya	Rural	19.0	53	60.0

Table 1: General information regarding the hospitals investigated in this study.

Group	Hospital Name	General	Pathogenic	Infectious	Sharps	Radioactive	Recyclable	Others	Total
A	Colombo General	3,000	260	1,000	30	3	150	0.5	4,444
	Angoda Mental	300	0	4	5	0	20	0	329
	Cancer Institute	320	70	10	10	25	56	4	495
	Castle	550	20	96	12	0	45	0	723
	De-Soysa	160	23	5	6.5	0	37	0	232
	Eye	450	0.3	66	4	0	4	0	524
	Kalubowila	150	85	50	26	0	130	0	441
	Lady Ridgeway (LR)	600	25	80	7	0	20	0	732
B	Angoda Fever	6	0.5	10	0.7	0	5	0	22.2
	Avissawella	500	45	47	2	0	60	0	654
	Homagama	420	54	60	12	0	45	0	591
	Moratuwa	35	0.25	6	0.3	0	7	0	48.6
	Premadasa Memorial Maligawatta	20	0	7	0.4	0	2	0	29.4
	Wetara	30	0.25	3	2	0	8.5	0	43.8
	Nawagamuwa	67	0.24	7	0.25	0	4	0	78.5
	Piliyandala	75	0.07	8	1	0	5.5	0	89.6
	Thalangama	40	0	2	4	0	2	0	48.0
	Aturugiriya	10	0	5	2	0	2	0	19.0

Table 2: Types of healthcare wastes generated from the hospitals.

Group	Hospital Name	Patients' Services	Theaters	Labs	Kitchens	Canteens	Staff Hostels	Pharmacies	Gardens
A	Colombo General	√	√	√	√	√	√	√	√
	Angoda Mental	√		√	√	√	√	√	√
	Cancer Institute	√	√	√	√	√	√	√	√
	Castle	√	√	√	√	√	√	√	√
	De-Soysa	√	√	√	√	√	√	√	√
	Eye	√	√	√	√	√	√	√	√
	Kalubowila	√	√	√	√	√	√	√	√
	Lady Ridgeway (LR)	√	√	√	√	√	√	√	√
B	Angoda Fever	√		√		√	√	√	√
	Avissawella	√	√	√		√	√	√	√
	Homagama	√	√	√	√	√	√	√	√
	Moratuwa	√		√		√		√	√
	Premadasa Memorial-Maligawatta	√			√			√	√
	Wetara	√	√	√				√	√
	Nawagamuwa	√		√				√	√
	Piliyandala	√	√	√				√	√
	Thalangama	√		√				√	√
	Aturugiriya	√						√	√

Table 3: Sources of waste generated.

Group	Hospital Name	Segregation as Category Wise with Color Code	Segregation into Four Category	Segregation only Sharps	No Segregation
A	Colombo General	√			
	Angoda Mental	√			
	Cancer Institute	√			
	Castle	√			
	De-Soysa	√			
	Eye	√			
	Kalubowila	√			
	Lady Ridgeway (LR)	√			
B	Angoda Fever		√		
	Avissawella		√		
	Homagama		√		
	Moratuwa			√	
	Premadasa Memorial-Maligawatta			√	
	Wetara			√	
	Nawagamuwa			√	
	Piliyandala				√
	Thalangama				√
	Aturugiriya				√

Table 4: Segregation at the various hospitals investigated in this study.

Group	Daily	Once in Two Days	Once in Three days	Weekly
A	6	0	2	0
B	3	2	4	1

Table 5: Healthcare waste collection patterns.

Group	Covered Fence and Locked Area	Cemented Floor with Roofs	Cemented Floor without Roofs	Covered Container	Open Container	Open Ground
A	0	1	0	0	5	2
B	3	0	0	0	0	7

Table 6: Healthcare waste storage conditions.

	Disposal Method	General	Pathologic	Infectious	Sharps	Radioactive	Recyclable	Others
A/B	Municipal Council	08-Jul		5/-				
	Disposal by Funeral Parlors		06-Jan					
	Open Dumping	-/2	-/1				01-Jan	
	Incineration in Site			1/-	4/-			2/-
	Incineration out site		1/-		2/-			
	Bury		-/5					
	Outsources					2/-	07-Jun	
	Open Burning	-/1		02-Sep	02-Sep		-/3	
	Burn and Bury			-/1	-/1			

Table 7: Ways of disposal for each category of wastes by the hospitals investigated.

Group	Applied and Rectifying	Planning to Applied	No Action
A	8	0	0
B	1	5	4

Table 8: Regulation adherence.

Group	Overalls	Boots	Hand Gloves	Face Masks
A	0	0	8	0
B	0	0	8	0

Table 9: Safety equipment offered by the hospitals.

23

Utilization of Pectinases for Fiber Extraction from Banana Plant's Waste

Sunita Chauhan* and Sharma AK

Kumarappa National Handmade Paper Institute (KNHPI), Ramsinghpura, Sikarpura Road, Sanganer, Jaipur, India

Abstract

Today, biotechnology is perceived as a revolution throughout the world. With biotechnology, certain crops have been developed that can withstand the brutalities of weather changes, helping poor farmers of the developing countries to retain their yield and increase their output manifold. Biotechnology has also made agriculture more competitive and sustainable by creating new non-food markets for crops. To exploit the vast potential of biotechnology involved in non-food plant-products, the present study was taken up to explore the possibilities of improving the fiber extraction process of banana plant with the help of commercially available pectinase enzyme. Waste biomass of banana plant is widely available in many countries and the fiber extracted from its pseudo stem has utility for diversified range of applications including the manufacture of good quality handmade paper. The enzymatic treatment of green stem and trunk of banana plant before extracting fiber with the Raspador machine has resulted into an improvement in the yield as well as the quality of fiber obtained. This may not only result into a better utilization of the waste biomass of banana plant but may also increase profitability of the banana cultivators besides providing a source of good raw material for making handmade paper.

Keywords: Banana psuedo stem; Waste biomass utilization; Fiber; Handmade paper; Pectinase; Raspador machine

Introduction

Banana plant is a valuable bioresource which is distributed in more than 120 countries, over an area of 48 lakh hectares, with an annual production of 99.99 million tons (Indian Horticulture Database, 2011). The banana plant is highly valued for its fruit, but it also yields vast quantities of bio-mass residues from the trunk and fruit bunch (raquis), which are discarded on the field or – in the case of raquis – at the fruit processing sites (packing for exports). Thus banana farming generates huge quantities of biomass most of which goes as waste due to non-availability of suitable technology for its commercial utilization. From these residues, good quality of fibers can be extracted along with numerous other plant components (juice) with bioconversion potential. India has about 8.3 lakh ha under banana cultivation producing approximately 51.18 million tons of pseudo stem waste per year. This can be profitably used for extracting approximately 3.87 million tons of fiber [1]. Banana fibers are mainly placed in the superimposed leaves which form the pseudostem and provide resistance to the plant. These fibers run longitudinally along the leaves [2,3].

There are two ways to extract banana fiber i.e. either manually by hand or by mechanically through Raspador machine. Therefore, two qualities viz. hand-extracted and machine-extracted banana fiber are available in the market. Machine-extracted fiber is the low-grade fiber and cheaper in cost while the hand-extracted fiber is good in quality with higher price. Hand-extracted fiber has been found suitable for making high-grade paper due to its high purity while due to the presence of adherent pith; the machine-extracted fiber produces inferior quality of the product. Although appreciable efforts have been made by various institutions like Krishi Vigyan Kendra and Khadi & Village Industries Commission (KVIC), Mumbai, India to develop improved versions of machines for fiber extraction but even the best available machine in the market is not able to take out the fiber in its purest form and lot of pith remains attached to the extracted fiber. This pith creates problems during its utilization for making specialty handmade papers and behaves as a dead load on the fiber thereby consuming lot of chemicals, resulting into poor quality of the product and raising concerns about environmental pollution. Therefore, a suitable technology is the need of the hour for improving the quality of machine-extracted fiber so

that it may become a good quality cellulosic fiber for making varieties of handmade paper as well as various other fibrous products, thereby promoting better utilization of this valuable bio resource.

There are several new technologies using enzymes able to modify fiber parameters, achieve desired properties, improve processing results and ecology in the area of bast fiber processing and fabric finishing. Enzymatic retting of flax, enzymatic cottonization of bast fiber, enzymatic hemp separation, enzymatic processing of flax rovings before wet spinning etc. create a new group of technologies supported by effective mechanical treatments [4]. A lot of work has been carried out on retting of bast fibers like flax, hemp etc. and similarly the enzyme treatment aspects of textile fibers has been studied extensively. However not many reports are available in the literature about using certain enzymes for extracting leaf fibers particularly from the banana plant.

Pectinases are believed to play a leading role in the processing of bast and leaf fibers, since 40% of the dry weight of plant cambium cells is comprised of pectin [5]. The enzymatic processing of plant fibers using pectinolytic enzymes result in no damage to the fibers and most importantly in addition to being energy conservative is environmentally friendly [6,7]. Pectinolytic enzymes from *Actinomycetes* are reported to be used for the degumming of ramie bast fibers [8]. Pectinase obtained from *Bacillus pumilus dcsr1* has been used for the treatment of dried and decorticated ramie fibers [9]. Jacob et al. [10] have reported scanning electron microscopic studies of the banana fibers treated with crude pectinase obtained from *Streptomyces lydicus* and showed that fiber cells were intact in the control while the cells were separated in

*Corresponding author: Sunita Chauhan, Scientist, Kumarappa National Handmade Paper Institute (KNHPI), Ramsinghpura, Sikarpura Road, Sanganer, Jaipur, India
E-mail: itsneeru@yahoo.com

the treated samples.

Keeping in view of above, the present study was taken up to explore the possibilities of improving the quality of machine-extracted banana fiber through pectinase treatment of green stem/trunk of banana plant and the waste banana leaves. Extraction of better quality banana fiber through enzymes may help in addressing the issues of global warming and other environmental concerns as well as generating an innovative outlet for sustainable development all over the world.

Experimental

Fiber extraction from waste banana leaves through enzymatic and non-enzymatic route

The waste banana leaves (broad having thick mid ribs) were procured from the vegetable market of Muhana Mandi Sanganer. In addition to this, small cut pieces of the branches of banana could also be obtained. For the fiber extraction studies, mid ribs were de-leaved and the small branches were cut vertically to divide into three layers.

Analysis of fiber yield

The yield percentage of extracted fiber was calculated using the formula:

Yield, % = $\dfrac{\text{OD Weight of the Extracted Fiber}}{\text{OD Weight of the Plant Part Used For Extraction}}$ x 100

Pulping of the banana fiber extracted from waste banana leaves

The banana fiber extracted at KNHPI using the Banana Fiber Extractor with and without treatment was subjected to enzyme treatment (0.5% enzyme at 40°C for 4 hours) followed by "Open Hot Digestion", the pulping procedure commonly utilized in the handmade paper sector (NaOH-8%, time-3.5 hours, temperature-100°C, bath ratio: 1:8). Black liquor was collected before thoroughly washing the cooked fibers with tap water. Then the cooked material was subjected to the beating to a CSF (Canadian Standard Freeness) of 300-400 ml using the Standard Test Methods (IS 6213, T-227). Actual time to attain the desired degree of beating i.e. CSF was also noted in all the cases so as to evaluate the effect of enzyme treatment on beating energy.

Characterization of black liquor collected

Black liquor collected was characterized for the parameters of interest viz. pH, Total solids and Residual Active Alkali (RAA). All of these parameters were tested in duplicates and the average values have been reported here.

The pH was determined at 30°C using the standard TAPPI Test method number T-625 CM-85.

Total solids (% w/w) were estimated by drying 10-20 g of black liquor in pre-weighed glass petri dishes in the oven at 102 + 2°C for an overnight. Then the dried contents were weighed to the constant weight. Petri dishes were cooled in desiccators before weighing. Calculation was done as per the given formula:

Total solids, % w/w = $\dfrac{\text{Weight of dried contents}}{\text{Weight of black liquor originally taken}}$ x 100

For determining the Residual Active Alkali in black liquor, 25 ml of it was taken in a small beaker and its pH was noted with pH meter. The liquor was then titrated with 0.1 N HCl to a pH of 7.0 and the RAA

was calculated using the formula given below. (Standard TAPPI Test method no. TAPPI T-625 CM-85).

R.A.A. (g/l as NaOH) = $\dfrac{\text{Milli equivalent of acid x 0.1x40}}{\text{Volume of black liquor taken (25 ml)}}$

Evaluation of physical strength properties of pulps obtained

All the pulps thus obtained from the extracted banana fibers were used for making hand sheets of 60 GSM as per the TAPPI Standard procedures using the British Sheet Former. These hand sheets were evaluated for the physical strength properties using standard TAPPI TEST Methods and/or ISO methods (Tensile Strength: T-411, Tear Strength: ISO-1924, Grammage (GSM): T-423) so as to ascertain the effect of enzyme treatment on the quality of extracted banana fiber. A total of three hand sheets were evaluated for each of the test parameters and values reported here is an average of them.

Processing of the waste obtained from fiber extraction

The left over waste of banana fiber extraction through the enzymatic and non-enzymatic route were cooked separately with 4% NaOH at boiling temperature for a period of 3-4 hours, the pulps thus obtained were washed thoroughly and subjected to the beating to a CSF (Canadian Standard Freeness) of 300-400 ml. Hand sheets of 60 GSM were prepared as mentioned in 3.5 above.

Fiber extraction from green stem and trunk of banana plant

To study the effect of enzyme treatments on fiber extraction from the green stem of banana and banana trunk, experiments were conducted at one of the fiber extraction facilities at Guntur, Hyderabad, Andhra Pradesh, India so as to utilize the opportunity of the availability of both the Raspador machine and the green stem/trunk together at a single location. For this, besides observing and understanding the routine process of fiber extraction using the Raspador machine, enzyme soaking of the banana stem and trunk was carried out under the optimized conditions separately in two different drums. Further, the enzyme liquors and the pieces of banana trunk and stem each were collected at an interval of 12 hours, 24 hours, 36 hours and 48 hrs. Then the fibers were extracted from those pieces with the help of Raspador for their further evaluation at KNHPI, Jaipur.

Analysis of enzyme liquors collected from fiber extraction

The enzyme liquors collected before fiber extraction from the treated banana stem and trunks at the intervals of 12, 24, 36 and 48 hours were analyzed for color and total solids. Color of the filtrates was determined by measuring absorbance at 465nm and converting it into the Platinum Cobalt Units (PCU) using the conversion factor (500 PCU=0.41 Absorbance). Total solids were determined by drying a measured amount of liquor in a petri dish kept inside the Hot Air Oven at 100±2°C for an overnight. lignin content in the enzyme filtrate was quantified according to TAPPI method T-222 by measuring absorbance at a wavelength of 280 nm using 20.2 l/g/cm as the extinction coefficient [11].

Analysis of fiber extracted from green stem and trunk of banana plant through CIRCOT Mumbai

Since the quantity of banana fiber extracted from enzyme treated stem and trunk were less enough to process them for handmade papermaking, the fiber strength was needed to be evaluated as such. Due to the non-availability of such facilities at KNHPI, Jaipur, all the extracted fiber samples were sent to the Central Institute for Research on Cotton Technology (CIRCOT), Matunga, Mumbai, India for their

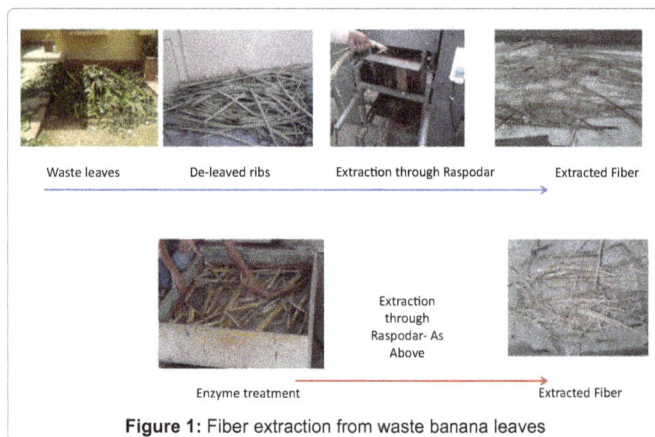

Figure 1: Fiber extraction from waste banana leaves

Parameters	De-leaved Ribs of Banana Plant		Banana Branches
	Natural/Untreated Control	Enzyme treated	Natural/untreated
Fiber yield,%	8.94%	23%	3.6%

Table 1: Fiber Extraction from Waste Banana Leaves

analysis.

Results and Discussion

Fiber extraction from waste banana leaves

For extracting fiber from the waste banana leaves, the mid rib portion was utilized. So, mid ribs were deleaved before processing them further as shown in Figure 1. The fiber yield obtained with and without enzyme treatment is given in Table 1 which indicates a great difference in the case of enzyme treated leaves as compared to the control case (23% vs. 8.94%). Table 2 shows the various parameters analyzed while processing the extracted banana fiber for making handmade paper. Actually, banana fiber obtained from banana ribs without enzyme treatment was less in quantity and having lot of dust, pith and waste along. So, the waste obtained during fiber extraction from untreated banana ribs was also cooked through open digestion using 4% NaOH for 3-4 hours to mix with it so as to enable its beating. Strength properties of the pulps prepared from waste left out of enzyme treated fiber extraction was found to be much poorer than that prepared from the waste left out of control fiber extraction. This implied that fiber extraction through enzymatic route resulted into lesser fibrous waste than through the untreated route (i.e. control) wherein lot of fiber was also lost into the waste generated down the equipment (Table 3).

Fiber extraction from green stem and trunk of banana plant

The fiber extraction process from banana trunk and banana stem has been shown in Figure 2 and 3 respectively. A pre-treatment of the green stem and trunk of banana plant with pectinases was found to be useful for fiber extraction process because in the fiber quality and yield obtained from banana trunk as well as banana stem was better with enzyme soaking than at the zero hours' stage which was equivalent to the conventional fiber extraction process. The quality of fiber extracted from banana trunk was better than the banana stem. The fiber yield varied with different periods of retention. In the case of banana stem, retention of 12 and 36 hours was found to be useful from the viewpoint of fiber yield (Table 4) but 36 hours retention could produce best quality fiber from the banana stem. Whereas in the case of banana trunk, a period of 24 hours was found to be the best for soaking it in the enzyme solution because it resulted not only in the maximum yield of fiber but also in the best quality of fiber (Table 4 and 5). Analysis

of colour and lignin in the collected effluents is presented in Table 6. Chauhan et al. [12] have also reported an increase in fiber yield and fiber quality through a prior treatment of *Calotropis procera* (Ankra twigs) with pectinase enzyme.

Analysis of strength properties of the extracted fiber through CIRCOT Mumbai

Physical strength properties of the extracted banana fiber were evaluated for the four important parameters of fiber quality viz. fiber strength, tenacity, fiber fineness and elongation through Central Institute of Research for Cotton Technology (CIRCOT), Mumbai.

Elongation percentage: Percent elongation at break is an important parameter of judging the quality of any fiber since it is the maximum possible extension of the fiber until it breaks. From the Table 6, it can be seen that the banana fiber extracted from the trunk always had better elongation % than that extracted from the banana stem. However, on comparing the fiber extracted from enzyme treated and untreated banana trunk, it was found to be higher in the treated stuff which may be due to more efficient removal of pectin by the pectinase treatment and increased proportion of cellulose during fiber extraction process through Raspodar [13]. But maximum elongation % could be achieved in the fiber extracted from the banana trunk after 24 hours of enzyme treatment while it took 36 hours of incubation for getting maximum value of elongation percentage in the case of banana stem. Further incubation till 48 hours resulted into its reduction from 2.4 to 1.8 (Table 6). Thus the retention period of enzyme treatment had a significant effect on the % of breaking elongation of banana fiber. Similar effect of retting period on elongation % has been reported by Resmina et al. [14] while treating banana pseudo stem with pectin decomposing bacteria and /or MgO.

Parameters	Fiber Extracted Without Enzyme Treatment Of Ribs	Fiber Extracted With Enzyme Treatment Of Ribs
Fiber yield, %	84.7%	93.2%
T.S in enzyme liquor, %	0.70%	0.44%
Pulping Of The Treated Fibers Through Open Digestion		
Pulp yield, %	18.24%	50%
T.S in the black liquor, %	1.25%	1.14%
pH RAA in the black liquor (gpl)	10.33 0.176gpl	10.37 0.224 gpl
Beating time and CSF	5.06 360 ml	3.20 370ml.
Physical Strength Properties of the pulps obtained		
Tensile index (Nm/gm)	41.8	48.85
Tear index (mN.m2/gm)	9.07	7.6
Folding Endurance, no.	448	244

Table 2: Enzyme Treatment of the Extracted Fibers

Parameters	Waste from untreated banana leaves	Waste from treated banana leaves
Pulp yield	78%	98%
T.S in the liquor	1.57%	2.03%
pH RAA in the liquor (gpl)	8.61 0.224	7.84 0.064
Beating time and CSF	7.22 410ml.	17.00 350ml
Physical Strength Properties of the pulps obtained		
Tensile index	13.55	1.05
Tear index	6.46	6.66
Double Fold	108	32

Table 3: Pulping Of Waste Obtained On Fiber Extraction through Open Digestion

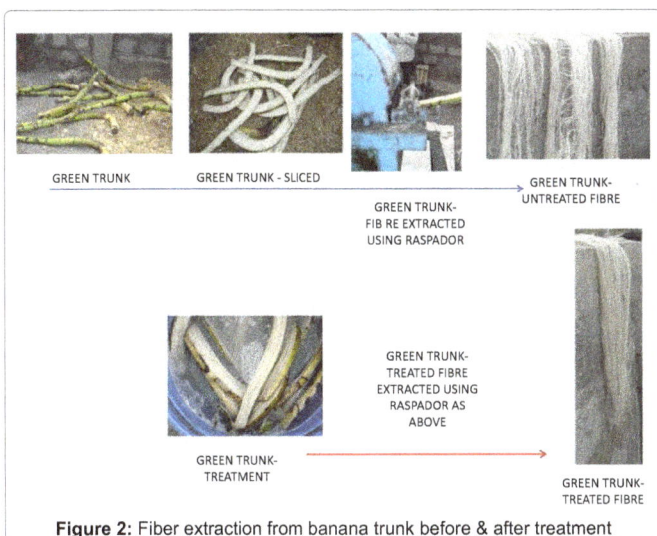

Figure 2: Fiber extraction from banana trunk before & after treatment

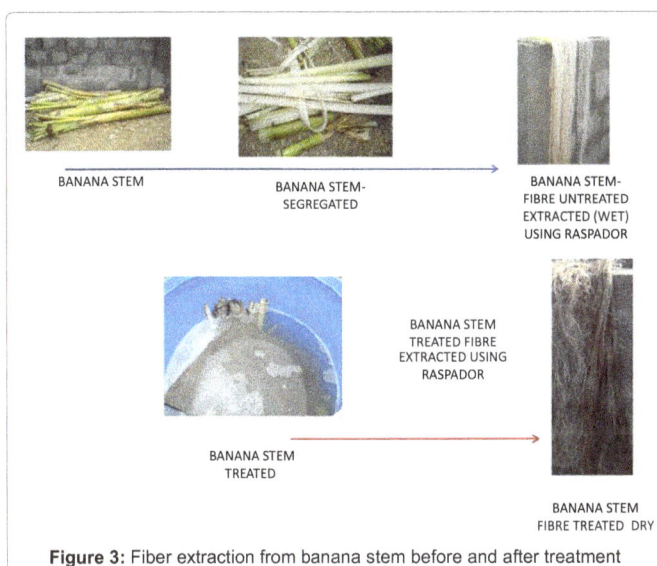

Figure 3: Fiber extraction from banana stem before and after treatment

Fiber strength: Fiber strength which is reported in the form of breaking strength, tensile strength and tenacity/intrinsic strength was also evaluated in all the extracted banana fibers. A significant increase in the value of this parameter in both the cases of banana trunk and banana stem as compared to the respective controls indicates a positive effect of pectinase enzyme treatment. However, the highest strength (432.4 gf Vs. 239.0 gf of control) was obtained after 36 hours of enzyme incubation in the case of banana stem while after even 24 hours of enzyme treatment in the case of banana trunk (401.7 gf Vs. 258.4). This also reflects about the efficiency of pectinase enzyme for improving the quality of extracted banana fibers because pectinases contribute to the breakdown of non-cellulosic substances like pectin materials and thereby separate the fibers from the core. Resmina et al. [14] have also reported the effectiveness of retting process on banana fiber by the quality of tensile strength and breaking elongation of the fiber but they had used pectin decomposing bacteria and MgO instead of the enzymes.

Fiber fineness: Fiber Fineness is also one of the most important fiber characteristics. Fibers exhibit a variety of cross sectional shapes and they also vary in section along their length and vary from fiber to fiber. Fiber fineness denotes the size of the cross sectional dimensions of the fibers. As the cross sectional features are irregular, direct determination of the area of cross section is difficult and often laborious. Some dimensional features such as swollen diameter, ribbon width etc. can be determined directly and sometimes used to specify the fineness of cotton fiber. The linear density or weight per unit length of the fiber is the more commonly used index of fineness. The linear density is called either the fiber weight per centimeter or hair weight per centimeter and is usually expressed in units of 10^{-8} g/cm or 10^{-5}mg/cm. In Tex system, the linear density of cotton fibers is expressed in terms of milli Tex which is the weight in milligrams of one kilometer length of fiber. It should be noted that it is quite possible to have fibers with identical linear densities but different cross sectional areas. For example, a fiber with a high density will have a smaller cross sectional area than a fiber of low density. Fineness of fibers (µg/inch) to a larger extent depends on the maturity of the fibers and to some extent is also influenced by the amount of the moisture present in the material [15]. Maximum value of fineness could be obtained after 36 hours of enzyme incubation in the case of banana stem while in the case of banana trunk the value of fiber fineness was almost equal in all the periods of enzyme incubation evaluated but it was of course more than the untreated/conventional case (Table 6).

Tenacity: Tenacity is the customary measure of strength of a fiber. It is usually defined as the ultimate (breaking) force of the fiber (in gram-force units) divided by the denier. Because denier is a measure of the linear density, the tenacity works out to be not a measure of force per unit area, but rather a quasi-dimensionless measure analogous to specific strength. In order to compare strength of two fibers differing in fineness, it is necessary to eliminate the effect of the difference in cross-sectional area by dividing the observed fiber strength by the fiber weight per unit length. The value so obtained is known as "Intrinsic Strength or Tenacity". Tenacity is found to be better related to spinning than the breaking strength. The strength characteristics can be determined either on individual fibers or on bundle of fibers. Mean single fiber strength determined is expressed in units of "grams/tex". As it is seen the unit for tenacity has the dimension of length only, and hence this property is also expressed as the "Breaking Length", which can be considered as the length of the specimen equivalent in weight to the breaking load. Since tex is the mass in grams of one kilometer of the specimen, the tenacity values expressed in grams/tex will correspond to the breaking length in kilometers. Tenacity was found to be maximum at an enzyme incubation of 24 hours in the case of the fiber extracted from both the banana stem and the trunk (Table 6).

The results obtained in the present study can be further co-related with some of earlier reports available in literature. Brahamkumar and Manilal [16] have studied SEM of bioextracted and physically extracted banana fibers and have reported cleaner and smoother surfaces in bioextracted fibers. Mooney et al. [17] have reported that the enzymatic hydrolysis of all the pectic and hemicellulosic materials results in a high yield, environmentally friendly produced pure cellulose and the production of individual fibers without the generation of Kink bands (resulting from the breaking and scotching process) generates fibers with much higher intrinsic fiber strength. Saleem et al. [18] have also reported positive effect of pectinase treatment on the mechanical properties of hemp fibers and the fiber reinforced polypropylene. The positive effect was told to be accomplished by enzymatic decomposition of middle lamella which caused separation of technical fibers into smaller bundles and single fiber cells. This in

Banana Stem					Banana Trunk				
Time interval	pH	Total solids (%)	Fiber extracted from 4 pieces of 700 gms each (i.e. nearly (2800gm)	Yield %	Time interval (hours)	pH	Total solids	Fiber extracted from 5 pieces of trunk (i.e. nearly 3000gm)	Yield %
Zero hours/ conven-tional process	5.53	0.016	20.370	0.73	Zero hours / conventional process	5.33	0.205	46.319	1.5
12 hrs.	6.16	0.014	24.358	0.87	12	6.44	0.263	55.902	1.9
24 hrs.	4.90	0.077	17.691	0.63	24	5.46	0.20	69.292	2.3
36 hrs.	4.64	0.089	23.015	0.82	36	5.76	0.29	52.062	1.73
48 hrs.	5.08	0.020	18.505	0.66	48	6.35	0.21	53.599	1.8

Table 4: Extraction of Banana Fiber from Banana Trunk and Stem

Parameters	Banana Stem					Banana Trunk				
	0 hrs	12 hrs	24 hrs	36 hrs	48 hrs	0 hrs	12 hrs	24 hrs	36 hrs	48 hrs
Fiber Strength (gf)	239.0	367.1	286.1	432.4	381.8	258.4	279.6	401.7	253.0	357.7
Elongation, %	1.4	1.6	1.2	2.4	1.8	2.6	3.2	4.0	3.8	4.0
Tenacity (gf/tex)	19.3	17.6	23.8	20.4	24.5	14.4	13.7	22.3	13.5	17.2
Fiber Fineness (Tex)	12.4	20.8	12.0	21.2	15.6	17.9	20.4	18.0	18.8	20.8

Table 5: Strength Properties of Banana Fiber Extracted From Banana Stem and Trunk through Enzymatic Routes at Guntur (Analyzed Through CIRCOT, Mumbai)

Banana Stem			Banana Trunk		
Sample	Lignin (gpl)	Color (PCU)	Sample	Lignin (gpl)	Color (PCU)
Zero hours/ conventional process	0.031	41.50	Zero hours/ conventional process	0.034	48.80
12 hrs.	0.015	31.71	12 hrs	0.02	53.60
24 hrs.	0.024	28.05	24 hrs	0.035	50.0
36 hrs.	0.036	90.24	36 hrs	0.051	142.7
48 hrs.	0.043	101.24	48 hrs	0.028	81.7

Table 6: Analysis of Color and Lignin of the Effluents Collected

turn resulted into improved tensile and flexural characteristics of the thermoplastic composites. Nalankilli et al. [19] have also reported efficacy of pectinase on removal of non-cellulosics but they have used cotton fibers. In comparison to the solvent and alkali scouring, they had reported lowest reduction in strength and elongation values of the enzyme scoured cotton fibers against the control fibers. Azzaz et al. [20] have reported that treatment of hand stripped and dried banana fibers with crude pectinase results into destruction of middle lamella which in turn effects the separation of fibers. Similarly, Chauhan and Sharma [21] have reported a positive role of enzyme treatment in improving the quality of pseudo stem fiber of banana plant but they have used the commercially available machine-extracted fiber of banana instead of the green plant parts.

Thus the present study/investigation has shed some light on the influence of pectinase enzyme treatment on fiber extraction from the banana plant's waste (leaves, pseudo stem and trunk). However, the time period of enzyme treatment must be judged carefully to get the best quality of fiber because an under-treatment makes the separation difficult while an over-treatment may weaken the fibers.

Conclusion

Thus, the overall investigation has been quite useful for solving out the basic problem of improving the quality of machine extracted banana fiber. KNHPI has also experienced good response during the field trial studies conducted in Southern India. During fiber extraction from the enzyme treated banana stem and trunk and evaluation of strength of the extracted fiber, it was found that an incubation period of 24 hours is sufficient for banana trunk but it requires an incubation of 36 hours for getting best quality fiber from banana stem. Adoption of the enzymatic route shall help in enabling the handmade paper manufacturers to utilize the machine extracted banana fiber that is available at cheaper

price than the hand-extracted banana fiber to produce a good quality handmade paper thereby improving their profitability. Further studies may be required to make the process of fiber extraction more cost-effective through enzymatic route. This may open great opportunity for employment generation for the economically backward youth and women particularly belonging to the rural areas of all the banana cultivating countries throughout the world besides addressing the problem of solid waste disposal generated from banana farming.

Acknowledgement

The authors gratefully acknowledge the financial support provided by the Directorate of Science & Technology, Khadi & Village Industries Commission (KVIC), Mumbai for conducting this study.

References

1. Preethi P, Balakrishna Murthy G (2013) Physical and Chemical Properties of Banana Fibre Extracted from Commercial Banana Cultivars Grown in Tamilnadu State. Agrotechnol S11:008.

2. Ortega Z, Beniltez AN, Manzon MD, Hernandez PM, Angula I, et al.(2010) Study of banana fiber as reingforcement of polyethylene samples made by compression and injection moulding. Journal of biobased materials and bioenergy 4: 114-120.

3. Ortega Z, Manzon MD, Soto P, Guinea I, Surrez L, et al. (2012) Use of banana fiber in composite parts for automotive sector. Proceedings of the 15th European conference on composite materisld, Venice, Italy.1-8.

4. Marek J, Vikter A, Marie B, Prokop S, Halgar F, et al. (2008) Enzymatic bioprocessing: new tool of extensive fiber source utilization. Proceedings of International conference on flax and other bast plants. 159-169.

5. Bajpai P (1999) Application of enzymes in the pulp and paper industry See comment in PubMed Commons below Biotechnol Prog 15: 147-157.

6. Gurucharanam K, Deshpande KS (1986) Polysaccharases of cruvalaria lunata-use in degumming of remine fibers. Indian phytopathol. 3385-3389.

7. C.Arunachalam, S.Asha (2010) Pectinolytic enzyme-a review of new studies. Advanced Biotech Journal 1-5.

8. Brühlmann F, Kim KS, Zimmerman W, Fiechter A (1994) Pectinolytic enzymes from actinomycetes for the degumming of ramie bast fibers. See comment in PubMed Commons below Appl Environ Microbiol 60: 2107-2112.

9. Sharma DC, Satyanarayan T (2005) A marked enhancement in the production of a highly alkaline and thermostable pectinase by *Bacillus pumilus dcsr1* in submerged fermentation by using statistical methods. Bioresour Technol 97: 727-733.

10. Jacob N, Niladevi KN, Anisha GS, Prema P (2008) Hydrolysis of pectin: an enzymatic approach and its application in banana fiber processing. Microbiological Research 163: 538-544.

11. Fengel, D, Wegener G (1989) Wood, chemistry, ultra structure, reaction. Walter De Gruyter, Berlin, Germany.

12. Chauhan S, Jain RK, Sharma AK (2013) Enzymatic Retting: A revolution in the handmade papermaking from Calotropis procera, in: R. C. Kuhad and A. Singh (Eds) Biotechnology for environmental management and resource recovery, Springer India. p77-88.

13. Skoglundh, M, Johansson H, Lowendahl L, Jansson, K., Dahl L, et al (1996) The quality of combine harvested fiber flax for industrial purposes depends on the degree of retting. Industrial crops and products 5: 67-78.

14. Resmina H, Naimah MS, Aziah H, Kiew YN, Konami Y (2011) Retting of musa sapientum psuedostem with pectin decomposing bacteria. Journal of Agricultural Science and Technology B1: 1238-1244.

15. Booth JE (1996) Principles of Textile Testing. Butterworth Heinmann, UK. p188.

16. Brahamkumar M, Manilal VB (2011) Banana pseudostem characterization and its fiber property evaluation on physical and bioextrateion. Journal of Natural Fibers 8:149-160.

17. Mooney C, Stolle Smits T, Schols H, Jong ED (2001) Analysis of retted and non-retted flax fibers by chemical and enzymatic means. J Biotechnol 89: 205-216.

18. Saleem Z, Rennebaun H, Pudal F, Grimm E (2007) Treating bast fibers with pectinase improves mechanical characteristics of reinforced thermoplastic composites. Composites Science and Technology 68: 471-476.

19. Nalankilli G, Sarvanan D, Govindraj N, Harish P (2008) Efficacy of solvent, alkali and pectinase on removal of non-cellulosics from cotton fibers. Indian Journal of Fiber & Textile Research 33: 438-442.

20. Azzaz HH, Murad HA, Khalif AM, Morsy TA, Mansour AM, et al (2013) Pectinase production optimization and its application in banana fiber degradation. Egyptian J. Nutrition and Foods 16: 117-125.

21. Chauhan S, Sharma AK (2014) Enzyme treatment in improving the quality of pseudo stem fiber of banana plant to use this bioresource for making handmade paper. International Journal of Fiber and Textile Research 4: 57-61.

24

Polyphenols Extracted from Olive Mill Wastewater Exert a Strong Antioxidant Effect in Human Neutrophils

Samia Bedouhene[1], Margarita Hurtado-Nedelec[2,3,4], Nassima sennani[1], Jean-Claude Marie[2,3], Jamel El-Benna[2,3*] and Farida Moulti-Mati[1]

[1]*Laboratoire de Biochimie appliquée et de biotechnologie, Faculté des Sciences Biologiques et des Sciences Agronomiques, Université M. Mammeri, BPN°17, RP 15000 Tizi-Ouzou, Algeria*
[2]*Inserm, U1149, CNRS-ERL8252, Centre de Recherche sur l'Inflammation, Paris, France*
[3]*Université Paris Diderot, Sorbonne Paris Cité, Laboratoire d'Excellence Inflamex , Faculté de Médecine, Site Xavier Bichat, Paris, France*
[4]*AP-HP, Unité Dysfonctionnement Immunitaire, Centre Hospitalo-Universitaire Xavier Bichat, Paris, F-75018, France*

Abstract

Olive mill wastewater (OMW) is produced seasonally by the olive oil-producing countries. A large amount of liquid waste results from olive oil extraction, with a very high organic load which renders it difficult to treat. Wastewater represents also a significant source of polyphenols which can be revalorized and used for medical or agro-alimentary purposes. The extraction of polyphenols will improve the wastewater biodegradation and reduce its phytotoxicity. The objective of this study was to extract polyphenols from OMW and to investigate their effect on reactive oxygen species (ROS) production by human neutrophil. Polyphenols were extracted from OMW by an established technique. Neutrophils, isolated from blood of healthy volunteers, were incubated with increased concentrations of polyphenol extract and ROS production was measured by luminol-amplified chemiluminescence and cytochrome c reduction techniques. Results show that the polyphenol extract from OMW inhibited phorbol-myristate acetate (PMA)-stimulated neutrophil ROS production as measured by the chemiluminescence assay. The polyphenols extract from OMW also inhibited neutrophil superoxide production as measured by the cytochrome c reduction assay; as well as H_2O_2 production as measured by flow cytometry. Also, the polyphenol extract reacted with pure H_2O_2 but did not affect superoxide anions production by the xanthine/xanthine oxidase enzymatic system. Our results show that polyphenols extracted from OMW exert a strong antioxidant effect and they could have an anti-inflammatory effect by inhibiting neutrophil ROS production and by scavenging hydrogen peroxide, thus limiting their toxic effects. OMW could be used to extract polyphenols for medicinal applications.

Keywords: Antioxidant; Olive mill wastewater; Polyphenols; Polymorphonuclear neutrophils; Reactive oxygen species

Introduction

The Mediterranean region is the first producer of olive oil in the world. Despite the known beneficial properties of olive oil on health, its production generates large amounts of by-products such as olive mill wastewater (OMW) [1], which constitutes an important environmental pollution product. The Mediterranean region accounts for 95% of the global OMW product in the world [2]. Phenolic compounds such as oleuropein, tyrosol, hydroxytyrosol and caffeic acid are abundant in OMW [3,4]. The presence of polyphenols in olive oil is believed to be responsible for its stability [5]. Olive polyphenols are also known to play a role in preventing chronic human diseases such as cardiovascular diseases and inflammatory diseases [6]. Several *in vitro* and *in vivo* studies confirmed that phenolic compounds from olive reduce the effect of oxidative stress associated with pathological disorders including atherosclerosis [7,8], cancer [9,10], inflammatory diseases [11-13], neurodegenerative diseases [14], and they have some anti-microbial and anti-viral properties [15-18].

Reactive oxygen species (ROS) such as superoxide anion ($O_2^{\cdot-}$), hydrogen peroxide (H_2O_2) and hydroxyl radical ($OH^\cdot$) are powerful oxidants produced by several enzymatic systems in the body [19]. Phagocytes such as neutrophils, monocytes and macrophages are major ROS producing cells [20]. Neutrophil activation leads to the production of ROS in a process called respiratory burst, mediated by a multi-component enzyme NADPH oxidase. ROS produced by this process are used by phagocytes to kill infectious agents. In contrast, over production of ROS may cause oxidative stress, leading to many deleterious effects for the organism [21]. In this context, polyphenol compounds, known to have an antioxidant effect, can be used to protect cell against oxidative damage and limit the risk of various degenerative diseases associated to excessive ROS production by phagocytes. While,

individual polyphenol compounds have been shown to display anti-oxidative effect, little is known about the effect of total polyphenol extracts from OMW [22-25]. In this work, we used OMW, from *chemlal* variety olive tree which is largely cultivated in Kabylia (Algeria), to extract polyphenols. Polyphenol extract were tested on ROS production by human neutrophils and enzymatic systems.

Materials and Methods

Materials

Ficoll, Dextran, cytochrome C, luminol, PMA (4b-phorbol-12b-myristate-a13-acetate), xanthine, xanthine-oxidase, H_2O_2, HRPO (horseradish peroxidase), 2′,7′ - Dichlorofluorescin diacetate (DCFH-DA), Dulbecco's Phosphate-buffered saline (PBS) and Hanks' balanced salt solution (HBSS) were purchased from Sigma– Aldrich Chemie GmbH (Steinheim, Germany). The different solutions were diluted in phosphate-buffered saline (PBS) immediately before use.

Extraction of polyphenols from olive mill wastewater

Polyphenols were extracted according to the method previously

*Corresponding author: Dr Jamel El Benna, Faculté de Médecine Xavier Bichat, 16 rue Henri Huchard, Paris, F-75018, France
E-mail: jamel.elbenna@inserm.fr

described [4]. Olive mill wastewater was collected from the region of Kabylia, the most important producer of olive oil in Algeria (Algeria). Liquid–liquid extraction of phenolic compounds with ethyl acetate was carried out on olive mill wastewater samples obtained from a continuous olive oil processing plant. OMW was acidified to pH 2 with HCl and washed with hexane in order to remove the lipid fraction, the mixture was vigorously shaken and centrifuged for 5 min at 3000 rpm. The phases were separated and washed successively twice. Extraction of phenolic compounds was then carried out with ethyl acetate: the washed OMW samples were mixed with ethyl acetate and vigorously shaken before centrifugation for 5 min at 3000 rpm. The phases were separated and the extraction was repeated four times successively. The ethyl acetate was evaporated using a speed Vac. The aliquots of polyphenol extract without solvent were frozen (-20°C) until utilization.

Determination of total polyphenol content

Total polyphenol compounds in OMW extract was determined by the Folin–Ciocalteu assay according to the method described by Singleton [26]. The absorbance was read at 725 nm in Uvikon 931 (Contron, Milano, Italy) UV–Vis spectrophotometer and all the experiments were performed in triplicate. The content of total polyphenol is reported as gallic acid equivalents (EqGA) by reference to standard curve. For experiments, polyphenols extract was prepared at 5 mg/mL in PBS then diluted in the assay at different concentrations.

Neutrophil preparation

Neutrophils were isolated from venous heparinized blood, freshly collected from healthy volunteers. We used Dextran (T500) to remove red blood cells, followed by centrifugation over Ficoll-Paque to remove mononuclear cells and hypotonic lysis of any remaining contaminating red blood cells as described previously [2]. Finally, the neutrophils were centrifuged and suspended in PBS before being counted.

Determination of cell viability

Isolated human neutrophils were incubated with polyphenols extract at concentrations of 0, 25, 50, 100, 200 and 250 μg/mL during 30 min. Cell viability was evaluated by the trypan blue exclusion test.

Measurement of ROS production by luminol-amplified chemiluminescence

Neutrophils (5 x 10^5/ 0.5 mL) were resuspended in HBSS in the absence or presence of polyphenol extracts (0, 25, 50,100, 200 and 250 μg/mL), in the presence of luminol (10 μM), for 15 min at 37°C, then

PMA (100 ng/mL) was added and chemiluminescence was evaluated with a luminometer (Auto Lumat LB953 model, EG & G Berthold). Light emission was recorded in counted photons per minute (cpm) during 30 min at 37° C. The percentage of inhibition was calculated in comparison to the control without polyphenol extract. To test the effect of polyphenol on H_2O_2, we use 0.003% H_2O_2 mixed to 5U of HRPO in the presence of luminol (10 μM) and increasing concentrations of polyphenol extract (0, 10, 20, 50, 100 μg/mL). Chemiluminescence was measured for 30 min as described above.

Measurement of superoxide production by the cytochrome c reduction assay

Neutrophils (1 x 10^6) were incubated during 10 min at 37°C with cytochrome c (1 mg/mL) and increasing concentrations polyphenol extract (0, 25, 50, 100, 200, 250 μg/mL), prior to stimulation with 100 ng/mL PMA. Superoxide anion production was determined by measuring the ferric cytochrome c reduction with a UVIKON 860 spectrophotometer at 550 nm during 10 min.

Xanthine-Xanthine Oxidase (XXO)-derived superoxide production was determined in 1 mL PBS containing xanthine (100 mM) and xanthine oxidase (1U), in the presence of cytochrome c and increasing concentrations of polyphenol extracts (0, 25, 50, 100, 200 and 250 μg/mL). Superoxide anion production was determined by measuring the ferric cytochrome c reduction with spectrophotometer at 550 nm over 10 min. Superoxide production values were calculated using the molar extinction coefficient of reduced cytochrome c ($2.1 x 10^4$ mol/L^{-1} cm^{-1}).

Measurement of intracellular H2O2 production by flow cytometry

Intracellular H_2O_2 production was measured using a flow cytometric assay. The method is based on that when 2′,7′ -dichlorofluorescin di-acetate (DCFH-DA) probe diffuses across neutrophils membrane , it is hydrolyzed by intracellular esterases to DCFH which remains trapped within the cells. In presence of H_2O_2, non-fluorescent intracellular DCFH is oxidized to highly fluorescent 2′,7′-dichlorofluorescein (DCF). PMNs (5 x 10^5 cells) in Hank's were incubated for 15 min with 2,7-DCFH-DA (100 mmol/l) in the dark and in a water bath at 37°C with gentle agitation. Thereafter, polyphenol extract was added at increasing concentrations (0, 25, 50, 100 and 200 μg/ mL) to the cells and further incubated for 10 min. Neutrophils were stimulated with PMA (100 ng/mL) over 15 min. The reaction was stopped at 4° for 15 min. Flow cytometric analysis was performed with a Becton Dickinson FACSCantoII (Immuno cytometry Systems). The fluorescence intensity

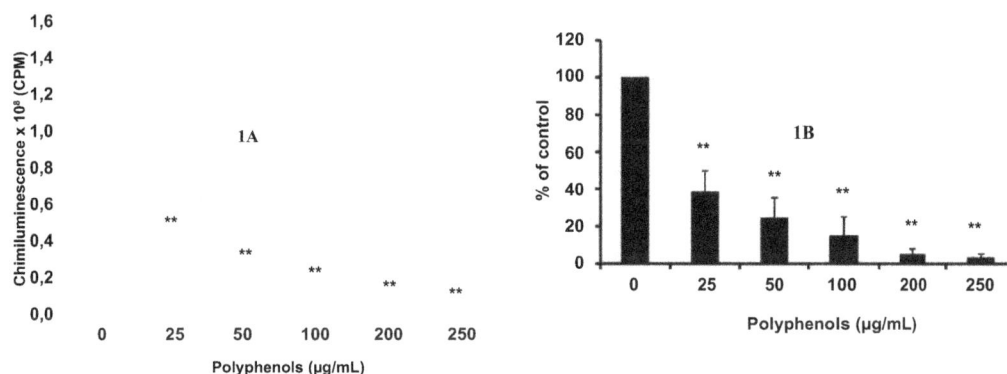

Figure 1: Effect of the polyphenols extract on ROS production by human neutrophils. Neutrophils were incubated with the OMW polyphenols extract at increasing concentrations and stimulated or not with PMA. Luminol-amplified chemiluminescence was measured during 30 min. Data are expressed as means ± SEM; n=3,* p < 0.05.

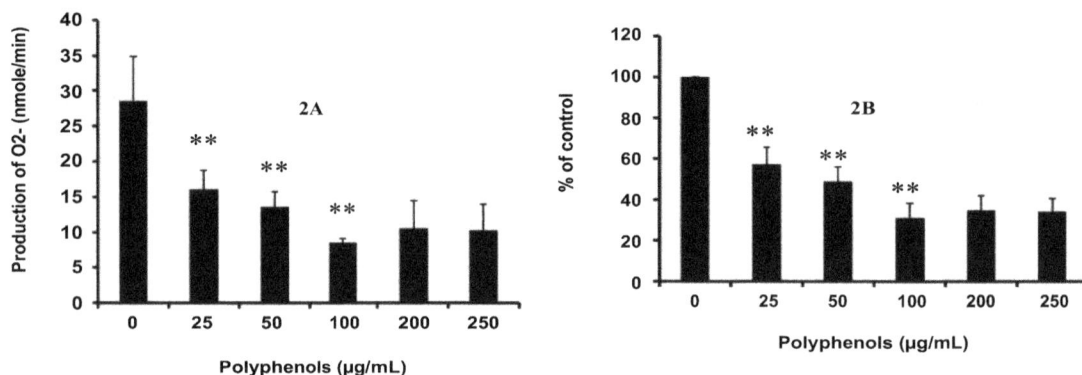

Figure 2: Effect of the polyphenols extract on superoxide anions production by human neutrophils. Neutrophils were incubated with the OMW polyphenols extract at increasing concentrations in the presence of cytochrome c and stimulated or not with PMA. Production of superoxide anions was monitored at 550 nm. Data are expressed as means ± SEM; n=3, * p< 0.05.

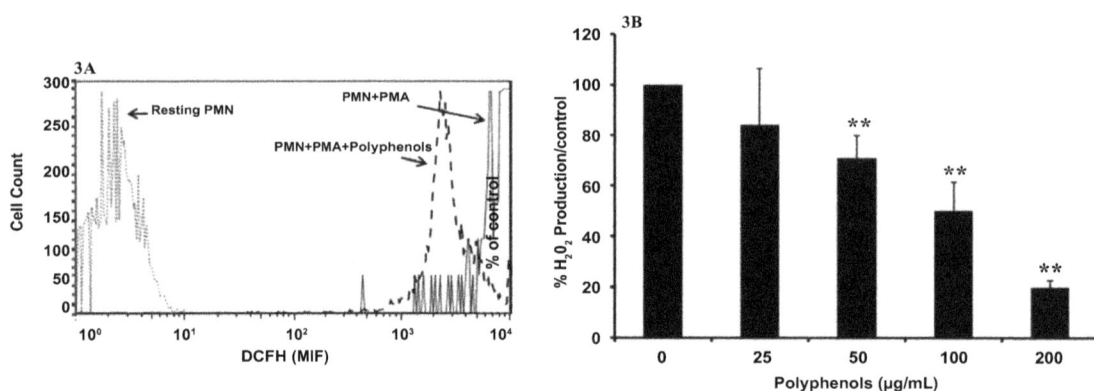

Figure 3: Effect of the polyphenols extract on DCFH-detected H2O2 production. Neutrophils were incubated with DCFH-DA at 37 °C at increasing concentrations of polyphenols extract. Flow cytometric analysis was performed with a Becton Dickinson FACSCantoII. Representative FACS profile of neutrophils stimulation and the effect of polyphenols extract (Counts: number of neutrophils; MFI: Mean fluorescence intensity). The results are calculated using a stimulation index (SI), namely the ratio of MFI of stimulated cells to that of unstimulated cells. Data are expressed as means ± SEM, n=3; * p< 0.05.

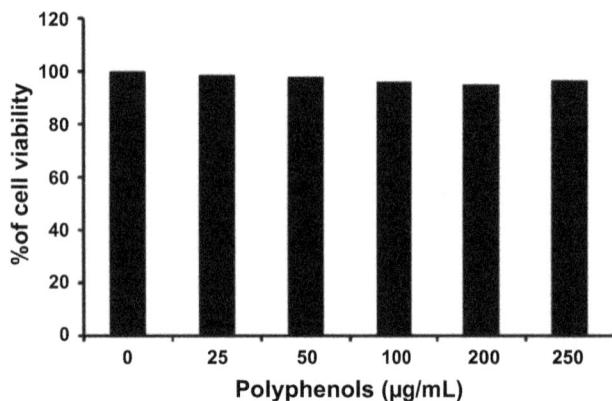

Figure 4: Effect of the polyphenols extract on neutrophil viability. Isolated human neutrophils were exposed to polyphenols extract at concentrations. Viability was evaluated with adding Trypan Blue, and blue cells were counted. % of viable cells was expressed compared to control conditions (without polyphenols). Results are expressed as mean+/- SEM, n= 3, * p< 0.05.

was measured simultaneously at wavelength of 488 nm for excitation and at 530 nm for emission. The data were analyzed using FACSDiva program and results are expressed as the mean fluorescence intensity (MFI) was used to quantitate the responses. The effect of polyphenol extract on H_2O_2 production was calculated by using a stimulation index (SI), namely the ratio of the MFI of stimulated cells versus that of unstimulated cells.

Statistical analysis

The data are presented as a percentage of the control and was calculated according to the equation: [% of control = (Response with polyphenol extract/Control Response) x 100]. Statistical analysis was established between controls and samples treated with polyphenols extract using student t- test.

Results:

Extraction of polyphenols from olive mill wastewater

Olive mill wastewater was collected from the region of Kabylia the most important producer of olive oil in Algeria. In this region, most of the olive trees are *chemlal* olive trees. Extraction of phenolic compounds was carried out on olive mill wastewater samples obtained from a continuous olive oil processing plant as described in material and method section. Total polyphenol compounds in OMW extract was determined by the Folin–Ciocalteu assay according to the method described by Singleton [26]. The content of total polyphenols is reported as gallic acid equivalents (EqGA) by reference to standard curve. After purification, we evaluated a content of polyphenol compounds in the

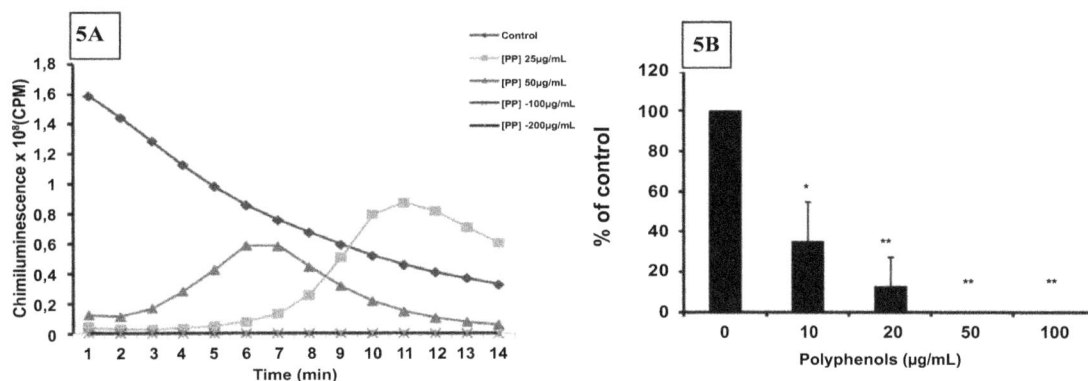

Figure 5: Effect of the polyphenols extract on luminol-amplified chemiluminescence in acellular system. H2O2 and HRPO were incubated with polyphenols extract at increasing concentrations. Luminol-amplified chemiluminescence was measured during 30 min. Data are expressed as means ± SEM, n=3, *p< 0.05.

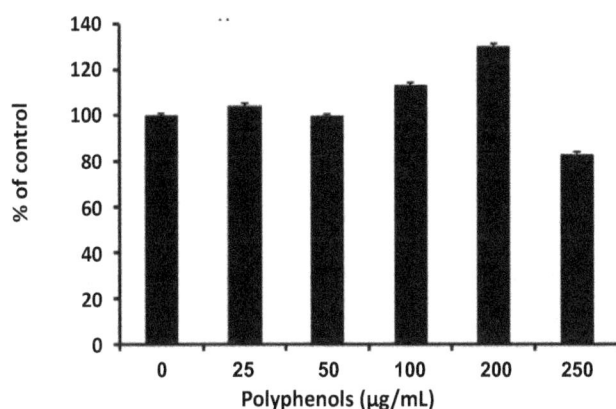

Figure 6: Effect of the polyphenols extract on superoxide anion production by the xanthine/xanthine oxidase cell free system. The polyphenols extract was incubated at increasing concentrations with xanthine oxidase, xanthine was added and superoxide was measured by the cytochrome c reduction assay at 550 nm. Results are expressed as the % of control without the extract. Data are expressed as means ± SEM, obtained in three independent experiments.

OMW extract by 0.148 mg EqGA/mg extract.

Effect of polyphenol extract from OMW on neutrophil ROS production

To evaluate the effect of the polyphenol extract on human neutrophil ROS production, freshly isolated neutrophils were treated with increasing concentrations of the polyphenol extract (0, 25, 50,100, 200 and 250 μg/mL), chemiluminescence was measured using luminol as probe. Luminol-amplified chemiluminescence detects multiple ROS, mainly superoxide anion, hydrogen peroxide and hypochlorous acid. The result shows that the polyphenol extract inhibited total ROS production by PMA-stimulated neutrophils (Figures 1A and 1B). This result suggests that the total polyphenol extract from OMW may scavenge ROS or affect the neutrophil NADPH oxidase activity or its upstream activation.

Effect of the polyphenol extract on superoxide production by neutrophils

To assess specifically the effect of the polyphenol extract on superoxide anion production, neutrophils were pre-incubated for 10 min at 37°C with cytochrome c and the polyphenol extract and then were stimulated with PMA. Superoxide anion production was determined by measuring the reduction of cytochrome c at 550 nm with a UVIKON 860 spectrophotometer. The result shows that the polyphenol extract inhibited the cytochrome c reduction (Figures 2A and 2B). This suggests that the extract may affect NADPH oxidase activity in neutrophils or could scavenge superoxide anions.

Effect of the polyphenol extract on H2O2 production by neutrophils

To test the effect of increasing concentration of polyphenol extract on intracellular production of H_2O_2 by neutrophil in response to PMA, we used DCFH probe which is fluorescent in presence of H_2O_2. We calculated the effect of polyphenol extract on H_2O_2 production by using a stimulation index (SI), namely the ratio of stimulated cells to that of unstimulated cells. The neutrophils were identified on the basis of forward and side scatter alone and analysed in combination with the DCFH probe. The fluorescence intensity decreased in neutrophils treated by polyphenols cells as compared to control cells, this effect was dependent on increasing concentrations of polyphenol extract. These results (Figures 3A and 3B) showed that the polyphenol extract inhibited intracellular production of H_2O_2, although at higher concentrations than those which inhibited chemiluminescence and cytochrome c reduction.

The polyphenol extract from OMW has no effect on human neutrophil viability

To verify that the inhibitory effect of the polyphenol extract was not due to its toxic activity on neutrophils, we examined its effect on cell viability using trypan blue exclusion assay. The result shows that after 30 min of neutrophils incubation with increasing concentrations of polyphenols, cell viability was greater than 95% (Figure 4). Thus cell viability was not affected by polyphenol extract.

Effect of polyphenol extract on H2O2 and superoxide anion in cell-free systems

To identify the specific target of polyphenols extract, pure H_2O_2 was incubated with increasing concentrations of the polyphenol extract and then the luminol-amplified chemiluminescence assay with HRPO (5U) was used to detect H_2O_2. Results show that polyphenols extract scavenges H_2O_2 at low concentrations in a dose dependent manner (Figures 5A and 5B). We also studied the effect of polyphenol extract on superoxide anion produced by the Xanthine/Xanthine Oxidase. Results show that polyphenols extract did not inhibit cytochrome c reduction (Figure 6).

Discussion

In this study, we extracted polyphenols from olive mill wastewater (OMW) from a specific variety of olive tree, the chemlal olive tree and tested its effects on ROS production by human neutrophils. Firstly, we used luminol-amplified chemiluminescence, a technique which detects the total ROS production. Results show that, in PMA stimulated neutrophils, the polyphenol extract significantly inhibits total ROS production in a dose-dependent manner. Secondly, using the cytochrome c reduction assay, a specific method to measure extracellular superoxide anions ($O_2^{\cdot-}$) production, we found that the polyphenol extract significantly inhibited neutrophil's $O_2^{\cdot-}$ production. Thirdly, using flow cytometry and DCFH probe, we showed that the polyphenol extract at low concentrations significantly inhibited the intracellular production of H_2O_2, in a dose-dependent manner. These effects were not due to a toxic effect of the polyphenol extract since cell viability was not affected. Interestingly, the polyphenol extract was able to react with pure H_2O_2 in vitro but not with $O_2^{\cdot-}$ produced by the xanthine/xanthine oxidase system. These results suggest that the polyphenol extract could have a double effect, 1) scavenging H_2O_2 and 2) inhibiting NADPH oxidase activity or its upstream activation. To check the second possibility, we tested the effect of the polyphenol extract on isolated neutrophil membranes containing activated NADPH oxidase. We found no effect on the isolated enzyme (data not shown). These results suggest that the polyphenol extract is able to affect NADPH oxidase activation, probably by interfering with the neutrophil signaling pathways involved in NADPH oxidase activation.

Our results show that OMW from *Chemlal* variety is a rich source of antioxidants such as phenolic compounds. Among the polyphenol compounds present in olive mill wastewater, only hydroxytyrosol and oleuropein showed a highly antioxidant effect [27-29]. Many recent human and animal studies have shown a spectrum of highly interesting bioactivities of polyphenols from olive, including antimicrobial activity, anti-cancer activity, anti-inflammatory activity and beneficial effect in cardiovascular diseases [3,8-10,12]. Also, these effects are well sustained by epidemiological studies on Mediterranean diet [30].

It has been found that polyphenol compounds are highly absorbed by human cells [31], and stored *in vivo* in some tissues such as prostate and breast tissues [30,32,33]. These data suggest that assimilated polyphenols can protect against H_2O_2 toxicity and exert beneficial effects on health. The total polyphenol extract could have a more effective effect *in vivo* due to synergistic action and effects of individual compounds.

It is well established that ROS generated by neutrophils and monocytes/macrophages are involved in inflammatory diseases such as inflammatory bowel diseases, cardiovascular diseases and rheumatoid arthritis [34].The polyphenol extract could be beneficial in these diseases due to its inhibitory action on neutrophil ROS production and also by scavenging extracellular H_2O_2 which is the most diffusible ROS. Thus the polyphenol can protect tissues of different organs from H_2O_2, limiting its bystander toxic effects. The scavenging of H_2O_2 polyphenols has also been reported *in vitro* by Ju et al. [35] and in neutrophils with another extract by Paula et al [36].

In conclusion, the polyphenol extract from OMW is a powerful hydrogen peroxide scavenger and a powerful inhibitor of neutrophil NADPH oxidase activation. These compounds present in OMW could contribute to the prevention of diseases in which ROS are involved. OMW could also be used as immunomodulatory compounds and a therapeutic adjuvant in the treatment of neutrophil-mediated inflammatory diseases as an alternative to synthetic antioxidants in pharmaceutical and agroalimentary process. As OMW is a significant source of phenolic compounds beneficial for health, its utilization as a resource of polyphenols can be valorized to reduce environmental waste [37]. Thus, the extraction of polyphenol compounds from OMW is a fundamental step before any biological degradation of this waste. OMW could be used as a source of antioxidant compounds with a clear impact on both health and environment.

References

1. Nafzaoui A (1991) Valorization des sous-produits de l'olivier. Options méditerranéennes, série séminaires 16 : 101-108.

2. Kapellakis IE, Tsagarakis KP, Avramaki C, Angelakis AN (2006): Olive mill wastewater management in river basins: A case in Greece. Agricultuural Water management 82: 354-370.

3. Czerwinska M, Kiss AK, Naruszewicz M (2012) A comparative study of the effects of oleuropein and oleacein on human neutrophil oxidative bursts and monocytes nitric oxide production. Food Chemistry 131: 940-947.

4. De Marco E, Sa Varese M, Paduano A, Sacchi R (2007) Characterization and fractionnement of phenolic compounds extracted from olive mill wastewaters. Food Chemistry 104: 858-867.

5. Chimi HJ, Cillard P and Rahmani M (1991) Peroxyl and hydroxyl radical scavenging activity of some natural phenolic antioxidants. JAOCS 68: 307-312.

6. Babich H, Visioli F (2003) In vitro cytotoxicity to human cells in culture of some phenolics from olive oil. Farmaco 58: 403-407.

7. Kastorini CM, Milionis HJ, Goudevenos JA, Panagiotakos DB (2010) Mediterranean diet and coronary heart disease: is obesity a link? - A systematic review. Nutr Metab Cardiovasc Dis 20: 536-551.

8. Ozsoy N, Candoken E and Akev N (2009) Implications for degenerative disorders: Antioxidative activity, total phenols, flavonoids, ascorbic acid, ß-carotene anda -tocopherol in Aloe vera. Oxidative Medicine and Cellular Longevity (Landes Bioscience) 2: 99-106.

9. Isik S, Karagöz A, Karaman S, Nergiz C (2012) Proliferative and apoptotic effects of olive extracts on cell lines and healthy human cells. Food Chemistry 134: 29–36.

10. Yang CS, Wang H, Li GX, Yang Z, Guan F, et al. (2011) Cancer prevention by tea: Evidence from laboratory studies. Pharmacol Res 64: 113-122.

11. de la Puerta R, Martínez-Domínguez E, Ruíz-Gutiérrez V (2000) Effect of minor components of virgin olive oil on topical antiinflammatory assays. Z Naturforsch C 55: 814-819.

12. Impellizzeri D, Esposito E, Mazzon E, Paterniti I, Di Paola R, et al. (2011) The effects of oleuropein aglycone, an olive oil compound, in a mouse model of carrageenan-induced pleurisy. Clin Nutr 30: 533-540.

13. Miles EA, Zoubouli P, Calder PC (2005) Differential anti-inflammatory effects of phenolic compounds from extra virgin olive oil identified in human whole blood cultures. Nutrition 21: 389-394.

14. Covas MI, Nyyssönen K, Poulsen HE, Kaikkonen J, Zunft HJ, et al. (2006) The effect of polyphenols in olive oil on heart disease risk factors: a randomized trial. Ann Intern Med 145: 333-341.

15. Bisignano G, Tomaino A, Lo Cascio R, Crisafi G, Uccella N, et al. (1999) On the in-vitro antimicrobial activity of oleuropein and hydroxytyrosol. J Pharm Pharmacol 51: 971-974.

16. Capasso R, Evidente A, Schivo L, Orru G, Marcialis MA, et al. (1995) Antibacterial polyphenols from olive oil mill waste waters. J Appl Bacteriol 79: 393-398.

17. Rodriguez-Vaquero MJ, Alberto MR, Manca-de-Nadra MC (2007) Antibacterial effect of phenolic compounds from different wines. Food Control 18: 93–101.

18. Yamada K, Ogawa H, Hara A, Yoshida Y, Yonezawa Y, et al. (2009) Mechanism of the antiviral effect of hydroxytyrosol on influenza virus appears to involve morphological change of the virus. Antiviral Res 83: 35-44.

19. Favier A (2006) Le stress oxydant, Intérêt conceptuel et expérimental dans la compréhension des mécanismes des maladies et potentiel thérapeutique. L'actualité chimique 11-12: 108-115.

20. Gougerot-Pocidalo MA, Elbim C, Dang PM, El Benna J (2006) [Primary immune deficiencies in neutrophil functioning]. Presse Med 35: 871-878.

21. Rotondo S, Rajtar G, Manarini S, Celardo A, Rotillo D, et al. (1998) Effect of trans-resveratrol, a natural polyphenolic compound, on human polymorphonuclear leukocyte function. Br J Pharmacol 123: 1691-1699.

22. Alvarado C, Alvarez P, Jiménez L, De la Fuente M (2005) Improvement of leukocyte functions in young prematurely aging mice after a 5-week ingestion of a diet supplemented with biscuits enriched in antioxidants. Antioxid Redox Signal 7: 1203-1210.

23. Léger CL1, Kadiri-Hassani N, Descomps B (2000) Decreased superoxide anion production in cultured human promonocyte cells (THP-1) due to polyphenol mixtures from olive oil processing wastewaters. J Agric Food Chem 48: 5061-5067.

24. Schaffer S, Müller WE, Eckert GP (2010) Cytoprotective effects of olive mill wastewater extract and its main constituent hydroxytyrosol in PC12 cells. Pharmacol Res 62: 322-327.

25. Visioli F, Bellomo G, Galli C (1998) Free radical-scavenging properties of olive oil polyphenols. Biochem Biophys Res Commun 247: 60-64.

26. Singleton VL, Rossi JA (1965) Colorimetry of total phenolics with phosphomolybdic-phosphotungstic reagents. Am J Enol Vitic; 16: 144–158.

27. O'Dowd Y, Driss F, Dang PM, Elbim C, Gougerot-Pocidalo MA, et al. (2004) Antioxidant effect of hydroxytyrosol, a polyphenol from olive oil: scavenging of hydrogen peroxide but not superoxide anion produced by human neutrophils. Biochem Pharmacol 68: 2003-2008.

28. Ju HY, Chen SC, Wua KJ, Kuo HC, Hseu YC, et al. (2012) Antioxidant phenolic profile from ethyl acetate fraction of Fructus Ligustri Lucidi with protection against hydrogen peroxide-induced oxidative damage in SH-SY5Y cells. Food Chem Toxicol 50: 492–502.

29. Paula FS, Kabeya LM, Kanashiro A, de Figueiredo AS, Azzolini AE, et al. (2009) Modulation of human neutrophil oxidative metabolism and degranulation by extract of Tamarindus indica L. fruit pulp. Food Chem Toxicol 47: 163-170.

30. Parzonko A, Naruszewicz M (2010) Silymarin inhibits endothelial progenitor cells' senescence and protects against the antiproliferative activity of rapamycin: preliminary study. J Cardiovasc Pharmacol 56: 610-618.

31. Visioli F, Poli A, Gall C (2002) Antioxidant and other biological activities of phenols from olives and olive oil. Med Res Rev 22: 65-75.

32. Shivashankara KS and Acharya SN (2010) Uptake of Polyphenols in Tissues. Bioavailability of Dietary Polyphenols and the Cardiovascular Diseases. The Open Nutraceuticals Journal 3: 227-241.

33. Scalbert A, Williamson G (2000) Dietary intake and bioavailability of polyphenols. J Nutr 130: 2073S-85S.

34. Manach C, Williamson G, Morand C, Scalbert A, Rémésy C (2005) Bioavailability and bioefficacy of polyphenols in humans. I. Review of 97 bioavailability studies. Am J Clin Nutr 81: 230S-242S.

35. Moreira MR, Kanashiro A, Kabeya LM, Polizello ACM, Azzolini AECS, et al. (2007) Neutrophil effector functions triggered by Fc-gamma and/or complement receptors are dependent on B-ring hydroxylation pattern and physicochemical properties of flavonols. Life Sci 81: 317–326.

36. Conner EM, Grisham MB (1996) Inflammation, free radicals, and antioxidants. Nutrition 12: 274-277.

37. Kalogerakis N, Politi M, Foteinis S, Chatzisymeon E, Mantzavinos D (2013) Recovery of antioxidants from olive mill wastewaters: a viable solution that promotes their overall sustainable management. J Environ Manage 128: 749-758.

25

Assessment of Environmental Flows for Various Sub-Watersheds of Damodar River Basin Using Different Hydrological Methods

Ravindra Kumar Verma[1], Shankar Murthy[2*] and Rajani Kant Tiwary[3]

[1]Fellow (Doctoral Programme), Environmental Engineering & Management Group, National Institute of Industrial Engineering, Mumbai, Maharashtra - 400087, India
[2]Associate Professor, Environmental Engineering & Management Group, National Institute of Industrial Engineering, Mumbai, Maharashtra - 400087, India
[3]Senior Scientist, Water Environment Division, Central Institute of Mining and Fuel Research (CIMFR), Barwa-Road, Dhanbad, Jharkhand-826015, India

Abstract

Environmental Flows (EFs) assessment is a global challenge involving a number of tangible and intangible segments of hydrology, hydraulics, biology, ecology, environment, socio-economics, and several other branches of engineering including the management of water resources. It has consequently led to the development of more than 240 methods available in literature. Required for the longevity of a river, EFs derived from a single method are usually not accepted. In the present study, the EFs variability was assessed using three hydrological methods: (i) Tennant, (ii) Tessman, and (iii) Flow Duration Curve (FDC) for various sub-watersheds of Damodar River Basin (DRB), located in the states of Jharkhand and West Bengal. The minimum and maximum range of magnitude, duration and frequency of flow estimated using these methods are recommended as EFs, which can be used by water resource managers for habitat protection, water supply planning and design, waste load allocation, reservoir design, future water resource and river health assessment in the basin.

Keywords: Environmental flows; Damodar river basin; Tennant; Tessman; Flow duration curve

Introduction

Assessment of EFs is of paramount importance in the present era of modernization/development leading to increasing hydrological alternation through dams and diversions and, in turn, modification to the natural conditions of stream flow and ecosystem as well due to increasing water extraction for meeting human demands for uses such as industry, agriculture, recreation, hydropower generation, domestic water supply [1-3]. It is estimated that more than 60% of the world's rivers are fragmented by hydrological alternation and modify the natural patterns of rivers or stream flow and this figure is projected to increase to 70% by 2025 [4]. The protection of aquatic resources against the impact of dam and water extraction in river is a challenging and elusive issue in sustainable river basin management. Now a day, EFs assessment has undergone a major paradigm shifts from a single hydrologic attribute (i.e. minimum flows) to a full range of flows (floods, average, and low flows) that account for seasonal and inter- and intra-annual variation in stream flow variability. In addition, its magnitude, timing, frequency, and rate of change also plays an important role in long-term sustainability of water resources and their proper utilization with river ecosystem in state of good health and other features (such as fish, wildlife habitat, environmental purposes, water quality) [5,6]. In last two decades, EFs is gaining more attention, because it is required for the longevity of a river, led to the development of more than 240 methods in published literature. Although [7] pointed out that none of developed methodology is considered as best and all methods have importance, depending on objectives for estimating EFs and hydro-geological condition of a watershed. Therefore, comparative approaches is ideal approach to provide an assessment of EFs in a river or stream because all categories of quantitative methods have importance and its use depends on the level of protection, environmental goals, and objectives of the study [8]. Further [8] pointed out that hydrological and hydraulic methods are useful in cases where there is a poor understanding of the ecosystem or where a high level of protection for an existing ecosystem is required. Based on this concept, a number of hydrological based comparative studies have been conducted throughout the world by researcher for different purposes. Table 1 shows only a few of them. In this view the objective of the study is assess EFs variability using three well known hydrological methods: i) Tennant, (ii) Tessman, and (iii) FDC in eight different sub-watersheds of the DRB, India.

Study Area

The DRB has area of about 23,170 sq. km. in the states of Jharkhand and West Bengal, India and lies between 22°15'N to 24°30' N latitude and 84°45' E to 88°30' E longitude Figure 1. The basin has two main rivers Damodar and Barakar, experienced exclusive anthropogenic activity and the riverbed is probably altered The water of the rivers is mainly used for agriculture, industry and domestic purposes and demand for water from these sectors would drastically increase in near future. Details of surface water availability, water use, and demand are summarised in Table 2. The basin hydrology is the product of its climate, geology, land use, topography and drainage systems. The flow response in the rivers is strongly influenced by the underlying geology and five storage dams that provide a measure of flow regulation. These dams were constructed under Tennessee Valley Authority (TVA) of United States of America project across Damodar and its tributaries at Tilaiya, Konar, Maithon, Panchet, Tenughat and a barrage at Durgapur and four are still proposed at Aiyar, Bermo, Balpahari, and Bokaro, so that the flow can be control in the lower valley and could be better utilized for industry, municipality, agriculture and other sectors. These dams as in case of other dams of the globe have affected the hydrological system of the rivers and altered the natural habitats of the river system. The basin experiences tropical climate; the winters are cold, summers

---

*Corresponding author: Shankar Murthy, Associate Professor, Environmental Engineering & Management Group, National Institute of Industrial Engineering, Mumbai, Maharashtra-400087, India
Email: murthyshanker@gmail.com

Reference	Purpose	EFs methods used	Finding
[9]	In-stream flow methods most often used in North America	IFIM, Tennant, Wetted Perimeter, 7Q10 flow, Aquatic Base Flow (ABF).	Comparing results provide useful update and widest application
[10]	In-stream flow methods used in Australia	Tennant, Flow duration (Q95& Q90), Constant yield.	Suggested different methods are suitable according to different conditions or objectives
[11]	Comparisons of hydrologically based in-stream flow methods in 70 rivers of Atlantic Canada	Tennant, 25% of the MAF, Monthly (Q50, Q90) flow, Aquatic Base Flow (ABF), 7Q10 flow.	Q_{90} and 7Q10 methods predict extremely low in-stream flows during winter and summer months; whereas, Q50 flows recommended for gauged and 25% of the MAF & ABF for ungagged basin.
[12]	Examine impact of temperature on low stream flow for 77 rivers in the Canadian Prairies and trends analysis	Seasonal 7-day low flow, Seasonal 25% of mean flow, Seasonal Q80, Monthly Q50 & Q90.	Decrease in the magnitude and an increase in the frequency of low flow results in poor water quality and negative impact on aquatic life in river, while temperature has an increasing tendency.
[13]	Assessment to fulfilled water requirement for industrial plant located in San and Brah river watershed, India.	50-, 75-, 90- percentile FDC	Only San river met the required water demand
[14]	To protect the fish habitat	MAF, Q50, Q90, 7Q2, 7Q10.	Q50 method provide high level and whereas Q90, 7Q2 and 7Q10 methods low level of in-stream flow in small river.
[15]	Assessment of EFRs in major Indian river basin.	Default FDC	Suggested 6 (From A to F) Environmental Management Classes (EMCs).
[16]	Link EFRs with EFs classes	FDC	Developed a software package named Global Environmental Flow Calculator (GEFC) for desktop assessment of EFs.
[17]	Assess and design EFs in the Brahmani-Baitarani river system, Odissa, India.	FDC	7Q10 FDC and 7Q100 FDC were appropriate methods for designed EFs during drought /low flow periods and normal precipitation years respectively.
[18]	Assess optimal EFs in Tungabhadra river, India	Tennant	Required more water improve water quality and livelihood support base of river ecosystem.
[19]	Computing the minimum water requirement to save biological activity in Safaood river, Iran.	Tennant, Q95 from FDC, Hydraulic	Q95 from FDC gave compatibility results with the rivers condition whereas; Tennant and Hydraulic methods gave overestimated results.
[20]	Comparisons of hydrological based methods for assess EFRs to maintain basic functions in Shahr chai river, Iran.	Tennant, FDC shifting, Low flow index, DRM, GEFC	FDC shifting and DRM methods are more reliable methods in compare to Tennant, 7Q10 and Q90 of AFDC for maintains basic river functions.

Table 1: Comparative studies based on hydrological based methods.

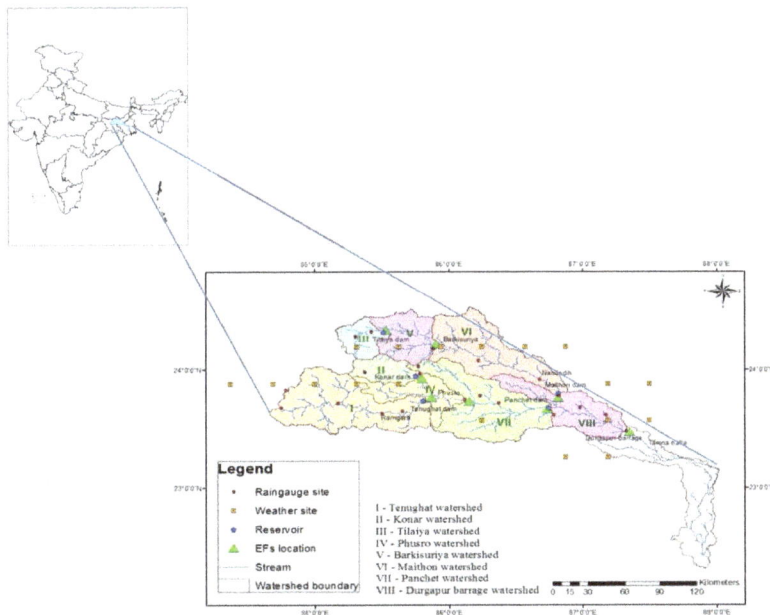

Figure 1: Showing eight sub-watersheds with EFs location.

are hot and the temperature difference between the two seasons is significant. Summers are usually very hot and dry with average 30°C, and during May-July month, temperatures can reach upto 48°C. Both rivers Damodar and Barakar are entirely rain fed. Mean annual precipitation over the whole basin varies from 1200-1400 mm (Barakar 1260 mm, Damodar 1272 mm, and lower valley 1329 mm). About 80% of mean annual runoff occurs during monsoon season from June to September. Mean annual precipitation varies from about 765 to 1850 mm. Generally, rainfall is occurring in April to August. The highest annual rainfall is 1650 mm in the southern part of the lower valley. The rainfall gradually decreases to less than 1050 mm in the northern part of the Barakar catchment.

Hydrological Methods Adopted to Assess EFs

Three hydrological methods: Tennant, Tessman, and FDC (traditional and stochastic) were used to assess EFs in eight sub-watersheds of the DRB. Details of Tennant, Tessman, and FDC methods can be found elsewhere, Whereas, described about stochastic FDC. Furthermore, the value of probability of exceedance equal to 95% (Q95) of FDC was chosen as "design EFs" in the DRB, because, the basin has extremely low flows during lean period and the ecosystem (flora and fauna) manages with the severity of flow from high to low flows very well.

Results and Discussion

Firstly, the Tennant method was applied to generate EFs requirements corresponding to different habitat condition in selected each watershed and recommended values are summarized in Tables 3 and 4. It can be noted that required values for optimum habitat status (60-100% of MAF) are maintaining in the DRB except in low flow season from April to July. Analysis shows that excellent habitat condition was maintain in downstream of Phusro, Maithion, and Panchet watersheds. On the other hand, reaches below Konar dam, Tilaya dam and, Barkisuriya did not satisfy required flow even 2.5% of the MAF throughout the year due to heavily flow regulation in Konar and Barakar watersheds

(Table 5). To make preliminary flow recommendations that take into consideration the needs of fish and other aquatic life, the Tennant method can be used in the DRB. However, it should be modified by adjusting the season of lowest flow to cover the period from April to July. Secondly, Tessmann method [22-24] was applied that mimic the natural flow on monthly basis and recommended values are shown in Figure 2. In the third method, FDCs of daily and 7-day were drawn for respective stations based on available period of record (POR). Figure 3 shows daily FDC at each monitored site. It has been observed that shape of computed daily FDCs at each sites were different, may be due to different in variability in precipitation, watershed characteristics, meteorological factors, urbanization and water abstraction or demand. The Damodar catchment is highly urbanized and having impervious surface which causes increase in storm water runoff and decrease in infiltration and ground water recharge.

Conclusions and Recommendation

The paper presents the preliminary EFs recommendations using three hydrological methods; (i) Tennant (ii) Tessman, and (iii) FDC in eight sub-watersheds in the DRB, which is the first attempt in this tropical river system. These methods are preliminary approaches, where insufficient ecological and hydraulic data are not available. Lack of eco-hydrological data makes it difficult to determine minimum flow thresholds and tipping points of different freshwater ecosystems across the world. However, in some cases, hydrological methods gave results that were in agreement with each other, but in other cases different approaches yielded different results and threshold values.

The flow recommendations in this study are often used as firsthand information, because neither a flow as magnitude computed by Tennant and Tessman and duration computed by FDC nor frequency obtained by stochastic FDC analysis has a consistent relationship to habitat or production across a range of stream geomorphology. It is therefore, likely to generate results with low confidence and monotonous and there is no provision to integrate other associated aspects, for instance-

Source	Amount (MCM)	Offstream User	Water demand (MCM/year)	
			In year 2012	In year 2021
DVC reservoir at 75% dependability (Konar, Tilaiya, Maithon, Panchet, Durgapur barrage)	4,855	Domestic	507.0	-
Less due to decrease in water holding capacity of reservoir	1,030	Industry	663	884
Net available	3,825	Agriculture	652.41	1948
Presently available from other sources	223			
Total water available	4,048			

Table 2: Water availability, withdrawal and demand in the DRB.

Station/its characteristics	TG dam	KN dam	TY dam	Phusro	Barkisuraiya	MT dam	PH dam	DB, Burnpur
River	Damodar	Konar	Barakar	Damodar	Barakar	Barakar	Damodar	Damodar
Locatitn	23°44' N 85°55' E	23°43' N 85°30' E	24°19' N 85°31' E	23°45' N 86°00' E	23°13' N 85°54' E	23°78' N 86°81' E	23°40' N 86°44' E	24°06 'N 86°13' E
Drainage area (km²)	3,393	997.1	984	5,352	2,681	6293.7	10,966	19,555
Annual runoff (Ha-m)	245,500	55,507	43,172	-	-	261,499	453,923	-
Average annual precipitation (cm)	132.08	132.08	111.76	-	-	114.17	114.17	132.08
Total dead storage capacity (Ha-m)	16,096	3,440	7,478	-	-	9,317	11,914	-
Spillway design discharge (cumec)	15,990	6,796	1,348	-	-	13,592	16,608	-

Table 3: Descriptions of selected EFs location characteristics in the DRB

SlNo.	EFs recommendation/ station	Period of record	MAF (cumec)	Flushing or Maximum	Outstanding		Excellent		Good		Fair or Grading		Poor or Minimum	Serve Degradation
				April to September @ 200%	April to September @ 60%	October to March @40%	April to September @ 50%	October toMarch @ 30%	April to September @ 40%	October to March @20%	April to September @ 30%	October to March @ 10%	October to September @10%	April to September @ < 10%
1.	Tenughat dam	1981-2010	69.83	139.66	41.89	27.93	34.92	20.95	27.93	13.96	20.95	6.98	6.98	< 6.98
2.	Konar dam	1981-2010	13.63	27.26	8.18	5.45	6.82	4.09	5.45	2.07	4.09	1.36	1.36	< 1.36
3.	Tilaya dam	1981-2010	10.37	20.79	6.22	4.15	5.19	3.11	4.15	2.08	3.11	1.04	1.04	< 1.04
4.	Barkisuriya	1981-2010	32.16	64.32	19.30	12.86	16.08	9.65	12.84	6.43	9.65	3.22	3.22	< 3.22
5.	Phusro	1988-2010	87.56	175.12	52.54	35.024	43.78	26.27	35.02	17.51	26.27	8.76	8.76	< 8.76
6.	Maithon dam	1981-2010	81.50	163.0	48.9	32.6	40.75	24.45	32.6	16.3	24.45	8.15	8.15	< 8.15
7.	Panchet dam	1981-2010	135.87	271.74	81.52	54.35	67.94	40.76	54.35	27.17	40.76	13.59	13.59	< 13.59
8.	Damodar bridge, Burnpur	1981-2010	168.03	336.06	100.82	67.21	84.02	50.41	67.21	33.61	50.41	16.80	16.80	< 16.80

			MAF (cumec)	Flow Indices from empirical 1day and 7-day POR FDC													
				Q10		Q17		Q40		Q50		Q75		Q90		Q95	
				1-day	7-day	1-day	7-day	1-day	7-day	1-day	7-day	1-day	7-day	1-day	7-day	1day	7-day
1.	Tenughat dam	1981-2010	69.83	187.73	187.74	128.4	133.71	6.37	8.48	5.70	6.25	5.10	5.24	4.68	4.86	4.39	4.51
2.	Konar dam	1981-2010	13.63	21.29	21.18	13.65	13.77	9.26	9.60	8.60	8.79	6.75	6.82	5.55	5.10	4.05	4.52
3.	Tilaya dam	1981-2010	10.37	21.87	21.76	20.2	18.88	9.72	9.54	4.16	4.92	0.075	0.38	0.25	0.35	0.011	0.34
4.	Barkisuriya	1981-2010	32.16	77.06	78.6	45.85	53.27	21.18	22.78	11.39	13.0	0.78	1.24	0.023	0.06	0.01	0.2
5.	Phusro	1988-2010	87.56	176.11	240.13	139.4	153.2	25.08	27.45	19.36	20.5	10.04	10.38	4.52	5.18	1.86	2.14
6.	Maithon dam	1981-2010	81.50	169.09	169.98	126.16	124.4	45.83	46.62	31.46	34.8	14.28	15.90	7.17	9.19	2.89	5.18
7.	Panchet dam	1981-2010	135.87	353.0	354.89	195.2	226.8	58.33	58.07	36.11	38.9	9.95	13.34	2.88	5.44	0.29	0.74
8.	Damodar bridge, Burnpur	1981-2010	168.03	473.12	478.0	282.5	305.1	46.8	52.25	24.0	26.9	4.49	4.86	0.22	1.10	0.22	0.23

Table 4: Comparison of recommended EFs in cumec in various sub-watersheds of the DRB assessed through Tennant, and traditional FDC

Sl. No	Sampling site	Q10 (cumec)				Q25 (cumec)			
		7Q10	7Q20	7Q50	7Q100	7Q10	7Q20	7Q50	7Q100
1.	Tenughat dam	99.79	133.71	192.59	449.04	7.31	25.76	80.86	240.38
2.	Konar dam	9.79	10.76	16.55	65.86	8.10	8.54	10.09	25.81
3.	Tilaya dam	10.52	13.19	18.68	82.49	2.43	4.91	15.49	25.21
4.	Barkisuriya	33.63	40.14	73.2	224.87	17.22	20.99	36.92	90.66
5.	Phusro	106.77	140.8	240.2	891.81	37.12	41.30	114.84	361.0
6.	Maithon dam	77.36	97.22	160.24	540.0	42.70	53.03	86.72	185.86
7.	Panchet dam	194.36	277.35	374.64	786.04	102.7	142.72	177.46	276.22
8.	Damodar bridge, Burnpur	72.29	195.11	507.81	1469.0	11.96	44.61	168.07	567.28
		Q50 (cumec)				Q75 (cumec)			
		7Q10	7Q20	7Q50	7Q100	7Q10	7Q20	7Q50	7Q100
1.	Tenughat dam	5.09	5.31	6.12	39.17	4.86	5.09	5.29	8.32
2.	Konar dam	5.67	6.02	8.38	22.59	4.4	5.65	7.10	10.99
3.	Tilaya dam	0.075	0.14	7.13	22.05	0.05	0.05	0.09	10.86
4.	Barkisuriya	2.49	5.0	13.92	30.59	0.04	0.06	1.25	12.36
5.	Phusro	10.2	15.09	20.34	37.79	3.25	7.25	10.25	24.45
6.	Maithon dam	13.75	23.97	37.68	62.88	7.97	12.91	17.16	29.40
7.	Panchet dam	34.87	57.87	60.62	71.25	15.6	14.25	21.1	22.77
8.	Damodar bridge, Burnpur	1.82	3.13	16.51	193.81	0.22	1.42	5.21	73.49

		Q90 (cumec)				Q95 (cumec)			
		7Q10	7Q20	7Q50	7Q100	7Q10	7Q20	7Q50	7Q100
1.	Tenughat dam	4.40	4.71	4.93	6.52	2.55	4.02	4.56	5.58
2.	Konar dam	3.47	4.56	6.02	10.88	3.40	4.51	5.9	10.88
3.	Tilaya dam	0.03	0.04	0.06	2.26	0.02	0.03	0.05	1.50
4.	Barkisuriya	0.01	0.02	0.10	2.99	0.01	0.02	0.10	1.98
5.	Phusro	1.87	2.89	7.47	21.74	1.23	1.3	6.90	19.31
6.	Maithon dam	2.89	7.28	11.6	20.47	1.67	6.35	8.20	17.23
7.	Panchet dam	9.80	6.22	10.22	14.19	2.76	5.21	8.68	10.44
8.	Damodar bridge, Burnpur	0.22	0.62	3.87	49.35	0.20	0.22	3.71	34.66

Table 5: Comparing the results for selected 7Q flow series in various return periods computed using SFDC analysis.

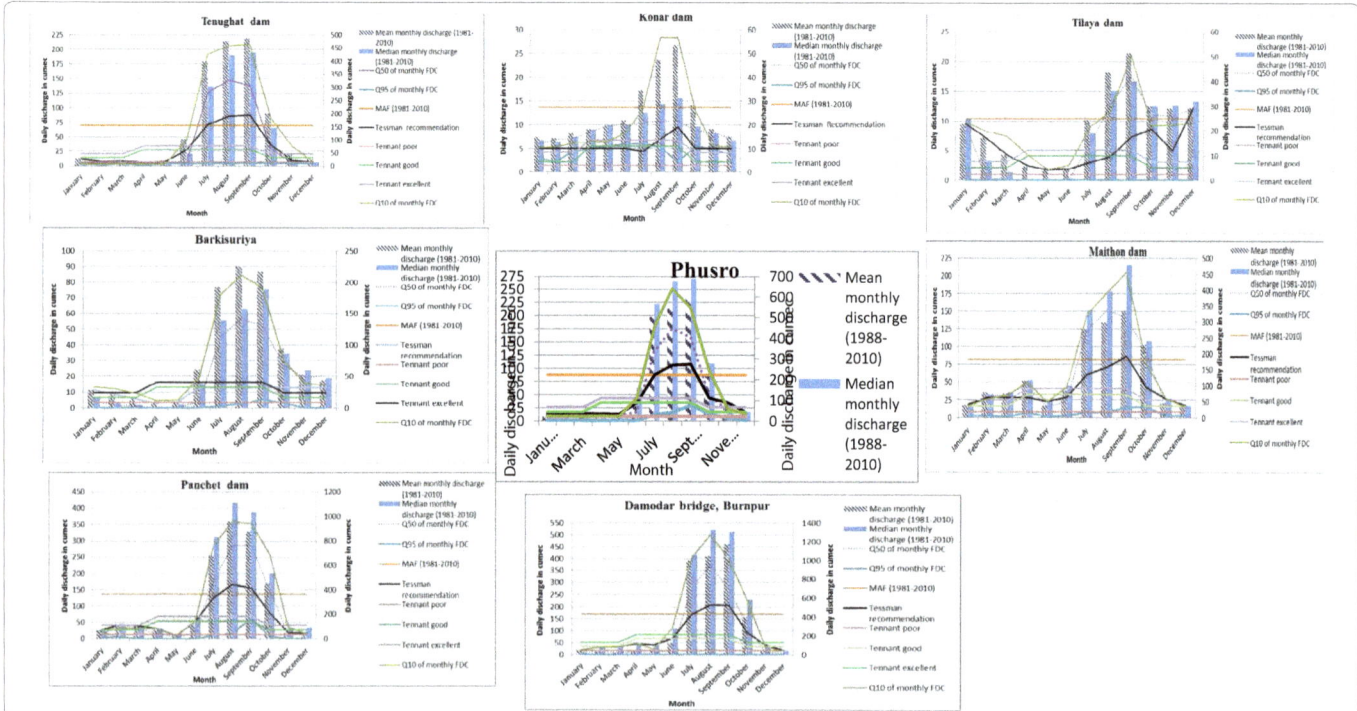

Figure 2: Estimated EFs in different watersheds of the DRB using Tennant method on monthly basis, Tessmann method and flow charecteristics at diffents % time exceedance (i.e Q10, Q50, Q95) on basis of monthly FDC.

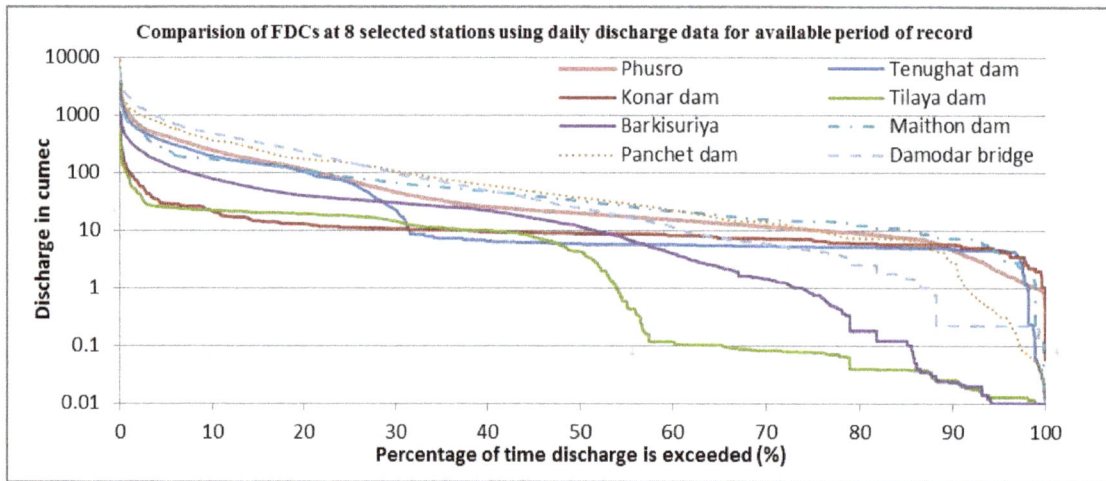

Figure 3: A graphical comparetive representation of daily FDC for selected stations in the DRB.

the ecology, biodiversity, riverine communities etc. Thus, there is a need for a better methodology that could be implemented and describes relationships between discharge and width, discharge and depth, and discharge and velocity. Hence, simple historical based approaches used in assessing EFs requirements are not found suitable for all types of watershed. However, the results obtained provide detailed information, which can be used for estimating the water supply, the water demand for both anthropogenic and ecological, and the amount available for withdrawal in near future regarding the planned anthropogenic alteration and its consequences. The values recommended for 7Q50 and 7Q100, will help water resource planner and decision makers to develop new water resource projects such as the design of storage facilities, assessment of water available for municipal, agricultural or industrial purposes and operating rules that satisfy EFs in DRB.

Acknowledgements

The authors are thankful to MRO office, DVC, Maithon for providing the historical discharge data to conduct this study. The authors would also like to thank Centre for Environmental Studies, NITIE for the financial support.

References

1. Crisp DT (1989) Some impacts of human activities on trout Salmo trutta populations. Freshwater Biology 21: 21-33.

2. King J, Brown C, Sabet H (2003) A scenario-based holistic approach to environmental flow assessment for rivers. River research and applications 19: 619-639.

3. Petts GE (2009) Instream flow science for sustainable river management. Journal of the American Water Resources Association 45: 1071-1086.

4. Revenga C, Brunner J, Henninger N, Kassem K, Payne R (2000) Pilot analysis of global ecosystems: freshwater systems. World Resources Institute.

5. Poff NL, Allan JD, Bain MB, Karr JR, Preste-gaard KL, et al. (1997) The natural flow regime. A paradigm for river conservation and restoration. Bio Science 47: 769-784.

6. Postel S, Richter BD (2003) Rivers for life: managing water for people and nature. Island Press Washington DC, USA.

7. Acreman M, Dunbar MJ (2004) Defining environmental river flow requirements-a review. Hydro Earth Sys Sci 8: 861-876.

8. Jowett IG (1997) Instream flow methods: A comparison of approaches. Regulated Rivers: Res Mange 13: 115-127.

9. Reiser DW, Wesche TA, Estes C (1989) Status of instream flow legislation and practices in North America. Fisheries 14: 22-28.

10. Karim K, Gubbels ME, Goulter IC (1995) Review of determination of instream flow requirements with special application to Australia. Water Resour Bull 31: 1063-1077.

11. Caissie D, El-Jabi N (1995) Comparison and regionalization of hydrologically based in-stream flow techniques in Atlantic Canada. Can J Civil Engg 22: 235-246.

12. Yulianti JS, Burn DH (1998) Investigating links between climatic warming and low stream flow in the Prairies region of Canada. Can J Water Resour 23: 45-60.

13. Pandey RP, Ramasastri KS (2003) Estimation of lean season water availability in streams with limited data. Journal of the Institution of Engineers. India, Civil Engineering Division 84:149-152.

14. Caissie D, El-Jabi N, Hébert C (2007) Comparison of hydrologically based instream flow methods using a resampling technique. Can J Civil Engg 34: 66-74.

15. Smakhtin VU, Anputhas M (2006) An assessment of environmental flow requirements of Indian river basins. International Water Management Institute, Colombo, Sri Lanka.

16. Smakhtin VU, Eriyagama N (2008) Developing a software package for global desktop assessment of environmental flows. Environ. Model. Software 23: 1396-1406.

17. Smakhtin V, Revenga C, Döll P (2004b) A pilot global assessment of environmental water requirements and scarcity. Water Int 29: 307-317.

18. Jha R, Sharma KD, Singh VP (2008) Critical appraisal of methods for the assessment of environmental flows and their application in two river systems of India. KSCE J Civil Engg 12: 213-219.

19. Babu KL, Kumara, BKH (2009) Environmental flows in river basins: a case study of river Bhadra. Current Science 96: 475-479.

20. Shokoohi A, Hong Y (2011) Using hydrologic and hydraulically derived geometric parameters of perennial rivers to determine minimum water requirements of ecological habitats (case study: Mazandaran Sea Basin-Iran.). J Hydrolog Process 25: 3490-3498.

21. Karimi SS, Yasi M, Eslamian S (2012) Use of hydrological methods for assessment of environmental flow in a river reach. Int J Environ Sci Technol 9: 549-558.

22. Sugiyama HV, Vudhivanich AC, Whitaker K, Lorsirirat (2003) Stochastic flow duration curves for evaluation of flow regimes of rivers. J Amer Water Resour Assoc 39: 47-58.

23. Tessman SA (1980) Environmental Use Sector Reconnaissance Elements of the Western Dakotas Region of South Dakota Study. Water Resources Research Institute.

24. Kumar SS, Sridhar KT (1998) GIS techniques for carrying capacity study of Damodar River Basin Central.

26

Effect of Used Motor Oil on the Macro and Micromechanical Properties of Crumb Rubber Modified Asphalt

Magdy Abdelrahman, Mohyeldin Ragab* and Daniel Bergerson

Department of Civil and Environmental Engineering North Dakota State University, USA

Abstract

The need to be more environmentally conscious has recently shifted toward the forefront of society. With this new focus on environmentally responsible behavior comes the practice of using recycled materials in construction when possible. Used motor oil (UMO) is an example of waste materials that can be utilized in numerous applications to alleviate its environmental disposal burden. In the current work, the effect of UMO on the internal structure of crumb rubber modified asphalt is investigated. This is carried out by employing rheological analysis including dynamic shear rheometer and microindentation testing. Rheological analysis was employed to determine the change in the phase angle (δ) as well as the complex modulus (G^*) of the UMO modified asphalts as well as the crumb rubber modified asphalt (CRMA), in addition, temperature sweep viscoelastic analysis was employed to investigate the change in internal structure of the produced modified asphalts. Microindentation analysis was utilized to determine the hardness and elastic modulus of the modified asphalt liquid phase. Microindentation tests served to simulate the effect of UMO on the behavior of the thin asphalt layer over the aggregate which has a thickness measured in microns. Results indicate that the utilization of UMO only as a modifier to asphalt severely deteriorates the macro and micro mechanical properties of the binder. Combining CRM with UMO as modifiers to asphalt had better results. It is suggested to use UMO at a rate of 3%, or less, by asphalt weight.

Keywords: Used motor oil; Crumb rubber modifier; Three dimensional network structure; Crumb rubber modified asphalt

Abbreviations: CRM: Crumb Rubber Modifier; UMO: Used Motor Oil; CRMA: Crumb Rubber Modified Asphalt; 3D: Three Dimensional

Introduction

Each year, two hundred million gallons of used oil are improperly disposed of [1]. In recent years, attention has been brought to the need to preserve the environment and its resources for future generations. This can be achieved through the utilization of waste materials, in addition to limiting the use of virgin products. Used motor oil (UMO) and crumb rubber modifier (CRM) are both waste materials that can be implemented in the paving industry. Although both materials have been investigated separately, up to this point no research has been dedicated to investigating the combined effect of such binder modifiers on the binder's rheological, internal and micromechanical properties.

CRM is an example of a recycled material that is incorporated into asphalt pavement. CRM is made from recycled tires. Asphalt is made up of continuous three- dimensional associations of polar molecules that are dispersed in a fluid of nonpolar or relatively low-polarity molecules [2]. Associations of different strengths are created by the polar functions within asphalt [3]. The typical viscoelastic properties for neat asphalts are the result of continuous formation and breakage of these associations under the effect of external factors such as shear stresses and temperature variations [2].

Research by Dedene et al. [4] and by Villanueva et al. [5] suggests that UMO can be used as rejuvenator for existing and recycled asphalt pavement and can enhance low temperature service performance.

In previous work by this research group it was verified that the existence of 3D network structures occurs in CRMA [6]. The effect of the developed three dimensional (3D) network structure in the liquid phase of CRMA at the end of the interaction time has been shown to be essential and deterministic to the enhancement of both the G^* and δ of CRMA [6]. However, the effect of addition of UMO to asphalt

that is either neat or modified with CRM needs further investigation. The interaction of CRM and asphalt results in a non-homogeneous mixture due to the fairly consistent physical shape of CRM when treated with the binder. Thus it is essential to understand how the UMO would alter the behavior of the asphalt-CRM mix. In another work by us, we investigated the environmental impact of UMO during the modification with asphalt [7]. We proved through air testing and batch leaching that interaction temperature, interaction time, and binder grade affect the amount of Benzene, Toluene, Ethylbenzene, and Xylenes (BTEX) leached from modified asphalt binder. In addition, we found that the binders modified with both CRM and UMO released less BTEX to both leachate and the air at the end of interaction time, indicating that CRM retains BTEX and prevents it from being released into the environment when used in conjunction with UMO [7].

The aim of this work is to investigate the effect of addition of UMO to neat as well as CRMA and also investigate how it affects the modified asphalt rheological (stiffness and elasticity), internal structure and micromechanical properties. This is achieved by investigating the change in the internal network structure of the modified asphalt through monitoring the change in the physical properties of asphalt through single point and temperature sweep viscoelastic tests as well as microindentation analysis. In addition, the utilization of microindentation serves to simulate the behavior of a thin layer or coating of asphalt such as that laid on the surface of aggregate.

*Corresponding author: Mohyeldin Ragab, Ph.D. Candidate, Department of Civil and Environmental Engineering, North Dakota State University
E-mail: Mohyeldin.ragab@ndsu.edu

Materials and Methods

Processing

Raw materials: In the current work, one asphalt binder was investigated in combination with one type of crumb rubber. The asphalt was a PG 64-22 based on the superpave grading system. The CRM was a cryogenic processed CRM from a mixed source of scrap tires. The CRM particle size was smaller than mesh #30 and larger than mesh #40, according to the US standard system.

Collected used motor oils have been tested for Benzene, Toluene, Ethyl-benzene, and Xylenes (BTEX) content and it was verified that the utilized UMO in the current experiments has less BTEX content than what is allowed by the United States Environmental Protection Agency (US-EPA) Maximum Contaminants Levels (MCLs) during leaching experiments and also from air tested samples above the interactions carried out in the lab.

Asphalt-CRM interactions: The interactions were conducted in 1 gallon cans, and a heating mantle connected to a bench type controller with a long temperature probe (12") was used to heat the material. A high shear mixer was used to mix the binder and crumb rubber. The amount of CRM was controlled to be either 10% or 20% of the initial asphalt binder weight in selected interactions. The UMO percentage varied from 3% to 9% of the final binder weight. Interactions were conducted for 120 minutes under one of two different temperatures (160 and 190°C) and a single mixing speed (30 Hz) for each temperature utilized. Samples were taken at 2, 30, and 120 minutes of interaction time and kept at -12°C to avoid any unwanted reactions. All interactions in this research were carried out under controlled atmosphere of nitrogen gas to prevent any oxidation. A specific coding for the samples was adopted in the current work, starting with the asphalt type, HU-64, followed by the interaction temperature, interaction speed, UMO percentage, and lastly CRM percentage. Table 1 illustrates the list of interactions utilized in this research work.

Characterization

Extraction of liquid phase: The liquid phase of CRMA was extracted by removing the non-dissolved CRM particles from the CRMA matrix. In this regard, the required amount of CRMA sample was heated to 165°C and drained through mesh #200 (75 µm) in the oven at 165°C for 25 minutes. The extracted liquid phase was stored at -12°C immediately to prevent any unwanted reactions.

Dynamic mechanical analysis: Dynamic Shear Rheometer from Bohlin Instruments CVO, (Worcestershire, UK) was used for viscoelastic analysis of neat modified asphalt samples and their liquid phase. Single point test and temperature sweep test were performed on the samples.

The single point test was performed on all modified asphalt samples before separation of the liquid phase at 64°C and 10 radian/sec using 25 mm diameter parallel plates. The temperature sweep test was performed on the liquid phase of modified asphalt at a temperature range of 10°C to 70°C with 6°C increments. 25 mm diameter plates were used for tests that conducted above 45°C, and 8 mm diameter plates were used for tests that were conducted below 45°C.

The gap between plates for CRMA samples was selected to be 2 mm, which is the minimum gap size that does not affect the results due the presence of CRM particle. For samples without CRM and for the liquid phase, the gap was selected to be 1 mm. For all tests that were conducted at temperatures below 45°C using the 8 mm diameter plates, the 2 mm gap size was selected, regardless of the type of the sample.

Preparation of CRMA microindentation samples: The procedure of asphalt sample preparation for indentation testing was adapted from literature [8,9]. The preparation of the CRMA liquid phase thin film involved the utilization of a glass slide surface that was covered with a high temperature resistant tape. A rectangular window of size 1.5x0.5" (38.1x12.7 mm) was made inside the high temperature resistant tape. Following that, CRMA was heated to 160°C and poured into the square window in the high temperature resistant tape. In order to have a smooth CRMA surface, the glass substrate coated with CRMA was placed in the oven at 160°C for 5 min. After that, the glass substrate was cooled down to room temperature. The CRMA film thickness ranges were controlled to be between 550–600 µm to insure that the measured values for hardness and elastic modulus are not affected by the glass substrate.

Microindentation tests procedure: In this research work, a FISCHERSCOPE HM2000S indenter was utilized for the indentation tests. Indentation tests were carried out using a tungsten carbide metal spherical tip of diameter 2mm that conforms with ISO 14577-3. All indentation tests were carried out under load control mode. The indentation load configuration was programmed to start with a constant loading rate followed by a holding period at maximum load and finishing with an unloading rate similar to the loading one. The maximum load was 5 mN. The loading and unloading times were 20 seconds. The dwell time at maximum load was 60 seconds to minimize the viscous effect on the unloading portion of the material [8-10]. All tests were carried out at ambient temperature. For each sample a minimum of 5 indentations were carried out to determine the hardness and elastic modulus. A minimum distance of 6 mm between two

#'	Code	Interaction temperature (°C)	Mixing speed (Hz)	UMO percentage	CRM percentage
1	HU-64-160-30-3% UMO	160	30	3	0
2	HU-64-160-30-3% UMO-10% CRM	160	30	3	10
3	HU-64-160-30-10% CRM	160	30	0	10
4	HU-64-160-30-9% UMO	160	30	9	0
5	HU-64-160-30-9% UMO-20% CRM	160	30	9	20
6	HU-64-160-30-20% CRM	160	30	0	20
7	HU-64-190-30-3% UMO	190	30	3	0
8	HU-64-190-30-3% UMO-10% CRM	190	30	3	10
9	HU-64-190-30-10% CRM	190	30	0	10
10	HU-64-190-30-9% UMO	190	30	9	0
11	HU-64-190-30-9% UMO-20% CRM	190	30	9	20
12	HU-64-190-30-20% CRM	190	30	0	20

Table 1: List of interactions.

indentations was utilized to avoid the pile up and sink-in effect for successive indentations and also to conform to the ASTM guidelines that require a distance of at least 6 indent radii from the previous indentation point. In this work the Oliver and Pharr method was utilized for the analysis of the indentation test data [11]. The Oliver and Pharr method not only accounts for the curvature in the unloading data but also provides a physically justifiable approach for determining the indentation depth to be used in conjunction with the indenter shape function to establish the contact area at peak load [11]. The analysis starts by defining a reduced modulus E_r that accounts for the effects of a non-rigid indenter on the load-indentation behavior as follows [12]:

$$\frac{1}{E_r} = \frac{1-v_s^2}{E_s} - \frac{1-v_i^2}{E_i} \dots\dots\dots\dots \quad ..(Eq.1)$$

where E_s=Young's modulus of the sample; v_s=Poisson's ratio of the sample; E_i=Young's modulus of indenter, v_i=Poisson's ratio of indenter; and E_r=reduced modulus.

The unloading portion of the indentation curve relates to the reduced modulus as follows [11,13]:

$$E_r = \frac{\sqrt{\pi}}{2} \frac{S}{\sqrt{A}} \dots\dots\dots\dots \quad (Eq.2)$$

where S=dP/dh=initial unloading stiffness and A=contact area.

To determine the initial unloading stiffness, Oliver and Pharr utilized curve fitting of the indentation depth versus loading/unloading data using the following power law function [11]:

$$P = \alpha(h - h_f)^2 \dots\dots\dots\dots \quad ....(Eq.3)$$

where h=any depth of penetration; h_f =unrecoverable or plastic depth; and α and m=constants.

In this regard, m is a power-law exponent that is related to the geometry of the indenter [8]. The initial unloading slope S is determined by differentiating Eq. 3 and evaluating the derivative at the peak load and displacement [8].

Based on Oliver and Pharr's approach, the total displacement h is given as [11]:

$$h = h_c + h_s \dots\dots \quad \dots\dots\dots(Eq.4)$$

where h_c is the vertical distance along which contact is made (contact depth) and h_s is the displacement of the surface at the perimeter of the contact [11]. The contact depth h_c can be determined from the experimental data by extrapolating the tangent line to the unloading curve at the maximum loading point down to zero loads [8,11]. h_s can be estimated from the intercept value for depth hi that relates to the contact depth h_c associated with the maximum loading point as follows [11]:

$$h_s = h_{max} - \varepsilon \frac{P_{max}}{S} \dots\dots \quad \dots\dots\dots\dots(Eq.5)$$

where ε=geometric constant and P_{max} is the maximum indentation load.

The hardness can be obtained as follows [11]:

$$H = \frac{P_{max}}{A} \dots\dots\dots\dots \quad (Eq.6)$$

where A=projected area of contact at the peak load.

Results and Discussion

Figure 1 illustrates the rheological properties; (a) Complex modulus (G*) and (b) phase angle (δ) for the samples interacted at 160°C with 3% UMo, 10% CRM or both modifiers.

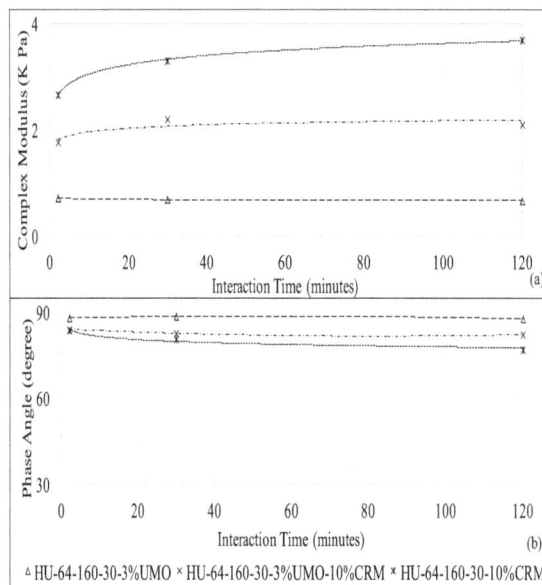

Figure 1: Rheological properties of the samples interacted with 3% UMO, 10% CRM, or both at 160C and 30Hz: (a) G* and (b) δ.

As shown in the Figure, the addition of 3% UMO to the asphalt resulted in deterioration in both the G* and δ. On the other hand, upon the utilization of 10% CRM only, enhancement in both the G* and δ was observed.

The combined use of 10% CRM and 3% UMO had a plateau effect for both the G* and δ values, but with values tending to slightly deteriorate at the end of interaction time (120 minutes). At such interaction temperature (160°C) and with the utilization of medium interaction speed (30Hz), the prevalent behavior observed in the swelling of the CRM particles with the low molecular weight fractions from asphalt and/or oil leading to minor enhancement of both the G* and δ [14]. On the other hand, the addition of UMO only to asphalt resulted in the disturbance of the asphalt's continuous three-dimensional associations leading to the observed deterioration in both G* and δ for the sample modified with 3% UMO only.

Figures 2a and 2b illustrates the rheological properties (G*) and (δ) for the samples interacted at 160°C with 9% UMO, 20% CRM, or both modifiers, respectively. A marked difference can be seen in the enhancement of both rheological parameters (G* and δ) for the sample with 20% CRM; this is due to the swelling of CRM and decrease of enter-particle distance [15]. In addition, for the sample with 20% CRM and 9% UMO, deterioration can be observed after 30 minutes of interaction time in both rheological parameters and continues until the end of the interaction time (120 minutes). This can result from the increased amount of UMO (9%) that significantly disturbed the asphalt's continuous three-dimensional associations, even in the presence of 20% CRM. This behavior of severe deterioration in both G* and δ was largely manifested in the sample with 9% UMO only.

Figure 3 illustrates the rheological properties; (a) Complex modulus (G*) and (b) phase angle (δ) for the samples interacted at 190°C with 3% UMO, 10% CRM, or both modifiers.

A distinctive behavior can be seen for the samples with 10% CRM and 3% UMO. There is continuous enhancement in both the G* and δ alongside the interaction time. Similar enhancements, with higher extent, can be seen for the sample with 10% CRM only. At such combination of moderate interaction temperature (190°C) and mixing speed (30 Hz) the interaction of CRMA involves not only the swelling of the CRM but also the occurrence of devulcanization processes of CRM that lead to the release of CRM components into the liquid phase of asphalt. This results in enhancements in the network structure of the CRMA whether UMO is present or not. On the other hand, the use of 3% UMO only resulted in deterioration in both the G* and δ as a result of the disturbance of the network structure of asphalt, as explained earlier.

Figures 4a and 4b illustrates the rheological properties (G*) and (δ) for the samples interacted at 190°C with 9% UMO, 20% CRM, or both modifiers, respectively.

As shown previously for the sample interacted at 160°C, a marked difference can be seen in the enhancement of both rheological parameters (G* and δ) for the sample with 20% CRM; this is due to the swelling of CRM as well as the occurrence of devulcanization at such interaction temperature (190°C) [16]. In addition, for the sample with 20% CRM and 9% UMO, unlike the behavior for the samples at 160°C, enhancements can be observed after 30minutes of interaction time in both rheological parameters and continue through the end of the interaction time (120 minutes). On the other hand, severe deterioration in both G* and δ was largely manifested in the sample with 9% UMO only.

Figure 5 illustrates the temperature sweep viscoelastic properties of the liquid phase for rheological properties (G*) and (δ) for the samples interacted at 160°C with 3% UMO, 10% CRM, or both modifiers after 120 minutes of interaction time, respectively.

In Figure 5, results show agreement with other studies which have shown that the CRM modification of asphalt mainly affects its high temperature properties and by decreasing the testing temperature the changes in physical properties reduce to a marginal level [5]. Results in Figure 5b show that the behavior of the δ lacks the presence of a plateau behavior, indicating lack of network structure in the liquid phase of modified asphalts [17,18].

Figure 6 shows the temperature sweep viscoelastic properties of liquid phase for rheological properties (G*) and (δ) for the samples interacted at 160°C with 9% UMO, 20% CRM, or both modifiers after 120 minutes of interaction time, respectively. It can be seen from the Figure that the behavior of the samples with UMO is almost similar with or without CRM addition. This indicates that the addition of the UMO significantly annihilated the network associations within the modified asphalt liquid phase.

Figure 7 shows the temperature sweep viscoelastic properties of the liquid phase for rheological properties (G*) and (δ) for the samples interacted at 190°C with 3% UMO, 10% CRM, or both modifiers after 120 minutes of interaction time, respectively. Results in Figure 7b show a distinctive behavior. Where, similarity can be observed between the behavior of both samples with 10% CRM only or 10% CRM and 3% UMO before 20°C and after 50°C. This indicates that the released components of the CRM in asphalt, in the presence of UMO, at certain interaction temperature (190°C and 30Hz) are capable of forming internal network in the asphalt matrix. The effect of such network was previously manifested by the enhancement in both G* and δ values of such sample as compared to the sample with 10% CRM only (illustrated in Figures 3a and 3b).

Figure 8 shows the temperature sweep viscoelastic properties of liquid phase for rheological properties (G*) and (δ) for the samples interacted at 190°C with 9% UMO, 20% CRM, or both modifiers after 120 minutes of interaction time, respectively.

Figure 8a, indicates that the G* values of the samples having UMO, either with or without CRM, are of similar values, in contrast with the samples with CRM only. As can be seen from Figure 8b, the behavior of the samples with 20% CRM and 9% UMO shows an intermediate trend between those that have 20% CRM only and those that contain 9% UMO only.

Figure 9 illustrates (a) the force vs. indentation depth profiles and (b) comparison of hardness and elastic modulus for the samples interacted at 160°C with 3% UMO, 10% CRM, or both modifiers, after 120 minutes of interaction time.

As can be seen from Figure 9a, the max load values show a continuous decrease during the dwell time, similar observations were recorded in the literature for the indentation of asphalt [8]. This was explained in terms of the decrease in contact area due to delayed (viscous) flow of asphalt binders at the indentation location [8]. Another reason is the minute scale load carrying capacity of the asphalt binders and binder softening which results in it being virtually impossible to keep the maximum applied load constant [8]. For the sample with only 3% UMO, the indentation depth was about 45 µm. For the sample with 10% CRM and 3% UMO the indentation depth was about 38 µm. However, upon utilizing 10% CRM only, further reduction in the indentation

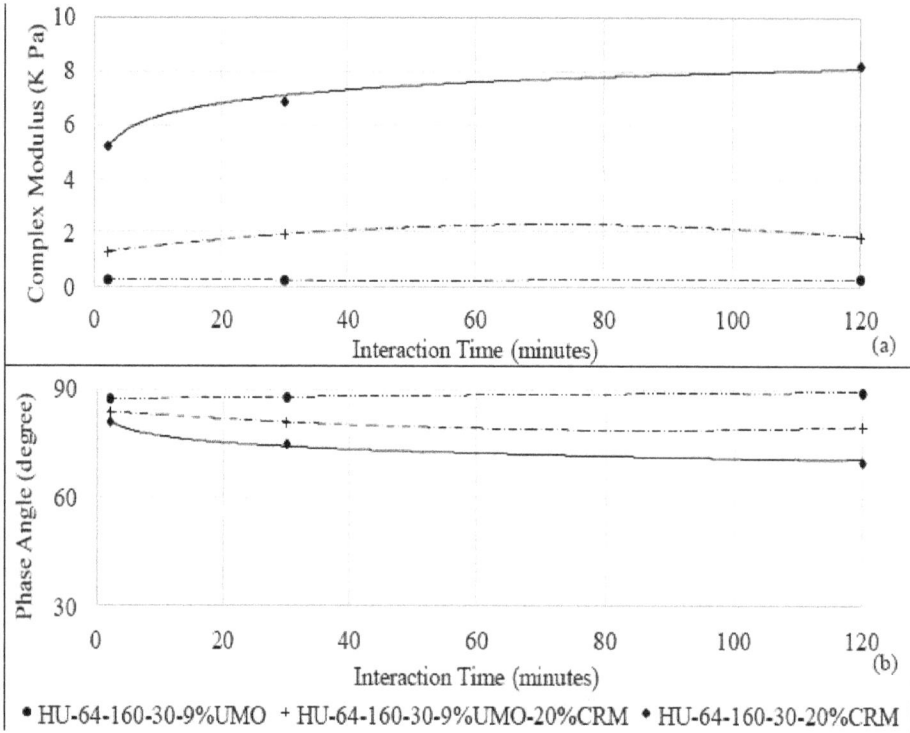

Figure 2: Rheological properties of the samples interacted with 9% UMO, 20% CRM or both at 160°C and 30Hz: (a) G* and (b) δ.

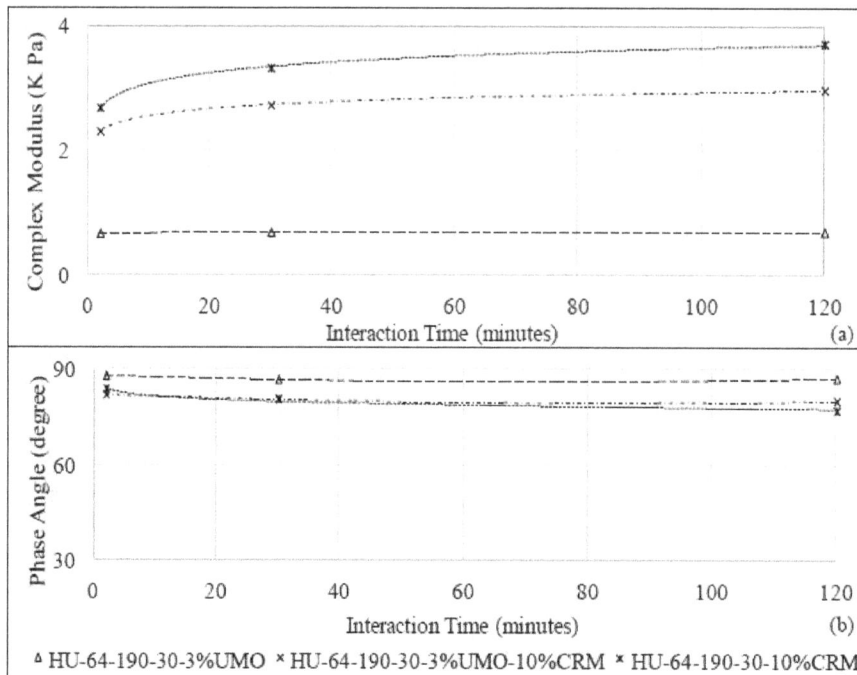

Figure 3: Rheological properties of the samples interacted with 3% UMO, 10% CRM or both at 190°C and 30Hz: (a) G* and (b) δ.

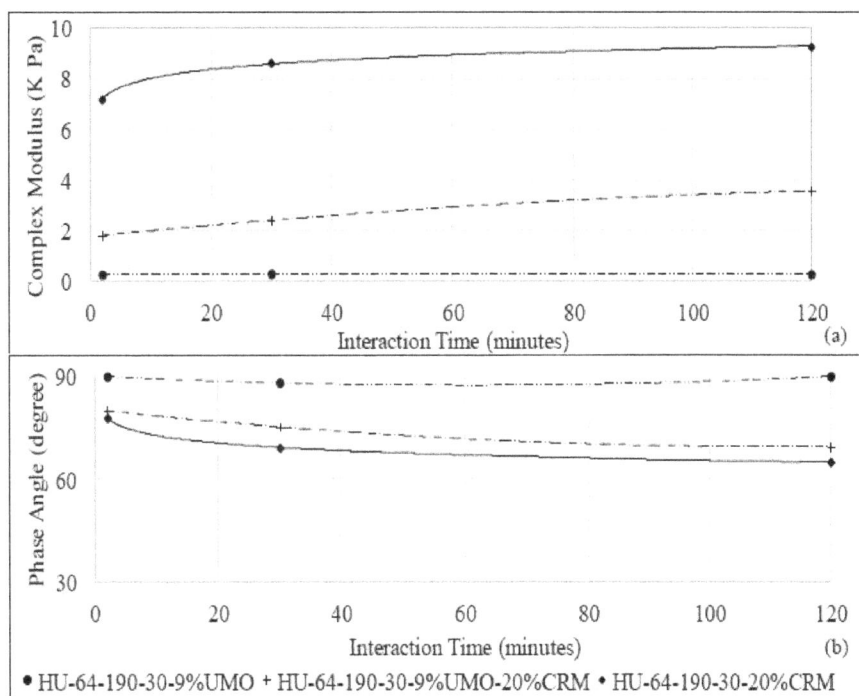

Figure 4: Rheological properties of the samples interacted with 9% UMO, 20%CRM or both at 190°C and 30Hz: (a) G* and (b) δ.

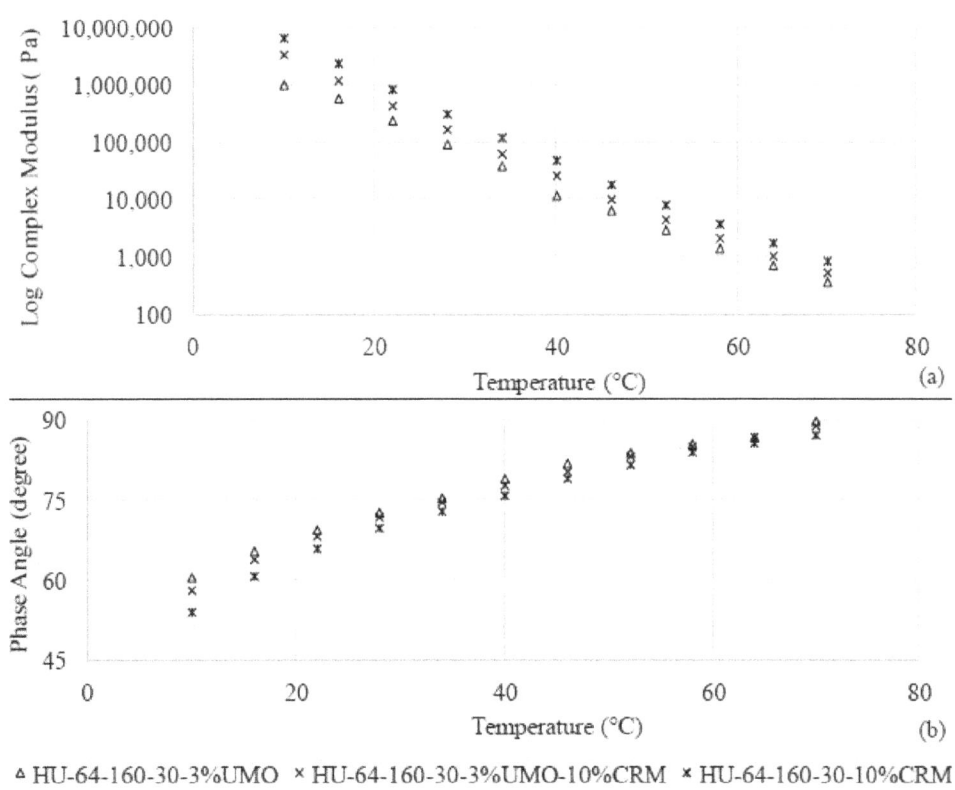

△ HU-64-160-30-3%UMO × HU-64-160-30-3%UMO-10%CRM × HU-64-160-30-10%CRM

Figure 5: Temperature Sweep Viscoelastic Properties of Liquid Phase samples interacted with 3% UMO, 10% CRM or both at 160°C and 30Hz: (a) G* and (b) δ.

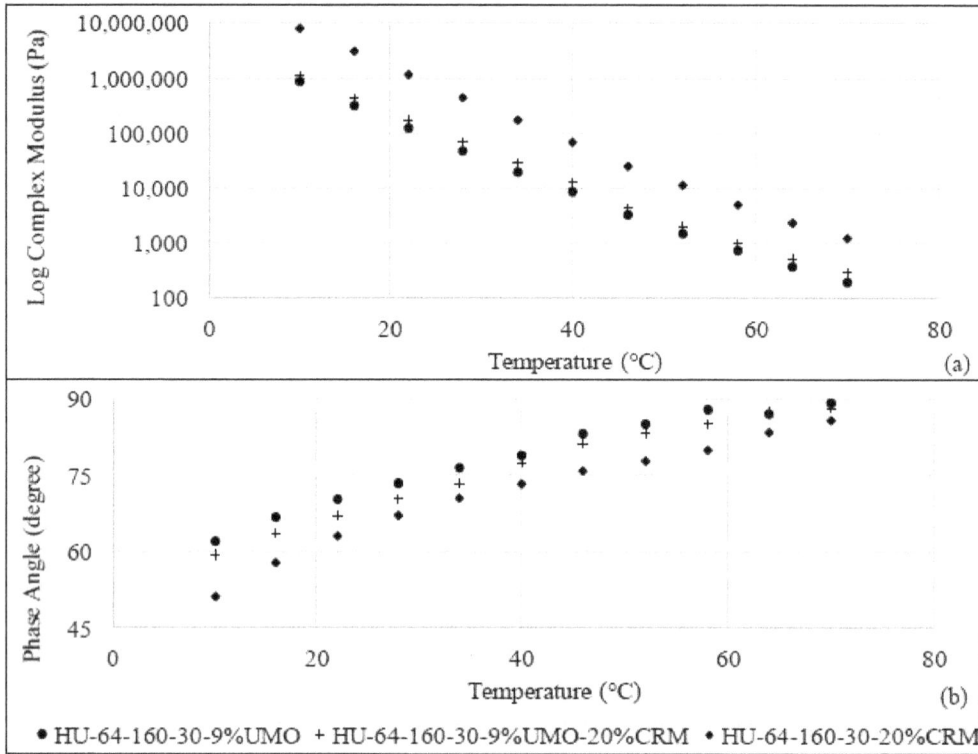

Figure 6: Temperature Sweep Viscoelastic Properties of Liquid Phase samples interacted with 9% UMO, 20% CRM or both at 160°C and 30Hz: (a) G* and (b) δ.

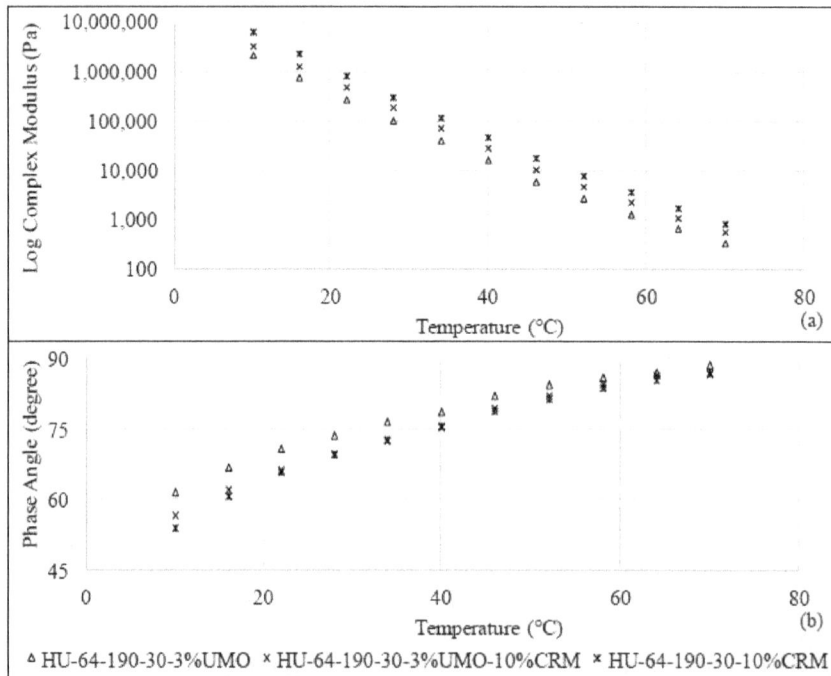

Figure 7: Temperature Sweep Viscoelastic Properties of Liquid Phase samples interacted with 3% UMO , 10% CRM or both at 190°C and 30Hz: (a) G* and (b) δ.

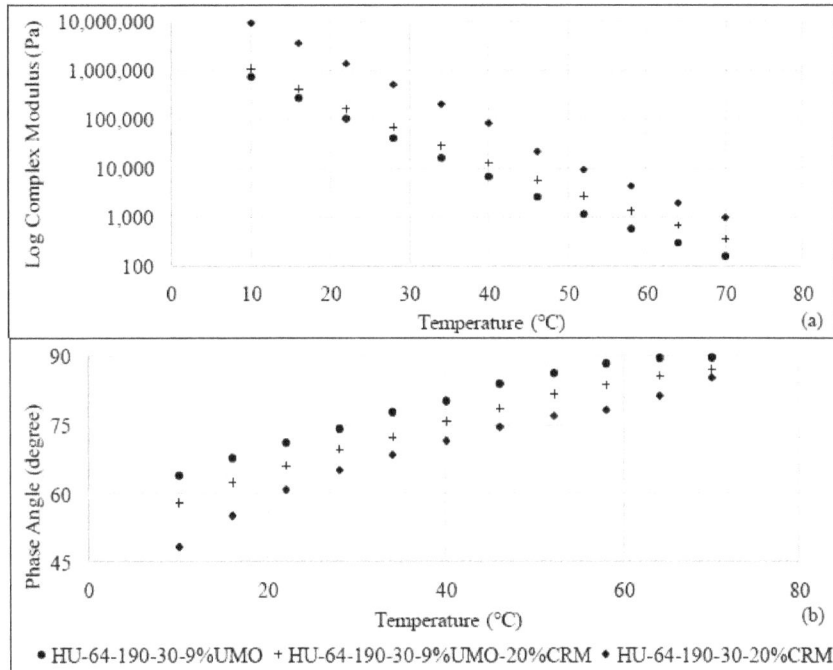

Figure 8: Temperature Sweep Viscoelastic Properties of Liquid Phase samples interacted with 9% UMO, 20% CRM or both at 190°C and 30Hz: (a) G* and (b) δ.

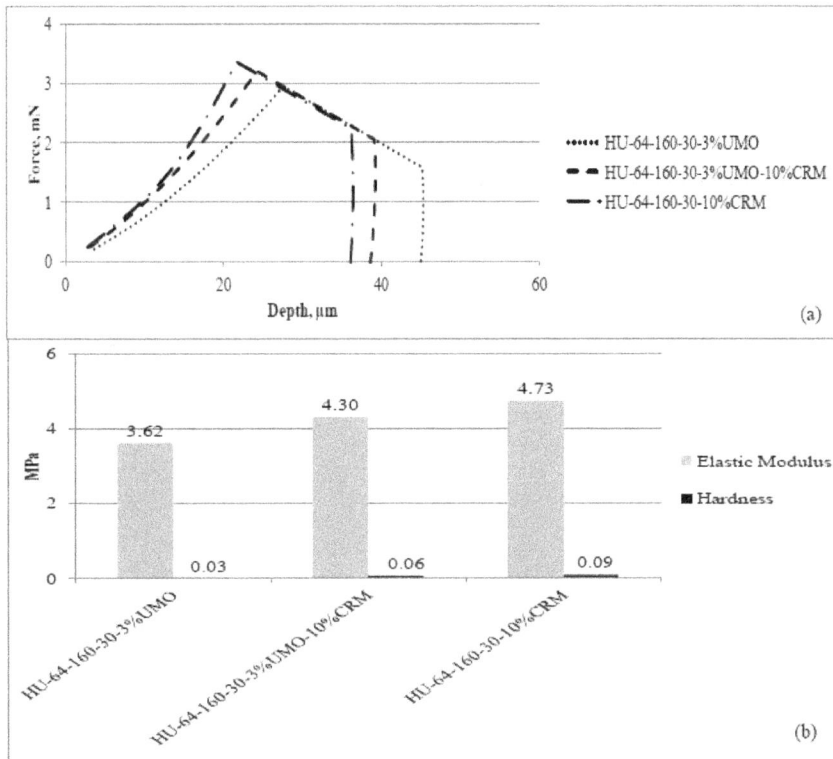

Figure 9: Micromechanical behavior of samples interacted at 160°C and 30Hz with 3% UMO, 10% CRM or both at 120 minutes interaction time: a) Force vs indentation depth profile, (b) Comparison of hardness and elastic modulus.

depth was recorded.

Figure 9b shows the hardness and elastic modulus values for the same samples. As shown in Figure 9b, a continuous increase in both the hardness and elastic modulus can be seen with the utilization of CRM with oil in asphalt and with CRMA only. The elastic modulus of samples was 3.62, 4.3, and 4.73 Mpa, for the asphalt samples modified with 3% UMO, 3% UMO with 10% CRM, and 10% CRM, respectively. On the other hand, the hardness of the samples was 0.03, 0.06, and 0.09 MPa. This indicates that at such a combination of interaction temperature (160°C) and interaction speed (30 Hz) the increase in the elastic modulus values occurs gradually with addition of CRM, whereas for the hardness, a major increase occurs when CRM is added to asphalt with UMO (almost double the values). This could be as a result of the higher absorption of the low molecular weight aromatics of asphalt by the CRM that lead to stiffer binder [14].

Figure 10 illustrates (a) the force vs. indentation depth profiles and (b) comparison of hardness and elastic modulus for the samples interacted at 160°C with 9% UMO, 20% CRM, or both modifiers after 120 minutes of interaction time.

Figure 10a shows a different trend than the samples interacted at 160°C with 30Hz with 3% UMO and 10% CRM. For the samples with 9% UMO only, the indentation depth is about 55 μm. However, upon addition of 20% CRM, a steep decrease in the indentation depth ranging around 45 μm is observed. On the other hand, for the samples with 20% CRM, the indentation depth was almost 23 μm. As illustrated in Figure 10b, an increase in both the hardness and elastic modulus values is evident after the addition of CRM. The elastic modulus was 2.9 MPa for the sample with 9% UMO and increased to 3.74, and 8.01MPa,

for the samples modified by 9% UMO with 20% CRM, and 20% CRM, respectively. The same trend was seen for the hardness values that were 0.009, 0.04, and 0.36 MPa, for the aforementioned samples.

Figure 11 illustrates (a) the force vs. indentation depth profiles and (b) comparison of hardness and elastic modulus for the samples interacted at 190°C with 3% UMO, 10% CRM, or both modifiers after 120 minutes of interaction time.

Figure 11a shows a similar trend to the samples interacted at 160°C with 30 Hz with 3% UMO and 10% CRM. For the samples with 3% UMO only, the indentation depth is about 40 μm. However, upon addition of 10% CRM, we observe a minor decrease in the indentation depth to be about 35 μm. On the other hand, for the samples with 10% CRM, the indentation depth was almost 30 μm. As illustrated in Figure 11b, an increase in both the hardness and elastic modulus values is evident after the addition of CRM. The elastic modulus was 4.08 MPa for the sample with 3% UMO and increased to 4.85, and 5.34 MPa, for the samples modified by 3% UMO with 10% CRM, and 10% CRM, respectively. The same trend was seen for the hardness values that were 0.05, 0.09, and 0.12 MPa, for the aforementioned samples. It should be noticed that, upon increasing the interaction temperature from 160°C to 190°C, and with the utilization of 10% CRM with 3% UMO, the values of the hardness and elastic modulus show enhancement for the same sample conditions. This is direct result of the development in the asphalts internal structure that was proven in the temperature sweep section.

Figure 12 illustrates (a) the force vs. indentation depth profiles and (b) comparison of hardness and elastic modulus for the samples interacted at 190°C with 9% UMO, 20% CRM, or both modifiers after

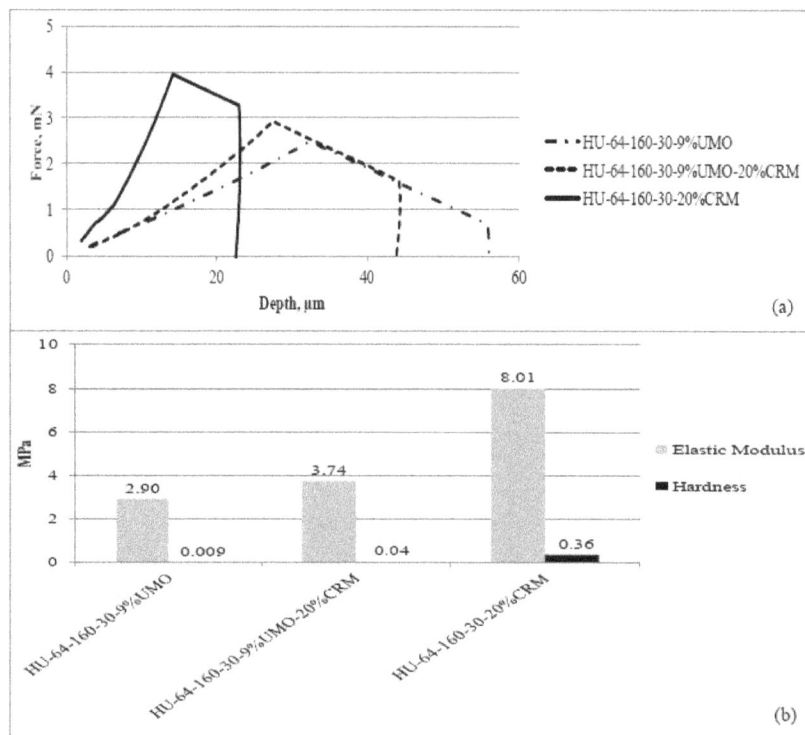

Figure 10: Micromechanical behavior of samples interacted at 160°C and 30Hz with 9% UMO, 20% CRM or both at 120 minutes interaction time: a) Force vs indentation depth profile, (b) Comparison of hardness and elastic modulus.

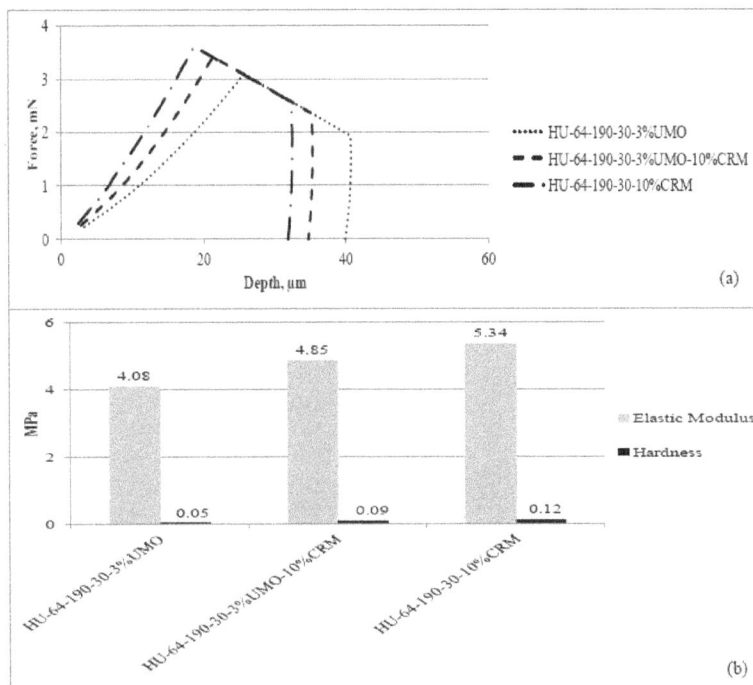

Figure 11: Micromechanical behavior of samples interacted at 190°C and 30Hz with 3% UMO, 10% CRM or both at 120 minutes interaction time: a) Force vs indentation depth profile, (b) Comparison of hardness and elastic modulus.

120 minutes of interaction time.

Figure 12a shows a different trend from that seen in the samples interacted at 190°C with 30 Hz with 3% UMO and 10% CRM. For the samples with 9% UMO only, the indentation depth is about 45 μm. However, upon addition of 20% CRM, we observe almost no change in the indentation depth ranging around 45 μm. On the other hand, for the samples with 20% CRM, the indentation depth was almost 20 μm. As illustrated in Figure 12b, a decrease in both the hardness and elastic modulus values is evident after the addition of CRM for the samples with UMO. The elastic modulus was 3.71 MPa for the sample with 9% UMO and decreased to 3.6, for the samples modified by 9% UMO with 20% CRM. However, a major increase in the elastic modulus (9.18 MPa) was recorded for the samples with 20% CRM only. The same trend was seen for the hardness values that were 0.004, 0.03, and 0.48 MPa, for the aforementioned samples.

Conclusions

This work investigated the effect of the addition of UMO to neat and CRMA at different interaction conditions. The changes in the macro and micromechanical properties of the modified asphalts were investigated. It was found that the utilization of UMO only as a modifier to asphalt severely deteriorates the macro and micro mechanical properties of the binder because the UMO disrupts the asphalt intermolecular associations and internal network structure, in the absence of CRM. Combining CRM with UMO as modifiers to asphalt had better results for the lower percentages of modifiers (10% CRM with 3% UMO) over the higher percentage (20% CRM with 9% UMO). A better balance between the light molecular fractions absorbed

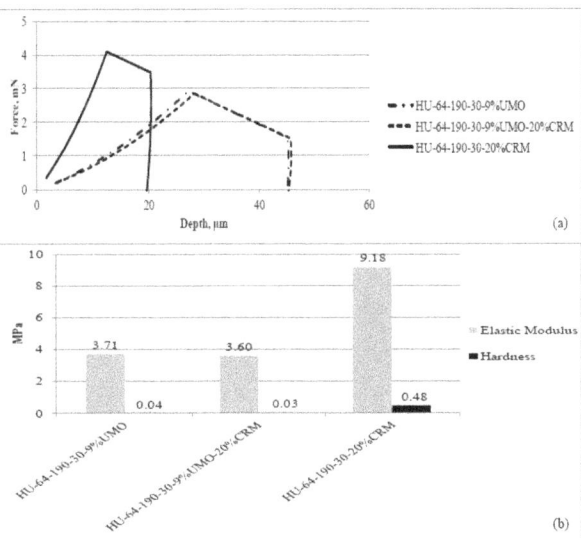

Figure 12: Micromechanical behavior of samples interacted at 190°C and 30Hz with 9% UMO, 20% CRM or both at 120 minutes interaction time: a) Force vs indentation depth profile, (b) Comparison of hardness and elastic modulus.

by CRM from asphalt and the compensated amounts from UMO was achieved with the utilization of 3% UMO and 10% CRM. When employing a combination of 10% CRM and 3% UMO as modifiers to the binder, the utilization of the low interaction temperature (160°C) for the synthesis of modified asphalt was not sufficient to efficiently develop the internal network structure of the modified asphalt that leads to enhancements in both the macro and micromechanical properties of the modified asphalt. Based on the current research work, it is suggested that when UMO is to be used in asphalt modification, it should be at a rate of less than 3%, and it should be combined with at least 10% CRM. The proposed synthesis conditions are an interaction temperature of 190°C and interaction speed of 30 Hz for a period of 120 minutes.

Acknowledgment

This material is based on the work supported by the National Science Foundation under Grant No. 0846861. Any opinions, findings, and conclusions or recommendations expressed in this material are those of the writer(s) and do not necessarily reflect the views of the National Science Foundation.

The writers would like to express their gratitude to Husky Energy for providing the asphalt used in this research work and also to Crumb Rubber Manufacturers for providing the crumb rubber used in this research work.

References

1. Agency EP. Wastes - Resource Conservation - Common Wastes & Materials. Available from: http://www.epa.gov/osw/conserve/materials/usedoil/oil.htm.

2. Wekumbura C, Stastna J, Zanzotto L (2007) Destruction and recovery of internal structure in polymer-modified asphalts. Journal of Materials in Civil Engineering 19: 227-232.

3. Stroup-Gardiner M (1995) Polymer Literature Review.

4. DeDene CD, You ZP (2014) The Performance of Aged Asphalt Materials Rejuvenated with Waste Engine Oil. International Journal of Pavement Research and Technology 7: 145-152.

5. Villanueva A, Ho S, Zanzotto L (2008) Asphalt modification with used lubricating oil. Canadian Journal of Civil Engineering 35: 148-157.

6. Ragab M, Abdelrahman M, Ghavibazoo A (2013) Performance Enhancement of Crumb Rubber-Modified Asphalts Through Control of the Developed Internal Network Structure. Transportation Research Record: Journal of the Transportation Research Board 2371: 96-104.

7. Bergerson D, Abdelrahman M, Ragab M (2014) Environmental Study of the Release of BTEX from Asphalt Modified with Used Motor Oil and Crumb Rubber Modifier. Int J Waste Resour 4: 165.

8. Tarefder RA, Zaman AM, Uddin W (2010) Determining hardness and elastic modulus of asphalt by nanoindentation. International Journal of Geomechanics 10: 106-116.

9. Tarefder RA, Faisal H (2012) Effects of dwell time and loading rate on the nanoindentation behavior of asphaltic materials. Journal of Nanomechanics and Micromechanics 3: 17-23.

10. Oyen ML, Ko CC (2007) Examination of local variations in viscous, elastic, and plastic indentation responses in healing bone. J Mater Sci Mater Med 18: 623-628.

11. Oliver WC, Pharr GM (1992) An improved technique for determining hardness and elastic modulus using load and displacement sensing indentation experiments. Journal of Materials Research 7: 1564-1583.

12. Stilwell N, Tabor D (1961) Elastic recovery of conical indentations. Proceedings of the Physical Society 78: 169.

13. Doerner MF, Nix WD (1986) A method for interpreting the data from depth-sensing indentation instruments. Journal of Materials Research. 1: 601-609.

14. Gawel I, Stepkowski R, Czechowski F (2006) Molecular interactions between rubber and asphalt. Industrial & Engineering Chemistry Research 45: 3044-3049.

15. Putman BJ, Amirkhanian SN (2006) Crumb rubber modification of binders: interaction and particle effects. Asphalt rubber 2006 conference, Palm Springs, USA.

16. Zanzotto L, Kennepohl GJ (1996) Development of rubber and asphalt binders by depolymerization and devulcanization of scrap tires in asphalt. Transportation Research Record: Journal of the Transportation Research Board 1530: 51-58.

17. Lu X, Isacsson U (2000) Modification of road bitumens with thermoplastic polymers. Polymer Testing 20:77-86.

18. Airey GD (2003) Rheological properties of styrene butadiene styrene polymer modified road bitumens. Fuel 82: 1709-1719

27

Not Too Little, Not Too Much and Shortcut: A Review on the Effectualness of Per Capita Pollutant Discharge Indicators

Yoshiaki Tsuzuki[1,2]*

[1]Engineering, Architecture and Information Technology (EAIT), The University of Queensland, St Lucia, QLD 4072, Australia
[2]Research Centre for Coastal Lagoon Environments (ReCCLE), Shimane University, Nishi-Kawatsu-cho, Matsue City, 690-8504, Japan

Abstract

In the fields of wastewater treatment planning, institutional and governance aspects are sometimes emphasised in developing and middle-developed countries. This paper summarises some important technical issues for the stakeholders of municipal wastewater treatment and water environment management. Pollutant removal efficiencies at wastewater treatment plants (WWTPs) are important but effluent water quality is sometimes emphasised. To achieve high pollutant removal efficiencies, maintaining a certain level of pollutant concentrations in the influent, or "Not Too Little" pollutant, is necessary. The second point is pollutant discharge from the river catchment should be "Not Too Much". Excess and rapid pollutant discharge increase in the catchment results in high costs and lengthy time periods for the river water environment to recover the original water environment conditions. The third point is that developing and middle-developed countries can use a "shortcut" or technological bypass by use of existing hard and soft measures to facilitate environmental improvements. This can be done with appropriate financial mechanisms. Pollutant discharge per capita (PDC) and pollutant load per capita flowing into water body (PLC$_{wb}$) are effective and efficient indicators that can be used to address these three concepts.

Keywords: Pollutant removal efficiency; Influent concentration; Monod-type equation; Pollutant discharge per capita; Inverted-U shaped curve relationship; Environmental Kuznets curve (EKC); Wastewater treatment planning

Introduction

In the fields of wastewater treatment planning in urban and peri-urban areas of economically developing and middle-developed countries, many researchers have conducted academic and practical purpose research over several decades [1]. Wastewater treatment management and planning is usually conducted with a primary requirement of effluent quality [2] and systematic considerations on the wastewater treatment systems have been conducted in developed countries [3]. Other aspects including economic, institutional and political, climatic, environmental, land availability-properties, socio-cultural and other-local aspects are secondary considerations, and lastly cost effectiveness is pursued. On the contrary, in developing countries, economic, institutional, political, and other-local aspects are usually the primary considerations. Normal pollutant concentration in the influent and effluent of wastewater treatment plant (WWTP) are typical by countries [4,5]. From the technological perspective, pollutant removal efficiencies are also important, as well as the requirement of effluent water quality [6,7].

From engineering perspectives, designing facilities with too small a capacity results in overflow of untreated wastewater to ambient water [2], and designing facilities with too large a capacity results in lower pollutant removal efficiencies [6,7]. Both centralised and decentralised municipal wastewater treatment systems have been developed based on socio-cultural conditions and available technologies [8]. Major decentralised treatment methods are Facultative Lagoons (FL) and Aerated Lagoons (AL), Anaerobic Lagoons (AnL), Aerobic Lagoons (AoL), Suspended Growth (SG), Sequencing Batch Reactor (SBR), Attached Growth (AG), Constructed Wetlands (CW) and simple and combined johkasou (SJ and CJ) [9,10]. Ecological wastewater treatment is a widely used term, and some of these technologies that apply biological and natural processes might be categorised as ecological wastewater treatment methods [6,11-14].

Technological, economic and institutional aspects of these ecological wastewater treatment methods have been studied many years. For example, technical and institutional aspects of constructed wetlands have been developed especially in developing countries [15-19]. A large percentage of municipal wastewater is still discharged without treatment especially in East and Southern Asia, Caspian Sea, Central and East Europe, West and Central Africa and Caribbean Regions [20]. From the sustainable engineering design concepts, sustainable design processes (SDPs) have advanced to integrated sustainable engineering design processes (ISEDP) [21]. When designing centralised and on-site WWTPs, the aim should be to attain high pollutant removal efficiencies in order to address "the needs of the present" and "their (future generations) own needs" World Commission on Environment and Development (WCED) [22]. When the design is not appropriate, it will take much time and cost to repair the damage to the environment [23].

A selection of wastewater treatment methods was systematically analysed in a case study in peri-urban and rural areas of South Africa and sustainability indicators were found to address all the sustainability dimensions [24], i.e., environmental, economic and socio-cultural, while other indicators and analysis methods such as energy analysis, Material Flow Analysis (MFA), economic analysis and Life Cycle Assessment (LCA), addressed limited aspects of the sustainability of wastewater treatment systems. Based on the results of their study, sustainability indicators are considered to be the most comprehensive tool to assess the sustainability of engineering systems. New paradigms

*Corresponding author: Yoshiaki Tsuzuki, Engineering, Architecture and Information Technology (EAIT), The University of Queensland, St Lucia, QLD 4072, Australia and Research Centre for Coastal Lagoon Environments (ReCCLE), Shimane University, Nishi-Kawatsu-cho, Matsue City, Japan
E-mail: y.tsuzuki@uq.edu.au, tsuzuki.yoshiaki@gmail.com

have been developed for sustainable wastewater treatment in developing countries [25,26].

A concept for planning of on-site and centralised wastewater systems has been developed from the aspects of pollutant discharge indicators, pollutant removal efficiency functions, and scenario-based pollutant load analysis [7,27]. Pollutant load, pollutant discharge and water quality should be quantitatively managed and planned from the planning phase of Wastewater Treatment Plants (WWTPs) especially in low- and middle-income countries because huge public investments have been directed into centralised WWTPs. These parameters should be the most basic and fundamental markers for the framework of planning of WWTPs. It should be noted that there are some WWTPs which miss the mark and are not efficiently operated and managed, e.g. smaller pollutant removal ratios because of smaller influent concentration.

The purpose of this paper is to describe some fundamental issues in wastewater treatment management and planning issues based on recent research and to discuss importance of these issues in the institutional and governance fields of wastewater treatment planning. In this paper, critical points on the planning are described based on these existing papers from the perspective of sustainability. Sustainability indicators include multiple aspects such as technology, economy and institutional [24,28]. This paper focuses on technological issues related to sustainability and discusses the meaning of results from several recent publications [5-7,26,27,29]. The research topics of these papers have been the necessity of maintaining a certain pollutant concentration or load in the influent of WWTPs [6,7], and a problem of a rapid and excess pollutant discharge increase in a catchment [30]. For a technological shortcut for municipal wastewater treatment systems, on-site treatment systems are considered to be cost effective in areas with smaller population density. A simple simulation on payment for the on-site WWTP under the conditions of moderate economic development with purchase-power parity based gross domestic product (PPP-GDP) in 2014 as US$ 2,000 was conducted in this paper to think about shortcuts or early development stages from an economics perspective.

For the three concepts presented in this paper, Not Too Much refers to restricting pollutant discharge into the catchment and corresponds to general directions for environmental preservation; however, Not Too Little suggests that WWTP influent should be maintained with a certain level of pollutants. Biological technologies which are applied in many centralised WWTPs need a certain level of pollutants for their operation. Development of a WWTP, especially a centralised system, needs a long-term and huge investment [31]. Shortcuts sometimes may need to utilise financial mechanisms to support the early stages of development. PDC and PLC_{wb} are efficient and effective indicators to address these three concepts. Biological (or biochemical) oxygen demand (BOD) is commonly applied in the three topics of this paper. BOD is considered to be an appropriate indicator for biological treatment processes and also for evaluating organic consumption capacity of ambient water. Regarding biological treatment processes, there are still problems with a mixture of non-biodegradable or toxic materials in the influent in some countries, however, such problems are beyond the scope of this paper.

Not Too Little: The Necessity to Maintain a Certain Pollutant Concentration or Load in Wastewater Treatment Plant (WWTP) Influent

The relationships between pollutant removal efficiency and influent

concentration can be expressed using the Monod-type equation 1. (Figure 1) [6,7,32]. The Monod-type equation has been originally developed for the growth rate of bacteria, and is applicable to pollutant removal efficiency at WWTPs. Organic carbon concentrations such as BOD and chemical oxygen demand (COD) are most applicable for the pollutants in this context [6]. Figure 1 shows that a certain pollutant concentration is necessary in WWTP influent to maintain large pollutant removal efficiency. Pollutant removal efficiency increases with an increase of pollutant concentration in the influent [6]. If the pollutant concentration is too small, pollutant removal efficiency decreases and sometimes becomes a negative value, i.e. effluent concentration is larger than influent concentration [7,25]. Small pollutant removal efficiencies can be found at some WWTPs, and these are mostly because of planning problems, i.e. the actual influent pollutant concentration and load is smaller than the planned pollutant load. There are several existing technological solutions for this problem [7]. In Figure 2, the relationship between pollutant influent concentration (C_{in}) and effluent concentration (C_{ef}) is illustrated more simply based on the following Eqs. 2 and 3 [6].

$$R_{BOD} = \frac{R_{BOD-MAX} \times C_{BOD,in}}{K_s + C_{BOD,in}} \tag{1}$$

Where R_{BOD} is the removal efficiency of BOD at a WWTP (%); $R_{BOD-MAX}$ is the maximum value of the removal efficiency of BOD (%); $C_{BOD,in}$ is the BOD influent concentration at a WWTP (mg/l); and Ks is the half saturation coefficient (mg/l).

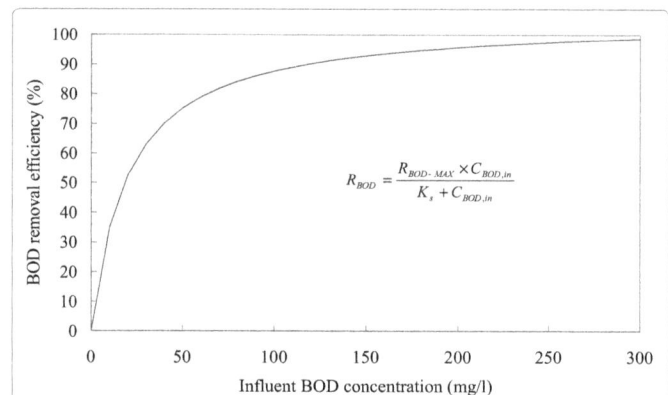

Figure 1: The Monod-type equation of BOD removal efficiency and influent BOD concentration which is obtained from management data from several ecological wastewater treatment plants (WWTPs) in Bangkok, Thailand (Modified from [6]).

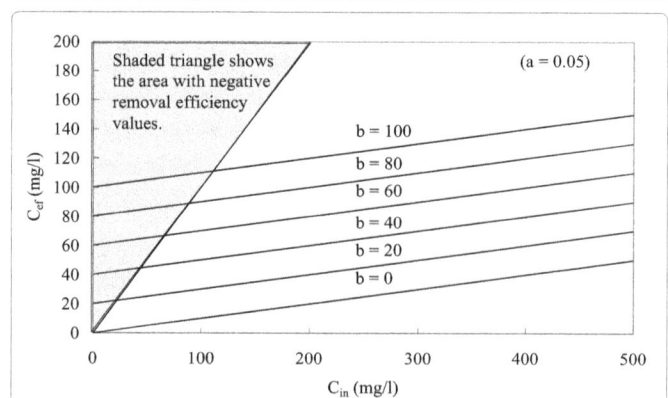

Figure 2: Relationship between influent and effluent concentrations of pollutant in wastewater treatment plants (Modified from [6]).

$$C_{ef} = a \times C_{in} + b \qquad (2)$$

where C_{ef} is the effluent pollutant concentration/load (g m^{-3} or g day^{-1}); C_{in} is the influent pollutant concentration/load (g m^{-3} or g day^{-1}); a is a coefficient (dimensionless); and b is a coefficient (g m^{-3} or g day^{-1}).

$$\text{Removal Efficiency (\%)} = (1 - \frac{C_{ef}}{C_{in}}) \times 100$$
$$= (1 - a - \frac{b}{C_{in}}) \times 100 \qquad (3)$$

The values of $R_{BOD-MAX}$ and Ks in Eq. 1 and a and b in Eqs. 2 and 3 can be obtained empirically based on management data of WWTPs or experimental data. Figure 2 shows effluent pollutant concentration increase with an increase of pollutant concentration in the influent. The shaded area shows negative removal efficiency, i.e. pollutant concentration in the effluent is larger than that in the influent.

These relationships will lead to one of the common understandings among wastewater professionals: it may be possible that total pollutant discharge to ambient water will decrease when more wastewater is collected and treated at WWTPs. Pollutant removal efficiency would increase under such conditions. This common understanding can be supported from the following contents in this paper, however, it may be also possible that some types of pollutants in the influent cannot be removed efficiently because of a design mismatch or design failure between the treatment plant itself and the piped collection system. Some methods to address design failures should be developed to address such cases [23].

The scenario-based analysis results show a decrease in total pollutant discharge per capita (PDC) with an increase in the percentage of wastewater treated at centralised WWTPs for Scenario 1 or for the percentage of population served with centralised WWTPs (%) for Scenario 2 (Figure 3) [27]. This analysis has been conducted in urban and peri-urban areas of Bangkok, Thailand, where both on-site and centralised wastewater treatment systems are applied. The scenario-based analysis results shown in Figure 3 have been derived from a set of PDC values which have been based on the material flow analysis (MFA) [7]. The x-axes or percentage values are different in the two

scenarios because existing mixture conditions of centralised and on-site treatment systems are considered to be maintained in Scenario 1 and centralised WWTPs areas and on-site WWTPs areas are considered to be divided in Scenario 2.

The values of pollutant discharge or pollutant load are expressed as pollutant discharge per capita (PDC) in Figure 3. When these percentages increase, pollutant load in WWTP influent increases and pollutant load in direct discharge to ambient water decreases. These parameters are measured using PDC. Pollutant discharge and pollutant load can be assessed as per capita values by using PDC. Pollutant discharge and pollutant load values have been converted to WWTP influent concentrations in the simulation analysis [27]. The alteration of WWTP effluent is not simple because pollutant removal functions are included in the analysis especially for Cases 2a and 2b. The increase or decrease of PDC for WWTP effluent depends on the explanatory variables. The increase in PDC for WWTP effluent is smaller compared to the other PDCs, and the total PDC decreases with an increase in the explanatory variables on the x-axes in both scenarios and all cases. In these investigations, costs were not taken into account. Inclusion of economic parameters and life cycle analysis (LCA) will lead to more comprehensive understanding of wastewater treatment planning [33]. These parameters are more sensitive to local natural and socio-economic conditions.

Equations 1–3 are based on normal operations of biological WWTPs for municipal wastewater. Inclusion of other wastewater in WWTP influent should be carefully considered and its dilution effects are beyond the scope of this paper. Note that dilution often does not decrease pollutant loads. Other aspects which should be further considered in the practical planning would be centralised and on-site wastewater treatment [7,27], treatment of black water and gray water [34] and source-separation [35].

Not Too Much: Excess Pollutant Discharge into the Catchment Causes Excessive Water Pollution

Natural conditions generally change in the long term and

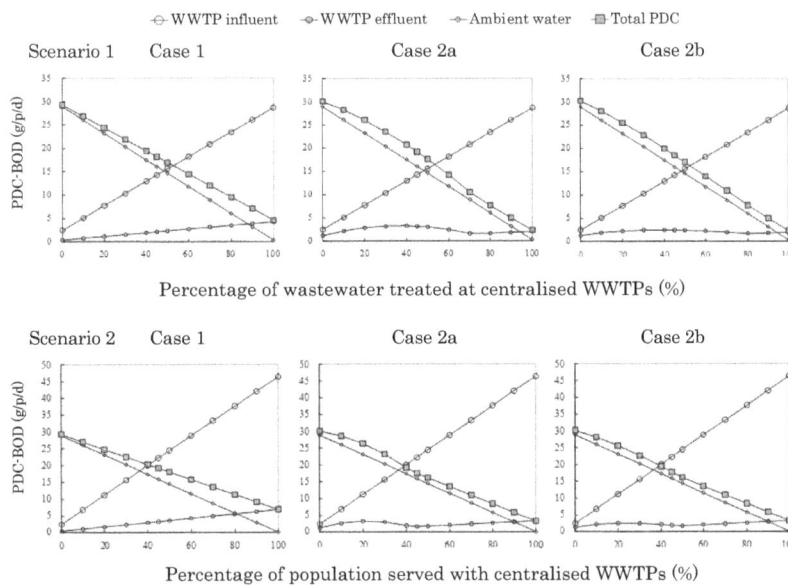

Figure 3: Estimation of pollutant discharge per capita (PDC) of BOD for two scenarios (Modified from [27]).

anthropogenic conditions change in the short term. In a natural water environment with no or minimum human effects, a certain relationship between pollutant discharge and ambient water quality should maintain some equilibrium conditions. The relationship between pollutant discharge and water quality in a natural water environment may be originally in the linear or first-order relationship, and perturbation or dynamic equilibrium change of stable conditions for the relationship have been observed under the rapid increase and excess pollutant discharge conditions in the Yamato-gawa River Catchment, Japan (Figure 4) [30]. Water quality stationarity alongside changes of pollutant discharge from the catchments has sometimes found [36,37]. Over a longer timeframe than anthropogenic condition changes, morphological or sea level changes also occur.

Many developed countries including Japan have experienced severe ambient water pollution especially from the 1960s to the 1980s during rapid economic growth (Figure 4) [30]. Because of such growth, urban development and industrialisation, BOD discharge in the Yamato-gawa River Catchment has increased rapidly in the late 1960s. National level environmental regulations including those protecting against water pollution have established in 1970, however, BOD discharge into the river basin has continued to increase until the late 1970s and BOD concentration in the river has deteriorated rapidly and became worse until the mid-1970s (Phase 1). In the late 1970s and 1980s, BOD discharge into the river basin has been still substantial, but BOD concentration has improved a little (Phase 2). In the 1990s, BOD discharge has decreased because of several measures to improve water quality (Phase 3). In the late 1990s, the relationship between BOD discharge and BOD concentration returned to that of the late 1960s. After that, BOD concentration improved with decreases in BOD discharge (Phase 4).

There should be a capacity in regards to pollutant discharge into the river basins. When pollutant discharge exceeds the capacity and increases rapidly, water quality deteriorates and a lot of time and expense would be necessary to recover to the original relationship of pollutant discharge and water quality. The grey line of Figure 4 supposes the original linear relationship between BOD discharge and BOD concentration. Rapid increase and excess BOD discharge has caused the relationship to drift away from the original linear relationship, and

the BOD concentration in Phase 1 has deteriorated to increase above the original relationship.

Shortcuts: Economic Development with Smaller Environmental Burden

In developing and middle-developed countries, many people are working to develop their technologies and economies. During rapid economic development periods, developed countries have experienced environmental pollution including water pollution, e.g. in the 1960s and 1970s in Japan. At that time, there were not effective and efficient measures to decrease environmental pollution, which cause severe water pollution (Figure 4). During the history of development of environmental friendly technologies in developed countries, appropriate technologies and methods have been developed to mitigate environmental pollution. Advanced technologies suitable for developing and middle-developed countries have also been developed [38-40].

For developing and middle-developed countries, shortcuts or bypasses of technology development should be theoretically possible (Figure 5). Many kinds of environmental friendly technologies and methods to improve river water quality have been developed over several decades, e.g. the soft measures to improve river water quality [41]. There would be institutional, governance, cost and application problems to actually introduce these technologies and methods into these countries [2], however, these countries are benefitting from the situations that socio-economic tools including cost estimation methods of centralised treatment system have been developed [42].

In the fields of water pollution, pollutant discharge per capita (PDC) would be one of the important indicators used to describe the magnitude of pollutant discharge from municipal wastewater in certain areas [5,7,43]. PDC is found to have an inverted-U shaped curve relationship with the economic development indicator, namely Environmental Kuznets Curve (EKC) relationship (Figure 5) [42]. The indicator, pollutant discharge per capita flowing into the water body (PLC_{wb}), is also applicable to assess the effect of pollutant discharge on ambient water quality [5]. "Shortcut" planning can be introduced for municipal wastewater treatment by applying PDC and PLC_{wb} Figure 5. PDC and PLC_{wb} address pollutant loading aspects and would be a part of a set of sustainable indicators together with indicators that address socio-economic aspects.

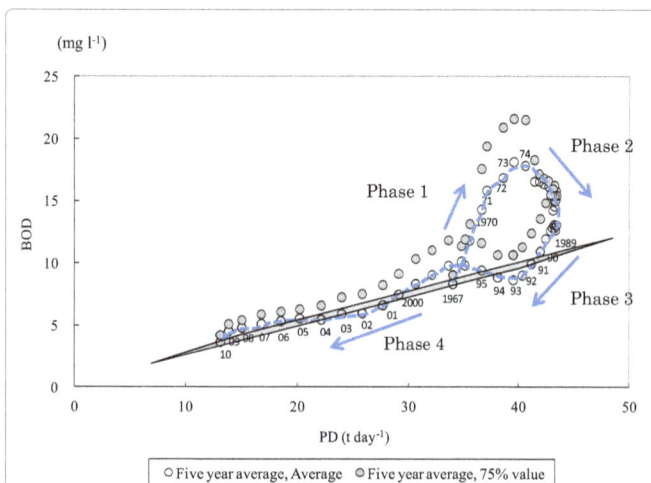

Figure 4: Relationship between BOD discharge and BOD concentration, five-year average of average and 75% value, in the Yamato-gawa River, Japan, during 1967–2010. The grey line indicates the original linear relationship between BOD discharge and BOD concentration (Modified from [30]).

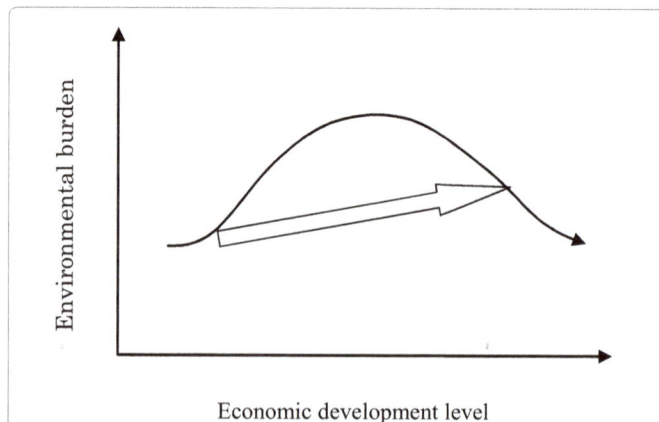

Figure 5: Relationship between economic development level and environmental burden. Inverted U-shaped curve is supposed. The white arrow shows the shortcut (Modified from [10]).

Simple simulation of income and payment for the on-site WWTPs showed longer time lags for the investments and utilisation of on-site WWTP with the deposit-and-pay method, compared to loans and 50% subsidies plus loans (Figures 6 and 7). Simulation conditions were assumed for purchase-power parity based gross domestic product (PPP-GDP) per capita, price of on-site WWTP for a family, the number of paid persons in a family, and several annual change rates as shown in Table 1. For the deposit-and-pay method, it will take 25 years to install and start to use the on-site WWTP because of the nominal price increase of an on-site WWTP (Figure 7). If the nominal price was fixed at US$ 5,000 because of technology development effects, installation of the on-site WWTP would be possible in 14 years. For both loan cases, the full-cost loan and the half-subsidies in 2014, the on-site WWTP could be used from 2014 and payment would be finalised in 20 years and 11 years, respectively. The simulation results showed the effectiveness of loans and subsidies for early development of the on-site treatment systems or shortcuts.

Conclusions

Among the three concepts presented above, two concepts state the importance of moderate conditions, i.e. "Not Too Little", suggesting that a certain pollutant inflow is necessary to maintain pollutant removal efficiency, and "Not Too Much", suggesting that rapid increase and excess pollutant discharge in a catchment cause severe

Parameter	Annual change rate (%)
PPP-GDP per capita	3
Consumer price index (CPI)	4
Percentage of income which is available for payment of on-site WWTP	5
Interest rate of loan for on-site WWTP construction	5
PPP-GDP per capita in 2014	US$ 2,000
Price of on-site WWTP for a family	US$ 5,000
The number of paid persons in a family	2 persons

Table 1: Simulation conditions on payment for on-site WWTPs.

water pollution. Another important concept is "Shortcut" or bypass of technologies and measures for environment improvement, which will be important for developing and middle-developed countries and involves applying both the existing and newly developed technologies. For the "Shortcut", financial mechanisms including subsidies and loans should be effectively utilised to introduce the measures to mitigate environmental pollution in the early stages of the development. In the municipal wastewater treatment and water environment fields, PDC and PLC_{wb} would be effective and efficient indicators to assess the magnitude of pollutant discharge and pollutant load.

Mutual and common understanding of scientific and technical issues is considered to enhance institutional and governance aspects of water environment preservation and improvement. The importance of maintaining moderate conditions for pollutant discharge and wastewater treatment and technological shortcut or bypass of environmental friendly technologies discussed in this paper would help enhance the mutual and common understanding among stakeholders of wastewater and water environment planning and management and sustainability in these fields.

Acknowledgments

This paper is based on several papers of the author which have been prepared with several researchers, students, engineers and government officers in several countries at that time. Some figures are modified from the existing literature after copyright permissions from Elsevier, the International Water Association (IWA), and Springer. Comments and suggestions from editors and reviewers have improved the quality of the paper. The manuscript English has been corrected by Jewel See Editing. The errors if any would be sole responsible of the author.

Conflicts of Interest

The authors declare no conflict of interest.

References

1. Mara DD (2004) Domestic wastewater treatment in developing countries. Earthscan UK and USA 217P.

2. Tsagarakis KP, Mara DD, Angelakis AN (2001) Wastewater management in Greece: experience and lessons for developing countries. Water Sci Technol 44: 163-172.

3. Yoo CK, Kim DS, Cho JH, Choi SW, Lee IB (2001) Process System Engineering in Wastewater Treatment Process. Korean Journal of Chemical Engineering 18: 408–421.

4. Taebi A, Droste RL (2004) Pollution loads in urban runoff and sanitary wastewater. Sci Total Environ 327: 175-184.

5. Tsuzuki Y, Koottatep T (2010) Municipal Wastewater Pollutant Discharge Indicator Estimation and Water Quality Prediction in Pak Kret District Bangkok Thailand. Journal of Water and Environment Technology 8: 51–75.

6. Tsuzuki Y (2012) Linking sanitation and wastewater treatment: from evaluation on the basis of effluent pollutant concentrations to evaluation on the basis of pollutant removal efficiencies. Water Sci Technol 65: 368-379.

7. Tsuzuki Y, Koottatep T, Sinsupan T, Jiawkok S, Wongburana C, et al. (2013a) A concept in planning and management of on-site and centralized municipal wastewater treatment systems, a case study in Bangkok Thailand I: pollutant discharge indicators and pollutant removal efficiency functions. Water Sci Technol 67: 1923–1933.

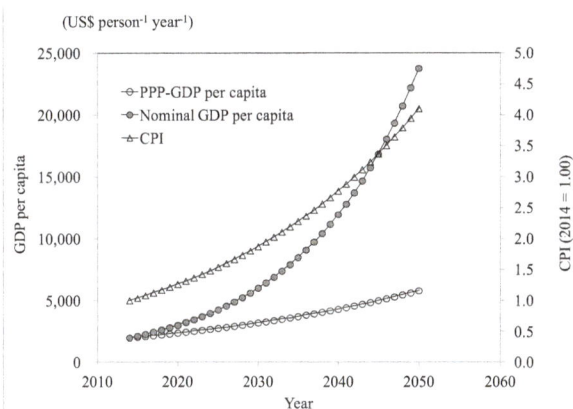

Figure 6: Chronological change of PPP-GDP per capita (2014 price), nominal GDP per capita, and consumer price index (PCI).

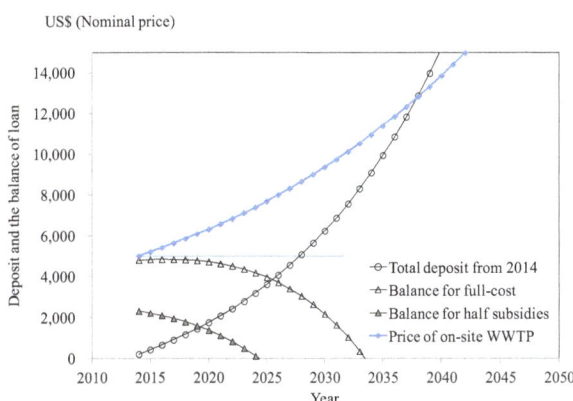

Figure 7: Total deposit from 2014, price of on-site WWTP, balances for loans of full-cost and half-subsides for on-site WWTP.

8. Massoud MA, Tarhini A, Nasr JA (2009) Decentralized approaches to wastewater treatment and management: applicability in developing countries. J Environ Manage 90: 652-659.

9. Gaulke LS (2006) On-site wastewater treatment and reuses in Japan. Proc. ICE Water Management 159: 103–109.

10. Tsuzuki Y (2006) An index directly indicates land-based pollutant load contributions of domestic wastewater to the water pollution and its application. Sci Total Environ 370: 425-440.

11. Burkhard R, Deletic A, Craig A (2000) Techniques for water and wastewater management a review of techniques and their integration in planning. Urban Water 2: 197–221.

12. Schmitt J, Chong J, Yap CB (2007) Water System Design for Stanford University Green Dorm: Final Report 104p.

13. Chen GQ, Shao L, Chen ZM, Li Z, Zhang B (2011) Low-carbon assessment for ecological wastewater treatment by a constructed wetland in Beijing. Ecological Engineering 37: 622–628.

14. Vera L, Martel G, Márquez M (2013) Two years monitoring of the natural system for wastewater reclamation in Santa Lucia Gran Canaria Island Ecological Engineering 50: 21–30.

15. Denny P (1997) Implementation of constructed wetlands in developing countries. Water Science and Technology 35: 27–34.

16. Koottatep T, Polprasert C (1997) Role of plant uptake on nitrogen removal in constructed wetlands located in the tropics. Water Science and Technology 36: 1–8.

17. Kivaisi A (2001) The potential for constructed wetlands for wastewater treatment and reuse in developing countries a review. Ecological Engineering 16: 545–560.

18. Koottatep T, Surinkul N, Polprasert C, Kamal AS, Koné D, et al. (2005) Treatment of septage in constructed wetlands in tropical climate: lessons learnt from seven years of operation. Water Sci Technol 51: 119-126.

19. Haberl R (1999) Constructed Wetlands: A Chance to Solve Wastewater Problems in Developing Countries. Water Science and Technology 40: 11–17.

20. Corcoran E, Nellemann C, Baker E, Bos R, Osborn D, et al. (2010) Sick Water? The central role of wastewater management in sustainable development: A rapid response assessment. United Nations Environment Programme UN-Habitat Grid-Arendal.

21. Gagnon B, Leduc R, Savard L (2012) From a conventional to a sustainable engineering design process different shades of sustainability. Journal of Engineering Design 23: 49–74.

22. World Commission on Environment and Development (WCED) (1987) Our common future. Oxford University Press.

23. Gupta P, Gupta S, Gandhi OP (2013) Modeling and evaluation of mean time to repair at product design stage based on contextual criteria. Journal of Engineering Design 24: 499–523.

24. Flores A, Buckley C, Fenner R (2008) Selecting Wastewater Systems for Sustainability in Developing Countries. 11th International Conference on Urban Drainage Edinburgh Scotland UK

25. Laugesen CH, Fryd O, Koottatep T, Brix H (2009) Sustainable Wastewater Management in Developing Countries: New Paradigms and Case Stories from the Field. American Society of Civil Engineers Reston VA USA 261p.

26. Tsuzuki Y (2014) Evaluation of the soft measures effects on ambient water quality improvement and household and industry economies. Journal of Cleaner Production 66: 577–587.

27. Tsuzuki Y, Koottatep T, Sinsupan T, Jiawkok S, Wongburana C, et al. (2013b) A concept in planning and management of on-site and centralized municipal wastewater treatment systems a case study in Bangkok Thailand II: Scenario-based pollutant load analysis. Water Science and Technology 67: 1934–1944.

28. Singhirunnusorn W (2009) An Appropriate Wastewater Treatment System in Developing Countries: Thailand as a Case Study PhD Thesis University of California Los Angeles 217p.

29. Tsuzuki Y, Koottatep T, Wattanachira S, Sarathai Y and Wongburana C (2009) On-site treatment systems in the wastewater treatment plants (WWTPs) service areas in Thailand: Scenario based pollutant loads estimation. Journal of Global Environment Engineering (Japan Society of Civil Engineers) 14: 57-65.

30. Tsuzuki Y (2013) Explanation of 47-Year BOD Alternation in a Japanese River Basin by BOD Generation and Discharge. Water Air & Soil Pollution 224: 1517.

31. Tsuzuki Y (2011) Chapter 6: Sanitation Development and Roles of Japan in Joel M McMann edn. Potable Water and Sanitation 266p Nova Science Publishers Inc New York 179–202.

32. Monod J (1949) The Growth of Bacterial Cultures. Annual Review of Microbiology 3: 371–394.

33. Loiseau E, Junqua G, Roux P, Bellon-Maurel V (2012) Environmental assessment of a territory: an overview of existing tools and methods. J Environ Manage 112: 213-225.

34. Ghunmi LA, Zeeman G, Fayyad M, Van Lier JB (2011) Grey water treatment systems: A review. Critical Reviews in Environmental Science and Technology 41: 657–698.

35. Tervahauta T, Hoang T, Hernández L, Zeeman G, Buisman C (2013) Prospects of Source-Separation-Based Sanitation Concepts: A Model-Based Study. Water 5: 1006–1035.

36. Eyre B (1997) Water quality changes in an episodically flushed sub-tropical Australian estuary: A 50 year perspective Marine Chemistry 59: 177–187.

37. Basu NB, Destouni G, Jawitz JW, Thompson SE, Loukinova NV, et al. (2010) Nutrient loads exported from managed catchments reveal emergent biogeochemical stationarity. Geophysical Research Letters 37: L23404.

38. Tanaka H, Takahashi M, Yoneyama Y, Syutsubo K, Kato K, et al. (2012) Energy saving system with high effluent quality for municipal sewage treatment by UASB-DHS. Water Sci Technol 66: 1186-1194.

39. Ushijima K, Ito K, Ito R, Funamizu N (2013) Greywater treatment by slanted soil system. Ecological Engineering 50: 62–68.

40. Sanguanpak S, Chiemchaisri C, Chiemchaisri W, Yamamoto K (2013) Removal and transformation of dissolved organic matter (DOM) during the treatment of partially stabilized leachate in membrane bioreactor. Water Sci Technol 68: 1091-1099.

41. Tsuzuki Y, Yoneda M, Tokunaga R, Morisawa S (2012) Quantitative evaluation of effects of the soft interventions or cleaner production in households and the hard interventions: a Social Experiment Programme in a large river basin in Japan. Ecological Indicators 20: 282-294.

42. Hernandez-Sancho F, Molinos-Senante M, Sala-Garrido R (2011) Cost modelling for wastewater treatment processes. Desalination 268: 1–5.

43. Tsuzuki Y (2008) Relationships between water pollutants discharges per capita (PDCs) and indicators of economic level water supply and sanitation in developing countries. Ecological Economics 68: 273–287.

28

Effect of Ruthenium Oxide/Titanium Mesh Anode Microstructure on Electrooxidation of Pharmaceutical Effluent

Vahidhabanu S, Abilash John Stephen, Ananthakumar S and Ramesh Babu B*

CSIR- Central Electrochemical Research Institute, Karaikudi – 630 003, India

Abstract

The present contribution investigates the influence of Ruthenium oxide (RuO_2) microstructure on titanium substrate for treatment of pharmaceutical effluent. RuO_2/Ti electrodes were prepared at two different sintering temperatures viz. 450°C and 550°C, and subjected to degradation studies on pharmaceutical effluent. Fourier Transform Infrared spectroscopy (FT-IR) was used for analysis of intermediates formed during degradation. The performance of these electrodes were presented and discussed on the basis of sintering temperatures. Electrodes prepared at 450°C and 550°C gave 84% and 96% color removal respectively. Chemical Oxygen Demand (COD) removal was found to be 68% and 79% for the electrodes prepared at 450°C and 550°C respectively. The surface morphology of these electrodes were identified and studied by Scanning electron Microscopy (SEM). X Ray Diffraction (XRD) patterns showed the presence of anatase phase TiO_2 at 550°C. The microstructural changes on sintering the catalytic coating caused a significant improvement in anode performance in electrodes sintered at 550°C . The electrodes are electrochemically active and stable, and are chemically inert under operating conditions.

Keywords: Pharmaceutical effluent; RuO_2/Ti anode; Electrooxidation; Microstructure; X Ray Diffraction; Scanning Electron Microscopy

Introduction

Pharmaceutical industries in and around Chennai, India produce a wide variety of products using both organic and inorganic substances as raw materials, thereby generating a large quantity of complex toxic organic liquid wastes with high concentrations of inorganic TDS. These wastes are highly toxic to biological life, and are usually characterized by high BOD, COD and COD: BOD ratio. In addition, wastes from drug manufacture also contain toxic components including cyanide [1]. Ground water quality is disturbed by penetration of pharmaceutical industry effluents. Advanced oxidation processes (AOPs), which rely on the generation of very reactive short lived hydroxyl radicals have been attempted to decontaminate pharmaceutical wastewater. Electrooxidation is one of the recent techniques which have been used in the treatment of pharmaceutical wastewater in an effective manner [2-5]. Industrial use of dimensionally stable anodes (DSA) for wastewater treatment has led to technological solutions, thus reducing operational and investment costs [6,7]. Such DSA-type materials have been used for the oxidation of model aqueous solutions containing non biodegradable organics, typically found in pharma effluents Recent DSA research emphasis focuses on titanium based anodes coated with a variety of oxide materials such as IrO_2, Ru_2, and SnO_2. DSA with RuO_2–TiO_2 coated titanium has been used widely and successfully as anode for chlor-alkali production and electro-oxidation of wastewater due to its good electro catalytic activity. RuO_2 is a good electro catalyst for oxygen evolution, in spite of limited service life [8]. The coating of RuO_2 by thermal decomposition of the precursors such as $RuCl_3$, can persist on titanium mesh surfaces for long durations and cause the activation of titanium anode through its pores. Surface activation by RuO_2 can occur only if the coating is sufficiently porous for diffusion of the ions involved during electro-oxidation. The coating also shows a high pseudo capacitance value due to the redox reactions of RuO_2. At 400°C, the normal firing temperature of $RuCl_3$ on titanium substrate, ruthenium penetrates deep into the bulk of the substrate and accumulates in the near surface region. The chemical stability and the electrochemical properties of RuO_2 layers are strongly dependent on their preparation process.

The Ti/RuO_2 system has received considerable attention due to its excellent stability, electro catalytic properties and prevention of Ti passivation by the oxide coating. A further advantage is the large number of Ru-oxidation states existing at the electrode surface in the potential region between the hydrogen evolution reaction, HER, and the oxygen evolution reaction, OER .The electrochemical oxidation of organic substances was attributed to OCl^-, OH^-, nascent oxygen and other reactive [9].

Titanium based anodes are unavoidable in the electro oxidation techniques. This type of anode is mostly coated with catalytic oxides such as Ruthenium oxide or iridium oxide to enhance the oxidation process of the anode [10,11]. The present work has the objective to study the effect of sintering temperature on RuO_2 coated titanium mesh anodes by means of preparing RuO_2 coating at two different temperatures. The coated electrodes are then employed for the treatment of pharmaceutical effluents [12-14]. The results obtained elucidate the effect of RuO_2 microstructural modifications at over Ti substrate at different sintering temperatures 450°C (Electrode A) and 550°C (Electrode B) on efficient electro-oxidation performance.

Materials and Methods

Materials

All chemicals employed were of the "AR" grade, and purchased from Merck India Ltd. Calibration of laboratory equipments were carried

*****Corresponding author:** Ramesh Babu B, CSIR- Central Electrochemical Research Institute, Karaikudi – 630 003, India
E-mail: brbabu2011@gmail.com

out with due accordance to accepted standards, and double de-ionized water was used while conducting all experiments. The pharmaceutical effluent samples were collected from a private pharmaceutical industry at Chennai, India. Table 1 provides the physico-chemical properties of effluent prior to electro-oxidation.

Anode fabrication

The anode substrate consisted of commercially available rectangular titanium mesh of 99% purity, with a 5 x 4 x 0.4 cm³ dimension. $RuCl_2.3H_2O$ solution was prepared in isopropanol, and used as the precursor. Post precursor application, the anode surface was dried and fired at two different temperatures, viz. 450°C (electrode A) and 550°C (electrode B) in the presence of air for a period of 10 min [15-17]. Surface pre-treatment of titanium substrate prior to RuO_2 coating was seen to be critical for the fabrication of stable electrodes. Etching of the substrate was done in HCl solution at 70°C for a 1-2 min time duration.

Electro oxidation procedure

About 200 ml of the effluent was taken in an electrochemical reactor. The anode was a catalytically coated titanium electrode and the cathode was a thin plate of stainless steel. The two electrodes were separated by a distance of about 2 cm. A current density of about 3 A/dm² was supplied between the electrodes. The sample to be analysed was pipetted out once in every one hour. The whole process was carried for 4 hours. The same experiment was carried out with both the electrodes electrode A and electrode B.

Physico-chemical characterization

The surface morphology of RuO_2 coatings at different temperatures was characterized using SEM (Bruker TESCAN) prior to and post electro-oxidation at 3 A/dm² for 4 hr. The coating from anodes A and B were scarped from the anode surface and ground well. The powder so obtained was used to determine the crystalline characteristics of the metal oxides by X-ray diffraction (XRD) analysis.

Electrochemical measurements

The electrochemical behaviour of RuO_2/Ti mesh anodes A and B prepared at two different temperatures were studied by cyclic voltmmetry. The anode mesh under study was employed as the working electrode, with an exposure area of 1 cm² in pharmaceutical effluent solution containing 1 g/L NaCl as supporting electrolyte. Saturated Calomel Electrode (SCE) and stainless steel were used as the reference electrode, and counter electrodes respectively. The cyclic voltammograms were recorded by sweeping the potential (Vs SCE) across the range of -0.8 to +1.2 V at a 5 mV/s scan rate. Studies were performed using a BIOLOGIC electrochemical analyzer.

COD Measurements

Measurements of COD in the pharmaceutical effluent were made at regular time intervals by acid solution precipitation method, as reported in our previous contribution [18]. COD acid and Potassium Dichromate

solutions were mixed with a pre-determined quantity of effluent solution. The mixture was placed in a COD digester (Spectroquant TR 420, Merck) at 148°C for 120 min. It was then titrated against ferrous ammonium sulphate (FAS) solution using Ferohin as the indicator. A change in color from red to brown indicated the end point. The COD was then calculated by the formula previously reported [18].

Results and Discussion

SEM observation

SEM observations were carried out for viewing variations in the morphology and structure of the two types of mesh catalyst coatings prepared at 450°C (Electrode A) and 550°C (Electrode B). Figs. 1 and 2 the SEM micrographs of the RuO_2 coated titanium electrodes prepared by conventional thermal decomposition procedure (Electrode A) prior to and post electro-oxidation at 3 A/dm² for 4 hr. Dried-mud cracks can be found in flat areas for both oxide electrodes, which is very typical for DSA type electrodes as a result of sintering process. A number of crystallite agglomerates are seen on the surface of electrode A. Some of the precipitates scatter in the flat area, while others are seen to be distributed along the ridges. This morphology with dried-mud cracks, flat areas and agglomerates is typical for RuO_2 coated titanium electrode derived from conventional thermal decomposition, as reported in previous literature [19].

The SEM images of electrodes prepared at 550°C (Electrode B) prior to and post electro-oxidation are provided in the Figures 3 and 4. It can be seen from these images that the size of RuO_2 oxide grains in the electrode B is much lower than that of electrode A. The agglomeration of crystallites is also found to be minimal in comparison to that of electrode A. This may be attributed to the formation of stable anatase phase RuO_2 on sintering at 550°C, resulting in the formation of small particles with high surface area, which is most suitable for electro catalytic oxidation process. There are number of pores on the surface of electrode B, contributing to the increased catalytic property and more reactive surfaces. This micro structural variation in electrode B contributes to more efficient treatment of pharmaceutical wastewater,

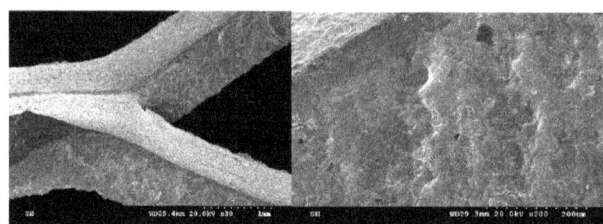

Figure 1: SEM image of RuO2/Ti electrodes sintered at 450°C (Electrode A) prior to electrooxidation.

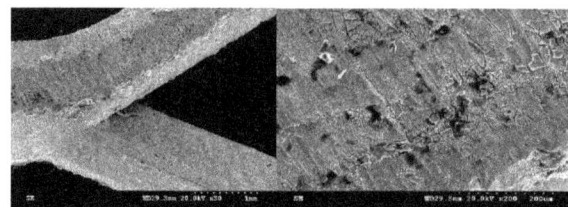

Figure 2: SEM image of RuO2/Ti electrodes sintered at 450°C (Electrode A) post electro-oxidation at 3 A/dm² for 4 hr.

S. no.	Parameter	Raw effluent
1	Color	Dark brown
2	pH	8.1
3	COD	1572 mg/l
4	BOD	965 mg/l
5	Total dissolved solids	215 mg/l

Table 1: Physico-chemical characteristics of raw pharmaceutical effluent.

Figure 3: SEM image of RuO2/Ti electrodes sintered at 550 °C (Electrode B) prior to electrooxidation.

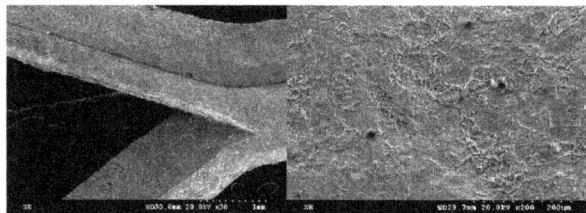

Figure 4: SEM image of RuO2/Ti electrodes sintered at 550 °C (Electrode B) post electro-oxidation at 3 A/dm² for 4 hr.

as depicted by further results presented in this report. The cracks that are formed on the surface of the electrode after electro-oxidation may be due to the evolution of chlorine gas and the evaporation of water vapor [20-25]. As observed from the Figures 2 and 4, the SEM images of electrodes A and B post electro-oxidation; it is evident that the electrode B is more stable than that of electrode A under operating conditions.

XRD analysis

XRD analysis was collected to help further characterize the coatings and highlight areas of similarity and difference between thermally decomposed electrodes prepared each at electrode A and at electrode B. The diffraction peaks were indexed in accordance to the standard data provided by JCPDS. Figure 5 (a) and 5(b) are the XRD patterns of RuO_2/Ti anodes prepared at 450°C (Electrode A) and 550°C electrode B. Diffraction peaks were observed at 25.32°, 37.56°, and 56.45°, corresponding to the anatase phase of RuO_2. The intensity of RuO_2 anatase diffraction peaks was seen to be higher in electrode B when compared to that of electrode A, which is an indication that the crystalline nature of anatase RuO_2 increases with increasing sintering temperature. This is illustrated by smaller sized grains and RuO_2 refined microstructure on the surface as electrode B, as illustrated by the SEM images. The porous surface nature results in an increased number of active sites, improving the electrocatalytic activity of electrode B for pharmaceutical effluent electro-oxidation. This is well revealed by results from COD removal and FT-IR studies. It is also important to mention here that the rutile phase of RuO_2 begins to form on sintering above temperatures of 600°C.

Cyclic voltammetry studies

Cyclic voltammetry studies (Figure 6) were carried out in order to compare the electrochemical stability of electrodes prepared at two different temperatures. The oxidation and reduction of pharmaceuticals in electrode B were found to occur at -0.45 V and 0.15 V respectively. It can be observed that durability of electrode B is much higher than electrode A. The difference in stability could be related to micro structural characteristics of the oxides in the coating. The oxides in electrode B are present in smaller crystallite sizes, resulting in enhanced cohesion between formed particles, and contribute to improvement in stability. Though it is a fact that stability could be altered by morphological and micro structural properties of the films, one has to take into account the effect of the real current density, i.e., the current normalised by the real surface area. In this sense, the real current density for the electrodes prepared at electrode A is lower which has lesser surface area as clearly indicated in Figure 6. Therefore, changing the preparation method could be considered as an approach to improve material performance. The SEM images showed that the RuO_2 films were totally damaged after electro-oxidation using electrode A (Figure 2).

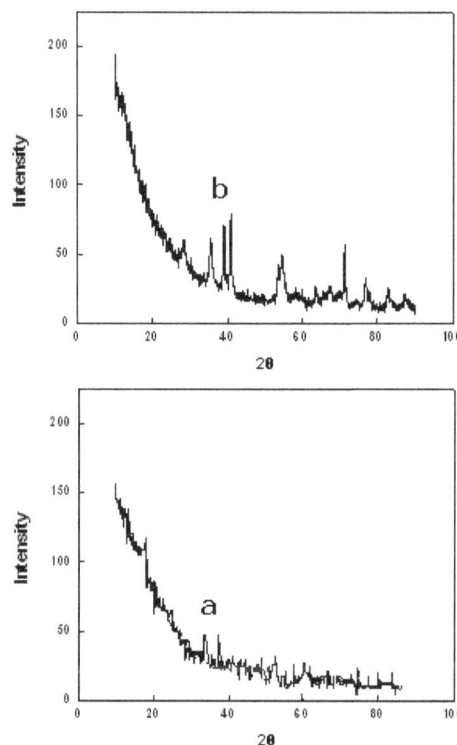

Figure 5 (a): XRD pattern of RuO2/Ti electrodes sintered at 450°C (Electrode A) prior to electrooxidation.
Figure 5 (b): XRD pattern of RuO2/Ti electrodes sintered at 550 °C (Electrode B) prior to electrooxidation

Therefore, dissolution of the RuO_2 occurs, which is associated with the formation of unstable ruthenium species, such as Ru(VI) and Ru(VIII) at this temperature. Erythromycin and amoxicillin undergoes oxidation at the peak potential of background response which indicates that erythromycin and amoxicillin is indirectly oxidized by some reactive species in the background electrolyte. This suggests that the oxidation of pharmaceuticals occur through indirect oxidation by hydroxyl or other oxidant reagent electro generated from the supporting electrolyte (mainly OCl⁻ ions in NaCl). Among all the results, the redox behavior and intensity of oxidation/reduction peaks was more pronounced in electrode B than that of electrode A. This observation can be attributed to high active surface area of the electrode B, which is rendered by smaller crystallite sizes and microstructural modifications occurring during sintering of electrodes at 550°C used in this study. The feasibility of electrode B for the electrochemical oxidation of pharmaceuticals is indicated by voltammetric observations.

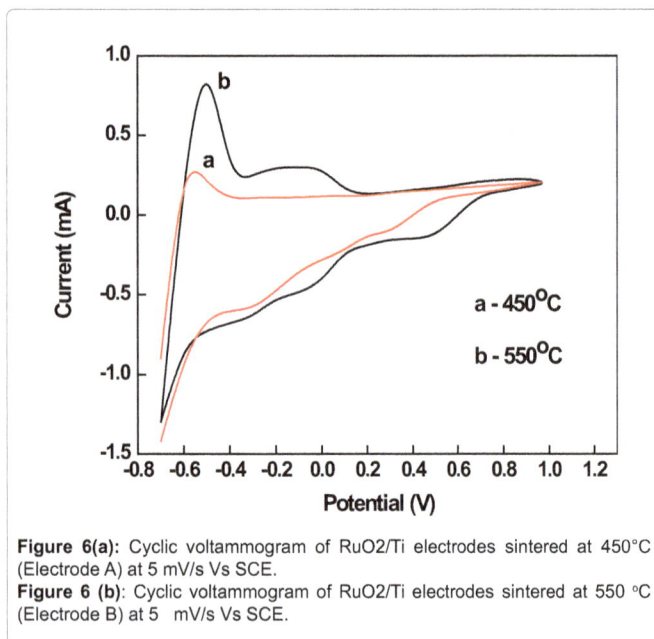

Figure 6(a): Cyclic voltammogram of RuO2/Ti electrodes sintered at 450°C (Electrode A) at 5 mV/s Vs SCE.
Figure 6 (b): Cyclic voltammogram of RuO2/Ti electrodes sintered at 550 °C (Electrode B) at 5 mV/s Vs SCE.

Electrode performance

Reduction in Chemical Oxygen Demand (COD) of the pharmaceutical effluent during electro-oxidation at a constant current density of 3 A/dm^2 was monitored at different time durations, and the results are provided in Figure 7. The rate of COD removal during the initial stages of the electrooxidation was observed to be high especially in anodes prepared at electrode B. Decrease in rate of COD removal at latter stages may be attributed to the formation of stable intermediates. At the end of electro-oxidation for four hours, the reduction in COD by electrode B was found to be significantly higher than that of electrode A. As seen from Table 2, the BOD, color and total dissolved solid reduction is also greater for electrode B than electrode A. Electrodes prepared at 450°C and 550°C gave 84% and 96% color removal respectively. COD removal was found to be 68% and 79% for the electrodes prepared at 450°C and 550°C respectively. The efficient electrooxidation performance of electrode A, when compared to that of electrode B can be attributed to the effect of microstructural modifications on the RuO$_2$ electrocatalytic coating, which is rendered by sintering at 550°C. The smaller grain size and greater porosity of stable anatase RuO$_2$ coating on the Ti mesh substrate results in the formation of a greater number of active sites for enhanced pharmaceutical effluent electro-oxidation.

FT-IR Spectral Studies

The FT-IR spectra for untreated and treated pharmaceutical effluent using electrode B were recorded on Thermo Nicolet Nexus 670 FT-IR spectrophotometer in the range between 400 and 4000 cm^{-1}. The samples were dispersed in pure KBr crystals to prepare pellets for analysis. From the spectra (Figure 8) it can be concluded that major structural changes might have occurred during the electrochemical oxidation process (electrodes prepared at electrode B). The functional group peaks covering the 700-1800 cm^{-1} region show that most of the peaks of raw effluent (8(b)) have disappeared or decreased in intensity in the spectrum of treated effluent (8(a)). Peaks at 3418, 1615, 1513, 1055 and 608 cm^{-1} correspond to N-H, C-N and carbonyl groups. The characteristic absorption peaks of alcoholic O-H groups were found at 3418 and 1398 cm^{-1}. The cyclic and acyclic ether linkage(C-O-C) absorbs at 1190 and 1118 cm^{-1}. After treatment, the intensities of amine

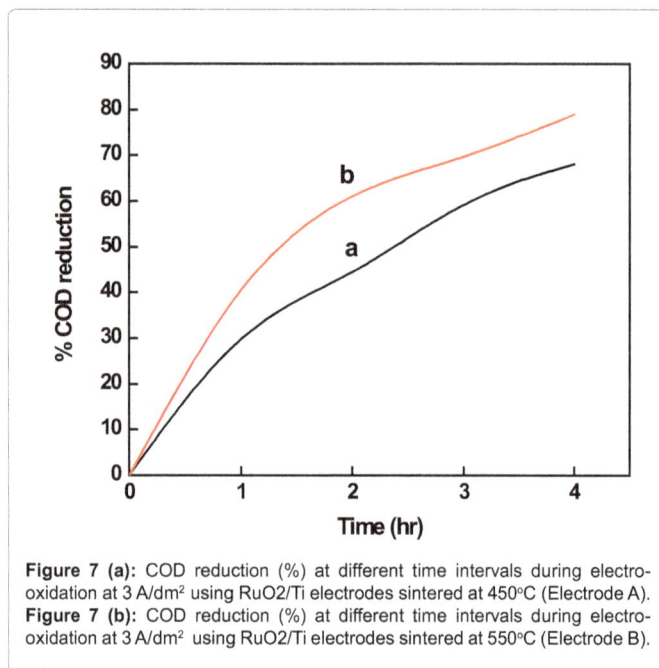

Figure 7 (a): COD reduction (%) at different time intervals during electrooxidation at 3 A/dm^2 using RuO2/Ti electrodes sintered at 450°C (Electrode A).
Figure 7 (b): COD reduction (%) at different time intervals during electrooxidation at 3 A/dm^2 using RuO2/Ti electrodes sintered at 550°C (Electrode B).

S. no.	Parameter	After electrooxidation (electrode A)	After electrooxidation (electrode B)
1	Color	Pale yellow	Colorless
2	pH	7.8	7.5
3	COD	504 mg/l	320 mg/l
4	BOD	185 mg/l	115 mg/l
5	Total dissolved solids	185 mg/l	163 mg/l

Table 2: Physico-chemical characteristics of pharmaceutical effluent after electrooxidation.

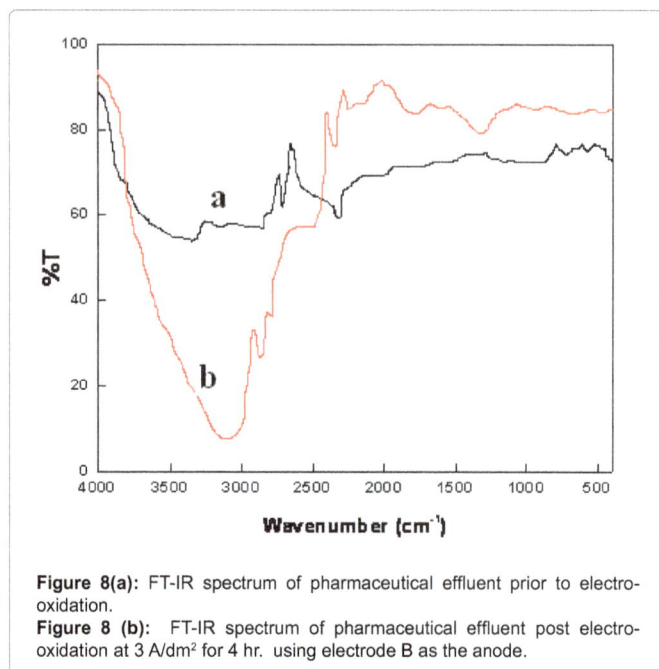

Figure 8(a): FT-IR spectrum of pharmaceutical effluent prior to electrooxidation.
Figure 8 (b): FT-IR spectrum of pharmaceutical effluent post electrooxidation at 3 A/dm^2 for 4 hr. using electrode B as the anode.

and O-H groups got reduced. This is mainly due to the fact that the organics present in pharmaceutical wastewater are converted into other secondary products. The reduction of peak intensity at 1190 and 1118 cm^{-1} may be due to the breaking up of cyclic ether linkage (C-O-C) and tetra hyropyran and cyclohexane rings are formed. Appearance of new peaks at 2926 and 2854 cm^{-1} and a low intense peak at 1445 cm^{-1} proves the formation of cyclohexane and tetra hyropyran ring in effluent. From the results, all characteristic absorption peaks appear reduced in the spectra. This indicates that the organics present in the effluent have converted into some other products such formic, oxalic and acetic acids. Further, these acids may get converted to CO_2 and water. The observations from FT-IR data point at the efficiency of RuO_2/Ti electrode B, prepared by sintering at 550°C for electro-oxidation of pharmaceutical effluent.

Conclusion

This contribution has reported the effect of RuO_2 microstructure on Ti substrate anodes with respect to efficient electrochemical oxidation of pharmaceutical effluent. Electrodes were prepared by sintering at two different temperatures viz. 450°C (Electrode A) and 550°C (Electrode B), and electro-oxidation was carried out at a current density of 3 A/dm^2 for 4 hr. Color removal of 84 % and 96 % were recorded electrodes prepared at 450°C and 550°C. COD removal was found to be 68 % and 79 % for the electrodes prepared at 450°C and 550°C respectively. Electrode B also exhibits more efficient redox behavior than that of electrode A, as depicted by results from cyclic voltmmetry. The efficient electro-oxidation performance of electrode A, when compared to that of electrode B can be attributed to the effect of microstructural modifications on the RuO_2 electrocatalytic coating, which is rendered by sintering at 550°C. This can be construed from the SEM and XRD results provided. The smaller grain size and greater porosity of stable anatase RuO_2 coating on the Ti mesh substrate results in the formation of a greater number of active sites for enhanced pharmaceutical effluent electro-oxidation. Thus, this contribution reports that preparation of RuO_2/Ti anode by sintering at 550°C would be an efficient technique for bringing about effective electro-oxidation of pharmaceutical effluent.

Acknowledgement

One of the authors Ms. S. Vahidhabanu thanks Department of Science and Technology, New Delhi for offering DST- Inspire fellowship.

References

1. Saritha P, Raj DSS, Aparna C, Nalini Vijaya Laxmi P, Himabindu V, et al. (2009) Degradative Oxidation of 2,4,6 Trichlorophenol Using Advanced Oxidation Processes –A Comparative Study. Water Air Soil Pollut 2: 169-179.

2. Han WQ, Wang LJ, Sun XY, Li JS (2008) Treatment of bactericide wastewater by combined process chemical coagulation, electrochemical oxidation and membrane bioreactor. J Hazard Mater 151: 306-315.

3. Mohan N, Balasubramanian N, Basha CA (2007) Electrochemical oxidation of textile wastewater and its reuse. J Hazard Mater 147: 644-51.

4. Yavuz Y, Koparal AS (2006) Electrochemical oxidation of phenol in a parallel plate reactor using ruthenium mixed metal oxide electrode. J Hazard Mater 136: 296-302.

5. Costa CR, Botta CMR, Espindola ELG, Olive P (2008) Electrochemical treatment of tannery wastewater using DSA electrodes. J Hazard Mater 153: 616-627.

6. Chatzisymeon E, Dimou A, Mantzavinos D, Katsaounis A (2009) Electrochemical oxidation of model compounds and olive mill wastewater over DSA electrodes: 1. The case of Ti/IrO2 anode. J Hazard Mater 167: 268-274.

7. Malpass GRP, Miwa DW, Miwa ACP, Machado SAS, Motheo AJ (2009) Study of photo-assisted electrochemical degradation of carbaryl at dimensionally stable anodes (DSA). J Hazard Mat 167: 224-229.

8. Pouilleau J, Devilliers D, Garrido F, Vidal SD, Mahé E (1997) Structure and composition of passive titanium oxide films. Mater Sci Eng B 47: 235-243.

9. Yang LX, Allen RG, Scott K, Christensen P, Roy S (2004) A comparative study of Pt-Ru and Pt-RuSn thermally formed on Ti support for methanol electrooxidation. J Power Sources 137: 257-263.

10. Yang LX, Allen RG, Scott K, Christensen P, Roy S (2005) A new Pt–Ru anode formed by thermal decomposition for the direct method fuel cell. J Fuel Cell Sci Technol 2: 104-110

11. Devilliers D, Mahé E (2010) Modified titanium electrodes Application to Ti/TiO2/PbO2 dimensionally stable anodes. Electrochim Acta 55: 8207-8214.

12. Klavarioti M, Mantzavinos D, Kassinos D (2009) Remediation of water pollution caused by pharmaceutical residues based on electrochemical separation and degradation technologies. Environment International 35: 402-417.

13. Chávez EI, Rodríguez RM, Cabot PL, Centellas F, Arias C, et al. (2011) Degradation of pharmaceutical beta-blockers by electrochemical advanced oxidation processes using a flow plant with a solar compound parabolic collector. Water Res 45: 4119-4130.

14. Boroski M, Rodrigues AC, Garcia JC, Sampaio LC, Nozaki J, et al. (2009) Combined electrocoagulation and TiO2 photoassisted treatment applied to wastewater. J Hazard Mater 162: 448-454.

15. L Lipp and D Pletcher (1997) The preparation and characterization of tin dioxide coated titanium electrodes. Electrochim Acta 42: 1091-1099.

16. Mahe′ E, Devilliers D (2001) Surface modification of titanium substrates for the preparation of noble metal coated anodes. Electrochim Acta 46: 629-636.

17. Yang LX, Allen RG, Scott K, Christenson PA, Roy S (2005) A study of PtRuO2 catalysts thermally formed on titanium mesh for methanol oxidation. Electrochim. Acta 50: 1217-1223.

18. Vignesh A, Siddarth AS, Gokul OS, Babu BR (2013) A novel approach for textile dye degradation by zinc, iron–doped tin oxide/titanium moving anode. Int J Environ Sci Technol 13762-013-0318-9.

19. Meen TH, Huang CJ, Chen YW, Ji LW, Diao CC, et al. (2008) Study of Different TiO2 Electrode Structures on Dye-Sensitized Solar Cell. Key Eng Mat 368-372: 1716-1719.

20. Chetty R, Scout K (2007) Catalysed titanium mesh electrodes for ethylene glycol fuel cells. J Appl Electrochem l37: 1077-1084.

21. Allen RG, Lim C, Yang LX, Scott K, Roy S (2005) Novel anode structure for the direct methanol fuel cell. J Power Sources 143: 142-149.

22. Babu BR, Venkatesan P, Kanimozhi R, Basha CA (2009) Removal of pharmaceuticals from wastewater by electrochemical oxidation using cylindrical flow reactor and optimization of treatment conditions, J Envin Sci Health Part A 44: 985-994.

23. Lim L, Chan K, Scott RG, Allen S, Roy, (2004) Direct methanol fuel cells using thermally catalysed Ti mesh. J Appl Electrochem 34: 929-933.

24. Rao MN, Datta AK (1987) Wastewater treatment. Oxford/IBH Publishing Co. New Delhi.

25. Turro E, Giannis A, Cossu R, Gidarakos E, Mantzavinos D, et al. (2011) Electrochemical oxidation of stabilized landfill leachate on DSA electrodes. J Hazard Mater 190: 460-465

29

Evaluation of Energy Conservation with Utilization of Marble Waste in Geotechnical Engineering

Nazile Ural*

Bilecik Seyh Edebali University, Department of Civil Eng., 11210, Bilecik, Turkey

Abstract

Nowadays one of the goals of all the countries has been to improve the international competitiveness and ensure a sustainable economy. This situation has become even more important with the increasing in population, consumption and rapid decreasing natural resources. Consumption of natural resources is created negative in terms of sustainable development with environmental problems. In addition, also, it is given the reason people's lives to reduce the energy needed to sustain a healthy and comfortable. Thus, recycling qualified wastes are a very substantial topic. In this study, the use of marble waste in geotechnical engineering, and the gains for energy to be obtained by recycling of its discussed.

Keywords: Stabilization; Solid waste; Geotechnical

Introduction

The industrial wastes can cause the environmental problem in the world. If we look at the issue in terms of sustainability, the reuse of these wastes is necessary. Especially, for many years large amounts of stone wastes are generated in the quarry processing plants. The environment, humans and economy are impacted significant because of these wastes. Especially, consumption increases with the increase of population in the world and with consumption our natural resources is rapidly declining. However, rapid economic growth, urbanization, rising population and increasing welfare leads to increasing amount of production waste. That large amounts of waste materials are required particularly in the construction industry, road construction, the construction of waste storage areas, and concrete manufacture makes recycling. Today, with the rapid growth of the industry is in excess of industrial wastes increased the performance of work on the evaluation of waste. Utilizing industrial waste provides both prevention of environmental pollution caused by waste and contribution to the national economy by using industrial waste in the construction industry.

Such industrial wastes as marble waste (pieces and dust), molding sand, fly ash, glass dust and sewage sludge are wastes that are environmentally harmful but contribute to national economy if they are recycled. Marble, mankind has lived in the area for centuries, it is particularly used because it is robust. Today, increasingly marble with the request of the construction sector development and better living people began to more consumption. The development of the marble industry and the rapid increase of the company has required an assessment of the marble waste.

There are many countries of the marble and stone quarries and marble the company in the world. Especially, Italy, Turkey, Spain, India and China are the top five dominant countries in terms of marble production [1]. A major part of production is consumed locally by producing countries, and only a small percentage of total production is exported. This fact indicates that local supply of marble remains less costly, while the transportation cost increases the price of exported marble products. Therefore, Afghan firms can win the domestic market with least effort. In addition to, Korai [2] was given world top ten marble exporters (China, Italy, Germany, Turkey, India, Japan, United States of America, Spain, Brazil and France) and importers (United States of America, China, Germany, Republic of Korea, Japan, Italy, France, United Kingdom, Netherlands and Belgium). However, Afghan and Pakistani are more the domestic market [1]. it is seems that in the

marble quarries of marble company emerged stone (marble) wastes in many countries. Constitute a significant portion of exports marble sector is growing rapidly in these countries. Sector grows, increases, the rate of waste left behind. Especially, in case uncontrolled spillage of waste, some environmental problems occur. Figure 1a is given image noise created by the marble waste, and Figure 1b is given waste from marble company, Bilecik, TURKEY. However, marble dust is left to nature in the form of aqueous sludge at the end of cutting process. Because of marble dust is uncontrollably left to nature, it damages to the environment that marble waste reduces water filtering capacity of the soil, prevents the development of vegetation, fills stream beds and contaminates water resources [3].

In bearing capacity weak soil, for the removal of existing soil, it is occured another area to transport, and instead of taking the high strength material handling, transportation, laying operations. The resulting new business processes, increase rental of construction equipment, fuel costs, labor costs, shipping charges, the unloading and storage costs. However, by adding additives to existing soil is provided the gain of said energy. By improving the application of lime, can provide up to 40% economy [4]. The contribution of the waste material, this will further increase the economic benefits. It is used marble dust and marble aggregate in geotechnical engineering. Marble dust is used very the department of building materials [5-10]. At the same time, department of transportation is used marble waste. Karaşahin and Terzi [11] investigated, the evaluation of waste marble dust covering layer. They said that local road in marble dust can be used as filler material in asphalt material. Seung and Fishman [12] studied a variety of aggregate wastes and fly ash were used in soil base. The experiments shown that waste aggregate and fly ash increased permeability and reduced plasticity was observed. Okagbue and Onyeobi [13] examined variation of geotechnical properties of three different red tropical soils by adding varying degrees of marble dust. Yhey were shown that

---

*Corresponding author: Nazile Ural, Bilecik Seyh Edebali University, Department of Civil Eng., 11210, Bilecik, Turkey
E-mail: nazile.ural@bilecik.edu.tr

Figure 1: (a) Waste storage, Bilecik, Turkey.

Figure 1: (b) Waste in marble company, Bilecik, Turkey.

plasticity and strength properties of red tropical soils were improved significantly with marble dust usage. Mishra et al. [14] indicated that marble waste might be mixed with soil used for road infrastructure and filling material. Zorluer and Usta [15] examined whether marble dust might be used for improving clay in swelling tests. They said that waste marble dust might be used for improving the soil. Sabat et al. [16] added varying degrees of fly ash and marble dust to a swelling soil. They determined that geotechnical properties of soils were improved significantly by increased amounts of fly ash-marble dust usage. Taspolat et al. [17] investigated the effects of marble dust used for layers of waste storage on freezing and thawing properties. They indicated that adding 10% and 15% marble dust to impermeable clay layers increased the soil strength against environmental conditions. Ural et al. [18] studied utilization of the waste marble in soil improvement of clayey soils and as a fine aggregate in concrete production. Physical, mechanical and physico chemical characteristics were determined on clayey soils with marble dust additive. The test results showed that some improvement occurred in behavior of clay soil. Also, the cutting waste of marble sludge will use as a filler material instead of fine aggregate in concrete production. Yıldız [19] evaluated marble dust and waste parts the availability of road-building. They said that in road construction, the use of marble dust and tracks may be appropriate and, in soil stabilization, with the use of dust waste, increased efficiency of at least 10 times. Gürer [20] and Lima [21] said that waste marble pieces can be evaluated as road base and sub-base material. Excess of that required in the construction industry raw materials is needed the use of raw materials and energy consumption by minimizing. Therefore, the use of recycled material in the road sub-aggregate material or as an additive

material in clayey soils has contributed to the significant amount of the national economy. This contribution emerges in material transport (shipping) or laying out as the work machine rental, fuel costs, labor costs, transportation costs, as the reduction of discharge-storage costs.

Formation of Marble Waste

Marble wastes are generated as a waste during the process of cutting and polishing. To come to the desired properly marble blocks, marble is cut into smaller blocks. Relatively large quantities of marble waste are becoming fragmented during this cutting (Figure 2). Marble and another stone industry produces large amounts of waste which causes environmental problems. Marble waste from the processing plant as part size can be as two different products. The first product is large-sized pieces of marble waste, the second product is dust waste maximum part size up to 2mm reach. Large-sized piece of marble wastes are disposed randomly nature show that the waste area or local government and very bad pollution demonstrate the image. Marble dust is settled by sedimentation and than, the marble dust is slurry obtained first and is left wet state at nature. In addition to forming dust in summer and threatening both agriculture and public health. Large size piece of waste that can be used as structural elements in the construction industry, the waste dust can be used directly in different industries. Part of waste concrete aggregate used as filling material, compacted road surface marble dust can be used in cement production, glass production, the film material in concrete and, as an additive material in soil improvement.

Evaluation of Energy savings using Waste of Marble

With population growth, is rapidly exhausted energy reserves of fossil fuels to meet a significant portion of its needs. The use of energy resources in an efficient manner is of great importance, due every day fossil fuels (coal, oil and natural gas) the reduction in the reserve. A significant amount of energy used recoverable, with energy-saving measures to do. Energy savings can be achieved in such as industry, buildings and transport sector to be taken measures. Energy savings are realized, more wise use of energy resources and to allow reduction of energy costs by making studies. In addition, it is of utmost importance, in terms of the reduction of environmental problems. Overall we can say the main energy saving methods in industry; improvement of combustion efficiency, waste water heat recovery, waste heat recovery from the flue gas, air to air heat recovery, compressor capacity control, limestone inhibitors and fleet of fuel-saving [22,23]. Hanieh et al. [24]

Figure 2: After the cutting of marble plates, marble waste. (a) Marble pieces

(b) Marble dust.

are given a flowchart for complete life cycle of the Stones. Lifecyle flowchart shows that tarting from the extraction process of the stones from rocks to using shaped stones in the building and possibility of recycling stones after buildings.

Energy saving to be obtained as aggregate of marble pieces and the marble dust by used as additives is obtained with the disappearance of the cost of extracted the rubble from the quarry using explosives, of preparation compressor or hands extract out of the quarry, of the installation of the wheel loader, of breaking with the crusher-sieved the classification, of overloading the transport vehicle and of the transportation to the construction site. Also transportation to the same equation will be used in natural and waste materials are the most important factor in moving away. Yıldız [19] said that for piece of marble waste for road construction in the area of 68.21km away, even if a quarry next to the construction site, the same cost gives with the use of natural materials to 68.21km from the use of waste brought to the piece of marble. Briefly, the marble using waste materials as aggregates or additives savings the energy used by the machinery and equipment used in quarries. Thus, the evaluation of the marble waste from industrial waste become the economy is contributing by providing both a reduction the environmental pollution and energy saving. Because a large number of factories in Bilecik City, Turkey; as example it can be give this city. Marble pieces was seen to be proper materials as sub-base/base materials according to Republic of Turkey Highways Technical Specification criteria. Consequently, it was made cost analysis and it was seen to suitable to be used the marble waste in order to reduce transport costs in the evaluation of the road construction in the region.

Conclusions

Natural stone such as marble cutting business by producing and processing enterprises are used natural resources as production inputs. But natural resources, can not be put in place again when consumed. Accordingly, these businesses should use resources efficiently and making production environment are also protected. As for environmental protection made of environmentally friendly production, minimization of harmful waste and recycling should be provided. The purpose of all these studies, To reduce the negative effects of the surrounding reducing waste levels,to reduce the quest for new sources of raw materials and provide energy savings.

In particular, a large amount of raw materials required by road, rail and environmental pollutants in the construction of structures such

as airport has industrial solid waste countries' potential to contribute to energy saving after making the necessary experiments using. The energy in a country's economic and social development in terms of direct influence, for use in any area of the material to be recycled it is also important in raising the living standards of humanity. In addition, the use of waste materials in construction industry will help to decrease environmental pollution and economic costs. Thus, utilization of marble waste different industries has a beneficial potential for sustainable construction technologies. The use of the replacement materials offer cost reduction, energy savings, arguably superior products, and fewer hazards in the environment.

References

1. Rassin AG (2011) A comprehensive study of marble industry in Afghanistan, Research & Statistics Department Afghanistan Investment Support Agency.

2. Korai MA, Hussain S, Abro A (2011) A report on marble & Granite, Trade Development Authority of Pakistan.

3. Öztürk M (2009) Environmental Pollution Resulting from Marble Cutting. Grand National Assembly of Turkey Environment Commission, Ankara, Turkey (in Turkish).

4. Akyarlı A, Kavak A, Atay S, Alkaya S (2009) Stabilization of Clayey Soil with Lime. TSE- Standard Economic and Technical Journal 48:108-112 (in Turkish).

5. Uğurlu A (1996) Effect of the use of Mineral Filler on the Properties of Concrete. 1. National Crushed Stone Symposium 303-323 (in Turkish).

6. Ünal O, Kibici A (2001) A Research Lsing Waste-Marble Dust in the Concrete Production. Turkey III Marble Symposium, Afyon Turkey 317-325 (in Turkish).

7. Agrawal V, Gupta M (2011). Expansive Soil Stabilization Using Marble Dust, International Journal of Earth Sciences and Engineering 4: 59-62.

8. Topçu IB, Bilir T, Uygunoğlu T (2009) Effect of waste marble dust content as filler on properties of self-compacting concrete. Construction and Building Materials 23: 1947-1953.

9. Corinaldesi V, Moriconi G, Naik TR (2010) Characterization of marble powder for its use in mortar and concrete. Construction and Building Materials 24: 113-117.

10. Taşdemir C, Atahan HN (1996) Effect of Microfiller Materials on the Mechanical Properties and Durability of Concrete. National Crushed Stone Symposium 251-265 (in Turkish).

11. Karaşahin M, Terzi S (2007) Evaluation of Marble Waste Dust in the Mixture of Asphaltic Concrete. Construction and Building Materials 21: 616-620.

12. Seung WL, Fishman KL (1993) Waste products as highway materials in flexible pavement system. Journal of Transportation Engineering 119: 433-449.

13. Okagbue CO, Onyeobi TUS (1999) Potential of marble dust to stabilise red tropical soils for road construction, Engineering Geology. Elsevier Science 53: 3 71-380.

14. Mishra AK, Mathur R, Goel P (2011) Marble slurry dust in roads- an apt solution for industrial waste. Jl. Of Highway Research Bulletin 65: 83-91.

15. Zorluer I, Usta M (2003) Stabilization of Soils By Waste Marble Dust, IV Marble ermer Symposium (Mersem '2003) Proceedings Book, Turkey, 305-311 (in Turkish).

16. Sabat AK, Behera SN, Dash SK (2005) Effect of Flyash- Marble Powder on the Engineering Properties of an Expansive Soil. Indian Geotechnical Conference 269-272.

17. Taşpolat LT, Zorluer I, Koyuncu H (2006) Effect of Freeze - Thaw Marble Waste Powder on impermeable clay layers. Electronic Journal of ConstructionTechnologies 2: 1–16.

18. Ural N, Karakurt C, Cömert AT (2014) Influence of marble wastes on soil improvement and concrete production. Journal of Material Cycles and Waste Management 16: 500-508.

19. Yıldız AH (2008) Evaluation of Road Construction Waste Marble Powder. PhD Thesis, Süleyman Demirel University,Turkey (in Turkish).

20. Gürer C (2005) Using Waste Marbles Within the Asphalt Pavements, Master Thesis, Afyon Kocatepe Üniversitesi.

21. Lima H (2002) Applicability of marble quarry waste in pavement layers, Proceedings of the Seminar, Appropriate use of natural materials in road, Ulann Baator (Mongolia).

22. Filik BÜ, Kurban M (2007) Effect of Boiler Efficiency in Industry and Energy Saving Analysis, EVK'2007 2. Energy Efficiency and Quality Symposium, Kocaeli, Turkey (in Turkish).

23. Kavak K (2005) Energy Efficiency in the world and Turkey and, Evaluation of Energy Efficiency in Turkish Industry, General Directorate of Economic Sectors and Coordination, Turkey, publication no: DPT: 2689(in Turkish).

24. Hanieh AA, AbdElall S, Hasan A (2013) Sustainable development of stone and marble sector in Palestine. Journal of Cleaner Production 84: 581-588.

30

E-Waste Trading Impact on Public Health and Ecosystem Services in Developing Countries

Ahsan Shamim[1*], Ali Mursheda K[2], and Islam Rafiq[3]

[1]Associate Professor of Environmental Science, Earth and Atmospheric Sciences Department, Metropolitan State University of Denver, CO, USA
[2]Adjunct Faculty, Biology Department, Metropolitan State University of Denver, CO, USA
[3]Soil, Water & Bioenergy Program Director, Ohio State University South Centers, Piketon, OH ,USA

Abstract

During recent year's accelerated global rise in Waste of Electric and Electronic Equipment (WEEE) and its indiscriminate disposal is becoming a foremost concern for human health and ecosystem services. With the rise in concerns on e-waste management and disposals practices, there are attempts to hold back e-waste generation and processing by a variety of regulatory instrument. Realistically there are substantial deficiencies in regulatory initiatives on worldwide trade, unlawful trafficking and improper handling of e-wastes. Currently, the center of attention on recent studies is primarily focusing on linkages of improper handling and consequent health effects on workers in the developing countries. Several studies emphasized on public health problems and reduced ecosystem services. An imminent concern of global calamity is expected, unless appropriate measures are not placed immediately into actions. These concerns demand a need to re-review the facts from recent research studies and suggest effective plans for collection, handling, disposal and remedy of e-wastes. An across-the-board review of available research and policy strategy is necessary to find a sustainable solution dealing with the global trafficking and trade of e-wastes.

Keywords: E-waste, trafficking, human health, recycling, regulatory compliance, Basel convention, Waste of electric and Electronic Equipment (WEEE), EU Directives.

Introduction

During last few decades, one special type of waste has raised a great concern in most of the developed- and in developing countries is the electronic waste. Currently, Waste Electrical and Electronic Equipment (WEEE) or e-waste is one of the fastest growing waste streams in the world [1]. Large quantities of e-waste represent an emerging environmental problem, as electrical and electronic waste equipment already constitutes more than 5% of municipal waste and is still growing fast in the developed countries [2]. Among the developed countries, United States produces largest volume of e-waste and it accounts for 1 to 3% of the total municipal waste generation. While in European Union (EU), the WEEE increases by16 to 28% every five years, which is three times faster than that of average annual municipal solid waste generation. One of the UNEP studies estimate that the total amount of WEEE generation in EU ranges from 5 to 7 million metric tons annually or about 14 to 15 kg per capita, which is expected to grow at a rate of 3 to 5% per year [1]. The 2012 UN report projected that by 2017 global e-waste will increase a further 33% from 49.7 million to 65.4 million metric tons annually [3].

The USEPA's report on e-waste statistics published in 2011 have reported that in 2009 there were 438 million new electronic products sold, 5 million short tons of electronic products were in storage, 2.37 million short tons of electronic products were ready for end-of-life management, and 25% of these tons were collected for recycling [4]. As fastest growing economies in the world, China and India are also experiencing exponential growth of Electrical and Electronic Equipment (EEE) consumption, which currently leads to a large volume of e-waste generation. In China, more than 5.1 million home appliances and 4.5 million personal computers are becoming obsolete each year [5]. In 2012 alone, China reportedly generated 11.1 million metric tons of e-waste as compared to 10 million tons of e-waste produced in the United States [6]. Besides domestic e-waste, being a developing country, over one million tons of e-waste from the U.S.,

Europe and other countries of the world are overflowing into China every year, taking advantage of the cheap labor costs and non-stringent environmental regulations [7]. It is assessed that one billion computers and accessories had been manufactured and subsequently discarded in 2008, and another one billion would be discarded worldwide within next 5 years [8].

The e-waste is a global environmental problem. Of the 20 million to 50 million metric tons of e-waste generated annually, it is estimated that 75 to 80% of that is shipped to developing countries especially in Asia and Africa for "recycling" and disposal [9]. Several studies have reported that a vast majority of the e-wastes is being exported to China and India. Moreover, other identified hot spots for e-waste destinations are Pakistan, Bangladesh, Ghana, Nigeria and Kenya. However, the handling and recycling techniques in these countries are often primitive and there is a little esteem for worker's safety or environmental protection [8,10], which are illegal under the Basel Convention of 1992 or any other existing national environmental legislations [11]. In contrast, the WEEE recycling in developing countries is a daisy chain of processes which are carried out in the informal economy. Informal economies constitute a considerable amount of the gross national product (GNP) of the developing or economically transitional countries [12].

The e-waste has raised great concerns as many components in these products are hazardous, toxic and non-biodegradable. A prolong

***Corresponding author:** Shamim A, Associate Professor of Environmental Science, Earth and Atmospheric Sciences Department, Metropolitan State University of Denver, CO, USA
E-mail: sahsan@msudenver.edu

exposure of these toxic constituents in the environment is almost definite to cause a long term concern for both human and ecosystem health. Based on environmental concerns and scientific findings, several EU countries banned e-waste from landfills in the 1990s [13]. The concern about the effects of chemical exposure from e-waste products and e-waste recycling on public health is increasing despite the paucity of scientifically valid research findings. Reported adverse effects of e-waste on public health include: fetal loss, prematurity, low birth weight, and congenital malformations, abnormal thyroid function and thyroid development, neuro-behavioral disturbances, and Geno toxicity [10,14].

The goal of this review paper is to assess the current status on generation and destination of WEEE, regulatory framework and its shortcomings, and likely human health and ecosystem effects summarized in related peer-reviewed research studies. This paper will also review the current sustainable e-waste management schemes, its outcomes, and possible solutions.

E-waste Generation, Disposal and Regulations

Growth of e-waste and future global forecast

Considerable global growth of e-waste has taken place in recent decades. According to UNEP reports, the estimated amount of annual global e-waste generation is around 20 to 50 million metric tons [15], about 1 to 3% of the global municipal solid waste production of 1636 million metric tons per year [16]. A recent UNU statistics indicates that the total annual global volume of WEEE – also referred to as e-waste – is soon expected to reach 40 million metric tons [17]. There are varying estimates as to the amount of domestic, regional, and global e-waste produced. According to StEP (Solving the E-waste Problem Initiative), the 2012 global generation of e-waste totaled 45.6 million metric tons [18]. On the other hand, the Global E-Waste Management Market report (2011 to 2016) projected that the global volume is expected to reach 93.5 million metric tons in 2016 from 41.5 million metric tons in 2011, at Compound Annual Growth Rate (CAGR) of 17.6% [19]. The USEPA has estimated a 5 to 10% annual increase in the generation of e-waste globally. Perhaps even more alarming is that only 5% of that amount is being recovered [4].

A WHO report recognized that the total amount of e-waste produced is exponentially increasing because of multiple and interlinked factors. Increasing consumer demand and a high obsolescence rate have led to frequent and unnecessary purchases of the EEE [20]. One of the major driving forces of growing e-waste problem is the short lifespan of most electronic products - less than two years for computers and cell phones [21,22]. Life span of computers has dropped in developed countries from six years in 1997 to just two years in 2005. Mobile phones have a lifespan of even less than two years [23]. The International Association of Electronics Recyclers (2006) projected that the current growth and planned obsolescence rates of the various categories of consumer electronics, somewhere in the proximity of 3 billion units which would be scrapped by 2010 or an average of about 400 million units a year. A recent UN University report estimated: by 2020 e-waste from old computers in South Africa and China will jump by 200 to 400% and by 500% in India compared to that of the 2007 levels. In 2012 alone, China reportedly generated 11.1 million metric tons of e-waste and the United States produced 10 million tons [6]. Even the least developed countries (LDC) like Senegal and Uganda are expecting e-waste flows from personal computers alone to increase 4 to 8-fold by 2020 [24]. Research data have shown that electronic waste in Europe is growing at three times the rate of other municipal wastes [25]. The reliable and precise

estimation on future e-waste generation varies from nation to nation, however, Khurrum et al. [26] analyzed e-waste problem providing an estimation of the amount of e-waste produced and recycled every year. The estimate leads us to believe that by the year 2015, over 500 million units of e-waste will be disposed-off and slightly over 113 million units are expected to be recycled. Recent rise in electronic sales eliciting great concern of continued increase in e-waste. An annual report of the Consumer Electronic Association (CEA) presented a healthy-looking growth for sale of consumer electronic products. The US consumer electronics industry made overall shipment revenues above $ 173 billion in 2008, while in July 2012 report estimates that industry sales will surpass $ 206 billion this year. Sales growth is projected to continue into 2013, when industry revenues will likely grow 4.5%, reaching $216 billion [27]. The global e-waste management market 2011-2016 report forecasted: The revenue generated from the e-waste management market is expected to grow from $ 9.15 billion in 2011 to $ 20.25 billion in 2016 at a growth rate of 17.2% from 2011 to 2016 [19]. According to a new report by Allied Market Research, the global e-waste Management Market would reach an amount of $ 49.4 billion by 2020, registering a Compound Annual Growth Rate (CAGR) of 23.5% during 2014 to 2020 [28]. It is quite apparent that in different reports there are discrepancies and lacking in the correctness of the forecast data, however, most of the reports revealed a just beginning global rise of e-waste. The increase in e-waste is the visible symptom of the "make, consume, and dispose" culture that has permeated the developed world and is now spreading across the developing countries [29].

Linkage of e-waste generation and GDP growth

The economics of e-waste production and consumption have been constructed directly correlated with the GDP growth in most of the countries which was illustrated graphically (Figure 1) in a paper presented by Robinson [10]. Lu et al. [30] validated the relationship between China's GDP per capita, urbanization rate and e-waste generated from 2001 to 2012 (Figure 2). It is apparent that per capita e-waste generation is even higher than per capita GDP and is almost double the urbanization rate, indicating that e-waste generation will create a big challenge to the whole country. All these projected growth in the electronics sector clearly paint the grim picture and raise apprehension of unmanaged and unprocessed e-wastes, unless appropriate recycling measures are undertaken.

Background of E-waste Recycling and Trade

In 1991, the first electronic recycling system was implemented in Switzerland beginning with the collection of used and obsolete refrigerators. Over the years, all other electric and electronic devices were gradually added to the system. A legislation followed in 1998, and since January 2005, it has been possible to return all electronic wastes to the sales points and other collection points free of charge. There are two established PROs (Producer Responsibility Organizations): Swiss Economic Association for the Suppliers of Information, Communication and Organizational Technology (SWICO), mainly handling electronic waste and Swiss Foundation for Waste Management (SENS), mainly responsible for electrical appliances. The total amount of recycled e-waste exceeds 10 kg per capita per year [31]. In the 1990s, governments in the EU countries, Japan and several US states set up e-waste 'recycling' systems, but many countries do not have the capacity to deal with the sheer quantity of e-waste they generated annually or with its hazardous nature [32].

In many cases, the cost of recycling e-waste exceeds the revenue recovered from materials especially in countries with strict environment

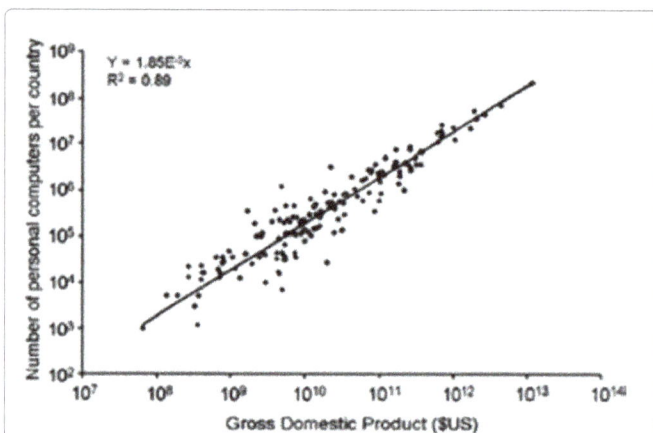

Figure 1: Number of PCs per country related to the country's GDP for 161 countries (Adapted from or Source: [10]).

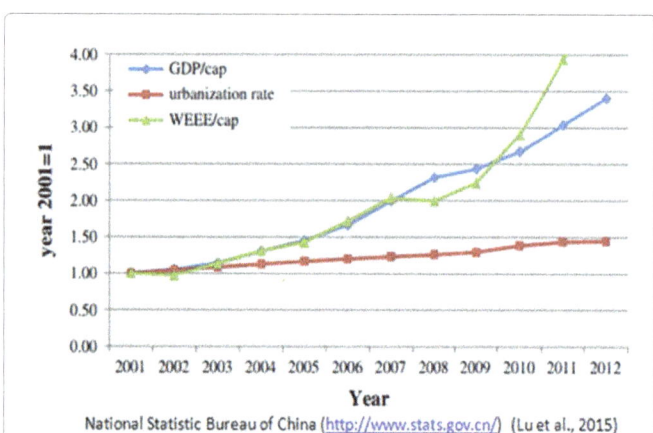

National Statistic Bureau of China (http://www.stats.gov.cn/) (Lu et al., 2015)

Figure 2: Urbanization rate in China with GDP/Cap, WEEE/cap Adapted from or Source: [30].

regulations. Therefore, e-waste mostly ends up dumped in countries where environmental standards are low or nonexistent and working conditions are poor. Historically, Asian countries have been a popular dumping ground for e-waste, but as regulations have tightened in these countries over time, this trade has moved to other regions of the world, particularly West Africa [33]. Several studies reported that one of the illegal recycling nucleus have grown in China over the years is a place named "Guiyu". The e-waste recycling reportedly began in Guiyu in the late 1980s. Laqiao is a town of 400,000 people in Taizhou and currently is the main e-waste recycling site in China. At least 10% of the population in Laqiao participates in e-waste recycling which first started in the 1970s [18,34]. A report by Toxics Link found that 70% of the e-waste collected at recycling units in New Delhi, India, was actually exported or dumped by developed countries, and about 50 to 80% of these e-wastes collected for recycling in the western US region are being exported to Asia [35,36]. About 90% of those e-wastes sent to China for recycling [35,36]. There are also e-waste recycling sites in Bengaluru and Delhi, India. In West Africa, e-waste recycling sites are located in Nigeria (Lagos) and Ghana (Accra, Agbogbloshie) [14,18,25].

The e-waste is informally processed in many countries of the world, but a high-volume of informal e-waste recycling has been reported in China, Ghana, Nigeria, India, Thailand, the Philippines, and Vietnam

[37]. Demand for e-waste recycling in Asia begins to grow when scrap yards found they could extract valuable metals such as copper, iron, silicon, nickel and gold, during the recycling process. A mobile phone, for example, is 19% copper and 8% iron [32]. Much of the informal e-waste recycling done in scrap yards and homes is done by children. The e-waste recycling in China is processed in the informal economy and constitutes a considerable amount of the gross national product (GNP) of the country [24]. Both the demand for recycled materials and the potential economic benefit are the main factors promoting the development of the disassembly (or recycling) industry for e-wastes [38]. Most of the literatures reported that informal waste recycling is carried out by poor and marginalized social groups who resort to scavenging and waste picking for income and survival [39].

Current Management Practices and Regulatory Framework

E-waste management practices

The existing management practices in US and Europe exerts greater economic impact on global trade and recycling due to generation of large volume of e-wastes. So far, legislation on WEEE is mainly driven by certain EU countries and their directive on WEEE. Most developing nations in the world are lagging behind in the development of similar regulations, in particular their enforcement [25]. In particular, various reports and studies by the mainstream Medias (e.g., Cbsnews.com, National Geographic, Scientific American)[cited in 40], environmental organization (e.g., Green peace [16], and researchers [26] have found primitive waste management practices in India and various countries in Africa and Asia. Existing e-waste recycling operations in Guiyu have gained a particular attention [40].

In the United States under most circumstances, e-waste can legally be disposed-off in a municipal solid waste landfill or recycled with little environmental regulatory requirements. The USEPA's 2011 report on e-waste shows that 2.4 million metric tons of e-waste were disposed in 2010 in the United States (Table 1). The USEPA report also elucidated that residential households store 5 times more computer products (by weight) than that in the commercial establishments. Approximately 2.37 million short tons of electronics are ready for end-of-life management, representing an increase of more than 120% compared to that in the 1999s [4]. As shown in Figure 3 that a large number of three major electronic devices were ready for end of life management in 2009, which eventually may add up the total volume as those are added in waste stream. Most recent report published by the USEPA in February 2014 shows; US generated 3.42 million metric tons of e-waste. Of this amount, only 29.2 % was recycled, (up from 10% in 2000). The rest 71% of the e-waste was landfilled or incinerated [43]. Data presented in Figure 4 shows that a slow growth in generation of e-waste as well rise in recycling practices by weight in 2012, i.e. 4.3% increase from previous year [42].

Increasing concerns about e-waste landfill disposal have led federal and state environmental agencies to encourage recycling in the United States. Although there may be limited data regarding how e-waste is managed, the consequences of e-waste export to the developing countries that handle it improperly are becoming increasingly evident. Reliable data regarding how much e-waste is generated, how it is managed, and where it is processed (either domestically or abroad) is largely unavailable. Because e-waste recycling is largely unregulated, virtually no data are available to track its fate [44]. The EU, Japan, South Korea, Taiwan and several US states have introduced legislation making producers responsible for their end-of-life products. The EU

	Total Disposed ** (tons)	Trashed (tons)	Recycled (tons)	Recycling rate (%)
Computers	423,000	255,000	168,00	40%
Monitors	595,000	401,000	194,000	33%
Hard Copy Devices	290,000	193,00	97,000	33%
Keyboards and Mice	67,800	61,400	6,460	10%
Televisions	1,040	864,000	181,000	17%
Mobile Devices	19,500	17,200	2,240	11%
TV peripherals*	Not included	Not included	Not included	Not included
Total (in tons)	2,440,000	1,790,000	649,000	27%
What's included here?				
Computer products include CPUs, desktops and portables.				
Hard copy devices are printers, digital copiers, scanners, multi-functions and faxes.				
Mobile devices are cell phones, personal digital assistants (PDAs), smartphones, and pagers				
*Study did not include a large category of e-waste: TV peripherals, such as VCRs, DVD players, DVRs, cable/satellite receivers, converter boxes, game consoles.				
**"Disposed" means going into trash or recycling. These totals don't include products that are no longer used, but which are still stored in homes and offices.				
Source: EPA, Adapted from [41]				

Table 1: E-Waste by the Ton in 2010 –Trashed or Recycled in USA.

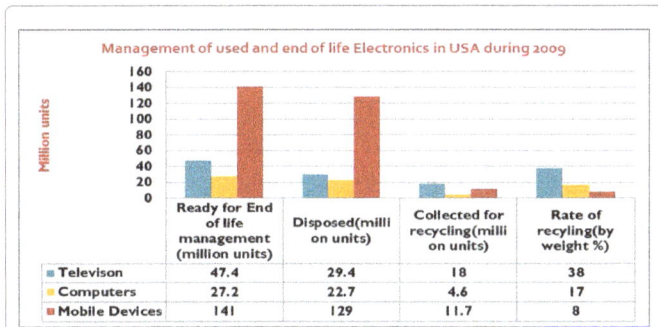

Figure 3: Management of used and end of life electronics in USA during 2009 [Data sources EPA].

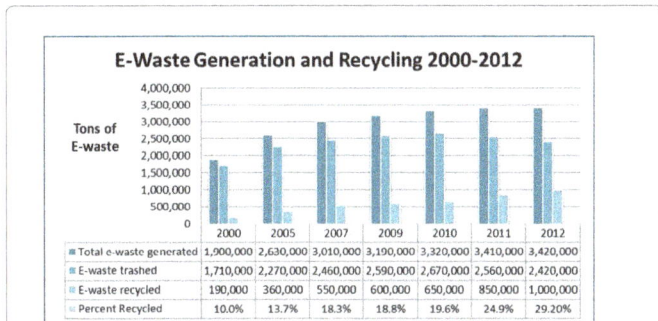

Figure 4: EPA data from "Municipal Solid Waste Generation, Recycling and Disposal in the United States, 2012 ," Feb 2014; These EPA numbers are for "selected consumer electronics" which include products such as TVs, VCRs, DVD players, video cameras, stereo systems, telephones, and computer equipment. [Adapted from 42].

has banned the use of certain hazardous substances in electrical and electronic products from July 2006, to facilitate safer recycling. However, the e-waste recycling sector in many Asian countries remains largely unregulated [23]. Restrictions on the use of certain chemicals are included in the EU Directive on Restrictions on Hazardous Substances – RoHS [45]. This directive has served as a useful guide for other developed countries, for example, China has recently drafted similar administrative measures [46].

International, Regional and national efforts to manage E-waste

Global management of WEEE falls under the Basel Convention on the Control of Trans- boundary Movements of Hazardous Wastes and their Disposal. The Convention was opened for signature on 22 March 1989, and enforced on 5 May 1992. Until 2006, the convention was ratified by 168 nations (Figure 5, Source Basel Convention). The sixth meeting of the Conference of the Parties to the Basel Convention (convened in 2002) recognized that the issue of e-waste recycling required urgent and in-depth supervision, particularly in the Asia-Pacific regional countries. This program was further strengthened at the ninth meeting of the Conference of the Parties to the Basel Convention in 2006, with the adoption of the Nairobi Declaration on the Environmentally Sound Management (ESM) of Electrical and Electronic Waste (decision IX/6). By this decision, the secretariat was requested to facilitate work and activities on the ESM of e-waste, focusing on the management needs of the developing countries and countries with economies in transition. The Secretariat of the Basel Convention, in consultation with selected countries in this region and the Basel Convention Regional Centers in China (BCRC China), Indonesia (BCRC-SEA) and the South Pacific (SPREP), developed a proposal for a pilot project on the ESM of e-waste products [47]. Under the Basel Convention, e-wastes are classified under Annex VIII entries

A1180, A1190, A1150 and A2010 and also under Annex IX as B1110. The e-wastes are characterized as hazardous wastes under the Convention when they contain reactive chemical components such as accumulators and other batteries, mercury switches, glass from cathode-ray tubes and other activated glass, PCB-containing capacitors or when contaminated with cadmium, mercury, lead or PCBs. Also, precious-metal ash from the incineration of printed circuit boards, LCD panels and glass waste from cathode-ray tubes and other activated glasses are characterized as hazardous wastes. To address the environmental issues related to the increasing trans-boundary movements of these wastes, and to ensure that their storage, transport, treatment, reuse, recycling, recovery and disposal is conducted in an environmentally sound manner, a proactive approach is essential. The plastics associated with e-wastes may need to be covered, under Annex II of the Basel Convention [48]. Despite the existence of these agreements and conventions, the transfer of WEEE from the United States, Canada, Australia, Europe, Japan and Korea to China, India, Bangladesh and Pakistan remains relatively high [5, 16, 49]. Although Basel Convention regulates e-waste, it does not ban a country's right to export it entirely.

In the United States, concerns regarding the potential impact of exporting e-waste for processing in the developing countries have led to increased scrutiny from members of the public and environmental organizations, as well as members of Congress. The US Government Accountability Office (GAO) stated that concerns have grown, that some U.S. companies are exporting these e-wastes to the developing countries, where unsafe and/or unregulated recycling practices can cause serious health hazards and environmental problems. Currently, U.S. regulatory controls do little to stem the export of potentially hazardous used electronics, primarily due to (a) Narrow scope of regulatory control (U.S. hazardous waste regulations do not consider most used electronic products, such as computers, printers, and cell phones, as hazardous, instead, under U.S. law, only exports of CRTs are regulated as hazardous waste) and (b) Regulatory controls easily circumvented (The export of CRTs from the United States in apparent violation of the CRT rule seems widespread, despite adoption of the CRT rule in 2006). The USEPA has done a very little to enforce the CRT rule (EPA has taken few steps to enforce the CRT rule since the rule took effect in January 2007) [50]. On May 21, 2009, Representative Gene Green introduced H.R. 2595, a bill that would amend the Solid Waste Disposal Act (42 U.S.C. 6921) to establish certain e-waste export restrictions. There have also been several congressional hearings on issues associated with e-waste management, one of which specifically addressed issues associated with e-waste exports [51]. In July 2011, a tri-organizational

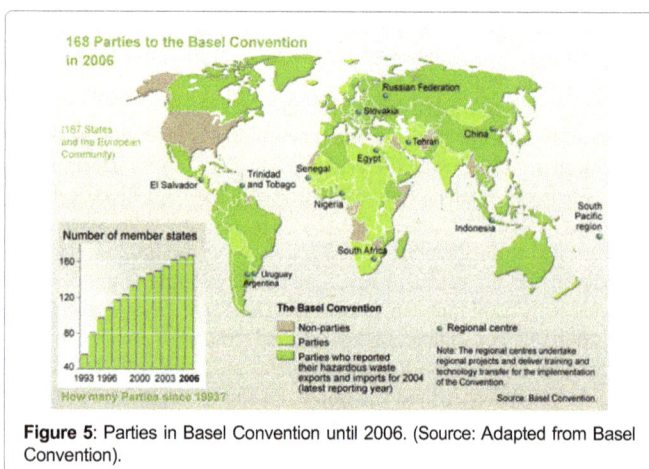

Figure 5: Parties in Basel Convention until 2006. (Source: Adapted from Basel Convention).

taskforce released the National Strategy for Electronics Stewardship ("NSES" or "National Strategy"), establishing an innovative, flexible, pragmatic, and yet unified framework to evolve electronics stewardship. The actions identified in the NSES provide a roadmap to ensure that electronics are designed, purchased, and managed in a more sustainable manner, help protect human health and the environment from harmful effects associated with the unsafe handling and indiscriminate disposal of used electronics, and simultaneously promote new and innovative technologies of the future [52].

Another larger producer of e-waste, the EU countries follow the WEEE Directive of the European Parliament and of the Council (2002/96/EC), which entered into force in 2003. Despite extensive legislation targeting the e-waste problem, experience in the first few years of implementation of the WEEE Directive has shown that it is facing difficulties. Less than half of the collected e-waste is currently treated and reported according to the Directive's requirements [53]. Given that the processing of WEEE involves a variety of problems, European Directive PE-CONS 2/12, which is due to enter into force on 15 August 2018, provides for the grouping of all EEE into six categories instead of the current used ten categories (which remain valid from August 2012 until August 2018) [54].

Presently, most developing countries are lacking in regulatory instruments, especially those are the recipients of most of the e-waste products. Among those major e-wastes informal processing hubs, China and few African countries are in the process of formulating regulatory framework to prevent future environmental degradation as a result of growing e-waste generation and processing. Recently, the Chinese government issued a variety of laws, regulations, policies, standards, and technical guidance to manage and control the EEE Production and WEEE recycling [55]. Beginning July 1, 2004, collection, storage and disposal of hazardous wastes including e-waste require business licensing [56]. Measures for the Control of Pollution from Electronic Information Products, which is the counterpart of the EU RoHS directive, restricted the use of six hazardous substances (i.e. Pb, Hg, Cd, Cr, PBB or PBDE) during the production, sale, and import of electronic information products in China destined for export [25,57]. However, there are few loopholes in this ordinance. For example, no deadline for the restriction has been fixed yet. Moreover, this regulation only applies to mainland China, not to Hong Kong or Macao; nevertheless, Hong Kong has already become the main receiving hub for used electrical and electronic products imported into China [55]. In 2008, the 'Administrative Measures on Pollution Prevention of WEEE' was enacted in order to prevent the pollution caused by the storage, transport, disassembly, recycling and disposal of e-waste. This policy also established a licensing scheme for e-waste recycling companies. Finally, in January 2011 the 'Regulation on Management of Recycling and Disposal of Waste Electrical and Electronic Equipment' was adopted. This regulation is similar to the EU WEEE Directive (Directive 2012/19/EU) as it makes e-waste collection and recycling mandatory [58].

Trade and Illegal Export beyond Regulatory Framework

The INTERPOL research by Pollution Crime Working Group (PCWG) has uncovered that there are huge potential for informal networks of criminals to profit from the illegal export of e-waste to the developing countries. The two most common methods of illegal export are mislabeling containers to conceal e-waste and mixing waste with a legitimate consignment, such as end-of life vehicles [59]. The primary driver of this trade is that e-waste contains valuable components, is easy to source and relatively cheap to ship, and the risk of being caught is generally low [60]. In East Asia and the Pacific, the illicit trade appears to be driven by recycling for metals to be used in manufacturing. Within the region, China is the main destination for e-waste, despite the fact that the country banned the import of used electronic and electrical equipment in 2000. Globally, it is estimated that 80% of e-waste is shipped to Asia (including India) – with 90% of that amount destined for China [61]. An investigation carried out by a UK-based NGO, the Environmental Investigation Agency (EIA), revealed criminal syndicates involved in trafficking e-waste. These groups were also involved in other crimes such as theft, human trafficking, fraud, drugs and firearms trafficking, and money laundering [62]. Most of the studies recognized that illegal trade of e-waste is predominantly driven by high economic return, and the globalization of the illegal e-waste trade has increased corporate or "white-collar" crime. East Asia plays a prominent role in the illegal trade of both e-waste and ozone depleting substances (ODS). The Asia-Pacific region is a major recipient of illicit e-waste. It is estimated that up to 10 million metric tons of e-waste are traded illegally into and around the region annually, with a potential value of at least US $ 3.75 billion. Considering that the global market for e-waste, including legal exports, has been predicted by the UNEP to be valued at around US $ 11 billion by 2009 [1], the scale of the estimate of US $ 3.75 billion of illegal e-waste in East Asia is reasonable [63]. A presence of the informal economy makes solid estimates of the value for the sector difficult. In another report by the UNEP using an estimate previously used by the INTERPOL of an average value of e-waste at US $ 500 per metric ton [59], the range of e-waste handled informally or unregistered, including illegally, amounts to US $ 12.5 to 18.8 billion annually [64]. In spite of national and international efforts, unaccounted e-waste in USA and EU countries is exported to developing countries. Although it is illegal in EU, such exports have been classified as legal recycling by the USEPA [16]. In the case of EU, despite strong legislations, a major source of e-waste which is illegally exported and dumped in developing countries. An estimated 75% of the e-waste generated in the EU, equivalent to 8 million metric tons a year, is unaccounted for [65]. It is believed that most of these 8 million metric tons were trafficked in several developing countries. The evolution of crime, even transnational organized crime, in the waste trading sector is a major threat. Whether the crime is associated with direct dumping or unsafe waste management, it is creating multi-faceted consequences that must be addressed urgently [64]. The first INTERPOL operation targeting the illegal trade of e-waste resulted in the seizure of more than 240 metric tons of electronic equipment and electrical goods and the launch of criminal investigations into some 40 companies involved in all aspects of the illicit trade [66]. Despite empirical data suffering from high uncertainties, the scale of the e-waste trade, its impacts across spheres, and its links to crime are difficult to contest. Although the crime itself often involves less structured and centralized groups than other crimes, the severity of its impacts and its relation with other crimes suggest the seriousness of this issue [58].

Environmental and Human Health Impact

Globalization of e-waste has adverse environmental and public health implications as the developing countries face economic challenges and lack the infrastructure for proactive management of hazardous wastes [67]. The degree of hazard posed to workers and the environment varies greatly depending on the individuals involved and the nature of operations. The short- and long-term effects of exposure to hazardous e-wastes are not fully understood, however, there are research conducted on the association between e-waste exposure and

higher levels of chemicals and metals in human-derived biological samples [18,68,69]. What is known is that the pollution generated by e-waste handling and processing brings about toxic or genotoxic effects on the human body, threatening the health not only of workers but also of current residents and future generations living in the local environment [70]. Most people are uninformed of the possible negative impact of the rapidly increasing use of electronic devices. When these products are placed in landfills or incinerated, they pose health risks due to the hazardous materials they contain [71]. Computers and display units contain a significant amount of diverse chemical compounds that are hazardous to human health if they are not disposed of properly. Monitors and televisions compose 40% of total lead and 70% of all heavy metals found in landfills. These heavy metals and other toxins that can leach into the soil and groundwater from landfills, evaporate into the air, and enter the air through incineration [72]. Many researchers have indicated that different types of chemicals and pollutants released from e-waste into the environment as a consequence improper handling and processing [21,35,69,73,74] and they can accumulate in the human body through various pathways. Inhalation of contaminated air and dust is believed to be one of most important pathways. Some pollutants such as PCDD/Fs, PCBs, PBDEs, PAHs, cadmium, chromium, lead, and arsenic have been identified in atmospheric particles in- and around the e-waste dismantling areas in China [21,69,75]. Long-range transport of pollutants has also been observed, which suggests a risk of secondary exposure in remote areas. Atmospheric pollution due to burning and dismantling activities seems to be the main cause of occupational and secondary exposure [25]. The BAN studies have identified a range of potential occupational safety hazards including silicosis, toxic exposure to dioxins, mercury and other metals and carcinogens through inhalation of fumes while processing e-waste or from local drinking water and food sources contaminated by e-waste by-products [76]. The potential adverse health effects of exposure to e-waste have been reviewed recently and may take account of changes in lung function, thyroid function, hormone expression, birth weight, birth outcomes, childhood growth rates, mental health, cognitive development, cytotoxicity, and genotoxity [18,37,77]. The toxicity of many individual substances found in e-waste is well documented in several studies, however, the toxicity of the mixtures of substances likely to be encountered through e-waste recycling is less well known. Heavy metals and halogenated compounds appear to have a major influence on public health [18,57]. Direct exposure entails skin contact with harmful substances, the inhalation of fine and coarse particles, and the ingestion of contaminated dust. Individuals who directly engage in e-waste recycling with poor protection incur high levels of direct, occupational exposure [18,34,78]. In most of the recycling operations involve burning the plastic coverings of materials to extract metals for scrap, openly burning circuit boards to remove solder or soaking them in acid baths to strip them for gold or other metals. Acid baths are then dumped into surface water which severely impact fresh water ecosystems. Despite the fact that uncontrolled open strong acid leaching of e-waste is officially banned and considered illegal, this practice is on-going in Guiyu, China. One such operation was identified during field sampling. Data presented in Table 2 have shown show comparison of e-waste processing site in China with some riverine systems in Australia [79] and the USA [80]. Data indicated that two rivers in Guiyu, Lianjian and Nanyang were considerably enriched with Cd, Co, Cu, Ni, Pb and Zn. The Pearl River Economic Zone is one of the largest light industrial bases in China, and "dissolved" metal concentrations in urban and rural beaches of the Pearl River were determined by Ouyang et al. [81]. It was found that Nanyang river of Guiyu was noticeably more polluted with Cd, Co, Cu, Ni, Pb and Zn

Locations	Cd (Mean ± SD)	Pb (Mean ± SD)	Co (Mean ± SD)	Cu (Mean ± SD)	Ni (Mean ± SD)	Zn (Mean ± SD)	Refs
Lianjiang, Guiyu	0.091 ± 0.010	1.48 ± 0.09	0.86 ± 0.096	7.80 ± 1.70	36.6 ± 6.2	30.6 ± 4.2	Coby S et al., 2007[82]
Nanyang, Guiyu	0.315 ± 0.032	1.81 ± 0.30	3.62 ± 0.82	50.8 ± 10.0	52.4 ± 7.6	106 ± 10	Coby S et al., 2007 [82]
Hawkesbury-Nepean River, Australia	0.045[a] (0.009-0.111)[b]	0.111[a] (0.027-0.321)	0.24 (0.16-0.35)	0.81a (0.20-2.13)	0.26 (0.18-0.39)	0.88[a] (0.21-2.37)	Markich and Brown, 1998[79]
St. Lawrence River Mouth, USA	0.022	0.0644	0.062	0.996	1.767	0.812	Gobeil et al., 2005[80]
Pearl River (Urban), China	0.127 (0.010-1.340)[b]	0.577 (0.10-2.64)[b]	0.418 (0.050-2.230)[b]	13.96 (1.20-95.41)[b]	28.57 (0.98-178.89)[b]	61.847 (1.74-706.25)[b]	Ouyang et al., 2006[80]
Pearl River (Rural), China	0.041 (0.01-0.13)[b]	0.429 (0.14-1.81)[b]	0.122 (0.04-0.34)[b]	5.358 (1.28-21.65)[b]	7.016 (0.35-29.78)[b]	6.008 (2.38-14.21)[b]	Ouyang et al., 2006[81]

Table 2: Comparison of dissolved metal concentration in freshwater at Guiyu, Australia and USA (µg/L) Data Source: Adapted from source [82].
[a] **Geometric mean,** [b] **Range**

than that of the urbanized region of the Pearl River, suggesting a significant discharge of these metals in Nanyang [82]. Figures 6 and 7 also show the higher concentration of heavy metal Pb, Cd, Co, Cu, Ni and Zn concentrations presented in the study by Coby et.al, and also data adapted from other studies cited in [82]. Several studies attempted to establish a link between (direct and indirect) children workers in e-waste facilities and their health impact. It is currently hard to estimate to what extent children work specifically on e-waste disposal sites; however, many studies reported that children's comprising a significant proportion of all workers on these sites. The difficulty in estimation results largely from a lack of data segregation for e-waste, as a considerable knowledge base exists for child laborers working as "scavengers" or "waste-pickers" [83]. One case study cited in [83] was led by Cuadra (2005) specifically investigated heavy metals exposure of child scavengers in the city of Managua, Nicaragua. Blood analysis of children, who worked as scavengers, showed that the children working at the waste disposal site had higher levels of lead in their blood compared with the non-working reference groups. Among the child workers at the waste disposal site, as many as 28% had blood lead levels greater than the community action level of 100 µg/L recommended by the Centers for Disease Control and Prevention (CDC) [84]. Two other studies have reported elevated body loadings of heavy metals [85] and persistent toxic substances in children and e-waste workers, respectively, in Guiyu [86]. Huo et al. [85] studied 165 children in Guiyu and 61 children in Chendian and measured their blood lead levels (BLLs). As expected, BLLs among Guiyu children were much higher than those in the children of Chendian (p<0.01). Among Guiyu children, 135 (81.8%) had BLLs >10 µg/dL, whereas 23 (37.7%) in Chendian (p<0.01) had high levels. Among 135 (81.8%) Guiyu children with elevated BLLs, 62% and 20% had BLLs >10 µg/dL and 20 µg/dL respectively, but lead levels >45 µg/dL were not found in any children's blood. The BLLs of children working in Guiyu increased somewhat with age (p<0.01); older children tended to have higher BLLs than the younger ones. The same study found no evidence for the association in lead concentrations or prevalence of elevated BLLs differentiated by sex (both at p>0.05). Compared to a study by Luo et al. [87] on children aged 1 to 5-year living in Shantou City (30 km from Guiyu), the average BLL was 7.9 (3.6 µg/dL; approx. 2-time lower than that in Guiyu children. In Guiyu, it is estimated that about 80% of the children suffer from respiratory

diseases [25]. Moreover, there has been a surge in cases of leukemia and high concentrations of lead in blood reported [57]. According to the China Labor Bulletin, e-waste recycling activities have contributed to elevated blood lead levels in children and high incidence of skin damage, headaches, vertigo, nausea, chronic gastritis, and gastric and duodenal ulcers [88]. Due to the existence of numerous e-wastes recycling sites in the Guiyu region is believed to be one of the most heavily chromium-polluted areas in China. Several research studies focused on levels of chromium concentrations among children and its consequent impacts on neonates. Li et al. [89] found that Umbilical cord blood chromium levels (UCBCLs) in neonates from Guiyu city had several folds higher than the normal values. A significant difference in UCBCLs in neonates was observed between Guiyu group and the control group. Based on their study, the authors concluded that, although the UCBCLs of neonates in 2007 had somewhat decreased compared with that of 2006, it was still a serious threat to neonates' health around the e-waste recycling areas [86]. In another study conducted by Xijin et al. [90] a total of 149 children from Guiyu and 146 from Chendian of the subjects enrolled in 2008 completed the questionnaires. The study found that the blood chromium levels in children corresponded to e-waste-related factors, such as occupation of either parent in, using the house as an e-waste facility, and residence adjacent to e-waste workshops. The median blood chromium in children, whose house was used as family workshops was 45.2 µg/L, whereas blood chromium in children's house not as a family workshop was 31.1 µg/L. Similarly, children to born to parents engaged in work related to e-waste recycling and residence adjacent to e-waste workshops, in comparison with those not related or adjacent to e-waste, had higher blood chromium levels. Besides the above mentioned ones, numerous studies imported negative impacts of heavy metals pollution on residents around the e-waste processing sites and workers. Study by Fu et al., [73] focused on resident dietary intake of heavy metals from rice sources and compared the tolerable daily intakes stipulated by the FAO/WHO standards [91] with the mean estimated daily intakes. Lead intake data was recorded (3.7 µg/day.kg.bw) slightly higher than the FAO prescribed tolerable intake of (3.6 µg/day.kg.bw). Another research conducted by Zheng et al. [92] indicated that daily intake of heavy metals from several food sources (chicken, fish, pork, rice and vegetables), house dust and groundwater. The study observed the

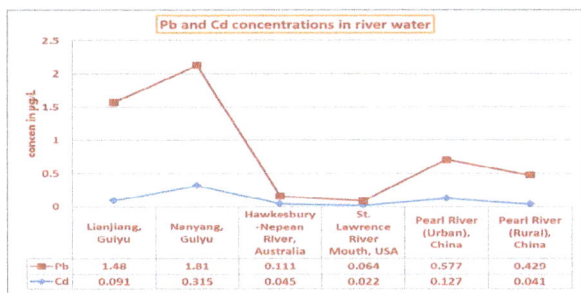

Figure 6: Comparison of Pb and Cd concentration in river waters Data Source: Adapted from source [82].

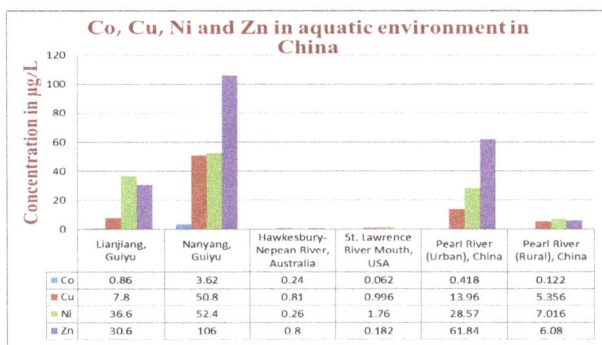

Figure 7: Comparison of Co, Cu, Ni and Zn concentration in river waters Data Source: Adapted from source [82].

higher daily intake, with serious health risks, particularly among the children. While Li et al. [93] evaluated the impacts of heavy metals on neonate's health; their study determined the levels of placental cadmium and cord blood cadmium. Among the 423 mothers included in the study from 2004 to 2007, a sample of 289 participants lived in Guiyu (exposed group) and 134 lived in Chaonan, located 10 km from Guiyu (control), and had never been exposed to e-waste pollution. The median placental cadmium was higher for Guiyu neonates than for the controls (3.61 vs. 1.25µg/L), with 25.6% of Guiyu subjects exhibiting a median cord blood cadmium that exceeded the safety limit defined by the WHO (5 µg/L), as compared with 14.2% of control neonates (p<0.01). In Guiyu, the mean placental cadmium was higher than that for controls (0.17 ± 0.48 vs. 0.10 ± 0.11 µg/g at p<0.05). The high levels of placental cadmium and cord blood cadmium were significantly associated with parents' occupational and environmental exposure to e-waste recycling pollutants. Therefore, elevated chromium levels in neonates were significantly correlated with improper e-waste recycling and disposal and its consequent effects on the environment and human health hazards. In their recent review paper Qingbin and Jinhui [94] reported that although many studies have estimated the potential daily intakes of the heavy metals in e-waste recycling sites, it should be noted that the use of data generated in surveys to estimate dietary exposure, inhalation, soil/dust ingestion and dermal exposure would likely overestimate the actual exposure. There are several other factors that can affect the daily intake. The environmental impacts associated with WEEE have translated directly into a serious public health threat. Many of these threats are already apparent in medical diagnoses and public health research. Some long-term risks may be yet to develop, and will still needs better understanding [83]. A large number of researches [85,89,95-107] have produced significant amount data on human

health impacts focusing on heavy metal levels in blood, impacts of chromium exposure in neonates, effect on chromosomal aberration, BFR and thyroid hormones, elevated body burdens of PBDEs, Dioxin and PCBs, lung of children, etc. A greater part of the research attempted to establish a likely link between human health impact and the contributing pollutants from e-waste processing practices. Even so, there are concerns among researchers and policy makers to establish the direct coherent link between data and health impacts. A group of biomedical researchers published review paper in the Lancet Global Health in 2013. After rigorous screening, a sample of 165 studies were considered and assessed for eligibility. Among those 23 papers reported associations between exposure to e-waste or waste electrical and electronic equipment and physical health, mental health, neurodevelopment, and learning outcomes. After comprehensive review of those studies, the group made a note of caution that few epidemiological data, weak associations, inconsistent findings across studies, and poor understanding of biological mechanisms preclude the establishment of a causal relation between exposure to e-waste and adverse health outcomes in the assessment of evidence by conventional epidemiological approaches. However, the widespread production and use of electronic and electrical equipment, the increasing contamination of the environment, and the persistence and bioaccumulation of these chemical components and their residuals warrant special consideration that e-waste is an emerging public health concern. Evidence suggests that WEEE is significantly increasing incidences of physical injuries and chronic disease, threatening not just workers but also current residents and future generations [37]. These opinions clearly imitate others views that there is a greater need to pursue cohesive and integrated research to establish a clear link between exposures in e-waste processing and human health impact.

Discussion

Challenges and opportunities ahead

In future, managing e-waste in a sustainable approach embodies a wide variety of challenges as well as opportunities for major stakeholders e.g., consumers, businesses and national governments. However, achieving sustainability goal is not a smooth horizontal path of transition but possible if pragmatic strategies entails awareness campaign, availability of technical tools and training programs, and regional and global cooperation of all actors. There is a greater and urgent need for political pledge for integration of existing fragmental approaches through committed financial mechanism. As stated in ILO report that, e-waste is a significant cross-cutting issue with global significance, and it therefore requires a cross-sectoral implementation. Many stakeholders are involved, including industry players, governments, customs authorities, regulatory agencies, intergovernmental organizations, non-governmental organizations and civil societies. What is needed is a range of interventions, international cooperation and goal-oriented actions on e-waste [83].

Recently, the "2013 Geneva Declaration on E-Waste and Children's Health" was published to raise awareness of human health risks by exposures to e-waste. Due to the broad scope and inherent global nature of these issues, appropriate solutions are challenging to find [108]. In contrast, Schluep et al. [24] acknowledged that although the current data presented are alarming, the situation could be improved rapidly by the implementation of more benign recycling techniques and the development and enforcement of WEEE-related legislation at the national level, including prevention of unregulated WEEE exports from industrialized countries. Several national, regional and global schemes are in place to keep in check the challenges of e-waste. Fewer options offer a step forward towards sustainable solution e.g., Extended Producer Responsibility (EPR), Waste Take back, efforts to control illegal trafficking and StEp program to better coordinate global management of e-waste.

EPR/E-waste takeback

Extended Producer Responsibility (EPR), as a principle, emerged in academic circles in the early 1990s. It is generally seen as a policy principle that requires manufacturers to accept responsibility for all stages in a product's lifecycle, including EoL management. There are three primary objectives of the EPR principle: Manufacturers shall be incentivized to improve the environmental design of their products and the environmental performance of supplying those products; Products should achieve a high utilization rate; and Materials should be preserved through effective and environmentally-sound collection, treatment, reuse, and recycling. Often EPR is narrowly defined as to be almost a synonym with a mandatory take-back system or some sort of financial responsibility. The establishment of feedback loops from the downstream EoL management into the upstream design phase is at the core of the EPR principle, and is what can distinguish EPR policies from the implementation of a mere take-back system [29]. Gregory et al. [109], proposed e-waste take-back system as means of solution, whose main functions are collection, processing, system management, and financing scheme. Meanwhile, several examples of current system models have been presented in the United States (such as California, Maine, and Minnesota), and Belgium, France, and Germany, in the EU. Even though some successful stories of e-waste take-back system currently exist, but several challenges still remain unresolved including (i) how to balance the harmonization between manufacturers and recyclers with respect to finance, operations, technologies, market, and so forth?, (ii) how to deal with different business models of stakeholders from various industries?, (iii) how to determine the number of policy in law, leaving others to be industrial standards?, and (iv) how to ensure that obligations are met by the stakeholders? [26]. Up until now, few developed countries took initiatives to embrace EPR for e-waste management with some success, while challenges still remains to achieve complete realization of EPR principles. Over 25 nations have some form of EPR program. EPR is most commonly applied for packaging waste, the most famous being Germany's packaging ordinance. However, batteries, electronic and electrical appliances and automobiles are also increasingly scrutinized under EPR programs. Electronic and electrical products are a major focus of EPR policies around the world [110] especially in Europe, where over the last couple of years, several countries have favored an EPR based e-waste policy. One of the holistic EPR strategies has implemented in Switzerland: over 10 years of success in a close loop system – very little waste into land-fill and 75% of the material was returned to the raw materials cycle. Figure 8 illustrates the circular flow of materials aims to optimize a closed loop material cycle, with the raw materials converted to finished EEE going through the retail and consumption stages and then at the end-of-life being collected and recycled to be put back into the production of new goods [111]. Some rapidly emerging economies, such as China, India and Indonesia have started to develop EPR programs even though these are generally not yet fully implemented and functioning. Malaysia and Thailand are also embarking the path towards EPR for e-waste, although these initiatives generally rely on voluntary participation of producers [112].

Emergence of StEP

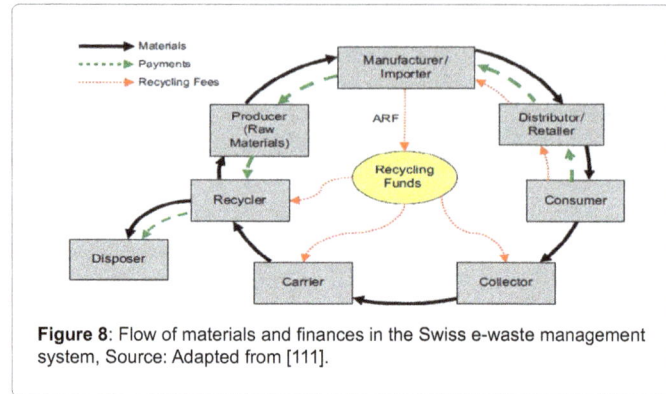

Figure 8: Flow of materials and finances in the Swiss e-waste management system, Source: Adapted from [111].

Another successful effort is the establishment of StEp, which became active in 2007, is coordinated by the UN University, the think-tank and research arm of the UN that hosts the Step Secretariat. Since commencement, the (StEP) initiative in solving the e-waste problem is well positioned to make a difference towards the development and practical implementation of sustainable solutions [6]. One of the pragmatic initiatives of StEp is to bring major stockholders on board e.g., IGO's, industry, governments, NGOs, environmental groups and academia to achieve common goals of developing and implementing e-waste strategies. The shared expertise and common vision of StEp's members focus on seven key areas: Reducing the materials used in manufacturing; Reusing equipment or components when practical; Refurbishing when possible; Recovering materials from obsolete equipment; Recycling the biggest possible level of materials; Developing policy recommendations for sustainable solutions and Administer trainings for key stakeholder groups. StEP's five Task Forces are advancing the e-waste agenda on many fronts. StEp takes a life-cycle approach to the global e-waste dilemma looking at the areas policy, redesign, reuse, recycle and capacity building. For 2014 and 2015, Step members have agreed to work on six projects to remaining Task Force work. Each project is led by two Step members, usually one representative from industry and one non industry actor [6]. Global initiatives have already been enacted at both the voluntary standard and regulatory levels in recognition of the importance in the responsible management of e-waste [113]. Over the past decade, China has made great advances to advocate better e-waste collection and recycling in both public and private sectors. There is a visible increase in domestic and foreign investments into recycling field, accompanied by encouraged transfer of international recycling technologies and western waste management principles. Under the auspices of EU, the UN-StEp program at different steps (e.g., capacity building, policy intervention) are undergoing in several Asian and African countries. At this outset as one of the largest producers and processor of e-waste, China necessitates to fill-up the loophole of their regulatory instruments and be proactive in collaborating with the global community to deal with massive e-waste related environmental issues. A schematic diagram of global cooperation is proposed in Figure 9. Minute details for the suggested global cooperation may be debatable, the contemplations suggested above for the life cycle of electronics has to be improved significantly to avoid an accelerated loss of scarce raw materials, emission of toxics into the environment, and most importantly to protect human health and ecosystem services.

Conclusion

E-waste is omnipresent. It is characterized by its unusual chemical composition and the difficulties associated with determining its mass

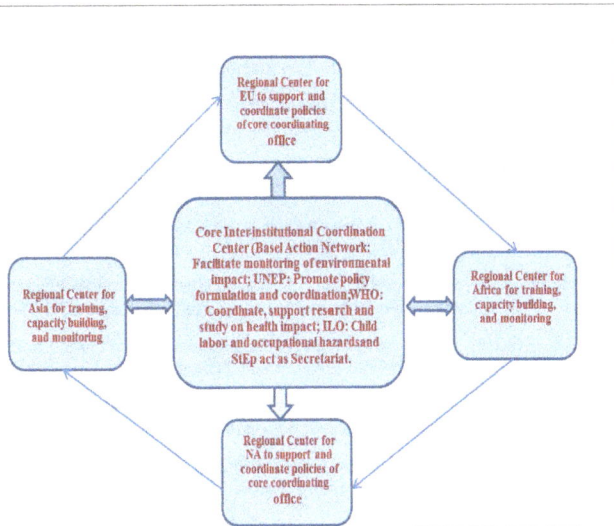

Figure 9: Schematic proposed plan for future collaboration among major stockholders.

and flux at both local and global scales. Contamination associated with e-waste has already caused considerable environmental degradation in the developing countries and adversely affected the health of the people who live in and around e-waste facilities [10]. The quantities of e-waste generated are predicted to grow substantially, both in industrialized countries and in the developing countries. The rich countries often legally or illegally divert this problem from their own backyards to the premises of the developing countries. The hidden flow of e-waste that results environmental damage in the backyards and scrapyards of the developing countries. It is quite perceptible from present and future scenarios of e-waste generation and hidden flow in the developing countries, mostly Asian and African continents will be adversely affected. There may be a shift of hub for informal recycling and processing from Guiyu, China to Ghana or Nigeria, however, challenges with impact on human health and ecosystem services will continue to exacerbate with hidden flow and improper recycling practices. One of the major challenges in achieving sustainable of e-waste management is to organize integration of formal and informal generation and process and establish better understanding on human health impact. As stated, some of the strategies and policy options are imparting piecemeal positive impacts but there are greater need to consolidation and integration. As observed in StEp green report that while the promotion and capacity-building of effective and efficient take-back solutions for End of Life (EoL) material is vital, the paper noted that there should also be a concerted effort to focus more on the reduction of e-waste volumes and the repair/reuse of EEE. Although several initiatives have been identified, especially by not-for-profits and the informal sector, there are few examples of public policy initiatives that have been successful [29]. It seems that there are enough missing links on policy initiatives among producers and recipient countries. Unless collaborative efforts are appropriated to consolidate cooperation among the major stockholders, sustainable e-waste management will not achieve complete success. To cultivate sustainable e-waste management practices: intensive awareness campaign; capacity building to prevent illegal trafficking; and proper technical training of formal e-waste processing practices must be ensured through collaborative approach. Poverty plays a crucial role in the exponential growth of informal sector. Also to prevent further growth of informal economy in e-waste management an economic tool to support growth of processed e-waste

material market must be established.

References

1. UNEP (2007) E-waste Inventory Assessment Manual volume I.

2. UNEP DEWA/GRID-Europe (2005) E-waste, the hidden side of IT equipment's manufacturing and use, in: Early Warning on Emerging Environmental Threats, (Chapter 5).

3. UNU, (2013) Solve the E-waste Problem (StEP), Massachusetts Institute of Technology (MIT), National Center for Electronics Recycling (NCER). World e-waste map reveals national volumes, international flows.

4. EPA 530-S-11-001, May (2011) "Electronics Waste Management in the United States.

5. Terazona A, Murakami S, Abe N, Inanc B, Moriguchi Y, et al. (2006), Current status and research on E-waste issues in Asia, J. Mater. Cycles Waste Management 8: 1-12.

6. StEP initiative, StEP Annual Report (2013/14).

7. Li J, Tian B, Liu T, Liu H, Wen X, et al. (2006) Status quo of e-waste management in mainland China, J. Mater. Cycles Waste Management 8: 13-20.

8. Ladou J, Lovegrove S (2008) Export of electronics equipment waste. Int J Occup Environ Health 14: 1–10.

9. Diaz-Barriga F (2013) Evidence-based intervention programs to reduce children's exposure to chemicals in e-waste sites. Discussion paper for WHO Working Meeting on e-waste and children's health. 1-90

10. B. Robinson (2009) E-waste: An assessment of global production and environmental impacts, Science of the Total Environment 408: 183-191.

11. UNEP (2009) Basel Convention on the Control of Transboundary Movements of Hazardous Wastes and their Disposal. United Nations Environment Programme.

12. Schneider F, Enste DH (2003) The Shadow Economy, an international survey. Cambridge: Cambridge University Press, UK.

13. CBC News, (2007) "How much e-waste do you dump? Cheryne's Diary".

14. Lundgren K (2012) The global impact of e-waste: addressing the challenge, International Labour Office (ILO), Geneva.

15. UNEP (2006) Call for Global Action on E-waste. United Nations Environment Programme.

16. Cobbing M (2008) Toxic tech: not in our backyard, uncovering the hidden flows of e-waste, Greenpeace International, Amsterdam.

17. UNU (2007) Review of Directive 2002/96 on Waste Electrical and Electronic Equipment (WEEE). Final Report to European Commission, Bonn.

18. Duffert C, Brune MN, Prout K Background document on exposures to e-waste. World Health Organization (WHO) Geneva, Switzerland.

19. Global E-Waste Management Market (2011-2016), Report Code: SE 1588, Retrieved from: marketsandmarkets.com, Publishing Date: August 2011.

20. Gagliardi D, Mirabile M. Overview of OSH issues related to the e-waste management: World Health Organization (WHO), Geneva, Switzerland.

21. Deng W J, Louie P, Liu W, Bi X, Fu J et al. (2006) Atmospheric levels and cytotoxicity of PAHs and heavy metals in TSP and PM2.5 at an electronic waste recycling site in southeast China Atmos. Environ. 40: 6945–55.

22. Macauley M, Palmer K, Shih JS (2003) Dealing with electronic waste: modeling the costs and environmental benefits of computer monitor disposal. Journal of Environmental Management 68: 13–22.

23. Brigden K, Labunska I, Santillo D, Allsopp M (2005) Greenpeace Research Laboratories, Department of Biological Sciences, University of Exeter, Exeter EX4 4PS, UK, Report on recycling of electronic wastes in China and India: Workplace & Environmental Contamination.

24. Schluep M, Hageluken C, Kuehr R, Magalini F, Maurer C, et al. (2009) Recycling – From E-waste to Resources, Report prepared by Swiss Federal Laboratories for Materials Testing and Research, Umicore Precious Metal Refining and the United Nations University.

25. Sepúlveda A, Schluep M, Renaud FG, Streicher M, Kuehr R (2010) A review of the environmental fate and effects of hazardous substances released from

electrical and electronic equipments during recycling: Examples from China and India, Environmental Impact Assessment Review 30:28–41.

26. Khurrum M, Bhutta Z, Omar A, Yang X (2011) Electronic Waste: A Growing Concern in Today's Environment. Economics Research International 11: 1-8.

27. Consumer Electronic Association [R] (2012) Corporate Report.

28. Global E-Waste Management Market - Size, Industry Analysis, Trends, Opportunities, Growth and Forecast, 2013-2020. Allied Market Research, Published: Jan-2015

29. Solving the E-Waste Problem (2015) Green Paper E-waste Prevention, Take-back System Design and Policy Approaches, ISSN: 2219-6579.

30. Lu C, Lin Z, Yongguang Z, Wanxia R, Mario (2015) An overview of e-waste management in China. Journal of Material Cycles and Waste Management 17: 1-12.

31. Electronic Waste, http://www.newworldencyclopedia.org/entry/Electronic_waste, (Cited in news article in world encyclopedia and retrieved on January 1, 2015)

32. Greenpeace International (2009) where dos e-waste end up?

33. Kuper J, Hojsik M (2008) Poisoning the Poor Electronic Waste in Gahana, Greenpeace International, Amsterdam, Netherlands.

34. Chan JKY, Wong MH (2012) A review of environmental fate, body burdens, and human health risk assessment of PCDD/Fs at two typical electronic waste recycling sites in China. Science of Total Environment 463-464:1111-23.

35. Schmidt C (2002) e-Junk explosion Environ. Health Perspect. 110:A188–A194.

36. Schmidt CW (2006) Unfair trade E-waste in Africa, Environ. Health Perspect. 114: A232–A235.

37. Kristen G, Fiona C G, Peter D, Marie B, Maria N (2013) Health consequences of exposure to e-waste: a systematic review. Lancet Glob Health 1: 350-361.

38. Gu Z, Feng J, Han W, Wua W, Fu J (2010) Characteristics of organic matter in PM$_{2.5}$ from an e-waste dismantling area in Taizhou, China, Chemosphere 80:800–806.

39. Wilson DC, Velis C, Cheeseman C (2006) Role of informal sector recycling in waste management in developing countries. Habitat International 30: 797-808.

40. Linda L (2010), Managing Electronic Waste: Issues with Exporting E-Waste, CRS Report for Congress, Prepared for Members and Committees of US Congress.

41. Electronic Take back Coalition (2012) E Waste Facts and Figures.

42. Electronic Take back Coalition (2014) E-Waste Facts and Figures.

43. EPA (2014) data from "Municipal Solid Waste Generation, Recycling and Disposal in the United States, 2012.

44. Robert Tonetti, (October 2007) EPA Office of Solid Waste, presentation materials, "EPA's Regulatory Program for E-Waste,".

45. European Union. EU Directive 2002/95/EC of the European Parliament and of the Council of 27 January 2003 on the restriction of the use of certain hazardous substances in electrical and electronic equipment.

46. National People Congress (2006) Administrative measures on the control of pollution caused by electronic information products.

47. Basel Convention, Decisions adopted by the Conference of the Parties to the Basel Convention on the Control of Trans-boundary Movements of Hazardous Wastes and their Disposal : COP 6 - 10 and ExCOPs , Compiled in 2012.

48. UNEP/CHW/OEWG/6/21, (2007) Open-ended Working Group of the Basel Convention on the Control of Trans-boundary Movements of Hazardous Wastes and Their Disposal Sixth session Geneva, 3–7 September 2007.

49. Exporting harm: the high-tech trashing of Asia (2002) Prepared by BAN and SVTC – The Basel Action Network and Silicon Valley Toxics Coalition.

50. GAO (2008), EPA Needs to Better Control Harmful U.S. Exports through Stronger Enforcement and More Comprehensive Regulation, GAO-08-1044.

51. House Committee on Foreign Affairs, Subcommittee on Asia, the Pacific and the Global Environment, "Exporting Toxic Trash: Are We Dumping Our Electronic Waste on Poorer Countries?" September 17, (2008).

52. UNEP, GSA and White House Council on Environmental Quality, (2014)

Interagency Task on Electronics Stewardship.

53. Computer Aid International (2010) WEEE Ver. 2.0: What Europe must do, Special Report Series: ICT and the Environment, Report 2.

54. Final Report (2014) ITU-D Study group 1, Strategies and policies for the proper disposal or reuse of telecommunication/ICT waste material.

55. Lin Weia, Yangsheng Liub (2012) Present status of e-waste disposal and recycling in China, Procedia Environmental Sciences 16: 506-514.

56. The State Council (2004) The measures for the Administration of Permit for Operation of Dangerous Wastes [Z].

57. Tsydenova O, Bengtsson M (2011) Chemical hazards associated with treatment of waste electrical and electronic equipment. Waste Management 31: 45-58.

58. EFFACE (2015), A case study on illegal e-waste export from the EU to China, European Union Action to Fight Environmental Crime.

59. INTERPOL, May (2009) Electronic Waste and Organized Crime: Assessing the link, Phase II report for the pollution crime working group.

60. INTERPOL, (2010) "International experts outline global strategy to tackle e-waste threat at INTERPOL meeting." Media Release 25 May.

61. Ongondo F O, Williams I D, Cherrett T J (2011) How are WEEE doing? A global review of the management of electrical and electronic wastes, Waste Management 31: 714-30.

62. Environmental Investigation Agency (2011) System failure: The UK's harmful trade in electronic waste.

63. UNDOC (2013) Transitional Organized Crime Threat Assessment- Asia and the Pacific, Chapter 9 Illicit trade in electrical and electronic waste (e-waste) from the world to the region.

64. UNEP (2015) A Rapid Response Assessment: Waste Crime Waste Risks Gaps in Meeting the Global Waste Challenge, ISBN: 978-82-7701-148-6.

65. The Times, "Britain's dirty little secret as a dumper of toxic waste", July 18, 2009 (Cited in SYSTEM FAILURE The UK's harmful trade in electronic waste, ENVIRONMENTAL INVESTIGATION AGENCY (EIA), 2011 Report).

66. INTERPOL (2013) INTERPOL operation targets illegal trade of e-waste in Europe, Africa. [Online] 25/02/2013.

67. Strategic Approach to International Chemicals Management (SAICM) (2009) Background information in relation to the emerging policy issue of electronic waste, paper presented at the International Conference on Chemicals Management, Geneva, 11–15 May (SAICM/ICCM.2/INF36).

68. Jing Yuan, Lan Chen, Duohong Chen, Huan Guo, Xinhui B, et al. (2008) Elevated serum polybrominated diphenyl ethers and thyroid-stimulation hormone associated with lymphocytic micronuclei in Chinese workers from an e-waste dismantling site Environ Sci Technol 42: 2195-2200.

69. Wen S, Yang F, Gong Y, ZhangX, Hui Y, Li J, et al. (2008) Elevated levels of urinary 8-hydroxy-2'-deoxyguanosine in male electrical and electronic equipment dismantling workers exposed to high concentrations of polychlorinated dibenzo- p-dioxins and dibenzofurans, polybrominated diphenyl ethers and polychlorinated biphenyls Environ. Sci. Technol 42: 4202-4207.

70. Qiang Liu, Jia Cao, Ke Qiu Li, Xu Hong Miao, Guang Li, et al. (2009) "Chromosomal aberrations and DNA damage in human populations exposed to the processing of electronics waste" Environmental Science and Pollution Research 16: 329-338.

71. Bina Rani, Upma Singh, Raaz Maheshwari, A K Chauhan (2012) Perils of Electronic Waste: Issues And Management Strategies, Journal of Advanced Scientific Research 3: 17-21.

72. GAO (2005) Electronic Waste – Observation on the Role of the Federal Government in Encouraging Recycling and Reuse.

73. Fu J, Zhou Q, Liu J, Liu W, Wang T, et al. (2008) High levels of heavy metals in rice (Oryza sativa L.) from a typical E-waste recycling area in southeast China and its potential risk to human health Chemosphere 71: 1269-75.

74. Wong M, Wu S, Deng W, Yu X, Luo Q, et al. (2007) Export of toxic chemicals: a review of the case of uncontrolled electronic-waste recycling, Environ. Pollution 149: 131-40.

75. Chen D, Bi X, Zhao J, Chen L, Tan J, et al. (2009) Pollution characterization and diurnal variation of PBDEs in the atmosphere of an E-waste dismantling region Environ. Pollution 157: 1051-1057.

76. BASEL ACTION NETWORK (2002) Exporting Harm: The High-Tech Trashing of Asia (Seattle, WA: Basel Action Network).

77. Chi X, Streicher-Porte M, Wang MYL, Reuter MA (2011) Informal electronic waste recycling: a sector review with special focus on China. WHO, The Geneva declaration on e-waste and children's health. Waste Management 31: 731-741.

78. Markich SJ, Brown PL (1998) Relative importance of natural and anthropogenic influences on the fresh surface water chemistry of the Hawkesbury- Nepean River, South-Eastern Australia. The Science of the Total Environment 217: 201-230.

79. Gobeil C, Rondeau B, Beaudin L, (2005) Contribution of municipal effluents to metal fluxes in the St. Lawrence River. Environmental Science and Technology 39: 456-464.

80. Ouyang TP, Zhu ZY, Kuang YQ, Huang NS, Tan JJ, et al. (2006) Dissolved trace elements in river water: spatial distribution and the influencing factor, a study for the Pearl River Delta Economic Zone, China. Environmental Geology 49: 733-742.

81. Coby SC, Wong, Nurdan S.,Duzgoren-Aydin, Adnan Aydin, Ming Hung Wong (2007) Evidence of excessive releases of metals from primitive e-waste processing in Guiyu, China, Environmental Pollution 148: 62-72.

82. ILO (2012) Report on The global impact of e-waste: Addressing the challenge, edited by Karin Lundgren, (Accessed on 12 March 2015).

83. Cuadra SN (2005) Child labour and health hazards: Chemical exposure and occupational injuries in Nicaraguan children working in a waste disposal site. Licentiate Thesis. Lund, Lund University, Faculty of Medicine.

84. Huo X, Peng L, Xu X J, Zheng L K, Qiu B, et al. (2007) Elevated blood lead levels of children in Guiyu, an electronic waste recycling town in China. Environmental Health Perspective 115: 1113-1117.

85. Bi XH, Thomas GO, Jones KC, Qu WY, Sheng GY, et al. (2007) Exposure of electronics dismantling workers to polybrominatetd diphenyl ethers, polychlorinated biphenyls, and organochlorine pesticides in South China. Environ. Sci Technol 41: 5647-5653.

86. China Labour Bulletin (2005) "The plight of China's e-waste workers", 15 Aug.

87. Li Y, Xu X, Liu J, Wu K, Gu C, et al. (2008) The hazard of chromium exposure to neonates in Guiyu of China. Science of Total Environment 403: 99-104.

88. Xijin Xu, Taofeek Akangbe Yekeen, Junxiao Liu, Bingrong Zhuang, Weiqiu Li, et al. (2015) Chromium exposure among children from an electronic waste recycling town of China, Environ Science Pollution Research 22: 1778-1785.

89. World Health Organization (2011) Evaluation of Certain Food Additives and Contaminants: Seventy-fourth Report of the Joint FAO/WHO Expert Committee on Food Additives.

90. Zheng J, Chen KH, Yan X, Chen SJ, Hu GC, et al. (2013) Heavy metals in food, house dust, and water from an e-waste recycling area in South China and the potential risk to human health. Ecotoxicol Environ Saf 96: 205-212.

91. Li Y, Huo X, Liu J, Peng L, Li W, et al. (2011) Assessment of cadmium exposure for neonates in Guiyu, an electronic waste pollution site of China. Environ Monitoring Assessment 177: 343-351.

92. Qingbin Song and Jinhui Li (2015) A review on human health consequences of metals exposure to e-waste in China. Environmental Pollution 196: 450-461.

93. Guo Y, Huo X, Li Y, Wu K, Liu J, Huang J, et al. (2010) Monitoring of lead, cadmium, chromium and nickel in placenta from an e-waste recycling town in China. Sci Total Environ 408: 3113-3117.

94. Guo Y, Huo X, Wu K, Liu J, Zhang Y, et al. (2012) Carcinogenic polycyclic aromatic hydrocarbons in umbilical cord blood of human neonates from Guiyu, China, Science of Total Environment 427-28: 35-40.

95. Han G, Ding G, Lou X, Wang X, Han J, et al. (2011) Correlations of PCBs, dioxin, and PBDE with TSH in children's blood in areas of computer e-waste recycling. Biomed Environ Science 24: 112-116.

96. Liu Q, Cao J, Li KQ, Miao XH, Li G, et al. (2009) Chromosomal aberrations and DNA damage in human populations exposed to the processing of electronics waste, Environ Science Pollution Research International 16: 329-338.

97. Liu J, Xu X, Wu K, Piao Z, Huang J, et al. (2011) Association between lead exposure from electronic waste recycling and child temperament alterations, Neurotoxicology 32: 458-464

98. Li Y, Xu X, Wu K, Chen G, Liu J, et al. (2008) Monitoring of lead load and its effect on neonatal behavioral neurological assessment scores in Guiyu, an electronic waste recycling town in China. J Environ Monitoring 10: 1233-1238.

99. Wang H, Zhang Y, Liu Q, Wang F, Nie J, et al. (2010) Examining the relationship between brominated flame retardants (BFR) exposure and changes of thyroid hormone levels around e-waste dismantling sites. Int J Hyg Environ Health 213: 369-380.

100. Wu K, Xu X, Liu J, Guo Y, Li Y, et al. (2010) Polybrominated diphenyl ethers in umbilical cord blood and relevant factors in neonates from Guiyu, China. Environ Sci Technol 44: 813-819.

101. Wu K, Xu X, Peng, Liua J, Guo Y, Huo X (2012) Association between maternal exposure to perfluorooctanoic acid (PFOA) from electronic waste recycling and neonatal health outcomes.Environ Int, 48: 1-8.

102. Xu X, Yang H, Chen A, Zhou Y, Wu K, et al. (2012) Birth outcomes related to informal e-waste recycling in Guiyu, China. Reprod Toxicol 33: 94-98.

103. Yuan J, Chen L, Chen D, Guo H, Bi X, et al. (2008) Elevated serum polybrominated diphenyl ethers and thyroid stimulating hormone associated with lymphocytic micronuclei in Chinese workers from an E-waste dismantling site. Environ Sci Technol 42: 2195-2200.

104. Zhang J, Jiang Y, Zhou J, Wu B, Liang Y, et al. (2010) Elevated body burdens of PBDEs, dioxins, and PCBs on thyroid hormone homeostasis at an electronic waste recycling site in China. Environ Sci Technol 44: 3956-3962.

105. Zheng G, Xu X, Li B, Wu K, Yekeen TA, et al. (2013) Association between lung function in school children and exposure to three transition metals from an e-waste recycling area. J Expo Sci Environ Epidemiol 23: 67-72.

106. Alabaster G, Ansong K, Bergman A, Birnbaum L, Noel M, et al. (2013) The Geneva Declaration on E-waste and Children's Health. World Health Organization.

107. Gregory F, Magalini R, Kuehr K, Huisman J (2009) "E-wastetake-back system design and policy approaches," Solving the e-Waste Problem (StEP), White Paper.

108. Fishbein B, (1998). EPR what does it mean? Where is it headed? Pollution Prevention Review 8(4): 43-55.

109. Khetriwal DS, Kraeuchi P, Widmer R (2009) Producer Responsibility for E-Waste Management: Key Issues for Consideration - Learning from the Swiss Experience Journal of Environmental Management 90: 153-65.

110. The State of Play on Extended Producer Responsibility (2014) Opportunities and Challenges, Issue Paper, Global Forum on Environment: Promoting Sustainable Materials Management through Extended Producer Responsibility (EPR), 17-19, Tokyo, Japan.

111. Solving the E-Waste Problem (2014) White Paper, Recommendations for Standards Development for Collection, Storage, Transport and Treatment of E-waste, June 2014.

112. Shinkuma T, Huong NTM (2009) The flow of e-waste material in the Asian region and a reconsideration of international trade policies on e-waste. Environmental Impact Assessment Review 29: 25-31.

113. Green Peace, Toxic Tech: Not in Our Backyard Uncovering the Hidden Flows of e-Waste, 2008.

31

Degradation of C.I. Reactive Dyes (Yellow 17 and Blue 4) by Electrooxidation

Vahidhabanu S and Ramesh Babu B*

CSIR- Central Electrochemical Research Institute, Karaikudi- 630 003, India

Abstract

Dyes are extensively used in textile industries which are responsible for different environmental problems. Conventional processes for effluent treatment are inefficient for the remediation of wastewaters containing toxic and bio recalcitrant organic pollutants. A large number of advanced oxidation processes (AOP's) have been successfully applied to degrade pollutants present in wastewaters. This paper examines the use of electro oxidation (EO) process for the recalcitrant of the textile dye effluent. The dye effluent containing C-I Reactive Yellow 17 and Blue 4 was treated using Ti/ RuO2 and stainless steel electrodes. The experimental study focused on the effect of supporting electrolytes such as NaCl and Na2SO4. The degradation process was enhanced appreciably by increasing the concentration of supporting electrolytes at optimum current density of 7 A/dm2 (pH=11). Efficiencies of COD reduction, color removal were also determined. It was established that NaCl was superior to Na2SO4 in terms of COD reduction as well as decolorization. The degradation was characterized by HPLC, FTIR and UV-Vis spectral analysis.

Keywords: Dye; Recalcitrant; HPLC

Introduction

Over 100,000 commercially available dyes exist and more than 7 $\times 10^5$ metric tonnes of dyes are produced worldwide annually [1]. The treatment of wastewater generated by the textile preparation, dyeing and finishing industry remains as a significant environmental pollution problem due to its huge quantity, variable nature and biologically difficult to degrade chemical composition. The main characteristics of effluent from the reactive dyeing process deserve particular attention due the relatively low dye fixation rate, high organic and inorganic content and high alkalinity [2]. Various physical and chemical processes are available for the treatment of textile wastewater, such as membrane filtration, coagulation- flocculation and sequential anaerobic and aerobic treatment, have been employed so far, however with limited success and or at unaffordable costs [3]. Now a days, advanced oxidation processes (AOP's) are available for the treatment of waste waters, like electro oxidation, wet oxidation, ozonization, photo catalytic degradation etc., [4,5]. Treatment using these processes is based on the production of hydroxyl radicals, highly reactive species, which promote oxidation of hazardous organic compounds.

Electrochemical technique offers high removal efficiencies and has lower temperature requirements compared to non-electrochemical treatment. In addition to the operating parameters, the rate of pollutant degradation depends on the anode material. When electrochemical reactors operate at high cell potential, the anodic process occurs in the potential region of water discharge, hydroxyl radicals are generated [6]. On the other hand, if chloride is present in the electrolyte, an indirect oxidation via active chlorine can be operatived [7-12], Naumczyk et al., [13] have demonstrated several anode materials, such as graphite and noble metal anodes successfully for the mediated oxidation of organic pollutants. In textile industries, synthetic dyes from residual dye baths are released to waste streams. It is estimated that up to 50% of the applied dye, depending on the type, can be lost in effluents during textile dyeing processes. Azo dyes, characterized by nitrogen to nitrogen double bonds (–N=N–), account for up to 70% of all textile dyes produced, and are the most common chromospheres in reactive dyes. Due to the characteristics of colored wastewaters containing reactive azo dyes, their treatment is rather difficult, especially by the non-electrochemical wastewater treatment methods based on adsorption and biodegradation. In order to assess the economic feasibility of EO to decolorize and at least partially oxidize reactive dye bath effluent, an evaluation of conception of electrical energy per volume of treated wastewater (KWh/l) was also conducted. During the dyeing process with reactive dyes, the addition of high concentrations of an electrolyte is necessary to obtain a better fixation and exhaustion. Generally an amount of 50- 80 g of litter of a salt is added as electrolyte, being NaCl or Na_2SO_4 the most common. Anastasios Sakalis et al., have also studied the use of different supporting electrolytes for the treatment of azo dyes by electrochemical method [14]. The main objective of the current work is to study the effect of concentration of supporting electrolytes for degradation of the dyes and optimize the operating conditions for economic feasibility.

Theory

Basically two different processes occur at the anode: 1. direct electrolysis and 2. indirect electrolysis. In direct electrolysis, anode has high electro-catalytic activity and oxidation occurs at the electrode surface. In indirect electrolysis, oxidation occurs via surface mediator on the anodic surface, where the oxidation occurs continuously. In indirect EO, chloride salts of sodium or potassium are added to the wastewater for better conductivity and generation of hypochlorite ions [15]. The reactions of anodic oxidation of chloride ions to form chlorine are given as below:

Anodic reactions

Indirect oxidation at anode surface

The generation of hypochlorous acid form chlorine (Eq. 1)

***Corresponding author:** Ramesh Babu B, CSIR- Central Electrochemical Research Institute, Karaikudi- 630 003, India
E-mail: brbabu2011@gmail.com

$$cl_2 + H_2O \rightarrow H^+ + Cl^- + HOCl$$

The generation and dissociation of hypochlorite ion (Eq. 2)

$$HOCl \longleftrightarrow H^+ + OCl^-$$

The hypochlorite ions act as main oxidizing agent for the degradation of dye effluents or other organic pollutants.

Direct oxidation at anode surface: A general scheme of electrochemical conversion of organic pollutants on RuO_2 coated Ti anode material is given below [16]: In first step, H_2O is discharged at anode to produce the hydroxyl radical, which get adsorbed on metal surface and form a complex $(RuO_2('OH))$.

The dissociation of H_2O, adsorptions of hydroxyl radical into RuO_2 (Eq. 3)

$$RuO_2 + H_2O \rightarrow RuO_2('OH) + H^+ + e^-$$

In the presence of organic pollutants, the active oxygen from hydroxyl radicals may involve in completely combustion of organics pollutants (Eq.4), and the active oxygen from higher oxygen radical can be involved in selective oxidation process and give respective products from organic pollutants (Eq.5) [17].

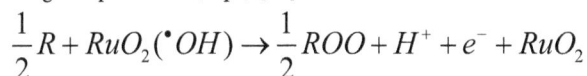

$$\frac{1}{2}R + RuO_2('OH) \rightarrow \frac{1}{2}ROO + H^+ + e^- + RuO_2$$

$$R + RuO_2('O) \rightarrow RO + RuO_2$$

In the presence of NaCl as supporting electrolyte, the chloride ion can also interact with active hydroxyl radical and give a hypochloride ion (Eq.6)

$$RuO_2('OH) + Cl^- \rightarrow RuO_2('OCl) + H^+ + 2e^-$$

The hydroxyl radical is also involved in the oxidation process of organic pollutants [18-20]. This reaction is a chain reaction, and continued until the formation of carbon dioxide and water (Eq.7-9).

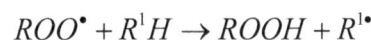

$$RH + ('OH) \rightarrow R^\bullet + H_2O$$

$$R^\bullet + O_2 \rightarrow ROO^\bullet$$

$$ROO^\bullet + R^1H \rightarrow ROOH + R^{1\bullet}$$

In the presence of Na_2SO_4 as supporting electrolyte, during the electrolysis, SO_2 is formed; it is a moderate reductant [14]. The degradation of organic pollutant or dye in the presence of Na_2SO_4 proceeds via direct redox reactions on the electrodes.

Cathodic reaction: Cathodic reaction is given in Eq.(10):

$$2H_2O + 2e^- \rightarrow H_2 + 2OH^-$$

The role of hypochlorite in electrochemical treatment of dye effluent via chlorine generation (Eq.11) is given below:

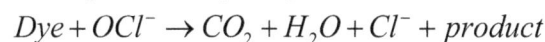

$$Dye + OCl^- \rightarrow CO_2 + H_2O + Cl^- + product$$

Materials and Methods

Materials

Reactive dye effluent containing C.I. Yellow 17 and Blue 4 was collected from a textile industry located in Tirupur, Tamilnadu, India. The structure of the dyes is given in Table 1. Commercially available Ti/RuO_2 and stainless steel were used as anode and cathode respectively.

S.No	Name of dye	Molecular structure
1	C.I. Reactive Yellow17	
2	C.I. Reactive Blue 4	

Table 1: Molecular structure of reactive dyes used in this study.

Two supporting electrolytes NaCl and Na_2SO_4 were used in this study. All the chemicals and reagents (analytical grade) were purchased from Merck and used as such.

Apparatus

The schematic diagram of experimental setup is shown in Figure 1. A 600 ml capacity glass beaker was used as a reservoir. Effluent was stored in the reservoir and it was brought into effect by batch system. A DC power supply was used. The main component of the experimental setup is electrochemical reactor. It consists of stainless steel cathode of size (7 cm x 5 cm x 0.2 cm) and an anode which is RuO_2 coated Ti mesh were used. Anode is placed as close as to cathode and the same was fixed rigidly on a PVC lid with the help of araldite. A magnetic stirrer was used for stirring the solution. Necessary provisions were made in the lid for sampling during the process.

Methods: A known quantity of effluent 400 ml was taken in each experiment. Either NaCl or Na_2SO_4 was added to the effluent as supporting electrolyte of various concentrations 1.5, 2.25 and 3.0 g/l at different applied current density (2, 5 and 7 A/dm^2). The pH of the effluent was adjusted to pH 11 by using 1 N NaOH and it was measured by using pH meter (Dot 491). Effluent was continuously stirred during the treatment using a magnetic stirrer. Experiments were carried out under batch conditions for maximum of 5 h. COD and color removal were determined periodically to know the extent of degradation of the effluent.

Extraction and HPLC analysis: Untreated and treated effluents were analyzed by the HPLC. Experiments were carried out on Shimadzu Instrument with UV detector at 254 nm. Each 5 ml of untreated and treated effluent was taken, centrifuged and filtered through a 0.45 μm membrane filter (Millipore make). The filtrate was then extracted by ethyl acetate. The extracted was dried over anhydrous Na_2SO_4 and evaporated to dryness in rotary evaporator. The obtained solid mass was dissolved in methanol and was analyzed by HPLC. A 20 μl sample was injected in to Octa Decyl Sinane (ODS -C18) column (4.6 mm ID x 250 mm length). Methanol of purity 100% was used as mobile phase with the flow rate of 1 ml/min for 10 minutes and UV detector was kept at 254 nm.

Extraction and FTIR analysis: Samples for FTIR analysis were prepared as mentioned in the HPLC analysis. Experiments were carried out on Thermo Nicolet Nexus 670 FT-IR spectrophotometer. The selected IR region was between 400 and 4000 cm^{-1}. To prepare pellets, the samples were mixed with spectroscopically pure KBr.

UV–Vis analysis: UV-Vis experiments were carried out using Varian UV-Vis-NIR-500 spectrophotometer. Untreated and treated effluents were analyzed by the UV – Vis spectrophotometer. Spectrophotometer was used together with a cell with a 1 cm optical path length to measure the UV spectra.

Analysis of COD

Analytical method of the COD of all samples was determined by the dichromate closed reflux method using Thermo reactor TR620-Merck. COD is generally considered as the oxygen equivalent to the amount of organic matter oxidizable by potassium dichromate. The organic matter of the sample is oxidized with a known excess of potassium dichromate in a 50% sulfuric acid solution. The excess dichromate is titrated with a standard solution of ferrous ammonium sulfate solution [21].

Determination of color

The selection of suitable wavelength in the spectrum can be made during the course of preparing of the calibration curve for the unknown samples. The particular wavelength which provides a maximum absorbance value will be considered as the best choice of wavelength. Reactive dye solutions show maximum absorbance at a wavelength of 430 nm. The UV–Vis spectra of the effluent was measured by using the spectrophotometer Spectroquant NOVA 60 at λ_{max} = 430 nm.

Color removal was calculated by using the following formula (1) [22]:

$$\text{Color Removal}\left(\%\right) = \frac{Abs_i - Abs_f}{Abs_i} \times 100$$

where, Abs is the average of absorbance values as it is maximum absorbency visible wavelength. Abs_i the value before electrolysis, Abs_f the value after electrolysis. Figures 5, 6 and 7 show the percentage of color removal with respect to time.

Determination of Instantaneous Current Efficiency (ICE)

Taking into account of the instantaneous current efficiency (ICE) of the electrolysis, it could be calculated using the following formula (2) [23]:

$$ICE = \frac{COD_t - COD_{t+\Delta t}}{8I\Delta t} FV$$

where, COD_t and $COD_{t+\Delta t}$ are the chemical oxygen demands at times t and $t +\Delta t$ (g O$_2$ l^{-1}), respectively, and I is the current (A), F the Faraday constant (96,487 C mol^{-1}) and V is the volume of electrolyte (l). Fig.8 shows the variation of ICE as a function of time at 7 A/dm^2.

Determination of Energy Consumption (EC)

Energy consumption is for the removal of 1 kg of COD (kWh kg–1 COD) was obtained by using the following formula (3):

$$EC = \frac{tUI/V}{\Delta COD} \times 10^3$$

Where, t is the electrolysis time (h), U the average electrolysis cell voltage (V), I the applied electrolysis current (A), V the simulated-wastewater volume (l), and ΔCOD the difference in COD (mg l^{-1}).

Results and Discussion

Effect of anode

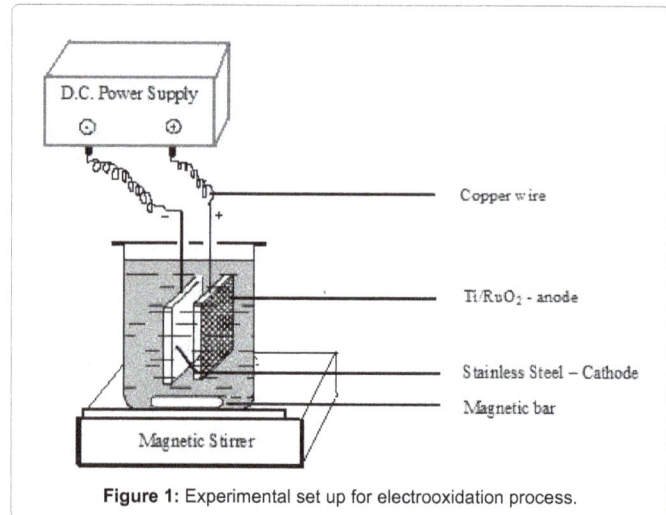

Figure 1: Experimental set up for electrooxidation process.

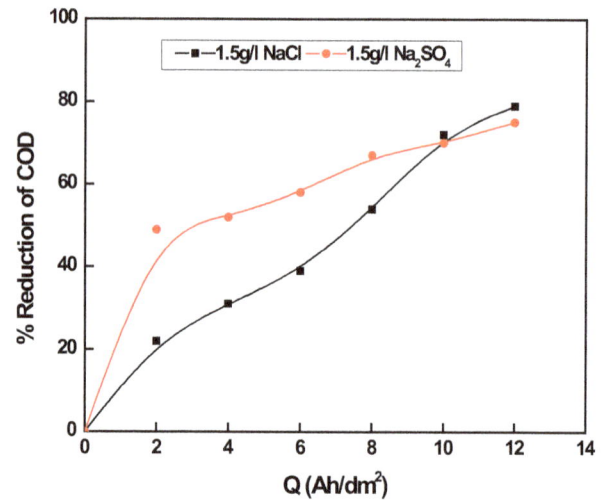

Figure 2: Percentage reduction of COD in different supporting electrolytes at 7 A/dm².

--◆-- 1.5 g/l NaCl

--●-- 1.5 g/l Na₂SO₄t

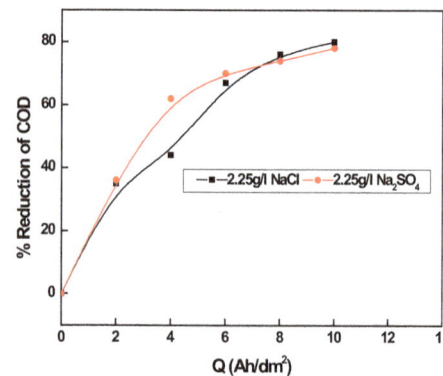

Figure 3: Percentage reduction of COD in different supporting electrolytes at 7 A/dm².

--◆-- 2.25 g/l NaCl

--●-- 2.25 g/l Na₂SO₄

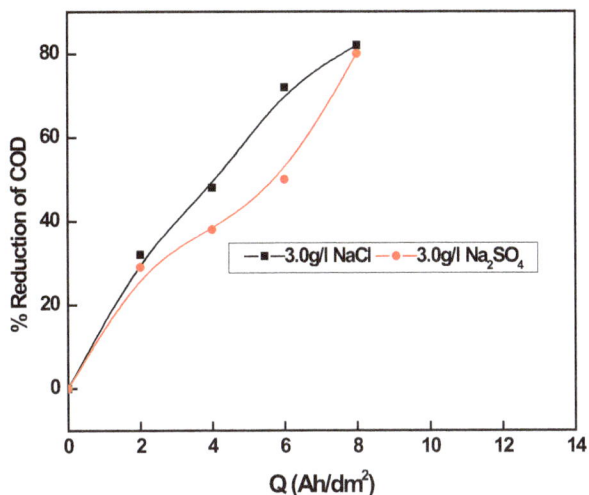

Figure 4: Percentage reduction of COD in different supporting electrolytes at 7 A/dm².

--◆-- 3.0 g/l NaCl

--●-- 3.0 g/l Na₂SO₄

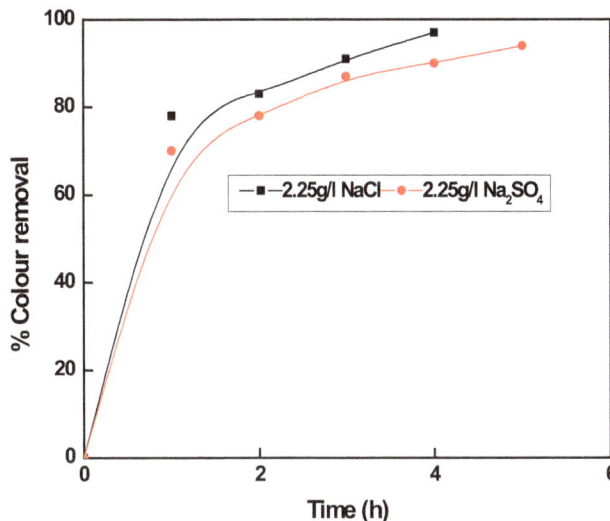

Figure 6: Percentage removal of color as a function of time at 7 A/dm².

--◆-- 2.25 g/l NaCl

--●-- 2.25 g/l Na₂SO₄

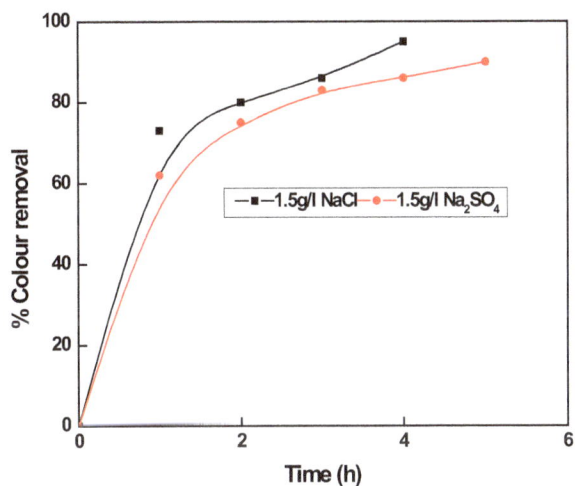

Figure 5: Percentage removal of color as a function of time at 7 A/dm².

--◆-- 1.5 g/l NaCl

--●-- 1.5 g/l Na₂SO₄

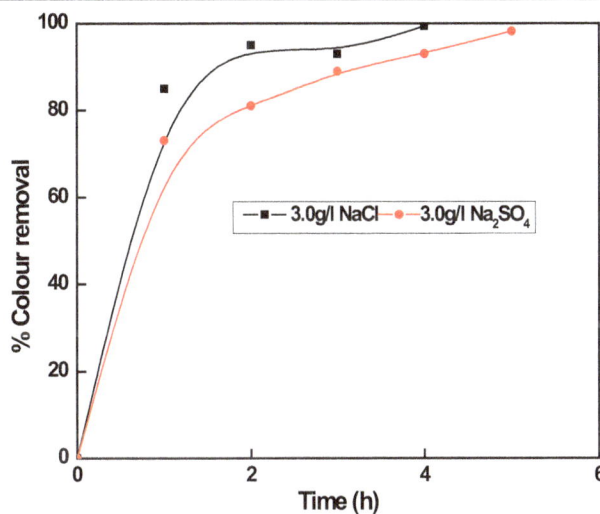

Figure 7: Percentage removal of color as a function of time at 7 A/dm².

--◆-- 3.0 g/l NaCl

--●-- 3.0 g/l Na₂SO₄

It is well known that the organic compounds are completely oxidized to carbon dioxide and water at electrode surface. The electrode Ti / RuO₂ has high electro catalytic activity which leads to simple fragments of complex organic compounds.

Effect of current density and supporting electrolyte

Figures 2-4 show the percentage reduction of COD in different supporting electrolytes. It was observed from the figures that reduction of COD was directly proportional to the applied current density. An optimum point 7 A/dm² was determined finally which gives a faster removal rate of COD. The easily oxidizable pollutants present in the effluent contribute to the decrease in COD under this condition. It is evident that the extent of degradation of reactive dye increases with

time. In the presence of NaCl, the dye was indirectly oxidized by perchloride ion, which was produced by hydroxyl radical. Dye was directly oxidized by hydroxyl or other oxidant reagent electro generated from the electrolysis. The removal of color and reduction of COD was directly proportional to concentration of electrolyte. The formation of perchloride ion and SO₂ depends upon the concentration of NaCl and Na₂SO₄ respectively. It is well known that the electrolysis of NaCl results in some very strong oxidants, such as free chlorine (Cl₂) and hypochlorite anions (ClO⁻) [24-26]. The supporting electrolyte Na₂SO₄ degrades the dyes due to the formation of SO₂, which is a moderate reductant. The decoloration of dye proceeds mainly via direct redox reactions on the electrodes, enhanced by indirect redox reactions by oxidants or reductants. Also it produces more ionic products which increase the ionic strength of the wastewater. Significant amount of $HOSO_3^-$ was identified in the treated wastewater, as a result of the degradation of the dye molecules, this ionic molecule fatherly degraded in non-ionic final products (SO_2/SO_3) [27]. It has been concluded from the results, that the efficiency of NaCl was found to be better than Na_2SO_4 regarding the degradation of C.I Reactive Yellow 17 and Blue 4.

Removal of color

Figures 5-7 show color removal efficiencies in different supporting electrolytes at 7 A/dm². The maximum color removal efficiency was obtained as 99.4% and 96.2% in NaCl and Na_2SO_4 respectively. The same trend was also observed in the reduction of COD.

Figure 8 shows the instantaneous current efficiency (ICE) as a function of time at 7 A/dm². From the figure, it was clearly observed that ICE was fairly good throughout the experiment in NaCl.

The energy consumption for the degradation of dyes was shown in Figures 9 and 10. Results indicate that energy consumption has been increased with increasing applied current density. 82% COD removal was obtained in 3.0 g/l NaCl at 7 A/dm² and energy consumption was 13 kWh/ kg of COD. At the same time, 80% of COD removal was obtained in 3.0 g/l Na_2SO_4 at 7 A/dm² and energy consumption was 16 kWh/ kg of COD Therefore, an optimum condition was determined to reduce COD to a maximum level and removal of color almost completely from the effluent with less energy consumption. An optimal current density of 7 A/dm² was obtained from this study. From the results, it was understood that, the process involving NaCl, completely removes the color of the dye with less consumption of electrical energy.

FT-IR

Figure 10 (a) shows the FT-IR spectra of effluent. Major peaks were obtained for untreated sample at 3363.34, 3282.40, 1637.20 and 663.38 cm^{-1}. The absorbance of peak at 3363.34 cm^{-1} was due to the stretching of N-H. The appearance of peak at 1636.47 cm^{-1} indicates the presence of aromatic C=C and also N=N groups. Figsure 10 (b & c) show the FTIR spectra of treated samples with 30 g Na_2SO_4 and 30 g NaCl respectively. It showed an absence of peak at 1600 cm^{-1} indicates the breakdown of

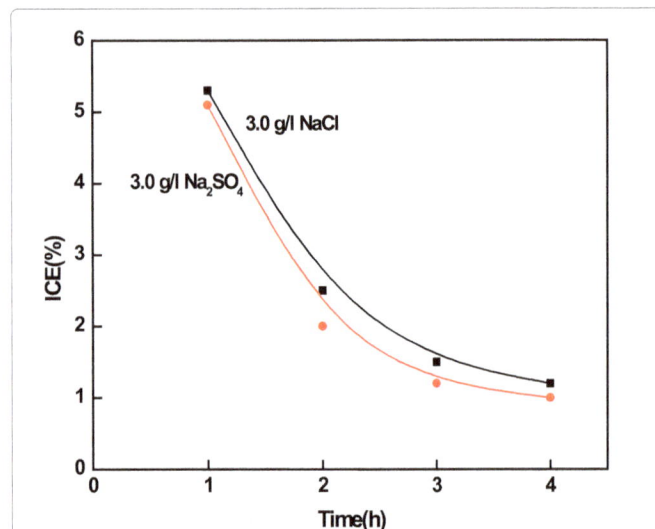

Figure 9: Energy consumption as a function of time in different supporting electrolytes at 7A/dm².

---◆-- 3.0 g/l NaCl

--●-- 3.0 g/l Na_2SO_4

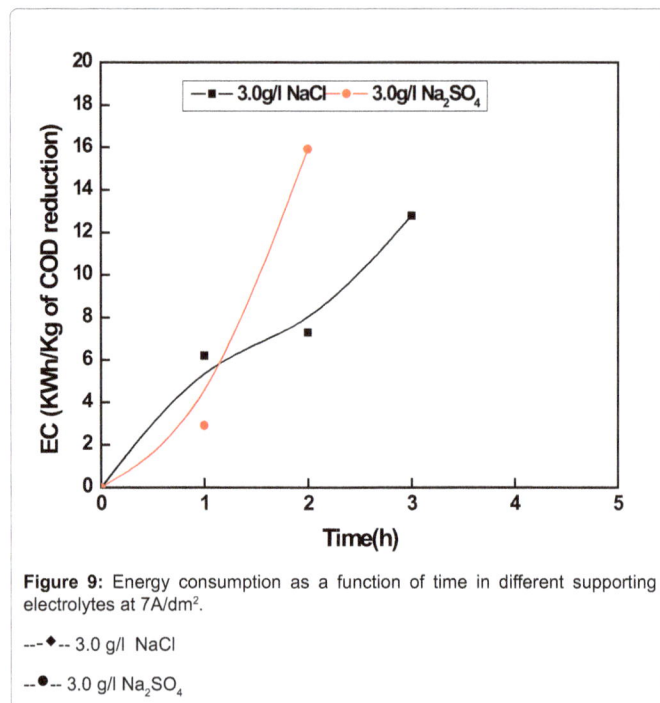

Figure 10: FTIR spectra of untreated and treated dye effluent during EO.
(a) Untreated effluent
(b) Treated with 3.0 g/l Na_2SO_4 at 7 A/dm²
(c) Treated with 3.0 g/l NaCl at 7 A/dm²

amine N-H functions and the absence of peak at 1636.47 cm^{-1} indicates breakdown of aromatic rings. Strong peaks at 686.38 cm^{-1} may be due to the presence of chlorine and hypo chlorite ion in the effluent.

HPLC

Figure 11(a) shows the chromatogram of untreated effluent, the major peak was obtained at 3.2 min (retention time). After treatment the intensity of peak was reduced drastically which was clearly shown in the Figure 11(b).

UV-Vis spectra

Figure 12 shows typical UV spectra for untreated and treated effluent. The spectrum of reactive dye in the visible region exhibits a

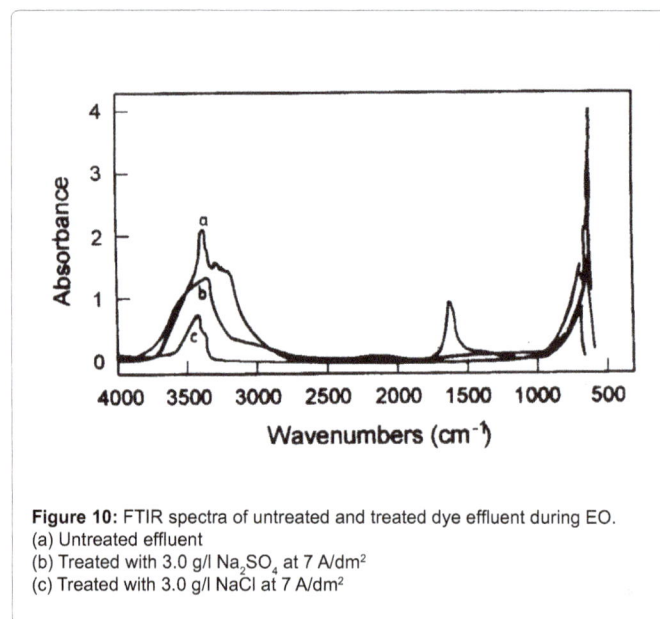

Figure 8: Variation of ICE as a function of time at 7A/dm².

--◆-- 3.0 g/l NaCl

--●-- 3.0 g/l Na_2SO_4

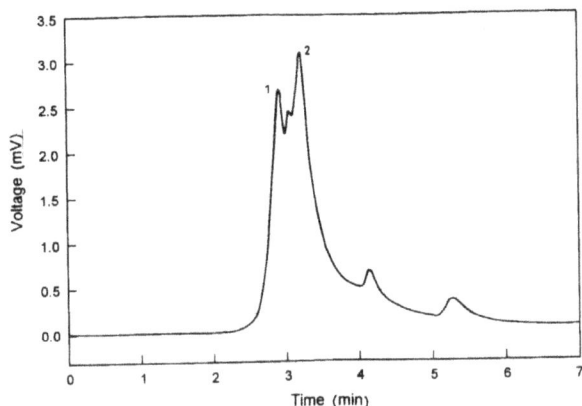

Figure 11a: The chromatogram of HPLC analysis.
(a) Untreated effluent

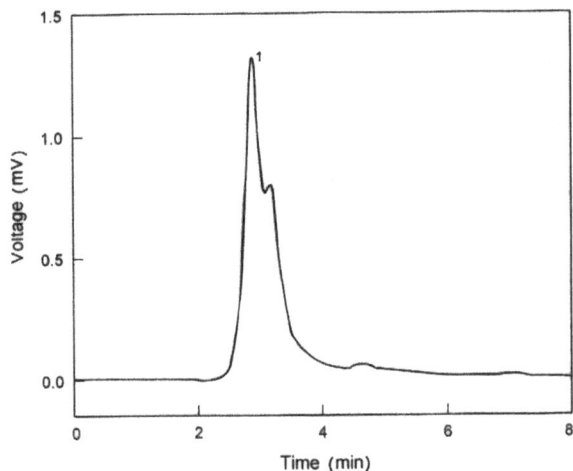

Figure 11b: Treated effluent.

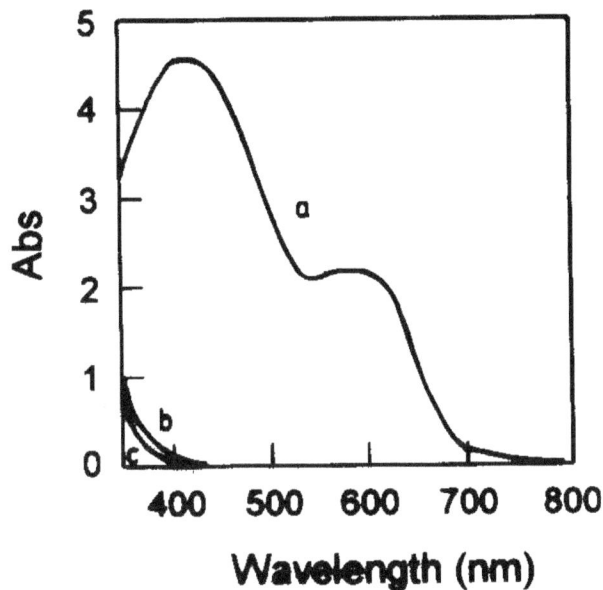

Figure 12: UV-Vis spectra of untreated and treated dye effluent during EO.
(a) Untreated effluent
(b) Treated with 3.0 g/l Na_2SO_4 at 7 A/dm^2
(c) Treated with 3.0 g/l NaCl at 7 A/dm^2

Parameters	Before treatment
pH	
Conductivity (m S)	
BOD (mg/L)	
TOC (mg/L)	
COD (mg/L)	
TDS (g/10 ml)	

Table 2: Characteristics of dyehouse effluent.

main band with a maximum at 430 nm. The decrease of adsorption peaks of reactive dye at λ_{max} = 430 nm which indicates a rapid degradation of reactive dye. The decrease in the intensity is due to the break of nitrogen double bond in the reactive dye, which is the most active site for oxidative attack. Complete discoloration of dye was observed after 3 h under the optimized conditions.

The characteristic band at 485 to 570 nm could be assigned to the n-* transition of azo group. The weak band below 350 nm could be attributed to the π-π* transition related to the aromatic ring attached to the azo group in the dye molecule. It is apparent that the intensity of characteristic band (545 nm) of dye solution was found to diminish gradually during the experiment and disappeared totally after treatment. The disappearance of the bands indicates the effective destruction of the azo.

Conclusion

EO process is a best technique for removal of color and COD for C.I. Reactive Yellow 17 and Blue 4. Degradation of the dye was better when NaCl was used as supporting electrolyte than Na_2SO_4 (Table 2) and it

is due to the formation of perchloride ion, which is used for indirect oxidation of dye. Maximum percentage removal of COD and color were 82 % and 99.4% at the optimum condition of 3.0 g/l NaCl and applied current density of 7 A/dm^2(pH= 11). Results are concurred with HPLC, FT-IR and UV-Vis spectral analysis. It is evident that EO process is a better technique for degradation of reactive dye in wastewater from textile industries.

Acknowledgement

One of the authors Ms. S. Vahidhabanu thanks Department of Science and Technology, New Delhi for offering DST- Inspirefellowship.

References

1. Rajagopalan S (1989) Pollution Management in Industries, Environmental Publication, Karad, India. 7: 31-33.

2. Corriea VM (1994) Characterization of textile wastewater - a review., Eviron. Technol. 15: 917-929.

3. Slokar YM, Marechal AM (1998) methods of discoloration of textile waste waters. Dyes pigments 37: 335-356.

4. Gutierrez MC, Crespi M (1999) A review of electrochemical treatments for colour elimination. J Soc Dyers Colourists 115: 342-345.

5. Lorimer JP, Mason TJ, Plattes M, Phull SS, Walton DJ (2001) Degradation of dye effluent. Pure Appl Chem 73: 1957-1968.

6. Simond O, Schaller V, Comninellis Ch (1997) Theoretical model for the anodic oxidation of organics on metal electrodes. Electrochim Acta 42: 2009-2012.

7. Kotz R, Stucki S, Carcer B (1991) Electrochemical wastewater treatment using high over voltage anodes. Part I. Physical and electrochemical properties of SnO2 anodes. J Appl Electrochem 21: 14-20.

8. Szpyrkowicz L, Santhosh NK, Neti RN, Satyanarayan S (2005) Influence of anode material on electrochemical oxidation for the treatment of tannery wastewater. Water Res 39: 1601-1613.

9. Rajkumar D, Song BJ, Kim JG (2007) electrochemical degradation of reactive blue 19 in chloride medium for treatment of textile dyeing wastewater with identification of intermediate compounds. Dyes and pigments 72: 1-7.

10. Szpyrkowicz L, Juzzolino C, Kaul SN (2001) A comparative study on oxidation of disperse dyes by electrochemical process, ozone, hypochlorite and Fenton reagent. Water Res 35: 2129-2136.

11. Daneshvar N, Sorkhabi HA, Kasiri MB (2004) Decolorization of dye solution containing Acid Red 14 by electrocoagulation with a comparative investigation of different electrode connections. Journal of Hazardous Materials 112: 55-62.

12. Fernandes A, Morao A, Magrinho M, Lopes, Goncalves I (2004) Electrochemical degradation of C.I. Acid Orange 7. Dyes and Pigments 61: 287-296.

13. Naumczyk J, Szpyrkowicz L, De Faveri M, Zilio Grandi F (1996) Electrochemical treatment of tannery wastewater containing high strength pollutants. Trans. I ChemE 74: 58-59.

14. Anastasios S, Konstantinos M, Ulrich N, Konstantinos F, Anastasios V (2005) Evaluation of a novel electrochemical pilot plant process for azodyes removal from textile wastewater. Chemical Engineering Journal 111: 63-70.

15. Rajeshwar K, Ibanez JG (1997) Environmental Electrochemistry, Academic Press, Inc.

16. Panizza M, Cerisola G (2004) Electrochemical oxidation as a final treatment of synthetic tannery wastewater. J Environ Sci Technol 38: 5470-5475.

17. Vlyssides A, Barampouti EM, Mai S (2004) Degradation of methyl parathion in aqueous solution by electrochemical oxidation. J Environ Sci Technol 38: 6125-6131.

18. Mohan N, Balasubramanian N, Ahmed Basha C (2007) Electrochemical oxidation of textile wastewater and its reuse. Journal of Hazardous Materials 147: 644-651.

19. Buso A, Balbo L, Giomo M, Farmia G, Sandona G (2000) Electrochemical removal of tannins from aqueous solutions. Ind Eng Chem Res 39: 494-499.

20. Mohan N (2000) Studies on electrochemical oxidation of acid dye effluent, Ph.D. Thesis, Anna University, Chennai, India.

21. Cleseeri LS, Greenberg AE, Eaton AD (1998) Standard Methods for Examinations of water and Waste water, 20th ed, American Public Health Association, American Water Works Association and Water environment Federation.

22. Neelavannan MG, Revathi M, Ahmed Basha C (2007) Photocatalytic and electrochemical combined treatment of textile wash water. Journal of Hazardous Materials 149:371-378.

23. Ahmed BC, Bhadrinarayana NS, Anantharaman N, Meera KM (2008) Heavy metal removal from copper smelting effluent using electrochemical cylindrical flow reactor. Journal of Hazardous Materials 152:71-78.

24. Pletcher D, Walsh FC (1990) Industrial Electrochemistry, 2nd ed., Chapman and Hall, New York, pp. 353-357.

25. Agadzhanyan SI, Romanova RG, Korshin GV, Kirpichnikov PA (1996) Physicochemical properties of sodium chloride solutions after electrochemical treatment in a flow electrolyzer. Russ J Appl Chem 69: 380-384.

26. Ananeva EA, Vidovich GL, Krotova MD, Bogdanovskii GA (1996) The study of a photochemical and electrochemical treatment of aqueous solutions of azo dyes. Russ J Electrochem 32: 936-938.

27. Sakalis A, Ansorgova D, Holcapek M, Jandera P, Voulgaropoulos A (2004) Analysis of sulphonated azodyes and their degradation products in aqueous solutions treated with a new electrochemical method. Int J Environ Anal Chem 84: 875- 888.

32

Biogas Potential from the Treatment of Solid Waste of Dairy Cattle: Case Study at Bangka Botanical Garden Pangkalpinang

Fianda Revina Widyastuti[1]*, Purwanto[1] and Hadiyanto[2]

[1]Environmental Science Master's Program, Diponegoro University Semarang, Indonesia
[2]Chemical Engineering Program, Technical Faculty, Diponegoro University, Indonesia

Abstract

Bangka Botanical Garden is an integrated cattle farm treating cattle solid waste as an energy source to produce biogas as fuel for gas stoves. At present, they make use of 132 kg of cattle waste from 5 cows and produce only 1 m³ of gas per day. This paper will discuss energy need, economic and environmental aspects of using cattle waste to produce biogas and the use of biogas to satisfy the need for electricity in the BBG farm. This study is descriptive in character. The data were collected through observation, measurement and interviews with informants. The biogas obtained could be used to provide lighting in the pens amounting to 60-100 W for 50 hours, as automotive fuel for a 1 HP engine for 17 hours, producing 39 kWh of electricity sufficient to cook three dishes for 40-48 servings. Producing 39.48 kWh of electricity per day, the generator could supply electricity in BBG farm using 35 light bulbs of 25 W each, switched on 12 hours per day. Thus, the electricity needed for lighting was 10 kWh per day. The milking machine needed 0.55 W per milking, or 1.1 W per day of two milking. The rest of the energy could be used to run water pumps, mowers and welding machines. The BBG farm needs to increase the efficiency of digester use by employing the inactive digester and improve the electricity installation for lighting.

Keywords: Biogas; Cattle; Energy; Potential

Introduction

Human beings need energy for their activity, and so does economy. Energy can be obtained from renewable natural sources and nonrenewable mineral sources, which takes a very long time to reproduce. Energy consumption in Indonesia has increased from time to time. Elinur et al. [1] stated that Indonesian crude oil reserves would be exhausted in 23 years, gas reserves in 59 years, and coal in 82 years.

According to the Mineral Resources Data and Information Center of the Ministry of Energy and Mineral Resources [1] unwise use of energy not only would exhaust the mineral resources but would also cause environmental pollution, through excessive emission of CO_2 and other green house effects. CO_2 emission, according to the Basic Scenario or Business as Usual (BAU) would increase to 1000 million tons in 2020 and would increase further to 2129 million tons in 2030. According to the scenario, the mitigation of CO_2 emission could reduce this to 706 million tons in 2020 and to 1219 million tons in 2030. The sources of CO_2 emission are coal (50.1%), natural gas (26%), and mineral fuel (23.9%). The industrial sector is the biggest contributor of CO_2 emission, followed by homes, transportation, commerce and farming, construction and mining.

Energy diversification is one of the solutions to the energy crisis threat in this country. Conservation could be done through energy saving and developing renewable energy sources, which of course, should be supported by government policies, which are environment-conscious. One of the renewable energy sources is biogas. Biogas can be produced from animal waste, tofu industry waste, organic waste from homes and traditional markets. Biogas has the potential to become an alternative renewable energy source in Indonesia, experiencing energy crisis, marked by the increasing scarcity and price of mineral fuel, resulting in high cost of electricity power production. Wahyuni [3] stated that biogas could create fire sparks with the power of 6400-6600 kcal/m³. The energy content of 1 m³ biogas equals that of 0.62 liter paraffin, 0.46 liter LPG, 0.52 liter diesel fuel, 0.08 liter gasoline, and 3.5 kg wood. Hanif's study [4] found that one cow could produce 25 kg waste. Thus, 411 animals could produce 10,275 kg waste with dry material content of 2,055 kg, which could produce 82.2 m³ of biogas per day. 1 m³ biogas could produce 4.7 kWh of electricity. Therefore, 441 cattle have the potential to produce 386.6 kWh of electricity per day. According to Hardianto et al. [5] there are three advantages of treating cattle waste to produce biogas, namely: 1) biogas can be used as alternative fuel instead of oil fuel or wood with a high heat quality, 2) the fluid sludge can be used as fertilizer, and 3) the solid sludge can be used as mixture of animal feed. According to Arifin et al. [6] in Saung Balong Pesantren (Islamic Boarding School), they had made a biogas pilot plant that produced 7 m³ of gas per day. The gas was used for the Pesantren's daily need, for cooking and lighting, and used to run a pure biogas generator producing 1,000-10,000 W (10 kW) using dual fuel system and had made biogas enrichment though absorption and putting biogas in tanks. The electricity produced by the biogas installation in Saung Balong Al-Barakah could reduce dependence on the electricity supplied by the state electricity industry (PLN).

The BBG farm is a farm inside the Integrated Farming Area (KUTBBG), owned or organized by *Yayasan Bangka Go Green* (Bangka Go Green Institute), located on the coast of Pantai Pasir Padi, Pangkalpinang, the Province of Bangka Belitung Islands. The BBG farm covers an area of 312 ha in the middle of the Ketapang Industrial and Harbor Area. KUTBBG is located in Bukit Intan Sub-Ditrict, the Kelurahan (Village) of Temberan. The daily activity of BBG is growing dairy and beef cattle, using the government supplied electricity

*Corresponding author: Fianda Revina Widyastuti, Environmental Science Master's Program, Diponegoro University Semarang, Indonesia
E-mail: fiandarevina@gmail.com

produced by the state electricity industry (PLN). The electric energy is used to run the water pumps, milking machines, lighting, workshop/welding, and mowers/choppers. Meanwhile, for cooking the BBG farm has used the biogas produced through cattle waste treatment.

The main problem faced by the farmers was that the electricity was often off. This frequent turning-off of electricity has negative consequences, as it disturbs the farmers' activity and has a bad effect on the electrical appliances. The switching off electricity also damages the electronic appliances. The BBG farm has two biogas installations, but only one unit is active, because the other unit is too far from the pens. The active biogas installation supplies energy for the farmers' cooking stoves. Every day the farmers only use 2 wheelbarrows or 132 kg of cattle waste.

Weight measurement of dairy cattle was done using 42 sample cattle and measurement was successful from 17 cattle, because the rest are too aggressive. Chest measurement was done using a chest measurement tape, because the farm does not have a cattle scale. According to the *Schoor formula*, Cattle weight (kg) = (chest measurement (cm) + 22)²: 100. Based on the results of measurements of bust 17 dairy cows gained weight average dairy farm in Bangka Botanical Garden is 432.07 kg / head as shown in Table 1.

The integrated farm area of Bangka Botanical Garden (KUTBBG) Pangkalpinang integrates plantation, fishery and animal husbandry. The initial concept of the KUTBBG planning applied the zero-waste system, where the animal waste was only used as fertilizer, while the leaves from the plantation were used for animal feed. Some of the animal waste was later processed to make compost and some to produce biogas for daily need of gas for cooking stoves. With 2 digester units and the ample availability of waste from the dairy cattle, the use of the animal waste potential had not been to the maximum. The problem was that one unit of the biogas installation was too far from the dairy cattle pens. If the digester in BBG farm were used efficiently, the dependence on the state supplied electricity could be reduced, thus saving electricity and fuel. Electricity needed in the BBG farm was 875 W, for 35 light bulbs of 25 W, 0.55 W for milking machine, for 1 AC Unit, for 1 mower/chopper, 1800 W for water pump, and for workshop/welding. This article will discuss the energy use problem, the economy

and environmental advantage of using cattle waste to produce biogas and the use of biogas to supply the electricity need of the BBG farm.

Study Method

The study was done at the farming complex as part of the Integrated Farming Area of Bangka Botanical Garden. The cattle population was 223, consisting of dairy cows, bulls and calves. We studied only 42 cattle, because they live in one pen, close to the biogas installation. Observing every step in raising the cattle. Secondary data were obtained through literature study and documents related to the Integrated Farming Area of BBG (KUTBBG). The pen janitors acting as informants consist of 4 persons from the dairy section, 2 persons from the bull section, and 6 from the calf section, and 1 nutritionist and 2 grass cutters, 2 persons from the composting section, 1 person from the biogas section. Biogas potential analysis was done by multiplying dairy cow waste mass and dry material content and converted to the amount of gas obtained (m^3) so that we got the daily biogas volume from 42 dairy cows.

Environment aspect analysis was done by calculating greenhouse gas emission formula quoted from the Indonesian Ministry of Environment, that is:

GG Emission = $A_i \times EF_i$

GG Emission = Greenhouse gas emission (CO_2, CH_4, N_2O)

A_i = type i material consumption, or the amount of i product.

EF_i = Emission Factor of i type material or i product.

The treatment of 1 ton CO_2 of greenhouse gas emission based on Kyoto Protocol requires 30 Euros. By using biogas, the BBG farm could help reduce the amount of greenhouse gas emission.

Results and Discussion

Biogas potential as energy source

Farming Mechanization Development Center in Serpong has studied the biogas use obtained from cattle waste for lighting and cooking stoves. Biogas is feasible technically and economically to use as energy source. The feasibility of biogas use in electricity generators has also been studied [7].

The BBG farm owns 223 heads of cattle, consisting of dairy cows, bulls and calves. As the cattle waste was not only treated in digester to produce biogas, but also composted to become fertilizer, so far the digester with a capacity of 4 m^3 has been using only 2 wheelbarrows or 132 kg of dairy cattle waste to produce gas. The produced gas was used only as fuel for cooking stoves and not as fuel for electric generator for lighting or other appliances using electricity. According the informant from the biogas section, biogas had been used to supply energy for generating electricity for lighting in the pens before the blockage and damage in the digester, resulting from the hard material in the feces, containing heavy copra residue that could not be removed and thus precipitated at the bottom of the digester. Up to now the digester has not been used any more to produce gas for lighting in the pens.

The number of dairy cows in the BBG farm is 42, weighing 450-500 kg. With the number of cattle in the farm, it has the potential to produce enough biogas as a source of energy to provide electricity and satisfy the fuel need in the husbandry section. Cattle waste is the most efficient material to produce biogas, because each 10-20 kg of waste per day can yield 2 m^3 of biogas. As the energy contained in 1 m^3 biogas is

No	CM (cm)	BW (kg)
1	188	441.00
2	180.5	410.06
3	187.5	438.90
4	164	345.96
5	203	506.25
6	186	432.64
7	175	388.09
8	190	449.44
9	172	376.36
10	181	412.09
11	178	400.00
12	194	466.56
13	188	441.00
14	202	501.76
15	197	479.61
16	196	475.24
17	173	380.25
Average weight (kg)		432.07

Table 1: Cattle Weight Measurement Result (*Source: Primary Data Measurement Result*, 2013).

2000-4000 Kkal, it is sufficient to provide cooking fuel need of a family (4-5 persons) for 3 hours [8].

From the chest measurement, we could roughly determine the weight of the dairy cows in the BBG farm is between 400 and 500 kg. According to Riliandi [9] a mature dairy cow can produce 25 kg of feces per day. Similar statements were made by Wiryosoehanto, Soedono and Solihat that a lactating dairy cow weighing 450 kg could produce around 25 kg of urine and feces per day. Therefore, we could assume that each dairy cow in the BBG farm produced 25 kg of waste per day.

Each dairy cow weighing 450 kg has the potential to produce 25 kg of feces and urine per day. According to the Farming Mechanization Development Center, Ministry of Agriculture's Research and Development Department (2008) quoted by Hanif [4] that 25-30 kg of cattle waste contained 20% dry material and biogas produced is 0.023 – 0.040 m³/kg of dry material. With the number of dairy cows in the BBG farm, 42 heads, the farm has the potential to produce 1,050 kg per day.

The total dry material content of cattle waste is 20% of the wet waste, 210 kgDW. According to the United Nations (1984) as quoted by Widodo [10], 1 kg of cow or buffalo waste can produce 0.023 – 0.040 m³ of biogas. Thus, the BBG farm's total potential to produce biogas from cattle waste is 8.4 m³ per day as shown in Table 2.

Environment aspect analysis

Indonesia is assumed to be the third largest contributor of greenhouse gas emission in the world, after China and the United States. In 2006, FAO produced a report titled "Livestock's Long Shadow" with the conclusion that animal husbandry was one of the main causes of global warming. The animal husbandry's contribution to global warming was 18%, bigger than that of transportation, which contributed 13.1% [11]. Meanwhile, world's husbandry contributed 37% methane and 65% dinitrogen oxide (IPCC 2001) [12].

Methane emission factor from dairy cow's digestion fermentation is 61 kg per head. We thus know that the BBG farm contributes 13,603 kg of methane emission or 340,075 kg CO_2e gas emission per year. If the waste produced by 42 cows were optimally used to produce biogas the methane emission reduction would be 2,562 kg of reducing CO_2e emission of 64,050 kg per year.

The treatment of 1 ton CO_2e emission following the GHG emission

	BBG Dairy Farm
Gas production benchmark per kg of waste (m³)	0.023 – 0.04
Cattle waste dry material content *)	20%
Daily dry waste (kg/day)	210
Daily biogas potential (m³/day)	8.4

Source: Processed primary data (2013) *): United Nations (1984) in Widodo (2004)

Table 2: Biogas Potential (m³/day) in Dairy Cow Solid Waste (Feces) in BBG Farm.

	BBG Dairy Cattle Farm
Daily biogas potential (m³/day)	8.4
Pen lighting (60 - 100 watt light bulb)(hour)	50
Energy source to run 1 HP engine (hour)	17
Electricity (kWh/day)	39
Cooking 3 dishes for 5 – 6 persons (servings)	40 - 48 servings

Source: Processed Study Data (2013)t

Table 3: Energy Conversion Based on Cattle Waste Potential (m3 Biogas) in Dairy Cattle Farming Activities in BBG Farm.

based on the standard Kyoto Protocol costs 30 Euros, when 1 Euro = Rp 12,000 (June 2013), the cost for 1 ton CO_2e emission treatment was Rp 360,000. Thus, if the dairy cow waste in the BBG were optimally used to produce biogas, there would be a cost saving in treating CO_2e emission amounting to Rp 23,058,000 per year.

Biogas use management

The calorie content of biogas makes it feasible for use in everyday activities, such as drying, lighting, and work requiring heat, like welding. As a fuel to run engines, biogas should be cleaned from the corrosive H_2S gas. There need to be some modification in the carburetor to make it feasible to use as engine fuel. The engines run by biogas fuel could be used in electric generators, water pumps and the like [13].

According to Suriawiria energy from 1 m³ of biogas could be used to provide energy for a 60-100 W light bulb for 6 hours, for cooking 3 dishes for 5-6 persons, and could run a 1 HP motor for 2 hours. It is known that 1 m³ of gas can be converted to 4.7 kWh of electricity [4,14].

Biogas produced from 42 dairy cows in the BBG farm could replace electric energy source to provide 60-100 W lighting for the pen for 50 hours, as fuel to move a 1 HP engine for 17 hours, producing 39 kWh of electricity and cooking 3 dishes for 40-48 servings. By producing 39.48 kWh/day, the BBG farm could use the biogas as electric energy source, with 35 light bulbs of 25 W for 12 hours/day. Lighting need is 10 kWh/day. The milking machine needs 0.55 W, thus 1.1 W for two daily milkings. The rest of the energy could be used to run the water pump, the mower/chopper and welders.

Conclusion and Suggestions

Processing dairy cow waste to produce biogas is feasible and highly recommended in the BBG farm, because besides reducing the negative environment impact, it could increase profit and save fuel and electric energy. The biogas potential produced from 42 dairy cows is 8.4 m³/day. The biogas could be used to provide 60-100 W lighting in the pen for 50 hours, to run a 1 HP engine for 17 hours, producing 39 kWh of electricity and to cook 3 dishes for 40-48 servings. At present the BBG farm use only 132 kg of animal waste, using waste from only 5 cows, producing only 1 m³ of gas per day. Therefore, there should be more efficient use of the existing digester, use more digesters, maximize the use of waste from 42 heads of dairy cattle, reinstall the electric installation to provide energy for 35 light bulbs of 25 W in the BBG farm complex, reuse the other digester by moving it closer to the cattle pens.

Expression of Gratitude

The writer, Fianda Revina Widyastuti would like to thank Pusbindiklatren Bappenas (Planning, Training, and Educational Development Center of the Indonesian National Planning Board) the owner and organizer of the Integrated Farming Area of the Bangka Botanical Garden, Pangkalpinang and everyone who has helped in the study process.

References

1. Elinur, Priyarsosno DS, Tambunan M, Firdaus M (2010) The Growth of Energy Supply and Consumption in the Indonesian Economy (Perkembangan Konsumsi dan Penyediaan Energi dalam Perekonomian Indonesia). Indonesian Journal of Agricultural Economics 2: 97–119.

2. Center for Data and Information on Energy and Mineral Resources (2010) Indonesia Energy Outlook 2010. Ministry of Energy and Mineral Resources.

3. Wahyuni S (2011) Assorted Produce Biogas from Waste (Menghasilkan Biogas dari Aneka Limbah). PT Agro Media Pustaka. Jakarta.

4. Hanif A (2010) Biogas Utilization Studies For Power 10 kw Farmers Mekarsari Dander village Bojonegoro Toward Energy Independent Village. Field Study of Power System Engineering Department of Electrical Engineering, Faculty of Industrial Technology Institute of Technology 'Sepuluh November.

5. Hardianto R, Wahyono DE, Andri KB, Hardini D, Setyorini D, et al. (2000) Technology Assessment Integrated Cropping Farming - Livestock in Dry Land. Proceedings of the Seminar and Exposure Research/Assessment BPTP East Java, 244-256.

6. Arifin M, Saepudin A, Santosa A (2011) Study Biogas as an Electrical Power Source in Pesantren Saung Balong Al_Barokah, Majalengka, West Java. Journal of Mechatronics, Electrical Power, and Vehicular Technology 2: 73-78.

7. Center for Development of Agricultural Mechanization Situgadung (2007) Scale Biogas for Household Electrical Generetor (Biogas untuk Generetor Listrik Skala Rumah Tangga). Journal of Agricultural Research and Development 29: 3-4.

8. Suriawiria U (2005) Reaping Biogas from Waste (Menuai Biogas dari Limbah).

9. Riliandi DK (2010) Studies use cow dung for biogas electricity generators, lighting and cook Nongkojajar Towards Self-energy village (district Tutur) (Studi pemanfaatan kotoran sapi untuk genset Listrik biogas, Penerangan dan memasak Menuju desa nongkojajar (kecamatan tutur).

10. Widodo TW, Nurhasanah A (2004) Technical Assessment and Potential Development of Biogas Technology in Indonesia (Kajian Teknis Teknologi Biogas dan Potensi Pengembangannya di Indonesia). Proceedings of the national seminar on agricultural mechanization. 189-202.

11. Food and Agriculture Organization (2006) *Livestock's Long Shadow*.

12. IPCC (2001) Climate Change 2001: Mitigation of Climate Change. Contribution of Working Group III to the Fourth Assessment Report of the Intergovernmental Panel on Climate Change Cambridge. University Press, Cambridge, United Kingdom and New York.

13. Widya SR, Muljatiningrum A (2011) Biogas from Animal Waste (Biogas dari Limbah Ternak) Nuansa Bandung.

14. Suhendra F (2008) The Usage of biogas Technology to Reduce Livestock Pollutant in Bali on Clean Development Mechanism. Mulya Tiara Nusa.

33

Household Waste Management in High-Rise Residential Building in Dhaka, Bangladesh: Users' Perspective

Tahmina Ahsan[1]* and Atiq Uz Zaman[2]

[1]School of Architecture and Built Environment, University of Adelaide, SA 5001, Australia
[2]Zero Waste SA Research Centre for Sustainable Design and Behaviour, School of Art, Architecture and Design, University of South Australia, Australia

Abstract

The Dhaka City Corporation (DCC) is primarily responsible for collecting and managing waste in Dhaka, Bangladesh. A significant amount of waste in Dhaka is not collected due to lack of infrastructure, funds and collection vehicles. Despites Dhaka's limited waste management service, community based door-to-door waste collection from households to local waste bins is considered as a success. Informal waste recycling systems is also highly effective in waste recycling and job creations for the poor. Even though both horizontal and vertical expansion is prevalent in Dhaka, there has been an increasing trend in vertical expansion of the city in recent years as horizontal expansion is not possible due to barriers such as the built up urban core and low lying flood plains. Very limited number of studies has been conducted on waste management system in high-rise residential buildings in Dhaka. Therefore, this study focuses on the waste management scenario of high-rise residential buildings in Dhaka. The study is particularly interested in the socio-demographic, cultural and environmental features in high-rise residential buildings. The study identified key areas in waste management systems in high-rise residential buildings which is the avenue for future studies for integrating waste management strategies in high density residential development in Dhaka. Based on the findings, the study recommends a further examination of the integration of waste management infrastructure in the high-rise residential building development in Dhaka.

Keywords: Waste management; High-rise residential building; Waste survey; Users' perspective

Introduction

Cities attract people for the various opportunities it can provide to the inhabitants in terms of income and services. Cities expand both horizontally and vertically to accommodate a huge number of people every year. This increasing population in cities produce an ever increasing amount of solid waste every day. Waste management is one of the most challenging and cost effective services that municipality around the globe offers to their citizen. The success of the waste management systems, both in developed and developing countries depends on a holistic waste management planning and adequate facilities Dhaka, the capital of Bangladesh is one of the densely populated cities in the world. Dhaka City Corporation (DCC) provides waste management services for 7 million people living in an area of 360 km² [1]. The DCC is responsible for collection and disposal of 3000-4000 tons of municipal solid waste from the city's 90 wards (local geo-administrative sub-division) every day. However, the DCC can collect and dispose only 40-50% of the total waste generated every day due to the lack of funds and infrastructure (Dhaka City Corporation, 2013). Only 14-17% of the total municipal budget is used for solid waste management which is approximately 0.5 USD per capita per year. As a result, the uncollected waste is primarily dumped illegally in the neighbourhood's streets, wastewater drains, ponds, lakes etc. or managed informally [2]. A significant amount of health and environmental problems are created by the improper waste management systems in Dhaka.

Due to the lack of waste management services from the local authority in Dhaka, private and community based waste management systems have been introduced recently. However, the recent waste management initiatives are facing various complex management challenges. The key challenges are storing, separation and collections of various types of waste from residential areas. Lack of common and uniform waste infrastructures is highly visible in housing sector in Dhaka which. In addition to this difficulty, residential buildings in Dhaka are of different types such as single-unit low-rise residences, multi-unit apartments, high-rise apartments and apartment complexes with both low-rise and high-rise apartments. Most of the residential housing in Dhaka does not meet the rules of the national building code. In addition, proper guideline on waste management systems is also absent in the national building code. This discrepancy creates additional challenges to the waste management authority.

As a mega city, Dhaka accommodates thousands of people every day. Hence, the number of residential buildings is increasing every year. Almost all available land area in Dhaka has now been used due to urbanization. As a result of limited horizontal land space in Dhaka, urban growth has now begun vertically through high-rise developments. Thus, the number of high-rise residential building is increasing significantly in Dhaka. It is important to understand that the people living in high-rise residential buildings would have different socio-economic conditions compared to other residential buildings in Dhaka. Hence, waste generation and management is also different in high-rise residential building compared to other housing types.

This study primarily focuses on exploring the existing waste management systems in high-rise residential building in Dhaka. Key challenges and opportunities will be identified in the study.

*Corresponding author: Tahmina Ahsan, School of Architecture and Built Environment, University of Adelaide, SA 5001, Australia
E-mail: tahmina.ahsan@adelaide.edu.au

Waste Management in Dhaka

The Dhaka City Corporation (DCC) is primarily responsible for collecting and managing waste in Dhaka. Waste management systems in Dhaka are primarily based on informal waste collection and recycling system. About 120,000 urban poor from the informal sector are involved in the recycling trade chain of Dhaka City. 15% of the total generated wastes in Dhaka (mainly inorganic) amounting to 475 tons/day are recycled daily [3]. Informal waste collection and management is not operated and controlled by the local authority (DCC for instance).

Waste is thrown on roadside and places other than dustbins due to lack of proper and sufficient waste infrastructure (bins) and collection systems in Dhaka. About 7146 cleaners are employed by DCC for street sweeping and collection of waste found in places other than dustbins, road side, open spaces, ditches etc. by hand trolley. DCC has 2,080 hand trolleys for primary collection of waste. DCC has 128 demountable container carrier trucks for collection of accumulated waste in 414 containers and 242 open trucks to collect waste from municipal bins at different locations [4].

A direct consequence of the poor performance of waste management services provided by the local government is the growth of community-based initiatives, private and non-government organizations, which are increasingly playing an important role in delivering conservancy services [5]. Inorganic waste such as paper, plastic, metal, glass and so on is mostly sorted at household level. Doo-to-door recyclables buyer locally known as 'feriwalla' buy sorted recycling waste. In the community based waste collection system, waste collector locally known as 'gariwalla' collects waste (mostly organic and mixed waste) from every household and transfers it to the designated roadside municipal collection points. Informal waste pickers locally known as 'tokai' further sort and collect recyclable materials from waste [1]. Waste is then transferred to the landfill site where another sorting and recycling activities are carried out by the informal recyclers. The following Figure 1 illustrates the waste flow in waste management systems in Dhaka.

Waste management expenditure in Dhaka

The quality and reliability of the waste management services depend

(a) waste storage   (b) door-to-door collection

(c) street-side waste bin   (d) waste transported to landfill site

Figure 1: Household's waste management system in Dhaka.

significantly on both infrastructure and operational budget to run the systems effectively. Waste management revenue expenditures in DCC is projected in 17.80 crore tk (USD 2.28 M) [1] which is only 8.68% of total revenue expenditures in DCC [6]. Total projected income of Dhaka North City Corporation (DNCC) in 2013 1984.53 crore tk and total projected expenditures is 1984.53 crore tk. Table 1 shows the comprehensive waste expenditures in Dhaka. Raising public awareness and promoting recycling through different awareness and training programmes are important. However, from the budget allocation on waste management systems in Dhaka, it is apparent that there is no fund for training programmes.

Government and non-government initiatives through public private partnership (PPP) are playing an important role in waste management development in Dhaka. The project 'Improvement of solid waste management in Dhaka city towards the low carbon society through enhancing waste transport capacity' has allocated about USD 2.04M for improving waste management systems in Dhaka. Most of the private solid waste management in Dhaka City is based on Public Private Partnership (PPP) mechanism. Dhaka North City Corporation fully privatizes its conservancy works in Uttara, Gulshan, Banani, Baridara, Mohakali and Tejgaon areas. The Private parties do the street sweeping, drain cleaning, door-to-door waste collection, dustbin cleaning and waste transportation and final disposal to landfill [6].

Similar studies in waste management systems in Dhaka

A number of studies have been done on waste management systems in Dhaka and study includes Afroz et al. in willingness to pay for waste management improvement in Dhaka city [7]. Facilitating people's participation in public–private partnerships for solid waste management in Dhaka [8]. In city governance and its impact in waste management system [5]. Enayetullah et al. in exploring waste management scenario in Uttara Model Town [9]. The effect of waste composting and landfill location [10]. The problem identification and nature of waste management problems in Dhaka [11]. Matter et al. in improving informal waste recycling through segregation of household waste [2]. In socio-cultural practices in household waste collection system from residential buildings [12]. In Waste management modelling in Dhaka [13]. Zaman studied in informal waste management in Dhaka and proposed an innovative social business model based on recycling materials [14].

Rationale and scope of the study

High-rise residential development is significantly important for the urban development in Dhaka. Hence, high-rise residential building must be the inevitable solution to cater for the urgently needed housing units in densely packed urban areas in Dhaka. Most of the private housing development is based on high-rise apartment housing. The government has planned the country's largest apartment project in Uttara to accommodate ever increasing population in Dhaka. The project proposed approximately 240 no's of 16 storied apartments in total 20,160 no of flats which would accommodate approximately more than 80,000 people in a single development project in Dhaka [15]. Combining all the high-rise development in Dhaka city the number would be significantly higher. Waste management system for high-rise residential buildings in particular, requires different infrastructure and management plans. Therefore, mega cities in developed countries such as Tokyo, Shanghai and New York have different waste management strategy and plan for high-rise residential buildings which is absent in the case of Dhaka. There is no such waste collection and management

[1] USD= 77.9 BDT as on 30th September 2013 currency conversion rate

Key Areas	Expenditures' Subsectors	Amount of expenditures (in Crore TK)	Reference section
Maintenance and protection	Solid waste workshop	1.5 (US$ 192,400)	subsection 3.7 (p-5)
Supply	Equipment	2.5 (US$ 320,650)	subsection 5.6 (p-5)
Urban Waste Management	Community waste management	0.3 (US$ 42,764)	subsection 16.1 (p-7)
	Privatised waste management	15 (US$ 1923,900)	subsection 16.2 (p-7)
	Special cleaning activities	1.5 (US$ 192,400)	subsection 16.3 (p-7)
	Landfill operation	1 (US$ 128,260)	subsection 16.4 (p-7)
	Subtotal:	17.80 (US$ 2.28M)	Section 16 (p4)
Infrastructure and development	Waste infrastructure (ward office)	0.5 (US$ 64,130)	subsection 4.12 (p-8)
	Waste management workshop	0.25 (US$ 32,065)	subsection 4.13 (p-8)
Environmental development	Dustbin/transfer station	1.5 (US$ 192,390)	subsection 5.4 (p-9)
	Landfill operation and development	5 (US$ 641,300)	subsection 5.7 (p-9)
Equipment and assets	Trucks/Motorcycles for waste management	5 (US$ 641,300)	Sub section 3 in sec 11 (p-10)
Public-private partnership project	Improvement of solid waste management in Dhaka city towards the low carbon society through enhancing waste transport capacity	15.91 (US$ 2.04M)	Subsection 1.11 (p-11)

Table 1: A comprehensive budget for waste management in Dhaka city (DNCC, 2013).

strategy in any high-rise residential building in Dhaka. Thus this study is primarily interested to explore the socio-demographic, socio-cultural and environmental features in waste management systems in high-rise residential buildings in Dhaka. The authors predict that the findings of this study would be beneficial for the local authority and city planner to integrate waste management planning in the high-rise residential buildings. The study primarily focuses on the socio-demographic, socio-cultural and environmental features of waste management systems in high-rise residential buildings in Dhaka.

Methodology and Method

A questionnaire survey has been used to explore the waste management scenario in high-rise residential buildings. The survey questionnaire consists of three different parts: socio-demographic, socio-cultural and environmental questions. A total of 117 households were surveyed during February-May 2013. A semi-structured questionnaire was used distributed randomly to occupants of high-rise residential building. The only controlled factor was the building height (according to Bangladesh National Building Code, A BUILDING more than 6 stories is considered as a high-rise building). Figure 2 shows the key indicators used in the questionnaire survey.

The questionnaire was based on the following basic questions:

i. Socio-demographic features

    a. Name/address

    b. Profession

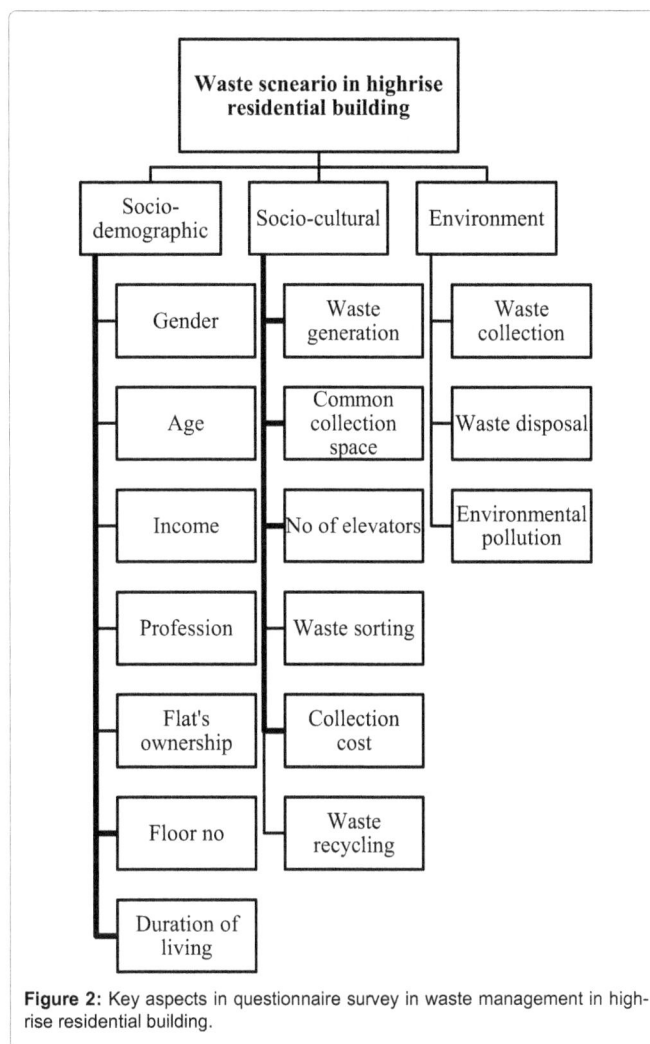

Figure 2: Key aspects in questionnaire survey in waste management in high-rise residential building.

    c. Gender

    d. Income group

    e. Flat ownership

    f. Floor level

    g. Duration of living

    h. Household size

i. No of elevators

ii. Socio-cultural features

    a. Any designated place for waste storing at household level

    b. No of bins used at household level

    c. Household waste collection provision

    d. Waste collection fees

    e. Waste management systems in surroundings areas

    f. Awareness program and training on waste management systems

iii. Environmental features

    a. Whether waste is collected separately

    b. Level of severity of waste problems

    c. Knowledge on waste separation

d. Willingness to recycle

e. Importance of proper waste management system

Results and Discussion

Socio-demographic features

Demographic features: As this study focuses on the waste management systems in high-rise residential building in Dhaka, it is important to understand the socio-demographic and economic characteristic of the participants living in the high-rise residential buildings. Table 2 shows the key characteristic of the survey participants'. The questionnaire was distributed to 117 (N=117) flats in 25 different high-rise residential buildings 62% of the respondents were male and 38% were female. About 70% of the participants live in their own apartment, 28% live in the rented and only 2% of them live in mortgaged apartments. Living in high-rise building in Dhaka is very expensive and the survey data also reflects the same. A majority (82%) of the participants belong to the higher income group (above 50,000tk/month) and only 27% were from the middle income (25-50,000tk) group. The average household size is 4.54. The average household size in the residential buildings of Dhaka calculated as 4.7 in 2010, quite close to the size found in this survey [16].

Participants' affiliations: Figure 3 shows the profession and affiliation of the participants.

Floor level surveyed: In this study, the building height (in this study the floor level) and the access to the flats are important factors in waste collection system. Figure 4 shows that, only 17 flats were located in buildings below six storeys and the remaining100 flats surveyed were located above six storeys. The highest floor level surveyed is the 19th floor (N=4). Most of the participants were from the 6th to 12th floor level.

No of elevators in the high-rise buildings: Accessibility for waste collection by collectors to the flats in the high-rise residential buildings is regarded as important. The high-rises do not have a service elevator and thus waste needs to be transported by the passenger elevators. An integrated waste collection system such as chute system is not mandatory for high-rise residential building as per the rules in Bangladesh National Building Code Therefore, waste is mostly collected on a door to door basis. It is obvious that every high-rise residential building has at least one or higher number of elevators. Figure 5 shows the Pareto diagram of the elevator number. From figure 5, 61 households have at least one elevator and 33 households have two, 18 households have three and 5 households have four elevators.

Duration of living: As shown in Figure 6, about 80% of participants have been living in their flats for 6-10 years which is comparatively long. The reason may be owing to the ownership of the flat, i.e. most of the participants own their flats. Community based waste collection systems is different in different wards or local sub-divisions. Long duration of flat occupancy has some benefits on overall waste management systems such as familiarity of wards based collection system, informal waste recycling and so on.

Socio-cultural features

Any designated place for waste storing at household level: Compared to developed cities, local government in Dhaka does not provide effective waste service to every household. Designated

Indicators		Frequency	Per cent
No of unit surveyed (N)		117	100%
Gender	Male	73	62%
	Female	44	38%
Housing Ownership	Own	82	70%
	Rent	33	28%
	Mortgage	2	2%
Income group	Middle income	21	18%
	Upper-middle income	31	27%
	Upper income	28	23%
	Upper-upper income	37	32%

Table 2: Key features of the participants.

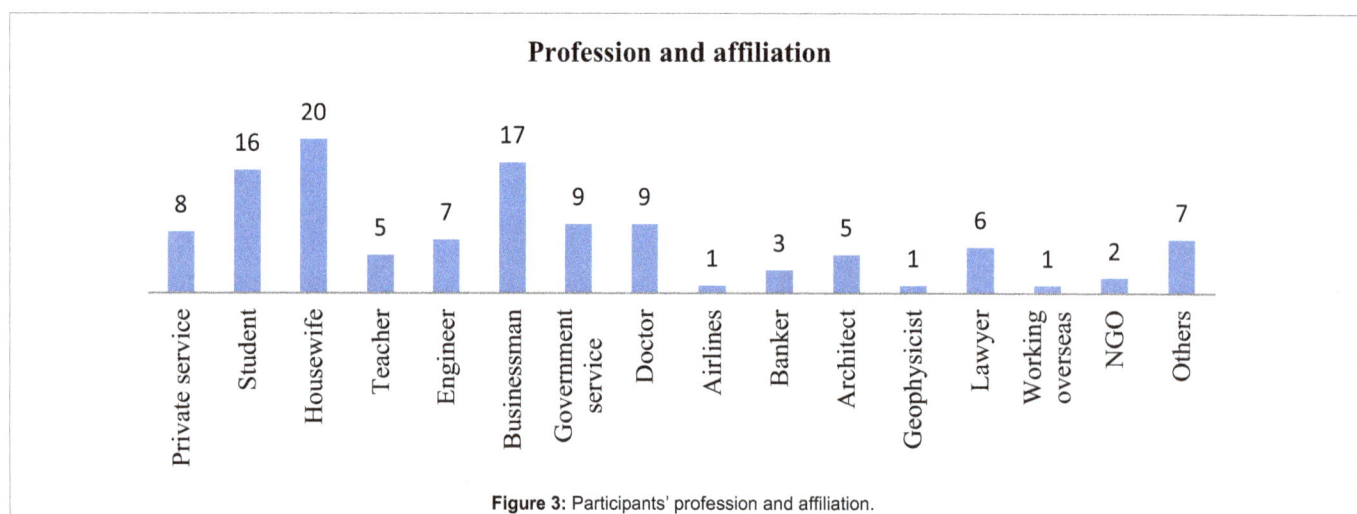

Figure 3: Participants' profession and affiliation.

Frequency of participants' floor level

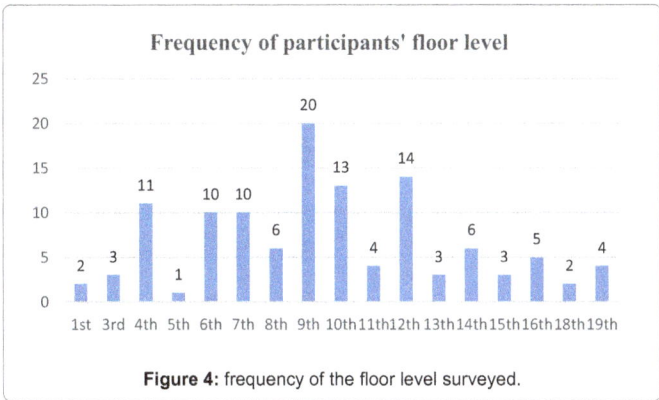

Figure 4: frequency of the floor level surveyed.

No of elevators in surveyed buidlings

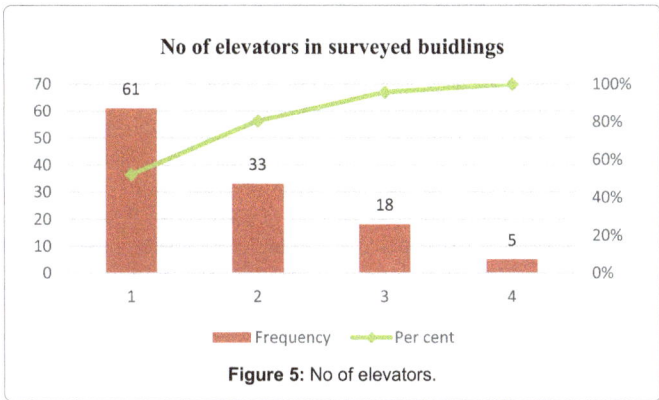

Figure 5: No of elevators.

Participants'occupancy period in flats

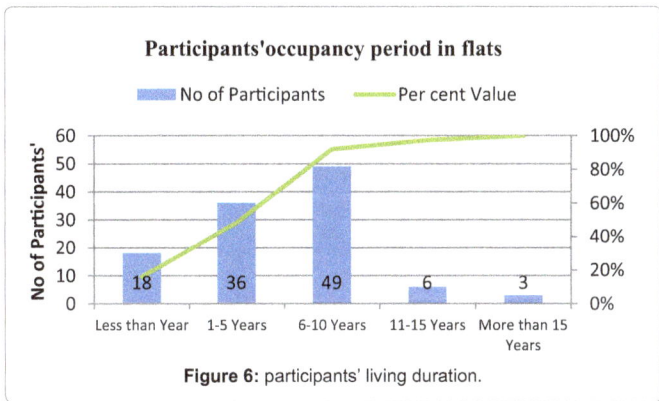

Figure 6: participants' living duration.

place for waste collection system is vital for effective waste collection and recycling systems. As shown in Figure 7, more than 97% of the households do not have the provision for common waste sorting place. Only 2.6% of the household have a common sorting place.

No of bins used at household level: Each of the 117 flats surveyed had only bin for collecting their waste. Waste from the individual bins of all the flats are emptied in the bin brought in by the waste collector during the door to door collection service. Even though in some apartment house, recycle bins were provided by the housing society, however they found unused and used inappropriately at the household level. Figure 8 shows the misuse of recycling bin in apartment in Dhaka.

Household waste collection provision: The local government in Dhaka does not collect waste at household level; waste is thus collected by other means and piled to the common garbage collection points such as roadside waste bins. Figure 9 shows the common way of waste

collection system in high-rise residential building in Dhaka. Wastes in 111 flats out of the 117 unit were collected by locally recruited collectors under the community based waste collection system. Only 5 flats had garbage chute and in one flat, wast was collected by a servant and transported to the local roadside waste bin.

Waste collection fees: Households waste is collected by community based door-to-door waste collection systems. Residents of the households bear the monthly collection fee. Most of the cases waste fee was included with the apartments' service charges. Approximately 50-200TK (USD 0.75-3) was allocated by every household for monthly waste collection systems, which was very little, compared to other basic services costs.

Waste management systems in surroundings areas: The high-rise residents were asked to rank the overall waste management systems in their neighbourhood. 62 participants (53%) out of 117 ranked waste management systems in the surrounding community as moderately good, 22 ranked very well and 9 ranked excellent. Only 23 were ranked

Designated storing place for household waste

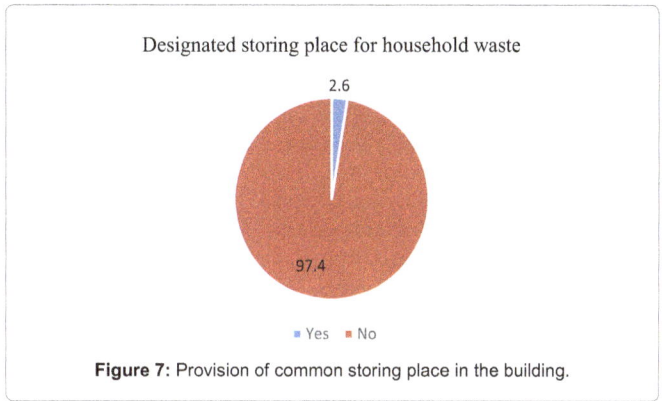

Figure 7: Provision of common storing place in the building.

(a) Cloths stored in bin (b) unused bins (c) paper in organic bin

Figure 8: Inappropriate use of recycle bins in apartment house in Dhaka.

Waste collection system from household

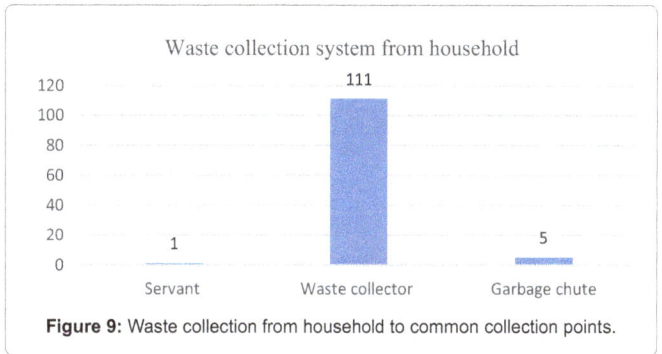

Figure 9: Waste collection from household to common collection points.

waste management systems as' somewhat good' to' poor'. Figure 10 shows the overall quality of waste management systems.

Awareness program and training on waste management systems: Success and effectiveness of waste management depends on the awareness and knowledge on waste management among local residents. However, lack of awareness program and funds makes it a seemingly difficult task to provide awareness to the local community. About 92.3% of the respondents indicated that no awareness program on waste was provided by local government. Only 7.7% of the respondents were aware of any kind of awareness program on waste provided by the local government. Figure 11 shows the percentage of waste management awareness programme provide in the surveyed households by the local government through any possible media.

Environmental features

Whether waste is collected separately: Sorting of waste at household level is very important for higher recycling efficiency. Provisions of formal waste recycling bins are thus important for sorting at household level. Only one household had separate waste collection system and the rest of 116 (99%) flats had no common separate waste collection system. Table 3 shows the provision of separate waste collection system in the surveyed flats.

Level of severity: Waste collection capacity in Dhaka is not enough to collect all the waste generated each day and hence, dumping of waste at roadside is a common problem in Dhaka. Figure 12 shows the level of severity of waste management systems in Dhaka. 46 participants (39%) out of 117 (N=117) ranked waste management system in Dhaka as very severe, 36 participants (30%) ranked as moderately severe, 32

participants ranked as somewhat severe and rest of 2 participants were ranked as not severe.

Knowledge on waste separation: Knowledge on waste sorting and recycling is vital for effective waste management services. The socio-economic condition of the participants of this survey is fairly good compared to the national average. Figure 13 shows the level of knowledge on waste separation at household level. Eight participants out of 117 claimed that they have excellent level of knowledge on waste separation. 42 ranked them as good, 41 ranked as moderate, 10 ranked as low and 15 participant said they really do not know about waste separation. Despite overall high level of waste separation by the informal sector, formal waste recycling is not high in Dhaka due to lack of waste infrastructure and absence of formal waste recycling system in place.

Willingness to recycling: The participant were asked whether they were willing to sort, separate and recycle if the bins or provision were provided by the local authority. Majority of the participant showed their willingness to sort waste at household level. From Figure 14, a total 7 participants out of 117 said they are willing to separate at any cost, 45 participants were willing to sort if they can afford the system, 41 participants were willing to sort but due to lack of infrastructure in place they cannot sort time at this moment. Only 10 participants were not willing to sort because they found the system costly and the remaining 13 participants were not willing to sort at all.

Importance of proper waste management system: Figure 15 shows participants' degree of importance on waste management systems. 37 participants (32%) out of 117 thought waste management system is extremely important, 47 participants (40%) though very important, 27 participants ranked as moderately important and only 6 participants ranked waste management as slightly important for the society.

Figure 10: Quality of waste management systems.

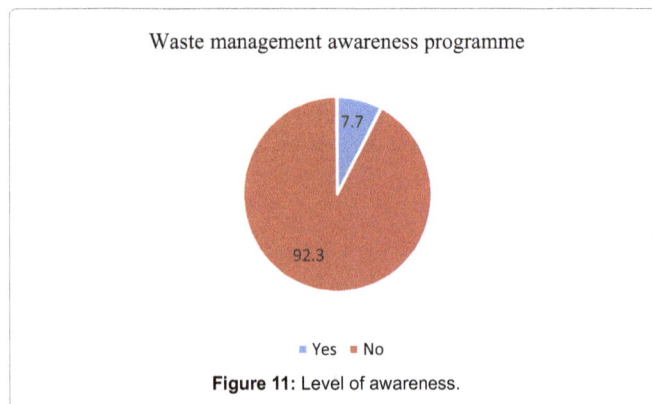
Figure 11: Level of awareness.

Separate waste collection system	Frequency	Per cent
Yes	1	.9
No	116	99.1

Table 3: Provision of separate waste collection system.

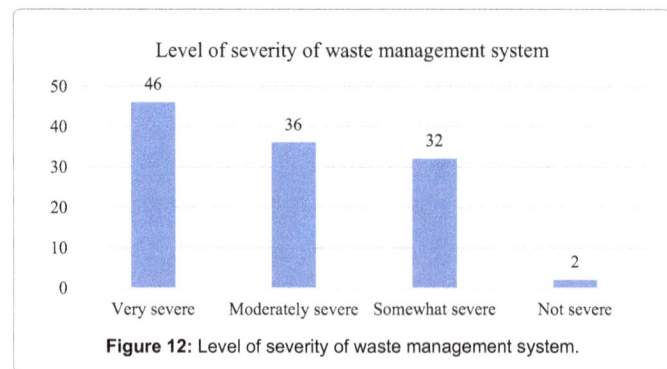
Figure 12: Level of severity of waste management system.

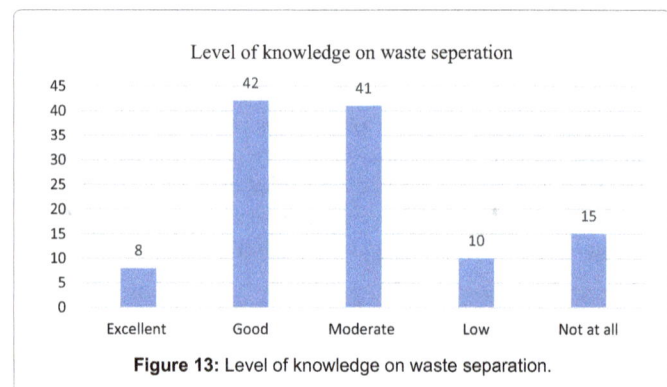
Figure 13: Level of knowledge on waste separation.

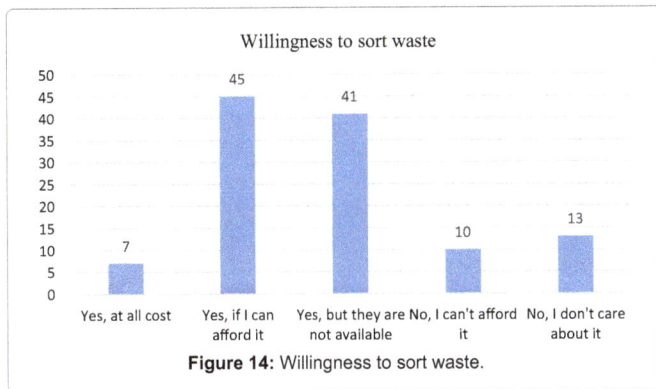

Figure 14: Willingness to sort waste.

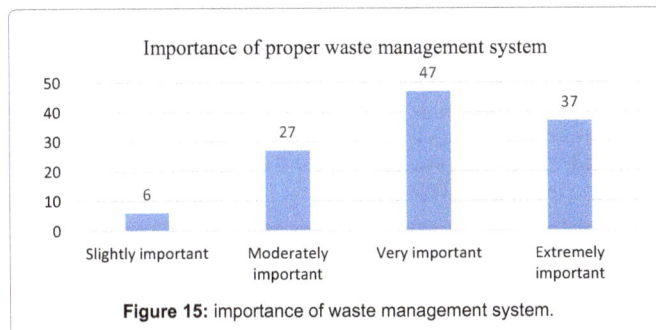

Figure 15: importance of waste management system.

Conclusion and Future Directions

Waste management systems and techniques vary depending on socio-economic, demographic, geographical and cultural differences. Despite poor waste collection system provided by the local authority in Dhaka, waste management systems in the high-rise residential buildings could be very efficient and different from the waste management systems in the other types of residential buildings for the following e key reasons: (i) income of the high-rise residential users is high compared to other types of residential building users', (ii) no of people live in per capita land area which affects the total households waste generation rate, (iii) fees for waste collection system which ensures frequent waste collection from households, (iv) no of elevation and access to the waste collectors which is not always obvious for every high-rise residential buildings in Dhaka, (v) designated and common waste sorting place to separate and recycle with high efficiency, (vi) total no of bins use for separation, (vii) awareness and recycle training programme and (viii) residents' willingness for recycling.

From the results of the survey findings it is evident that the residents of the high-rise buildings are higher-medium to upper income group people with higher social and economic status. As a result, they are also more aware of social issues such as waste problems in Dhaka. People from the surveyed apartments (N=117) in Dhaka shows a great sign of willingness to recycle right if the provisions and infrastructure are in place. They are also willing to bear the extra waste management cost. So now it is up to government to provide such infrastructure and service to the users.

The study identified some key areas in waste management systems in high-rise residential building which can be further studied to analyse the need for holistic waste management strategies for high-density residential development in Dhaka.

An extended study can be done to identify the provisions for the integration of waste management infrastructure in high-rise residential buildings. Long term waste collection and management strategies in high-rise residential buildings in Dhaka can be studied more comprehensively.

Acknowledgement

The authors are very grateful to the respondents who participated in the survey. The authors thank two anonymous referees for their insightful comments.

References

1. Dhaka City Corporation (DCC) (2013) Solid Waste Management.

2. Matter A, Dietschi M, Zurbrügg C (2013) Improving the informal recycling sector through segregation of waste in the household: The case of Dhaka Bangladesh Habitat International 38: 150-156.

3. Ministry of Environment and Forests (MoEF) (2010) National 3 R strategies for waste management. Government of the People's Republic of Bangladesh.

4. Islam S (2013) Solid Waste Management in Dhaka City

5. Bhuiyan SH (2010) A crisis in governance: Urban solid waste management in Bangladesh. Habitat International 34: 1 25–133.

6. Dhaka North City Corporation (DNCC) (2013) Waste management budget in Dhaka.

7. Afroz R, Hanaki K, Kiyo HK (2009) Willingness to pay for waste management improvement in Dhaka city Bangladesh. Journal of Environmental Management 90: 492-503.

8. Ahmed SZ, Ali SM (2006) People as partners: Facilitating people's participation in public–private partnerships for solid waste management. Habitat International 30: 781-796.

9. Enayetullah I, Sinha AHMM, Khan KH, Roy KS, Kabir SM, et al. (2006) Report on baseline survey on solid waste management in UttaraModel Town, unpublished results.

10. Hai, Faisal I, Ali M, Ashraf A (2005) Study on Solid Waste Management System of Dhaka City Corporation: Effect of Composting and Landfill Location. UAP Journal of Civil and Environmental Engineering 1: 18-26.

11. Hasan S (1998) Problems of municipal waste management in Bangladesh: An inquiry into its nature, Habitat International 22: 191-202.

12. Shah E (1999) Community based separation at source in Bangalore, India.

13. Sufian MA, Bala BK (2007) Modelling of urban solid waste management system: The case of Dhaka city, Waste Manag 27: 858-868.

14. Zaman AU (2012) Developing a Social Business Model for Zero Waste Management Systems: A Case Study Analysis. Journal of Environmental Protection 3: 1458-1469.

15. Rajuk (2013) Uttara Apartment Project: project information.

16. Bangladesh Bureau of Statistics (BBS) (2001) Population Census Dhaka Bangladesh.

34

Transaction Costs (Tcs) Framework to Understand the Concerns of Building Energy Efficiency (BEE) Investment in Hong Kong

Queena K Qian[1]*, Steffen Lehmann[2], Abd Ghani Bin Khalid[3] and Edwin HW Chan[4]

[1]*Endeavour Australia Cheung Kong Fellow, Center for Sustainable Design and Behaviour (sd+b), University of South Australia, Australia*
[2]*Professor of Sustainable Design and Behaviour, School of Art, Architecture and Design Director, Zero Waste Research Centre, The University of South Australia, Australia*
[3]*Professor, Faculty of Built Environment, Universiti Teknologi Malaysia, Malaysia*
[4]*Professor, Building and Real Estate Department, The Hong Kong Polytechnic University, Hong Kong S.A.R., China*

Abstract

Factors, such as split incentive, information asymmetry, opportunistic behavior, ill-informed users, and institutional transitions, etc., incur different levels of Transaction costs (TCs) and affect the stakeholders' willingness to take part in building energy efficiency (BEE). A better understanding of the nature and structure of TCs is essential to improve the market mechanisms for BEE investment. It covers three dimensions of TCs: specific investment, frequency and uncertainty. The paper provides a framework to understand BEE barriers in general and the TCs concerns of stakeholders in particular. In-depth interviews questions are designed to be conducted with the real estate representatives and architects, using a case of Hong Kong, where real estate developers are chosen to be the study object as they are the initiative and dominate force. The study focuses on how to smooth BEE transactions and lessen TCs involved. It indicates that TCs are the key factors impeding BEE market penetration, and will provide references to design a governance structure as well as to design policy packages to promote BEE.

Keywords: Transaction costs (TCs), Building Energy Efficiency (BEE), Real estate developer, Hong Kong

Introduction

Buildings account for 40% of global energy consumption and nearly one-third of global CO_2 emissions Levine et al. [1]. New buildings that are energy-inefficient are being built every day, and millions of today's inefficient buildings will remain standing in 2050 [2]. Moreover, the energy usage of buildings is growing rapidly as more people move into modern homes and acquire amenities such as heating, cooling, and refrigeration. Large and attractive opportunities exist to reduce buildings' energy use at lower costs and higher returns than in other sectors. Compared to developed economies, developing countries in general lack the incentive and technical knowhow to pursue sustainability [3]. There is an urgent call for the developing countries to raise their awareness and contribute their efforts on BEE development so as to combat the climate change and address the environmental concerns [4].

Up to 50% of all energy is consumed by buildings, including the development of materials, construction, and operation. In Hong Kong, for example, buildings consume over half of all energy and about 89% of electricity, mainly for air-conditioning, which is the source of roughly 17% of all Hong Kong's greenhouse gas emissions [5-7]. In practice, improving BEE is complicated due to the many parties and factors involved: the government, the market, the end-users, many practitioners, a range of technologies,, and a variety of cultures. It would be helpful for governments to know how to oversee BEE development most efficiently. BEE studies, though complicated, are necessary for improving energy efficiency and must involve more than just improving technology. Reports [2,8] show that with currently available technology, the energy-efficiency level could be increased by 40%, yet this does not happen. There must be some underlying reasons that call for the attention and collaboration among the key players of governing institutions, based on multi-disciplinary studies that consider economics, politics, society, technology, and so forth.

Economic theories suggest that market structure and performance is determined by the ease of entry to and exit from a market [9]. Chiang found that the institutional environment in Hong Kong led to the market concentration of the construction industry [10]. Building contractors compete intensely over cost reductions rather than technology improvements. According to the Hong Kong Consumer Council's study, the local property development market was also highly concentrated [11]. It is still true that only the large developers with superior financial resources can remain active in the sector. Under such market situations, the key market players have little incentive to venture into the new business of green building. Compared with conventional building, the entry barrier to the BEE market is higher because of the new information, expertise, new technology, and financial risk involved. If there is asymmetric information about quality standards or mandatory requirements that are not imposed on the market, the opportunistic behavior of most market players may make them continue to produce conventional buildings [12].

From the new institutional economics perspective, when Transaction Costs (TCs) are too large, they inhibit exchange, production, and economic growth. The functioning of TCs under alternative institutional arrangements is also crucial to the workings of markets [13-17]. From the transaction cost economics, energy efficiency is a coordination and incentive problem rather than one of utility maximization [18]. This view also emphasizes that policy

***Corresponding author:** Queena K Qian, Endeavour Australia Cheung Kong Fellow, Center for Sustainable Design and Behaviour (sd+b), University of South Australia, Australia
E-mail: kun.qian@fulbrightmail.org

interventions and different institutional structures may lower TCs and provide net social benefits [8,18,19]. The situation calls for a thorough study focusing on how to smooth transactions for market stakeholders in energy-efficient development, with the aim of lessening the TCs involved in BEE transactions.

Figure 1 illustrates the key issues of this research for which a critical review of the literature is provided to develop a clear understanding of how they relate to one another. The consolidated issues are summarized to help develop the research questions and propositions.

This research mainly focuses on how to smooth transactions among the market stakeholders in energy efficiency development in order to realize the energy-saving target. The study does not focus on any particular type of building technology, but rather on how to marketize energy-efficient residential buildings to be more acceptable to market stakeholders. It is to lessen the TCs due to the barriers to BEE and to propose policy packages accordingly. Its intention is to identify key areas where policy initiatives can help address the market's needs for BEE with follow-up empirical study.

Literature Review

TCs approach

"Without the concept of transaction costs, which is largely absent from current economic theory, it is my contention that it is impossible to understand the working of the economic system, to analyze many of its problems in a useful way, or to have a basis for determining policy" [20].

Transaction costs, in Coase's original formulation, refer to "the cost of using the price mechanism" or "the cost of carrying out a transaction by means of an exchange on the open market" [21,22]. In Demsetz's study, "TCs may be defined as the cost of exchanging ownership titles" [23]. Gordon consolidated definition of TCs as the expense of organizing and participating in a market or implementing a government policy is the definition used in this study [24]. A number of transaction-cost issues arise with respect to the development and implementation of BEE incentive schemes. Adapting this definition is in line with the work of other authors who treat TCs and administrative costs as essentially interchangeable terms [25]. As Coase explains, "In order to carry out a market transaction, it is necessary to discover who it is that one wishes to deal with, to inform people that one wishes to deal and on what terms, to conduct negotiations leading up to a bargain, to draw up the contract, to undertake the inspection needed to make sure that the terms of the contract are being observed, and so on" [22]. Thus, there is reason to consider changing institutions, formal and informal rules and their enforcement arrangements, to the extent that these influence the nature of transactions and thus their costs.

Transaction cost economics (TCE) argues that markets and organizations provide alternative means of organizing economic activities and that the choice between them depends upon a number of factors, including the relative magnitude of TCs [26,27]. In common with orthodox economic theory, TCE explains the behavior of individuals rather than social structures and assumes these individuals to be rational actors in that they seek out opportunities to improve economic efficiency. This research applies TCE to study the underlying reasons why the market is reluctant to accept BEE by choice. The findings help establish the study's discussion on how to choose a particular governance structure to solve the existing problem.

In empirical studies, a direct measurement of TCs is simply the economic value of resources used in locating trading partners and executing transactions. Another common measurement of TCs is the difference between the prices paid by the buyers and received by the sellers. Some studies focus more on secondary costs than on direct costs per se. For example, Williamson Ian TCE is primarily interested in the secondary costs of negotiation and enforcement. Some are concerned with the cost of government regulations imposed on market entry and transactions, which either reduces the size of the market or eliminates the market altogether. In this study, the key TCA independent variables for measuring the preference of developers' for BEE investment are asset specificity (or specific investment), uncertainty (economic, market and policy uncertainty), and frequency. Asset specificity refers to durable investments that are undertaken in support of particular transactions. These specific investments represent sunk costs that have a much lower value outside of these particular transactions [27]. Uncertainty refers to three aspects: economic uncertainty, market uncertainty and policy uncertainty. Frequency refers to how often the buyers make purchases in the market [27]. Figure 2 is a TCA Model developed for this study to help understand developers' preferences for BEE investment with the consideration of TCs. Three measurement indicators for TCs items in this study are money, time, and worry. Propositions will be developed, and a set of interviews will be conducted with real estate developers and their representatives to determine the importance of TCs.

BEE and its barriers

With socioeconomic progress, more market stakeholders are getting involved in the building sector and are dedicated to their own business interests. Real estate developers intend to do no more than

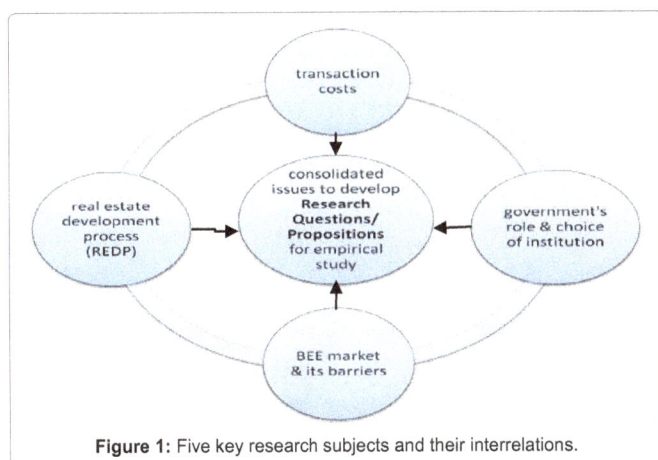

Figure 1: Five key research subjects and their interrelations.

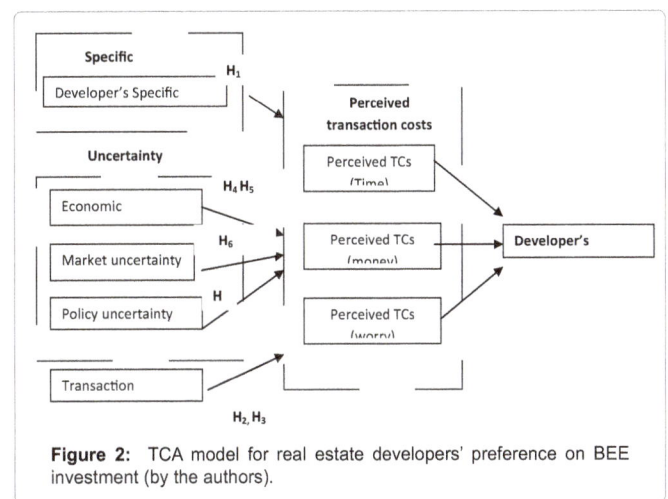

Figure 2: TCA model for real estate developers' preference on BEE investment (by the authors).

obey the basic requirements of the law and regulatory policies to minimize the increasing costs engendered by the extra work entailed by mandatory energy regulations. Contractors also want to avoid these extra tasks, because they require special expertise and specialized equipment that they do not typically possess. Manufacturers of BEE products want regulations to be still stricter to create greater demand. Building-design institutes will not be greatly influenced by the new policies but are apt to succumb to the demands of developers because of the nature of their relationship with them. However, these interests have not yet been fully expressed by the stakeholders themselves. These conflicting interests are the main source of the risks of and barriers to BEE development.

The number of barriers is enormous – according to some estimates, they are higher in the building sector than in any other sectors [8,28]. In this context, a barrier refers to a mechanism that inhibits decisions or behavior that appear to be both energy and economic efficient. In particular, barriers are claimed to prevent investment in cost-effective energy-efficient technologies [29]. The terms "barrier" and "market barrier" were introduced by researchers using engineering-economic models to study the technical and economic potential for energy efficiency. Often little interest in investments with high rates of return led to postulate that such BEE investments were inhibited by various barriers. This study has developed a framework for analyzing BEE barriers and TCs incurred based on earlier work [29,30-32]. The interview questions for the empirical study are designed based on this literature review framework (Figure 3).

Barriers relating to the BEE market (Developers and End-Users) and government's role

This study focuses on two main market players involved with energy efficient buildings – the real estate developers and the end-users,

who are at the two ends of the delivery of energy-efficient buildings. It makes sense to believe that these two players are so interrelated in the market that any concerns that hinder them from investing in BEE will eventually keep the transaction from happening. Therefore, this part of the analysis is to determine how their interactions affect their willingness to invest.

Promoting BEE requires that government and all parties in the market work together. By and large, the government agencies concerned with energy efficiency end up confining themselves to providing publicity and information. The government needs to play the role of a moderator to make it convenient for the market to embrace BEE. The growth of the BEE market requires a politically friendly environment with the appropriate combination of government intervention and flexibility; it also needs a well-designed institutional structure to encourage investment and change the business culture. The government's role is mainly to set out a good foundation (the well-organized institution) and a clear domain (clear of constraints, but also some flexibility) for the BEE market stakeholders.

The hypotheses and the interview questions from BEE barriers- an overview

Design of the interview question was based on the discussions on BEE barriers, to address three major theoretical dimensions of TCs: specific investment, frequency, and uncertainty. Seven hypotheses regarding these three aspects were developed, and related open questions about the interviewees' opinions were designed to test each of them (Table 1). Table 1 summarizes the barriers discussed in literature that the authors take reference to develop the hypotheses for empirical study. The hypotheses and the interview questions were developed based on the literature review and pilot interview with a few experts in industry and academia. The relations between the three dimensions,

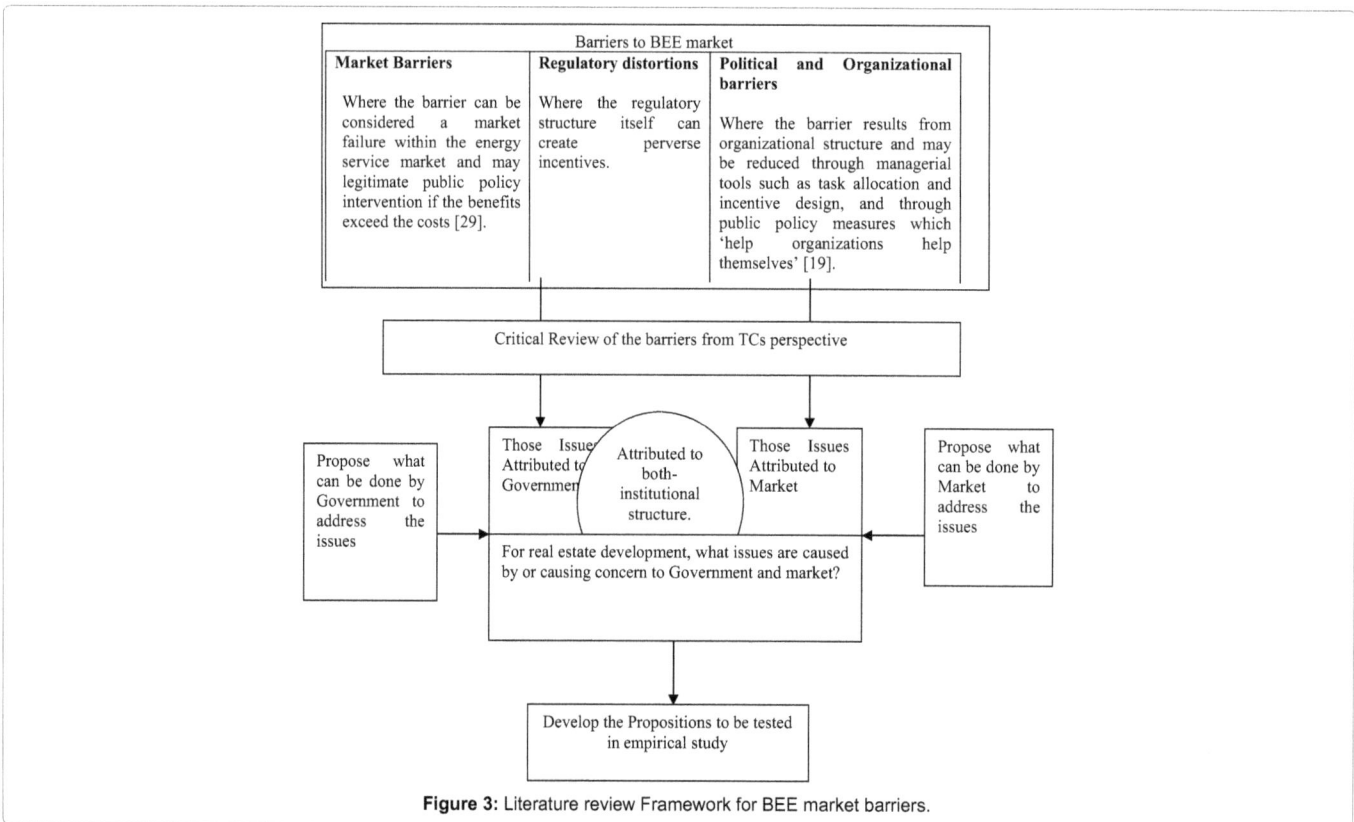

Figure 3: Literature review Framework for BEE market barriers.

Barriers	Common Claims	TCs considerations of BEE- developed to Hypotheses
Risk aversion	• BEE investment represents a higher technical or financial risk. The business & market uncertainty encourages short time capital return. • To replace familiar technologies and partnerships with new but more efficient ones is difficult due to risks, including economic fluctuations, policy instability, possible delay, and litigation, which should average out across the entire society and yield a positive economic return.	
Hidden costs and benefit	• BEE potential may be overestimated by failing to account for the reduction in utility associated with BEE technologies and other additional costs. • The hidden costs and benefits are not captured directly in financial flows, including costs associated with securing the energy efficient solution and risks associated with the replacement technology. • TCs are often high due to the fragmented structure of the building sector with its many small owners and agents. • New technologies may not be compatible with existing sockets. • The indirect benefits of improved energy efficiency, such as reduced air pollution and improved health, are often ignored.	**Specific investment:** H1: Dividing the transactions of the real estate development process into smaller established stages helps government to better understand the process and make policies with a more focused emphasis on the different stages of transaction to promote BEE more efficiently.
Imperfect information	• Lack of information about the possibilities and techniques for and potential of energy-efficient solutions is a major barrier, especially in developing countries. • TCs of information acquisition may be high due to quality and credibility.	**Frequency:** H2: There is a positive relationship between the size of the company and the TCs in BEE projects. H3: There is a positive relationship between the frequency of BEE investments and the TCs incurred in developing BEE projects.
Negative externality	• Non-BEE buildings consume more energy and release more carbon emissions, which are the negative externalities and need to be taken into account and be fairly apportioned to keep the end-users and developers from losing the motivation to further invest in BEE.	**Uncertainty- economic** H4: The economic context (upturn or downturn economic transition) affects the concerns of the real estate developers about BEE investment. H5: Changes in economic conditions call for the attention of government to adjust BEE policies as necessary to seize BEE development opportunities.
Access to capital	• BEE investments may be avoided if market stakeholders have insufficient capital through internal funds, and has difficulty in raising additional funds, due to internal capital budgeting procedures, investment appraisal rules and the short-term/instable incentives. • Higher capital costs raise the uncertainty and opportunity costs to the stakeholders, especially if the investment is financed by a mortgage or other loan. Besides, BEE investment would normally require a longer payback period, which increases business risk.	
Public goods	• BEE would prevent society from consuming extra energy and releasing unnecessary pollution, and have a collective effect society as a whole would benefit from, and many would benefit as free riders. BEE itself creates a lack of interest in itself as a business initiative. • The availability of BEE information is also a public good. The public requires a large flow of extra information to have confidence in breaking its routine to invest in BEE. Market stakeholders need to have public and transparent information about technology.	
Bounded	• Owing to constraints on time, cost, and the ability to process information, imitated knowledge of the stakeholders on BEE investment, they either have an irrationally high expectation for the BEE investment	**Uncertainty- market** H6: The end-users' variable expectations about BEE increase market uncertainty to the developers (e.g.., they may misinterpret a focused group as the end-users of their final products).
rationality	• Return and/or payback period or more interested in pursuing other short-term alternatives or to neglect the small cost savings from the energy efficiency improvement. • Changing behavior or lifestyles is very difficult. A lack of awareness and information about the opportunities and low costs of energy savings are a related problem. Energy subsidies are considered to be one of the most important BEE barriers in developing countries.	
Split incentives	• The cost and benefit of BEE is bear by different parties and difficult to appropriate among the investors. • The developers are reluctant to invest unless someone is going to pay for it. Similarly, utilities have no direct interest in measures for reducing their clients' energy use.	
Regulatory distortions	• The regulatory structure can create perverse incentives in form of regulatory bias, under-priced energy, building codes, and subsidies to established energy technologies. The building codes are not updated in a timely fashion, which inhibits technology innovation and interferes with efficient construction; inconsistency confuses the market and creates obstacles to achieve economies-of-scale for BEE.	**Uncertainty- policy** H7: The earlier the stage of BEE policy implementation, the greater the real estate developers' concern about TCs.
Political and Organizational barriers	• A lack of government involvement in promoting BEE due to inadequate enforcement structures and institutions; inappropriate government intervention that distorts business activities; the inflexibility of local governments; an insufficient number of qualified personnel; the lack of a long-term energy conservation mechanism; a lack of credible third-party agencies; the slow pace of institutional reform; worries about social stability; policies or programs that are incompatible with one another; resistance from interested parties; legal & urban-planning constraints; weak investment culture; weak managerial supervision of manpower & organization; problems with multi-institutional collaboration & coordination; local governments' resistance to change; and corruption.	

Simon [36] , Stern and Aronson [37], Koomey [31], Golove and Eto [19], Alam [38], Deringer [39], Shove [40], Wrestling [41], Evander [42], Sorrell [29], Chappells and Shove [43], Carbon Trust [44], Vine [45], Koeppel [8].

Table 1: Barriers to BEE and their TCs considerations to develop the hypotheses for empirical interviews.

seven hypotheses (H), and fifteen interview questions (Q) are listed in Tables 2-4 below. The Remarks explain how the interview questions relate to the hypotheses.

Design of the Interview Questions from the Hypotheses- In Details

Questions for Specific Investment

Specific investment in BEE increases the workloads of developers and the resources they need, which increases their concerns as they decide whether to make a BEE investment.

Hypothesis (H1) proposes that in securing a detailed understanding of the BEE elements, such as technologies, appliances, or inputs for specific investments, it is better to break down the real estate development process, by the difficulty of retrofits (Q2), by the type of buildings (luxury or low-price), (Q3). The purpose of these questions is to determine whether the policies can be designed for a highly specific group with effective incentives for securing investments in BEE.

Q1 is to elicit the underlying reasons and the approximate limit (as percentage of the development budget) that the developers would be willing to invest in BEE without incentives from the government. The purpose is to see if the government could create a business environment conducive to BEE with any market interventions. Q4 addresses one of the most notorious features of BEE – misplaced interests – in order to understand how they affect the current situation, determine what about them concerns developers, and determine what, if any, resolution is called for by the market. These five interview questions collectively address, from different perspectives, the issues raised by Hypothesis H1. The open question format allows the interviewees to talk freely about their concerns in a wider context.

Questions for frequency of BEE investment

The frequency of BEE transactions is another dimension that affects TCs. How frequently the developers invest in BEE may affect their concerns differently. The TCs may thus change accordingly (H3).

Q7 and Q8 are the two questions that address the relationship between the level of concerns about BEE investment and the frequency of BEE transactions. The nature of this relationship may help governments design different policies to encourage investment by frequent and occasional investors by taking into account their different concerns.

The size of the company and the size of the project also affect investors' capacity to invest and, therefore, the frequency with which they do so (H2). Big companies may have different concerns and strategies than smaller ones when it comes to BEE investments (Q6). To integrate green features into bigger projects may have different impacts in terms of TCs, compared to smaller ones (Q5).

To understand how changing concerns are a function of the size of the RED company or project and the frequency of BEE investment requires knowing market segmentation according to both size and frequency. This information allows government to design and specify incentives for more focused groups.

Questions for uncertainty

Uncertainty about BEE investments is one of the general features of TCs that causes real estate developers worry. Uncertainty is examined in this study from three perspectives: economic uncertainty, market uncertainty, and policy uncertainty.

What is the impact of economic transition on the BEE development (to the developer – H4; to the government – H5)? Is it a challenge or an opportunity? How do the developers' concerns change in an economic

Q1	What are the reasons that make developers willing to invest in new BEE technology without government incentives? What price difference (% of development cost) would be acceptable?
Q2	Uneven emphasis on incentives: What facilities/building elements are more expensive/difficult to be retrofitted, if not installed in the first place? Should they be emphasized in incentives to promote BEE investment?
Q3	For developers investing in BEE, what are the different concerns of investing in luxury buildings and in lower-priced buildings? Why?
Q4	There is misplaced benefit between the people who pay and who gain from BEE. To address this problem, would rental/selling-price differences help?

Table 2: Specific Investment – relating to H1.

Q5	Will the size of the project affect the developers' concerns about BEE investment?
Q6	Will the size of the RED company affect the developers' concern about BEE investment?
Q7	How does the frequency (e.g., regular, occasional, or at one-time) of developers' BEE investments affect their concerns about BEE investment?
Q8	Would the developers' concerns change if they invested in BEE projects more frequently? Why?

Table 3: Frequency: relating to H2-H3.

Economic uncertainty	
Q9	At times of economic transition, what new challenges or opportunities arise for investments in BEE? How do shifts in the economy change the developers' major concerns (neutral, positive, or negative) and in which aspects?
Q10	When the direction of the economy shifts, how might developers integrate green features into original investments to increase market competitiveness?
Q11	What role should government play in BEE promotion (more intervention or less intervention in a recessionary economy)?
Q12	What BEE promotions or incentive could government introduce in times of economic change that would be less upsetting to the market players' normal activities?
Market uncertainty	
Q13	Occupants' behavioral differences may lead developers to produce different BEE/GB at different performance levels. What is your view?
Q14	Will concerns about social classes (different education levels, experiences, financial ability to enjoy the benefits of BEE) affect the developers' concern about BEE investment?
Policy uncertainty	
Q15	Would a new incentive and a currently mature incentive affect the developers' concerns about BEE differently? In other words, encountering BEE incentives, would the developers have more concerns during the early or later stage of the implementation of the incentive? How are they different?

Table 4: Uncertainty: relating to H4-H7.

downturn or upturn? What should government be alert to during such periods and how can it develop the most effective policies to promote BEE accordingly? These are the main issues that are addressed in interview questions Q9- Q12.

The market also creates many uncertainties for developers. They may be hesitant to invest in BEE due to a lack of confidence in estimations of market demand. The end-users' expectations and concerns about BEE may be better known, so that both the developers and the government could seize the opportunity to promote BEE. This brings H6 onto the horizon. Q13 and Q14 are designed to detail the behavior and concerns of the market end-users about BEE by segmenting the customers so that the real estate developers might have a more confident business strategy and so that the government can design its incentive policies to cater to more focused groups based on a better understanding of the needs and concerns of both end-users and developers.

Policy also affects uncertainty during different implementation stages. This uncertainty affects the worries and enthusiasm of the market variously, thus affecting the effectiveness of the policies themselves. The policy uncertainty is based on the assumption that the timing of the policy's introduction is a major factor in causing uncertainty for the real estate developers (H7). Q15 is designed to elicit information about how the stage at which the policy is implemented affects the real estate developer's concerns, which gives government information that lets it have market concerns in mind as it implements policy at different points in the process.

Discussions on the Role and the Partnership of Government and Business in Promoting BEE

As an authority to set up institutions and design policies, government is more able to improve its own efficiency and internal decision-making than to improve its external counterpart, the market. Government should adopt a clear national policy to improve energy efficiency through a coherent package of policy measures. Policy mechanisms alone will not work and market forces by themselves will not achieve the potential for energy efficiency. Because the spread of energy efficiency improvements cannot be left to the market, there has to be an emphasis on policy-assisted, market-oriented mechanisms for promoting energy efficiency.

To determine the most needed policies to improve BEE in a particular society requires an in-depth understanding of the expectations of the market and government. Most policymakers regard energy efficiency principally as an environmental or social issue, rather than an economic one. Hence, policies are designed with inadequate consideration of the needs of market stakeholders and not pay enough attention to the necessity that businesses accept them. Government tends to pay more attention to the environmental consequences of energy consumption, and business enterprises may care more about their technical and financial ability to make changes, their potential economic benefits, and so forth. Detailed negotiation and greater understanding between government and the market stakeholders is needed to reach a win-win outcome.

Only when both the end-users and the developers appreciate the benefits of energy efficiency building will they create a business channel for BEE products and the BEE market. Each of the barriers discussed above provides an opportunity for policies to address, but it will involve simple matches of one policy to one barrier. It will require a careful selection and combination of a set of policy instruments to overcome these existing barriers. How, then, do we choose among so

many policy instruments? Economic theory, along with careful analysis of BEE barriers, provides guidance for matching policies to barriers.

Follow-Up Studies: Empirical Research on BEE to Be Conducted In Hong Kong

Why is Hong Kong chosen as a case study?

Hong Kong is a suitable choice for this study, as it is economically well-developed regions with free markets and fairly educated professionals for green building and energy efficiency. The GDP per capita at current market prices in 2007 in Hong Kong was U.S. $41,110 (IMD World Competitiveness Yearbook, 2007). Construction as a share of total GDP has been in the range of 5-7% in Hong Kong in recent years [33]. Harnessing solar energy through solar cells, sun-shading devices, low-emissivity glass, energy-efficient air-conditioning systems, and building-space planning and orientation are common design considerations for BEE in Hong Kong. Hong Kong relies more on voluntary effort, and there are several green groups, such as the Professional Green Building Council and the Green Council, promoting the voluntary use of BEE. The HK-BEAM and other green-label programs are accepted assessment tools promulgated by voluntary bodies in the past decade. In recent years, the Hong Kong government has begun to take an active part in driving BEE initiatives [34,35].

Why in-depth interview the real estate developers?

In-depth interviews were conducted with the executives and architects who work in big real estate development firms in Hong Kong to solicit their views on issues regarding BEE investment. The interviewees to be selected are top managers or directors from the top 6 real estate development companies, who actively worked in major real estate development firms or architectural firms, which covered 80% of real estate activities in Hong Kong. As the decision-makings and strategic plans for the real estate development- whether BEE or not, and market expectations/ concerns to BEE, are only done by people who are senior and stay high position.

The interviews to the high profile practitioners will reflect their preference to current BEE development in practice, which directly and indirectly reflect their will if and how to achieve the BEE decision-makings, and have a very heavy weight to influence the other stakeholders in the BEE market and overall BEE promotions. Therefore, the findings of the interviews are to get the perspectives of real estate developers and to check the assumptions and findings about BEE market barriers in the literature review. It is to hope to provide a reference for designing rational policy.

Conclusions

This study develops a methodology framework of the TC theories to be tested in the real world interviews in an empirical study of Hong Kong. This study has adopted a holistic approach to studying the barriers to BEE investments and has focused on TCs in particular. It provides a review of diverse literatures, including those on building energy efficiency, TCs, and real estate development. This research has comprehensively analyzed the market barriers to BEE and TCs incurred from the perspectives of the developers. The overall methodology framework is theoretically significant with the original data from a case study in Hong Kong to bring a thorough understanding of BEE market. The results will definitely help understand the real market concerns in terms of TCs regarding BEE development. It helps the policy makers to understand when, to whom, where, and how to design the policies that are in favor to the real need of the market. It, therefore, ensures the success of BEE implementation.

Acknowledgement

The work described in this paper was supported by joint research grant from The Hong Kong Polytechnic University, and University Technology Malaysia (UTM), Malaysia. Special thanks to the Endeavour Research Fellowship Program and Center for Sustainable Design and Behaviour (sd+b), University of South Australia for the supports.

References

1. Levine M, Urge-Vorsatz D, Blok K, Geng L, Harvey D, et al. (2007) Residential and commercial buildings. In Climate Change 2007 Mitigation Contribution of Working Group III to the Fourth Assessment Report of the Intergovernmental Panel on Climate Change 389-437.

2. WBCSD (2009) Energy Efficiency in Building: Business realities and opportunities. World Business Council for Sustainable Development.

3. Ugwu OO, Haupt TC (2007) Key performance indicators and assessment methods for infrastructure sustainability a South African construction industry perspective. Building and Environment 42: 665-680.

4. Qian QK (2012) Barriers to Building Energy Efficiency (BEE) Promotion: A Transaction Costs (TCs) perspective. The Hong Kong Polytechnic University.

5. Civil Exchange (CE) (2008) Submission on a proposal on the mandatory implementation of the building energy codes.

6. Environmental Bureau (EB) (2008) Policy and consultation papers: a proposal on the mandatory implementation of the building energy codes.

7. Chan EHW, Qian QK, Lam PTI (2009) The Market for Green Building in Developed Asian Cities the Perspectives of Building Designers. Energy Policy 37: 3061-3070.

8. Koeppel, Urge-Vorsatz D (2007) Assessment of policy instruments for reducing greenhouse gas emissions from buildings.

9. Baumol W, Panzer JC, Willig RD (1982) Contestable markets and the theory of industry. Structure Harcourt Brace Jovanovich.

10. Chiang YH, Tang BS, Leung WY (2001) Market structure of the construction industry in Hong Kong. Construction Management and Economics 19: 675-687.

11. Hong Kong Consumer Council (HKCC) (1996) How competitive is the private residential property market?

12. Akerlof G (1970) The market for lemons: quality uncertainty and the market mechanism. Quarterly Journal of Economics 84: 488-500.

13. Cheung SNS (1998) The transaction cost paradigm Presidential Address Western Economic Association. Economic Inquiry 36: 514-521.

14. Coase RH (1988) The Firm the Market and the Law. University of Chicago Press.

15. Benham A, Benham L (1997) Property rights in transition economies: A commentary on what economists know.

16. North DC (1990) Institutions institutional change and economic performance. Cambridge and New York: Cambridge University Press.

17. North DC (1991) Institutions. Journal of Economic Perspectives 5: 97–112.

18. Levine MD, Koomey JG, McMahon JE, Sanstad AH, Hirst E (1995) Energy Efficiency Policy and Market Failures. Annual Review of Energy and the Environment 20: 535-555.

19. Golove WH, Eto JH (1996) Market barriers to energy Efficiency: a critical reappraisal of the rationale for public policies to promote energy efficiency. Energy & Environment Division USA: 1-46.

20. Coase RH (1998) Our Challenge and Goals.

21. Coase RH (1937) The nature of the firm. Economics 4: 386-405.

22. Coase RH (1961) The problem of social cost. Journal of Law and Economics 3: 1-44.

23. Demsetz H (1968) The cost of transacting. Quarterly J of Econ 82: 33-53.

24. Gordon RL (1994) Regulation and Economic Analysis: A critique over two centuries. Topics in Regulatory Economics and Policy Series 16.

25. McCann L, Colby B, Easter KW, Kasterine A, Kuperan KV (2005) Transaction cost measurement for evaluating environmental policies. Ecological Economics 52: 527-542.

26. Williamson OE (1979) Transaction cost economics: the governance of contractual relations. Journal of Law and Economics, 22: 233-261.

27. Williamson OE (1985) The Economic Institutes of Capitalism. New York.

28. Parry ML, Canziani OF, Palutikof JP, Linden PJ, Hanson CE (2007) Climate Change: Impact Adaptation and Vulnerability. Contribution of Working Group II to the Fourth Assessment Report of the Intergovernmental Panel on Climate Change.

29. Sorrell S, O'Malley E, Schleich J, Scott S (2004) The Economics of Energy Efficiency: Barriers to Cost-Effective Investment., Edward Elgar Publishing Limited.

30. Robert GH, Carman JM (1983) Public Regulation of marketing Activity: Part I Institutional Typologies of Market Failure. Journal of Macromarketing 3: 49-58.

31. Koomey JG (1990) Energy Efficiency in New Office Buildings: An Investigation of Market Failures and Corrective Policies. Energy and Resources Group Berkeley.

32. Jaff AB, Stavins RN (1994) The energy-efficiency gap What does it mean? Energy Policy 22: 804-810.

33. Raftery J, Anson M, Chiang YH, Sharma S (2004) Regional overview. The Construction Sector in Asian Economics London.

34. Chan EH (2000) Impact of major environmental legislation on property development in Hong Kong: 273-295.

35. Chan EH, Lau SS (2005) Energy conscious building design for the humid subtropical climate of Southern China Green Buildings Design: Experiences in Hong Kong and Shanghai Architecture and Technology Publisher China: 90-113.

36. Simon HA (1960) The New Science of Management Decision. Harper & Bros New York.

37. Stern P, Aronson E (1984) Energy Use: The Human Dimension.

38. Alam M, Sathaye J, Barnes D (1998) Urban household energy use in India efficiency and policy implications. Energy Policy 26: 885-891.

39. Deringer J, Iyer M, Huang YJ (2004) Transferred Just on Paper? Why Doesn't the Reality of Transferring / Adapting Energy Efficiency Codes and Standards Come Close to the Potential? ACEEE Summer Study on Energy Efficiency in Buildings.

40. Shove E (2003) Comfort, cleanliness, and convenience: the social organization of normality. Berg Publishers Oxford and New York.

41. Wrestling H (2003) Performance Contracting. Summary Report from the IEA DSM Task X within the IEA DSM Implementing Agreement. International Energy Agency Paris.

42. Evander AG, Sieboock, Neij L (2004) Diffusion and development of new energy technologies: lessons learned in view of renewable energy and energy efficiency end-use projects in developing countries Lund: International Institute for Industrial Environmental Economics.

43. Chappells H, Shove E (2005) Debating the future of comfort: environmental sustainability, energy consumption and the indoor environment. Building Research and Information, 33: 32-40.

44. Carbon Trust (2005) The UK Climate Change Programme: Potential Evolution for Business and the Public Sector.

45. Vine E (2005) An international survey of the energy service company (ESCO) industry. Energy Policy 33: 691-704.

Pollution of Freshwater *Coelatura* species (*Mollusca: Bivalvia: Unionidae*) with Heavy Metals and its impact on the Ecosystem of the River Nile in Egypt

Faiza M El Assal*, Salwa F Sabet, Kohar G Varjabedian and Mona F Fol

Department of Zoology, Faculty of Science, Cairo University, Giza, Egypt

Abstract

The Knowledge of heavy metal concentrations in aquatic species is important with respect to genetic variation and extinction of some species and loss of biodiversity in the ecosystem of rivers and lakes. We used random amplified polymorphic DNA-polymerase chain reaction (RAPD-PCR) to examine genetic differentiation among *Coelatura* species collected from the River Nile, at two polluted locations (El-Kanater, Qalyoubyia governorate and Tura, Cairo governorate, Egypt) and the impact of heavy metal pollution on the genetic structure of these species (*C. aegyptiaca, C. prasidens, C. canopicus, C. gaillardoti* and *C. parreyssi*). RAPD PCR was carried out using five random primers (UBC 476, UBC 477, UBC 478, UBC 479 and UBC 487) that provided strong amplifications. The RAPD- PCR analysis between any given pair of species, based on the number of bands, showed natural differences or polymorphism among the *Coelatura* species under investigation. The greatest number of PCR fragments was found with primers UBC 478 and UBC 479 (6-7 bands), while less fragments were obtained with primers UBC 476, UBC 477 and UBC 487 (2-4 bands)

Primers UBC 477 and UBC 479 clearly distinguished the five studied *Coelatura* species into only three species, *C. aegyptiaca, C. parreyssi* and *C. canopicus* and primer UBC 478 showed DNA alteration concerning *C. parreyssi, C. gaillardoti* and *C. canopicus.*

Genetic diversity was also measured as the percentage of polymorphic bands for each primer.

The dendograms and the similarity index (D) showed, also, that the five studied species could be classified into only three species, *C. aegyptiaca, C. canopicus* and *C. parreyssi*

The concentration of six heavy metals (copper, cobalt, nickel, manganese, lead and iron) was determined in the soft parts of the *Coelatura* species to assess the impact of heavy metal pollution on their genetic variation. Metal concentrations in the tissues were found to be higher than the permissible limits, indicating that heavy metals might play an important role in the genetic variation of *Coelatura* species by inducing DNA damage and alteration of the genetic pattern as well as they may be the cause of the extinction of some species and the loss of biodiversity in the ecosystem of the freshwater ecosystem.

Keywords: *Coelatura; C. aegyptiac;, C. prasidens; C. canopicus; C. gaillardoti; C. parreyssi;* Taxonomy, RAPD-PC; Genetic variation; Heavy metals; Pollution

Introduction

Pollution of the freshwater environments by heavy metals due to increased action of flowing out discharge from various industries has received considerable attention in that it is able to influence freshwater organisms, leading to modify their genetic diversity [1-7]. Pollution affects adversely organisms and could be the cause of the genetic variation of some species.

Metal exposure was found to lead to various types of DNA damages and alteration of genetic patterns within populations [8,9] and also, DNA damage may indicate levels of metal toxicity.

In Egypt, *Coelatura* species showed great argument on their taxonomy, and their number ranged from 1 to 14 species in various studies [10-12] which consequently lead to questionable taxonomy. Therefore, in the present study, we used RAPD-PCR method to resolve the conflict on the taxonomical status of some *Coelatura* species from the River Nile in Egypt and to discuss the effect of metal pollution in this respect.

On the other hand, *Unionidae* are declining at a catastrophic rate worldwide. They are threatened by a number of factors among which industrial and human activities inducing environmental pollution [13],

pointing toward impending mass extinction. The significant loss of biodiversity may permanently alter ecosystem functioning in rivers and lakes as well as alter the rate of ecological processes [14].

Metal pollution of freshwater sources appears to be the main cause of the endangerment of freshwater mussels which are endangered nowadays worldwide and it is possible that high amounts of metals are toxic and could be a contributing threatening factor [15]. Therefore, it is important to estimate the accurate levels of trace elements in some mussels' species (*Coelatura* species as example) to assess the impact of heavy metals on their genetic variation and on the loss of biodiversity in the ecosystem of the River Nile in Egypt.

*Corresponding author: Faiza M El Assal, Department of Zoology, Faculty of Science, Cairo University, Giza, Egypt, 12613
E-mail: faizaelassal@yahoo.com

Material and Methods

Collections of samples

The *Coelatura* species *(C. aegyptiaca, C. prasidens, C. gaillardoti, C. canopicus and C. parreyssi)* were collected from the River Nile at two localities, known to be heavy metal polluted (El-Kanater, Qalyoubia Governorate and Tura, Cairo Governorate). Samples were monthly and randomly collected, for one year, from September 2009 to August 2010, then transported to the laboratory, sorted and maintained under the same conditions of food and temperature

Genetic study

DNA extraction and RAPD-PCR analysis: Samples of the *Coelatura* species were dissected and their soft parts were preserved in 100% ethyl alcohol at -20°C until used. Total genomic DNA was extracted from frozen ethanol-preserved mantle using Qiagen Dneasy tissue kit (Valencia, CA, USA) according to the manufacturer's manual. Seven primers were used in the present work, which were previously used in the bivalve RAPD-PCR [11,16,17].

476: 5`- TTG AGG CCC T – 3`

477: 5` -TGT TGT GCC C – 3`

478: 5` - CGA GCT GGT C – 3`

479: 5` - CTC ATA CGC G – 3`

483: 5` - GCA CTA AGA C– 3`

486: 5` - CCA GCA TCA G – 3`

487: 5` - GTG GCT AGG T – 3`

Only five primers worked out (UBC 476, UBC 477, UBC 478, UBC 479 and UBC 487). Amplifications were performed by modifying the protocol reported by Williams et al (1990) [18]. The 25 μl mixture contained 25 ng of template DNA, 1.5 unit of Taq Polymerase, 10 mMdNTPs, 10 pM primer, and 2.5 μl of 10x PCR buffer. (Dream Taq Green PCR MasterMix (2X) (Fermentas). Each amplification reaction was performed using a single primer and repeated twice to verify band autosimilarity [19].

Amplifications were performed in T-personal thermal cycler (Techne, TC-3000G), programmed for 45 cycles of 94°C for 1 minutes., 35°C for 1 minute., and 72°C for 1 minute. An initial denaturation step (3 minutes, 94°C) and a final extension holding (10 minutes, 72°C) were included in the first and last cycles, respectively.

Ten μl of the reaction products were resolved by 2% agarose gel electrophoresis at 85 volt in 1x TAE (Tris-acetate-EDTA) buffer. The gel was stained with ethidium bromide and photographed by a digital camera under UV transilluminator. For the comparison of the amplified products, population-specific fragments were detected using Gene Ruler 1 kb Plus DNA Ladder from Fermentas.

Molecular data analysis: Molecular data analysis was carried out using gel documentation system (SynGene, GeneTools - File version: 4.02.03, France), for the dendogram and calculation of similarity index of each primer between the studied *Coelatura* species. RAPD amplification products were scored as 0/1 for absence / presence of homologous bands [20] and analyses carried out using the NTSYS PC2.0 software [21].

Similarity coefficient matrix was calculated using Jaccard similarity algorithm [22] for RAPD markers. Dendograms were constructed using the UPGMA method, Unweighed pair-Group Method with arithmetical algorithms Averages [23]. Genetic diversity was also measured as the percentage of polymorphic bands. The percentage of polymorphic RAPD loci was calculated for each species, as well as the mean and overall value for all species and each primer.

Heavy metal analysis

Water and sediment analysis: Water and sediment samples collected from the two studied regions were analysed to determine the concentrations of heavy metals, using atomic absorption spectrophotometer model A-Analyst 100 Perkin Elmer. Metals analysed were Copper (Cu), Cobalt (Co), Nickel (Ni), Manganese (Mn), Lead (Pb), and Iron (Fe).

Tissue analysis [24]: Mussels were dissected and the soft parts were excised on clean tared pieces of plastic. Wet weights were then determined by the method of Johanson et al. [25]. Tissues were dried to constant weight, at room temperature, for 24 hours, removed from the plastic pieces and placed in 1.5 ml washed micro centrifuge tubes.

To each tube, 5 ml of piperidine (mole/litre) was added, the tubes were then cooled to room temperature, after which 2 ml of 61% (V/V) $HCLO_4$ was added to the precipitate. After 10 minutes, 7 ml of deionized water was added and mixed. Fifteen minutes later, the tubes were centrifuged for one minute, at 10,000 r.p.m. in a microcentrifuge (Beck Man/ Model J-2, 21). Supernatants were added in aliquots for analysis; using an atomic absorption spectrophotometer, model A-Analyst 100 Perkin Elmer instrumentation laboratories.

Single cuvette attached to an aspiration pump was used to minimize handling of samples and absorption of each ion was integrated for 2 seconds. Metals measured in tissues were Cu, Co, Ni, Mn, Pb and Fe.

The transfer factor (TF) in mussel tissues from the aquatic ecosystem, which include water and sediments, was calculated according to Kalfakakour and Akrida-Demertzi [26] and Rashed [27] as follows:

TF= Mtissue/ Msediment or Mwater

Mtissue = the metal concentration in mussel tissue, Msediment = the metal concentration in sediment and Mwater = the metal concentration in water.

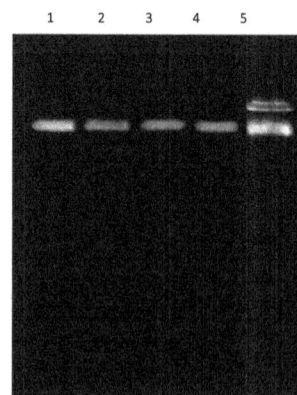

Figure 1: Agrose gel of extracted DNA from the mantle of the five *Coelatura* species under investigation.
1. *C. parreyssi*
2. *C. aegyptiaca*
3. *C. gaillardoti*
4. *C. canopicus*
5. *C. prasidens*

Statistical analysis

A software computer program SPSS Version 19 was used to test the significance differences between mean values of the different parameters in the studied mussels. One - way ANOVA and MANOVA were employed to find the difference in the ecological analysis at a probability level P>0.05 for insignificant results and P<0.05 and P<0.0001 for significant results.

Results

Genetic studies

Individual amplifications of agarose gel extracted DNA from the mantle of the five studied Coelatura species (Figure 1) were performed using the five primers UBC 476, UBC 477, UBC 478, UBC 479 and UBC 487, in order to determine the genetic relationship between them.

RAPD PCR carried out using the five primers provided strongly amplified fragments (Figures 2-6).

Genetic variability was observed among the studied Coelatura species. The greatest number of PCR fragments was found with primers UBC 478 and UBC 479 (6-7 bands), while less fragments were obtained with primers UBC 476, UBC 477 and UBC 487 (2-4 bands). The RAPD- PCR analysis was based on the number of bands that were different between any given pair of species (Table 1). Analyses showed

natural differences (polymorphism) among Coelatura species under investigation.

Primers UBC 477 (Figure 3a) and UBC 479 (Figure 5a) gave monomorphic bands with C. Parreyssi and C. gaillardoti and as well as with C. aegyptiaca and C. prasidens. While, C. canopicus revealed some polymorphic bands (Figures 3a and 5a). Primer UBC 476 gave similar results for C. aegyptiaca and C. prasidens as with primers UBC 477 and UBC 479, but it showed monomorphic bands for C. parreyssi, C. gaillardoti together with C. canopicus (Figure 2a).

However, Primer UBC 478 (Figure 4a) showed DNA alteration concerning C. parreyssi, C. gaillardoti and C. canopicus. This DNA alteration might have resulted from mutation or rearrangements at or between oligonucleotide primer binding sites in a genome. Primer UBC 487 (Figure 6a) revealed monomorphic bands for all five studied Coelatura species.

Genetic diversity was also measured as the percentage of polymorphic bands for each primer (Tables 2 and 3). 9.26 % of the bands were polymorphic among the five studied primers. Except primer UBC 487 which revealed no polymorphism, the other primers produced 1 to 6 polymorphic bands. Some RAPD fragments were found to be unique; 1 in C. parreyssi and C. gaillardoti and 3 in C. canopicus (Table 3).

Considering the similarity index (D) of the Coelatura species (Tables 5-9), utilizing RAPD-PCR markers, species were considered

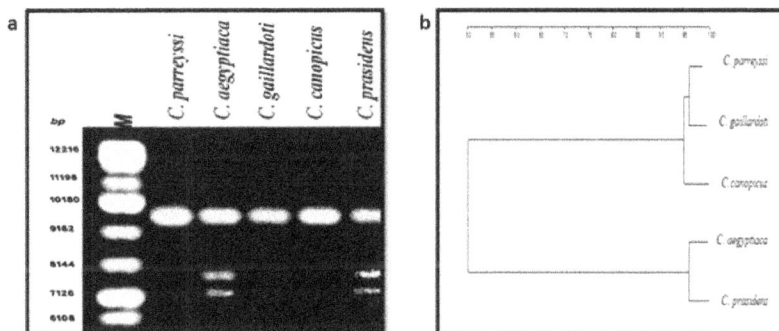

Figure 2: (a) RAPD-PCR profiles of the five Coelatura species under investigation using primer (UBC 476), and M: 1kb DNA marker shows one monomorphic band for all studied species and revealed 2, other monomorphic band for C. aegyptica and C. prasidens.
(b) Dendrogram of primer UBC 476 demonstrating the relationships of the five Coelatura species under investigation, based on compiled data set shows that C. parreyssi, C. gaillardoti identical species and C. canopicus are closed one also, C. aegyptiaca and C. prasidens similar species.

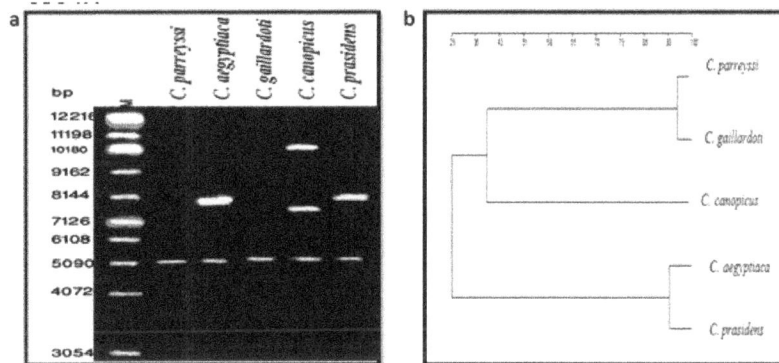

Figure 3: RAPD-PCR analysis of the five Coelatura species under investigation using primer (UBC 477)
(a) Gel electrophoresis showing amplification profile of samples. M: 1kb DNA marker.
(b) Dendrogram demonstrating the similarity relationship between the five Coelatura species under investigation.

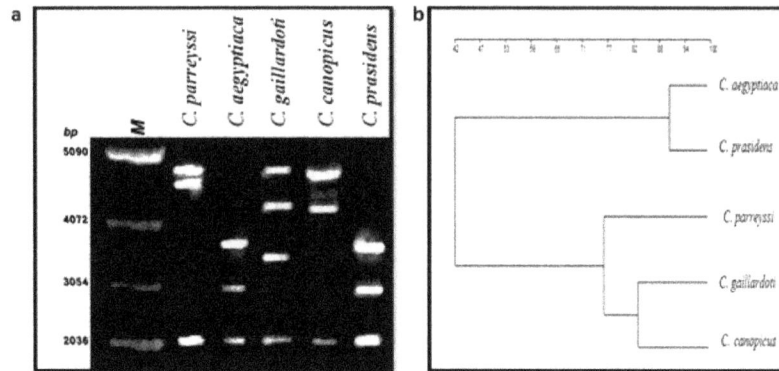

Figure 4: RAPD-PCR analysis of the five *Coelatura* species under investigation using primer (UBC 478)
(a) Gel electrophoresis showing amplification profile of samples. M: 1kb DNA marker.
(b) Dendrogram demonstrating the similarity relationship between the five *Coelatura* species under investigation.

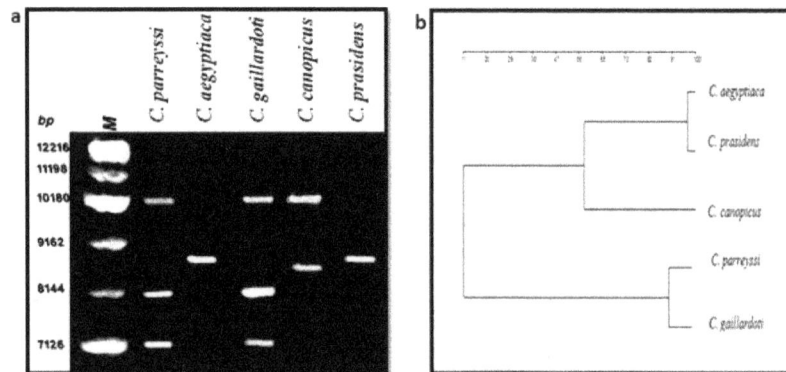

Figure 5: RAPD-PCR analysis of the five *Coelatura* species under investigation using primer (UBC 479)
(a) Gel electrophoresis showing amplification profile of samples. M: 1kb DNA marker.
(b) Dendrogram demonstrating the similarity relationship between the five *Coelatura* species under investigation.

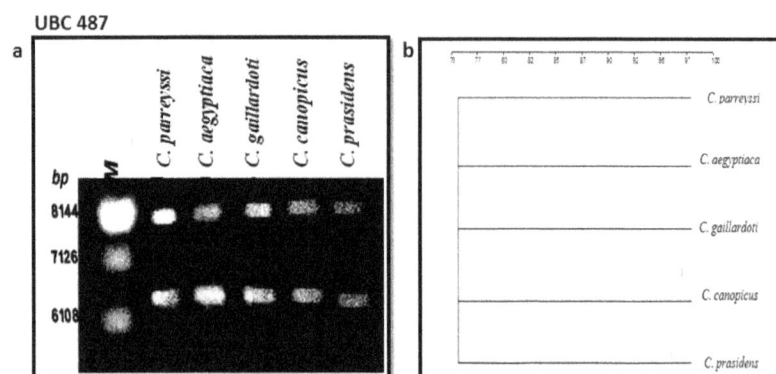

Figure 6: RAPD-PCR analysis of the five *Coelatura* species under investigation using primer (UBC 487)
(a) Gel electrophoresis amplification profile of samples M: 1kb DNA marker.
(b) Dendrogram demonstrating the similarity relationship between the five *Coelatura* species under investigation.

similar when the (D) value between two species is equal or close to 1. While, when (D) is distant from 1, the two species were regarded as separate species.

The similarity index (D) between *C. aegyptiaca* and *C. prasidens*, using all studied primers, was close to 1 (0.90-0.97), thus they were the closest species, and were considered one species, *C. aegyptiaca*. While, it was distant from 1 between these two species and the other studied species, except for primer UBC 487 which showed no polymorphism. Also, the (D) value, using the primers UBC 476 and UBC 478, was close to 1 between *C. parreyssi*, *C. gaillardoti* and *C. canopicus*. However, using

Primers/Species	C. parreyssi	C. aegyptiaca	C. gaillardoti	C. canopicus	C. prasidens
			UBC 476 (Figure 1a)		
1	Band 1 at ~ 9671.16 bp	Band 1 at ~ 9677.96 bp	Band 1 at ~ 9732.50bp	Band 1 at ~ 9739.76 pb	Band 1 at ~ 9759.88bp
2		Band 2 at ~ 7766.91 bp			Band 2 at ~ 7830.31 bp
3		Band 3 at ~ 7261.65 bp			Band 3 at ~ 7302.69 bp
			UBC 477 (Figure 2a)		
1	Band 1 at ~ 5157.41bp	Band 1 at ~ 8004.24 bp	Band 1 at ~ 5265.15bp	Band 1 at ~ 10284.48 pb	Band 1 at ~ 8133.93 bp
2		Band 2 at ~ 5167.11 bp		Band 2 at ~ 7655.77bp	Band 2 at ~ 5265.15 bp
3				Band 3 at ~ 5265.15 bp	
			UBC 478 (Figure 3a)		
1	Band 1 at ~ 4939.03bp	Band 1 at ~ 3788.28 bp	Band 1 at-4952.17 bp	Band 1 at ~ 4890.24bp	Band 1 at ~ 3716.02 bp
2	Band 2 at ~ 4737.70 bp	Band 2 at ~ 3094.70 bp	Band 2 at ~ 4398.09 bp	Band 2 at ~ 4359.32bp	Band 2 at ~ 3061.36 bp
3	Band 3 at ~ 2154.60 bp	Band 3 at ~ 2158.54 bp	Band 3 at ~ 3554.15 bp	Band 3 at ~ 2131.12bp	Band 3 at ~ 2158.54 bp
4			Band 3 at ~ 2158.54 bp		
			UBC 479 (Figure 4a)		
1	Band 1 at ~ 10195.41bp	Band 1 at ~ 8834.00 bp	Band 1 at ~ 10272.82 bp	Band 1 at ~ 10319.55bp	Band 1 at ~ 8825.74 bp
2	Band 3 at ~ 8112.54 bp		Band 2 at ~ 8174.51 bp	Band 2 at ~ 8654.18bp	
3	Band 4 at ~ 7153.63 bp		Band 3 at ~ 7188.33 bp		
			UBC 487 (Figure 5a)		
1	Band 1 at ~ 8093.80 bp	Band 1 at ~ 8093.80 bp	Band 1 at ~ 8193.97 bp	Band 1 at ~ 8106.52 bp	Band 1 at ~ 8156.88 bp
2	Band 2 at ~ 6542.64 bp	Band 2 at ~ 6542.64 bp	Band 2 at ~ 6556.28 bp	Band 2 at ~ 6583.66 bp	Band 2 at ~ 6529.02 bp

Table 1: Bands Pattern in *C. parreyssi*, *C. aegyptiaca*, *C. gaillardoti*, *C. canopicus* and *C. prasidens* using the five primers.

Primer	Total number of bands	Monomorphic	Polymorphic	% of polymorphism
UBC 476	3	1	2	66.7
UBC 477	4	1	3	75
UBC 478	7	1	6	85.7
UBC 479	5	0	5	100
UBC 487	2	2	0	0

Table 2: Total number of bands (monomorphic, polymorphic and percentage of polymorphism) of each primer, in *Coelatura* species under investigation.
*The repeated bands in all species are counted once.

Bands	Total	C. prasidens	C. canopicus	C. gaillardoti	C. aegyptiaca	C. parreyssi
Amplified	54	11	11	11	11	10
Monomorphic	49	11	8	10	11	9
Polymorphic	5	0	3	1	0	1
Unique	5	0	3	1	0	1
% of polymorphism	9.26	0 %	27.3 %	9.1 %	0 %	10 %

Table 3: Total number of bands for all studied primers (monomorphic, polymorphic, unique) and percentage of polymorphism, revealed by RAPD markers among the five studied *Coelatura* species.

Species	C. parreyssi	C. aegyptiaca	C. gaillardoti	C. canopicus
C. aegyptiaca	0.53			
C. gaillardoti	0.96	0.51		
C. canopicus	0.94	0.52	0.96	
C. prasidens	0.49	0.96	0.49	0.47

Table 4: Similarity index (D) of the Egyptian *Coelatura* specie susing UBC 476 primer.

Species	C. parreyssi	C. aegyptiaca	C. gaillardoti	C. canopicus
C. aegyptiaca	0.29			
C. gaillardoti	0.96	0.28		
C. canopicus	0.39	0.30	0.40	
C. prasidens	0.39	0.94	0.38	0.11

Table 5: Similarity index (D) of the Egyptian *Coelatura* specie susing UBC 477 primer.

the primers UBC 477 and UBC 479, (D) was distant from 1. *C. parreyssi* and *C. gaillardoti* were the most closely associated species and may be considered one species, *C. parreyssi*. While, *C. canopicus* was somewhat distant and may be regarded as distinct species or subspecies.

The dendrogram analyses, using primers UBC 476, UBC 477 and UBC 479 (Figures 2b, 3b and 5b) confirmed the results obtained with the RAPD profiles and those of the (D) value. *C. aegyptiaca* and *C. prasidens* were the closest species, as well as are *C. parreyssi* and *C. gaillardoti*. While, *C. canopicus* was a separate species. The dendogram using primer UBC 478 showed the same result for *C. aegyptiaca* and *C.*

Species	C. parreyssi	C. aegyptiaca	C. gaillardoti	C. canopicus
C. aegyptiaca	0.55			
C. gaillardoti	0.78	0.43		
C. canopicus	0.73	0.35	0.83	
C. prasidens	0.47	0.91	0.39	0.30

Table 6: Similarity index (D) of the Egyptian *Coelatura* specie susing UBC 478 primer.

Species ar	C. parreyssi	C. aegyptiaca	C. gaillardoti	C. canopicus
C. aegyptiaca	0.02			
C. gaillardoti	0.90	0.00		
C. canopicus	0.35	0.58	0.30	
C. prasidens	0.01	0.97	0.008	0.58

Table 7: Similarity index (D) of the Egyptian *Coelatura* specie susing UBC 479 primer.

Species	C. parreyssi	C. aegyptiaca	C. gaillardoti	C. canopicus
C. aegyptiaca	0.91			
C. gaillardoti	0.91	0.92		
C. canopicus	0.84	0.85	0.92	
C. prasidens	0.74	0.90	0.77	0.85

Table 8: Similarity index (D) of the Egyptian *Coelatura* specie susing UBC 487 primer.
^ All studied species show high similarity index (D), ranging from 0.74 to 0.92.

Species	C. parreyssi	C. aegyptiaca	C. gaillardoti	C. canopicus
C. aegyptiaca	0.91			
C. gaillardoti	0.91	0.92		
C. canopicus	0.84	0.85	0.92	
C. prasidens	0.74	0.90	0.77	0.85

Table 9: Similarity coefficient matrix of all primers calculated by NTSYS of the Egyptian *Coelatura* species.

prasidens, while some difference was revealed concerning *C. parreyssi*, *C. gaillardoti* and *C. canopicus*. The two latter species were the most related species and *C. parreyssi* was somewhat distant (Figure 4b).

According to the Similarity coefficient matrix of all primers (Table 9), the highest (D) value (0.55 and 0.67) was between *C. gaillardoti* and *C. parreyssi* and between *C. prasidens* and *C. aegyptiaca*. While, the lowest D-value (0.12) was recorded between *C. gaillardoti* and *C. aegyptiaca*, *C. canopicus* and *C. aegyptiaca* and between *C. prasidens* and *C. parreyssi*. This confirms that *C. aegyptiaca* and *C. prasidens* are similar and *C. gaillardoti* and *C. parreyssi* are also similar, while *C. canopicus* is different.

Heavy metal analysis

Nile water and sediment analysis: The mean values of the concentrations of the trace elements measured in the water and sediment of the studied areas (El-Kanater and Tura regions), are given in Table 10.

There was no significant difference ($P>0.05$) recorded in Cu and Fe measured in the water or sediment between both sites, while, significant difference ($P<0.0001$) in the concentration of Pb in the sediment and water between the two localities, was recorded. Also, Co concentration in the water showed significant difference ($P<0.0001$) between both sites. Ni and Mn revealed, too, significant difference ($P<0.0001$) in the sediment between the two regions.

Trace elements recorded in the Nile water of both regions were in the permissible levels for Cu, Mn and Fe, while the levels of Co, Pb and Ni exceeded these levels (Table 10).

The concentrations of the different studied heavy metals in water of the two locations were in the following decreasing order:

Tura region: Fe>Co>Ni >Mn>Pb> Cu

El-Kanater region: Fe>Co>Ni >Mn>Pb> Cu

Metal concentrations in the sediment of the two locations were in the following sequence:

Tura region : Fe>Mn>Ni >Pb>Co>Cu

El-Kanater region: Fe>Mn> Co>Ni>Cu>Pb

Tissue analysis

There was a great variation in the amount of the trace elements accumulated in the different soft parts of the studied mussels (Tables 11-14).

In general, significant difference was recorded in the concentration of the studied heavy metals in the different soft parts, between the studied *Coelatura* species ($P<0.05$, $P<0.0001$), at the two localities under investigation, except in some instance.

Heavy metals analyzed in all tissues of the three studied *Coelatura* species exceeded the permissible levels according to WHO [28] and FAO/WHO [29]. The calculated transfer factor (TF) in the different tissues from water and sediments at the two localities is shown in Tables 15-17. Results show that the TF of the sediments was greater than that of water.

Discussion

Human exploitation of world mineral resources and advances in industrialization has resulted in high levels of heavy metals in the environment [30-32]. The aquatic bodies near the industrial and urban areas are more able to accumulate such metals, causing hazardous impact on the freshwater fauna. The impact of metals on different bivalve populations revealed that those inhabiting environments contaminated by heavy metals exhibited a higher allelic diversity [33]. DNA damage and genetic diversity in aquatic animal populations induced by chemical contaminants have been successfully detected using RAPD method [34-37].

RAPD-PCR analysis proved to be helpful in estimating genetic variations among species [11,38]. Analyses of the RAPD-PCR showed natural differences or polymorphism among the *Coelatura* species under investigation, and distinguished them to only three species namely, *C. aegyptiaca*, *C. parreyssi* and *C. canopicus* which were also confirmed by dendograms and (D) values. In fact, thorough revision of genus *Coelatura* was needed by applying molecular techniques to reveal the current concept that it represents a lumped species complex, as claimed by Ortmann [39], Graf [40] and Graf and Cummings [12].

The present study shows that *C. aegyptiaca* and *C. prasidens* are closely related and could be considered as one species, *C. aegyptiaca*, which is the type species of the genus *Coelatura*. Also, *C. parreyssi* and *C. gaillardoti* are closely related and are considered as the same species, *C. parreyssi*, which has advantage over *C. gaillardoti* because of nomenclature priority [41]. On the other hand, *C. canopicus* is somewhat distant from the other studied species and may be considered a separate species or a subspecies. Finally, the similarity coefficient matrix and the UPGMA dendrogram of all primers confirmed that the five *Coelatura* species under investigation should be classified into three species only namely, *C. aegyptiaca*, *C. parreyssi* and *C. canopicus* (Figure 7). Thus, assessing the genetic diversity of populations could

Metals / Parameters	Fe	Mn	Ni	Co	Cu	Pb
Sediment at Tura	303.26 ± 60.7	198.7 ± 12.2	5.32 ± 0.43	3.9 ± 0.72	3.34 ± 0.2	5.24 ± 0.53
Sediment at El-Kanater	245.08 ± 20.87	234.67 ± 10.6	3.64 ± 0.48	3.7 ± 0.47	3.2 ± 0.19	2.7 ± 0.4
P value*	P > 0.05	*P < 0.0001	*P < 0.0001	P > 0.05	P > 0.05	*P < 0.0001
Water at Tura	0.26 ± 0.03	0.052 ± 0.01	0.06 ± 0.012	0.22 ± 0.13	0.022 ± 0.008	0.05 ± 0.01
Water at El-Kanater	0.14 ± 0.076	0.04 ± 0.004	0.05 ± 0.005	0.09 ± 0.007	0.03 ± 0.01	0.032 ± 0.008
P value*	P > 0.05	P > 0.05	P > 0.05	*P < 0.0001	P > 0.05	*P < 0.0001
Permissible levels of water	1	0.4	0.02	0.001-0.002	2	0.01

Table10: Concentration of heavy metals (in ppm) in water and sediment of the River Nile at El-Kanater and Tura regions, and the permissible levels in water according to the WHO (2008, 1996).
*Significant at P< 0.0001 and insignificant at P <0.05

Heavy metals	Tura region			P- value	El-Kanater region			P- value	*Permissible levels in mg/kg
	C. aegyptiaca	C.canopicus	C. parreyssi		C. aegyptiaca	C.canopicus	C. parreyssi		
Lead	7.3 ± 0.80	5.93 ± 0.78	13.1 ± 3.1	***P < 0.0001	3.73 ± 0.24	0.25 (0.00025 g)	12 ± 1.3	***P < 0.0001	9.1 ± 0.5
Copper	13 ± 1.7	13.65 ± 2.93	8.34 ± 2.28	***P < 0.0001	9.35 ± 0.83	3 (0.003 g)	6.4 ± 1.23	***P < 0.0001	8 ± 0.7
Cobalt	6 ± 1	4.57 ± 2.18	5.36 ± 2.2	P > 0.05	0.83 ± 0.1	-	5.5 ± 0.67	***P < 0.0001	3.1 ± 0.2
Nickel**	6.79 ± 1.26	4 ± 2.58	0.54 ± 0.18	***P < 0.0001	3.64 ± 0.52	0.5-1.0 (0.0005-0.001 g)	3.84 ± 0.7	P > 0.05	3.17 ± 0.54
Manganese	164.4 ± 17.9	532.39 ± 136.56	425.2 ± 12.79	***P < 0.0001	223.8 ± 24.2	2-9 (0.002-0.009 g)	306.1 ± 86.3	P > 0.05	263 ± 37.2
Iron	654 ± 85.4	568.66 ± 12	472 ± 15.7	***P < 0.05	278.2 ± 24	43 (0.043 g)	453.3 ± 200.2	P > 0.05	339.3 ± 34.3

Table 11: Mean concentrations of the heavy metals in the foot of C. aegyptiaca, C. parreyssi and C. canopicus in g /kg at Tura and El-Kanater regions ± standard deviation.
*Permissible levels of heavy metals according to FAO/WHO, (1999).
**Permissible levels of Ni according to WHO (1989).
***Significant at P < 0.05, P < 0.0001 and insignificant at P > 0.05

Heavy metals	Tura region			P- value	El-Kanater region			P- value	*Permissible levels in mg/kg
	C. aegyptiaca	C.canopicus	C. parreyssi		C. aegyptiaca	C.canopicus	C. parreyssi		
Lead	0.25 (0.00025 g)	9.4 ± 1.1	4.5 ± 1	***P < 0.0001	2.59 ± 0.7	7.63 ± 2.13	3.3 ± 0.76	***P < 0.0001	3.8 ± 0.74
Copper	3 (0.003 g)	3.99 ± 0.4	3.94 ± 0.3	P > 0.05	3.9 ± 0.34	7.15 ± 0.89	7.22 ± 1	***P < 0.0001	10.4 ±1.24
Cobalt	-	5.1 ± 0.7	3.8 ± 0.83	***P < 0.0001	2.57 ± 0.55	14.56 ± 2.4	11.69 ± 1.52	***P < 0.0001	7 ± 1
Nickel**	0.5-1.0 (0.0005-0.001 g)	2.5 ± 0.6	4.3 ± 0.4	***P < 0.0001	6.13 ± 1.3	7.9 ± 0.69	11 ± 2.5	***P < 0.05	11.57 ± 3.2
Manganese	2-9 (0.002-0.009 g)	178 ± 58.4	376.7 ± 43	***P < 0.0001	649 ± 82.9	568.2 ± 104.7	475.3 ± 70.7	***P < 0.0001	392.94 ± 19.8
Iron	43 (0.043 g)	356.5 ± 36.2	511.86 ± 40	***P < 0.0001	638.4 ± 54.6	473.8 ± 47.26	481.6 ± 74.27	***P < 0.0001	238.8 ± 31.7

Table 12: Mean concentrations of the heavy metals in the mantle of C. aegyptiaca, C. parreyssi and C. canopicus in g /kg at Tura and El-Kanater regions ± standard deviation.
*Permissible levels of heavy metals according to FAO/WHO, (1999).
**Permissible levels of Ni according to WHO (1989).
***Significant at P < 0.05, P < 0.0001 and insignificant at P > 0.05.

Heavy metals	Tura region			P- value	El-Kanater region			P- value	*Permissible levels in mg/kg
	C. aegyptiaca	C.canopicus	C. parreyssi		C. aegyptiaca	C.canopicus	C. parreyssi		
Lead	0.25 (0.00025 g)	7.27 ± 1.1	5.33 ± 1.1	***P < 0.0001	4.85 ± 0.56	2.54 ± 0.58	2.94 ± 0.62	P > 0.05	2.35 ± 0.78
Copper	3 (0.003 g)	5.9 ± 0.44	5.32 ± 0.7	P > 0.05	5.6 ± 1.2	10.98 ± 1	13.45 ± 1.96	***P < 0.05	11.9 ± 0.98
Cobalt	-	9.3 ± 1.1	7.3 ± 0.76	***P < 0.0001	8.5 ± 1	3.77 ± 1.2	5.45 ± 0.79	***P < 0.0001	9.99 ± 1.5
Nickel**	0.5-1.0 (0.0005-0.001 g)	6.67 ± 0.44	5.94 ± 1.2	***P < 0.0001	3.6 ± 0.7	11 ± 0.47	9.7 ± 1.2	***P < 0.05	11.2 ± 0.86
Manganese	2-9 (0.002-0.009 g)	447 ± 26.54	120.75 ± 41	***P < 0.0001	136.56 ± 5.25	142.78 ± 25.75	498.1 ± 72.8	***P < 0.0001	282.02 ± 17.63
Iron	43 (0.043 g)	261.8 ± 25.49	224.81 ± 30.66	***P < 0.0001	155.98 ± 12.32	144.49 ± 24.78	105.83 ± 81.9	P > 0.05	152.47 ± 23.375

Table 13: Mean concentrations of the heavy metals in the gills of C. aegyptiaca, C. parreyssi and C. canopicus in g /kg at Tura and El-Kanater regions ± standard deviation.
*Permissible levels of heavy metals according to FAO/WHO, (1999).
**Permissible levels of Ni according to WHO (1989).
***Significant at P < 0.05, P < 0.0001 and insignificant at P > 0.05.

Heavy metals	Tura region C. aegyptiaca	C.canopicus	C. parreyssi	P- value	El-Kanater region C. aegyptiaca	C.canopicus	C. parreyssi	P- value	*Permissible levels in mg/kg
Lead	0.25 (0.00025 g)	3.3 ± 0.6	3.44 ± 0.2	***P<0.0001	5.1 ± 0.54	8.45 ± 0.98	7.44 ± 1.1	***P< 0.05	6.6 ± 1.3
Copper	3 (0.003 g)	8.4 ± 0.7	6.56 ± 0.44	***P<0.0001	4.75 ± 0.24	4.9 ± 0.47	6.4 ± 0.89	***P<0.0001	7.16 ± 1
Cobalt	-	1.63 ± 0.3	1.26 ± 0.3	***P<0.05	1.1 ± 0.07	1.1 ± 0.4	2.3 ± 0.54	***P< 0.05	2.55 ± 1.17
Nickel**	0.5-1.0 (0.0005-0.001 g)	7.3 ± 1.5	5.8 ± 0.99	***P<0.0001	3.93 ± 0.6	2.7 ± 1.3	6.1 ± 0.5	***P<0.0001	7.4 ± 1
Manganese	2-9 (0.002-0.009 g)	70 ± 14.82	115.4 ± 15.35	***P<0.0001	153.5 ± 33	628.9 ± 53.7	358 ± 120.8	***P<0.0001	333.3 ± 33
Iron	43 (0.043 g)	758.7 ± 22.4	669.45 ± 27.35	***P<0.0001	453.2± 17.4	235.2 ± 93.9	911.69 ± 85	***P<0.0001	152.88 ± 13.17

Table 14: Mean concentrations of the heavy metals in the digestive tissues of *C. aegyptiaca*, *C. parreyssi* and *C. canopicus*in g /kg at Tura and El-Kanater regions ± standard deviation.
*Permissible levels of heavy metals according to FAO/WHO, (1999).
**Permissible levels of Ni according to WHO (1989).
***Significant at $P < 0.05$, $P < 0.0001$and insignificant at $P > 0.05$.

Site Heavy metals	Tura Pb	Cu	Co	Ni	Mn	Fe	El kanater Pb	Cu	Co	Ni	Mn	Fe
Water/Foot	0.007	0.002	0.043	0.01	0.003	0.0003	0.008	0.003	0.01	0.01	0.0002	0.0005
Water/Mantle	0.013	0.002	0.037	0.01	0.0001	0.0008	0.012	0.008	0.004	0.01	0.0001	0.0002
Water/Gill	0.02	0.002	0.026	0.01	0.0002	0.0013	0.006	0.005	0.001	0.01	0.0003	0.0009
Water/Digestive tissue	0.008	0.003	0.1	0.01	0.0002	0.0013	0.006	0.006	0.008	0.01	0.0003	0.0003
Sediment/Foot	0.72	0.3	0.7	0.76	1.2	0.46	0.7	0.3	4.4	1.0	1.0	0.9
Sediment/Mantle	1.4	0.3	0.6	0.44	0.5	1.3	1.0	0.8	1.4	0.6	0.36	0.4
Sediment/Gill	2.2	0.3	0.4	0.46	0.7	2.0	0.6	0.6	0.4	1.0	1.7	1.6
Sediment/Digestive tissue	0.8	0.5	1.5	0.7	0.6	2	0.5	0.7	3.4	0.93	1.53	0.54

Table 15: Mean transfer factor (TF) of the different heavy metals in the different soft parts of *C. aegyptiaca* (g/kg dry weight) and in Nile water samples (mg/l) from Tura and El-Kanater regions.

Site Heavy metals	Tura Pb	Cu	Co	Ni	Mn	Fe	El kanater Pb	Cu	Co	Ni	Mn	Fe
Water/Foot	0.004	0.002	0.05	0.1	0.0001	0.0004	0.002	0.005	0.002	0.01	0.0001	0.0003
Water/Mantle	0.007	0.003	0.02	0.01	0.0001	0.0004	0.003	0.008	0.002	0.02	0.0002	0.0004
Water/Gill	0.02	0.002	0.07	0.01	0.0004	0.0014	0.004	0.005	0.001	0.01	0.0001	0.0005
Water/Digestive tissue	0.006	0.004	0.24	0.02	0.0001	0.0009	0.009	0.004	0.006	0.01	0.0006	0.0002
Sediment/Foot	0.4	0.4	0.7	9.6	0.7	0.64	0.2	0.5	0.7	0.95	0.77	0.54
Sediment/Mantle	0.7	0.5	0.3	0.65	0.35	0.64	0.7	0.8	0.7	1.46	1.32	0.7
Sediment/Gill	2.1	0.3	1.0	0.5	1.4	1.3	0.4	0.5	0.4	0.55	0.53	0.94
Sediment/Digestive tissue	0.6	0.7	3.5	1.97	0.32	1.3	0.8	0.4	2.3	0.5	3.36	0.32

Table 16: Mean transfer factor (TF) of the different heavy metals in the different soft parts of *C. parreyssi* (g/kg dry weight) and in Nile water samples (mg/l) from Tura and El-Kanater regions.

Site Heavy metals	Tura Pb	Cu	Co	Ni	Mn	Fe	El kanater Pb	Cu	Co	Ni	Mn	Fe
Water/Foot	0.008	0.001	0.057	0.02	0.0001	0.0004	0.003	0.004	0.003	0.02	0.0002	0.0004
Water/Mantle	0.015	0.003	0.022	0.01	0.0001	0.0004	0.007	0.008	0.002	0.01	0.0001	0.0003
Water/Gill	0.017	0.001	0.05	0.01	0.0001	0.002	0.006	0.006	0.001	0.01	0.0003	0.0006
Water/Digestive tissue	0.007	0.003	0.1	0.01	0.0001	0.0002	0.009	0.005	0.007	0.01	0.0003	0.0002
Sediment/Foot	0.9	0.2	0.9	1.33	0.37	0.53	0.3	0.4	1.2	1.16	0.9	0.72
Sediment/Mantle	1.6	0.5	0.3	0.5	0.42	0.63	0.6	0.8	1.0	0.85	0.62	0.48
Sediment/Gill	1.8	0.2	0.7	0.54	0.4	0.33	0.5	0.6	0.5	0.6	1.94	1.1
Sediment/Digestive tissue	0.7	0.5	1.7	0.87	0.56	0.33	0.8	0.5	2.9	0.63	2.0	0.37

Table 17: Mean transfer factor (TF) of the different heavy metals in the different soft parts of *C. canopicus* (g/kg dry weight) and in Nile water samples (mg/l) from Tura and El-Kanater regions.

be a valuable addition to more traditional tools for determining the effects of environmental pollution on aquatic ecosystems as confirmed by Nevo et al. [42], Bickham and Smolen [43] and Nadig et al. [34].

Primer UBC 478 showed DNA alteration concerning *C. parreyssi*, *C. gaillardoti* and *C.canopicus*. The gain/loss of RAPD bands may be related to DNA damage, mutation or structural rearrangements induced by genotoxic agents affecting the primer sites [37]. Mutation may be due to quantitative or qualitative changes or rearrangement of

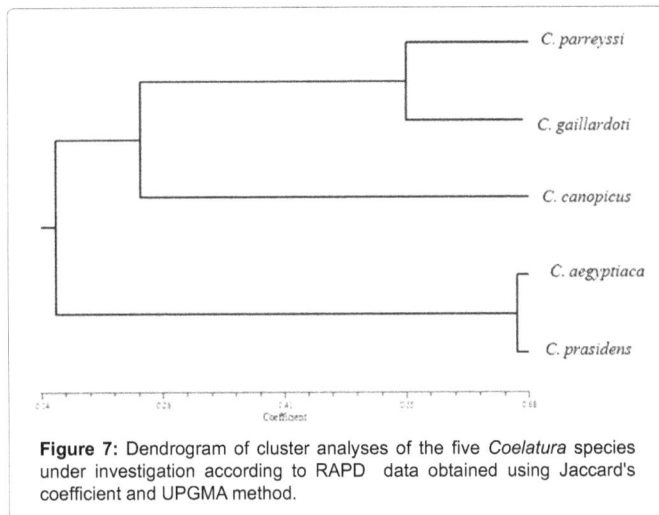

Figure 7: Dendrogram of cluster analyses of the five *Coelatura* species under investigation according to RAPD data obtained using Jaccard's coefficient and UPGMA method.

the genetic material, most probably due to metal (Pb, Mn, Cu, Fe, Co and Ni) pollution of the environment, recorded in the two localities of the study. The concept that genetic patterns within populations may be altered by exposure to contaminants was reported by Bishop and Cook (1981) [8], Klerks and Weis [9], Abdul-Aziz [44] and Giantsis et al. [38]. Also, the latter authors, examining genetic differentiation and potential impact of heavy metals pollution, using RAPD markers, observed a loss in genetic variability of *Mytilus galloprovincialis* population. They concluded that metal pollution appears to have played an important role in shapping pattern of genetic diversity and differentiation among Greek *M. galloprovincialis* population. Yap et al. [45] found that heavy metal contamination was a main causal agent for the genetic differentiation of *Pera viridis* in Peninsula Malaysia. The evidence for pollutant to induce genotoxicity has been also determined by several authors [33,38,46-50]. Neeratanaphan et al. [51] postulated that understanding the effect of heavy metals on genetic variability is fundamental for preservation. They detected genomic DNA modification such as damage and structure variation in the freshwater snail *Filopaludina martensi* affected by lead and cadmium

Forni [52] and Reid et al. [53] reported Cu, Fe, Cd and Ni as mutagenic agents. These metals have the tendency to bind to phosphates and wide variety of organic molecules including base residues of DNA, which can lead to mutations by altering structures of DNA [54] or modifying the genetic diversity of populations. Also, exposure of mussels in the field to water polluted by different mixtures of genotoxic contaminants was reported by Izquierdo et al. [55] to induce DNA alterations, leading to genetic variation among species and populations. Thus, the conflict in the taxonomy of *Coelatura* species in the different studies is most probably due to the environmental pollution with heavy metals among other factors. Heavy metals analysed in all tissues of the studied *Coelatura* species exceeded the permissible levels according to WHO (1989) [28] and FAO/WHO [29].

The numbers of threatened aquatic species and species extinctions increase at an alarming rate [56]. Molluscs are one of the most threatened major taxonomic groups worldwide [57]. Within this group, the unionids are highly threatened throughout their distribution [58] and are declining globally due to alteration in habitat, decline and extinction of fish host populations, pollution and environmental changes, pointing toward impending extinction. They are the most imperiled group of species and many species became extinct in several parts of the world including Egypt, while others are threatened or endangered. The loss

of benthic biomass may result in large scale alterations of freshwater ecosystem processes and functions [59].

Little information is available about the effect of frequent exposure to metals on mussels; it is possible that higher metal amounts than required could be a contributing factor to the extinction of some mussel species and genetic variation of some other species. In fact, the mussel fauna in Egypt is threatened due to heavy metal pollution of the River Nile water and sediment among other factors.

All species of genus *Unio* which were used to live in the River Nile are today extinct [10]. Only fossils were recorded by these authors from El Fayoum, Komombo, Idfu and Isna i.e from Upper Egypt. Although, some investigators [60,61] have referred to living *Unio* specimens in the River Nile in Lower Egypt. But, this is uncertain and needs to be thoroughly revised and the occurrence of *Unio* species in Egypt is still doubtful.

In general, studies on heavy metals are important in two main aspects, the public health point of view and the aquatic environment conservation. Heavy metals are present in the aquatic environment where they can accumulate along the food chain. Moreover, small amounts of absorbed heavy metals are either stored in a metabolically available form for essential biochemical processes or detoxified into metabolically inert forms and held in the body either temporarily or permanently [62-64]. Thus, determination of chemical quality of aquatic organisms, particularly the content of heavy metals is extremely important.

References

1. Hochwald S, Bauer G (1988) Gutachten zur Bestandssituation und zumschutz der Bachmuschel Unio crassus (Phil.) Fischer und Teichwirt, 12: 366-371.

2. Ashraf W (2005) Accumulation of heavy metals in kidney and heart tissues of Epinephelus microdon fish from the Arabian Gulf. Environ Monit Assess 101: 311-316.

3. Al-Weher SM (2008) Levels of heavy metals Cd, Cu and Zn in three fish species collected from the Northern Jordan Valley, Jordan. Jordan Journal of Biological Sciences 1: 41-46.

4. Bala M, Shehu RA, Lawal M (2008) Determination of the level of some heavy metals in water collected from two pollution- Prone irrigation areas around Kano Metropolis. Bayero Journal of Pure and Applied Sciences 1: 36-38.

5. Obasohan EE, Oronsaye JAO, Eguavoen OI (2008) A comparative Assessment of the heavy metal loads in the tissues of a common catfish (Clarias gariepinus) from Ikpoba and Ogba Rivers in Benin City, Nigeria. African Scientist 9: 13-23.

6. El-Assal F, Fol M (2011) Corbicula species (Bivalvia: Veneroida), from the river Nile in Egypt as bioindicators of pollution. Environmental Pollution Ecological Impacts Health Issues and Management 82-91.

7. Kamaruzzaman BY, MohdZahir MS, Akbar John B, Jalal KCA, Shahbudin, S, et al (2011): Bioaccumulation of some heavy metals by green mussel Perna viridis (Linnaeus, 1758) from Pekan, Pahang, Malaysia. International Journal of Biological Chemistry, 5: 54-60.

8. Bishop JA, Cook LM (1981) Genetic consequences of man-made change. Academic Press, London, UK. p209.

9. Klerks PL, Weis JS (1987) Genetic adaptation to heavy metals in aquatic organisms: a review. Environ Pollut 45: 173-205.

10. Ibrahim AM, Bishai HM, Khalil MT (1999) Freshwater molluscs of Egypt: publication of National Biodiversity unit, No.10. Egyptian Environmental Affairs Agency, Dept. of Nature protection.

11. Sleem SH, Ali TG (2008) Application of RAPD-PCR in taxonomy of certain freshwater bivalves of genus Caelatura. Global Journal of Molecular Sciences, 3: 27-31.

12. Graf DL, Cummings KS (2007) Preliminary review of the freshwater mussels (Mollusca: Bivalvia: Unionoida) of Northern Africa, with an emphasis to the Nile. Journal of the Egyptian German Society of Zoology 53: 89-118.

13. Williams JD, Warren ML, Cummings KS, Harris JL, Neves RL (1992) Conservation status of freshwater mussels of the United States and Canada. Fisheries 18: 6-22.

14. Hastie LC, Cosgrove PJ, Ellis N, Gaywood MJ (2003) The threat of climate change to freshwater pearl mussel populations. Ambio 32: 40-46.

15. Wang N, Ingersoll CG, Ivey CD, Hardesty DK, May TW, et al. (2010) Sensitivity of early life stages of freshwater mussels (Unionidae) to acute and chronic toxicity of lead, cadmium, and zinc in water. Environ Toxicol Chem 29: 2053-2063.

16. Ibrahim AM, Aly RH, Kenchington E, Ali TG (2008) Genetic polymorphism among five populations of Pinctada radiata from the Mediterranean coast in Egypt indicated by RAPD-PCR technique. Egyptian Journal of Zoology 50: 467-477.

17. Yousif F, Ibrahim A, Sleem S, El Bardicy S, Ayoub M (2009) Morphological and genetic analyses of Melanoides tuberculata populations in Egypt. Global Journal of Molecular Sciences, 4: 112-117.

18. Williams JG, Kubelik AR, Livak KJ, Rafalski JA, Tingey SV (1990) DNA polymorphisms amplified by arbitrary primers are useful as genetic markers. Nucleic Acids Res 18: 6531-6535.

19. Pérez T, Albornoz J, Domínguez A (1998) An evaluation of RAPD fragment reproducibility and nature. Mol Ecol 7: 1347-1357.

20. Abdellatif KF, Khidr YA (2010) Genetic diversity of new maize hybrids based on SSR markers as compared with other molecular and biochemical markers. Journal of Crop Science and Biotechnology 13: 139-145.

21. Rohlf FJ (1998) NTSYSpc Numerical taxonomy and multi variate analysis system, version 2.02c. Exeter Software, New York, USA.

22. Jaccard P (1908) Nouvelles recherches sur la distribution florale. Bull Soc Vaudoise Sci Nat 44: 223–270.

23. Sneath PHA, Sokal RR (1973) Numerical taxonomy. Freeman, San Francisco, USA.

24. Murphy VA (1987) Method of determination of Sodium, Potassium, Calcium, Magnesium, Chloride, and Phosphate in the rat choroids plexus by Flame Atomic Absorption and Visible Spectroscopy. Anal Biochem 161: 144-151.

25. Johanson CE, Reed DJ, Woodbury DM (1976) Developmental studies of the compartmentalization of water and electrolytes in the choroid plexus of the neonatal rat brain. Brain Res 116: 35-48.

26. Kalfakakour V, Akrida-Demertzi K (2000) Transfer factors of heavy metals in aquatic organisms of different trophic levels 1: 768-786.

27. Rashed MN (2001) Monitoring of environmental heavy metals in fish from Nasser Lake. Environ Int 27: 27-33.

28. World Health Organization (WHO) (1989) Heavy metals-environmental aspects. Environment Health Criteria.No. 85. Geneva, Switzerland.

29. FAO/WHO (Food and Agriculture Organization of the United Nations/ World Health Organization) (1999) Evaluation of certain food additives and contaminants. Expert Committee on Food Additives. WHO Technical Report Series, No. 884, p7.

30. Haggag AM, Marie MAS, Zaghloul KH (1999) Seasonal effects of the industrial effluents on the Nile catfish; Clarias gariepinus. J. Egypt. Ger. Soc. Zool., 28: 365-391.

31. Salah El-Deen MA, Khalid HZ, Gamal O, Abo-Hegab S (1999) Concentrations of heavy metals in water sediment and fish in the River Nile in the industrial area of Helwan. Egypt. J. Zool. 32: 373-395.

32. Zaghloul KH, Omar WA, Mikhail WZ, Abo-Hegab S (2000) Ecological and biochemical studies on the Nile fish Orechromis niloticus (L.), cultured in different aquatic habitats. Egypt. J. Zool., 34: 379-409.

33. Moraga D, Mdelgi-Lasram E, Romdhane MS, El Abed A, Boutet I, et al. (2002) Genetic responses to metal contamination in two clams: Ruditapes decussatus and Ruditapes philippinarum. Mar Environ Res 54: 521-525.

34. Nadig SG, Lee KL, Adams SM (1998) Evaluating alterations of genetic diversity in sunfish populations exposed to contaminants using RAPD assay. Aquatic Toxicology 43: 163-178.

35. Krane DE, Sternberg DC, Burton GA (1999) Randomly amplified polymorphic DNA profile-based measures of genetic diversity in crayfish correlated with environmental impacts. Environmental Toxicology and Chemistry 18: 504-508.

36. Atienzar FA, Cordi B, Donkin ME, Evenden AJ, Jha AN, et al (2000) Comparison of ultraviolet-induced genotoxicity detected by random amplified polymorphic DNA with chlorophyll fluorescence and growth in a marine macroalgae, Palmaria palmate. Aquat Toxicol 50: 1-12.

37. Atienzar FA, Venier P, Jha AN, Depledge MH (2002) Evaluation of the random amplified polymorphic DNA (RAPD) assay for the detection of DNA damage and mutations. Mutat Res 521: 151-163.

38. Giantsis IA, Kravva N, Apostolidis AP (2012) Genetic characterization and evaluation of anthropogenic impacts on genetic patterns in cultured and wild populations of mussels (Mytilus galloprovincialis) from Greece. Genet Mol Res 11: 3814-3823.

39. Ortmann AE (1920) Correlation of shape and station in freshwater mussels (naiads) Proceedings of the American Philosophical Society 59: 269-312.

40. Graf DL (1998) Sympatric speciation of freshwater mussels (Bivalvia: Unionoidea): a model. American Malacological Bulletin 14: 35-40.

41. Philippi RA (1848) Abbildungen und Beschreibungen never Oder wenig gekannter conchylien, II and III Kassel.

42. Nevo E, Noy R, Lavie B, Beiles A, Muchtar S (1986) Genetic diversity and resistance to marine pollution. Biological Journal of the Linnean Society 29: 139-144.

43. Bickham JW, Smolen MJ (1994) Somatic and heritable effects of environmental genotoxins and the emergence of evolutionary toxicology. Environ Health Perspect 102 Suppl 12: 25-28.

44. Abdul-Aziz KK (2012) The Physiological Status and Genetic Variations of the Bivalve, Pinctada radiata Affected by Environmental Pollution. Research. Journal of Pharmaceutical, Biological and Chemical Sciences 3: 277-291.

45. Yap CK, Cheng WA, Ong CC, Tan SG (2013) Heavy Metal Concentration and Physical Barrier are Main causal agents for the Genetic Differentiation of Pera viridis in Peninsular Malaysia. Sains Malaysiana 42: 1557- 1564.

46. Belfiore NM, Anderson SL (2001) Effects of contaminants on genetic patterns in aquatic organisms: a review. Mutat Res 489: 97-122.

47. Coughlan BM, Hartl MGJ, O'Reilly SJ, Sheehan D, Morthersill C, et al (2002) Detecting genotoxicity using the Comet assay following chronic exposure of Manila clam Tapes semidecussatus to polluted estuarine sediments. Mar Pollut Bull 44: 1359-1365.

48. De Wolf H, Blust R, Backeljan, T (2004) The population genetic structure of littorina littorea (Mollusca: Gastropoda) along population gradient in the scheldt Estuary the Netherlands using RAPD analysis. Sci. Total Environ 325: 59-69.

49. Liyan Z, Ying H, Guangxing L (2005) Using DNA damage to monitor water environment. Chinese J Oceanol Limnol 23: 340-348.

50. Gaikwad SS, Kamble NA (2014) Heavy Metal Pollution of Indian Rivers and its Biomagnifications in the Molluscs. Oct Jour Env Res 2: 67-76.

51. Neeratanaphan L, Sudmoon R, Chaveerach R (2014) Genetic erosion in the freshwater snails filopalodina martensi is affected by lead and cadmium. Applied Ecology and Environmental Research 12: 991- 1001.

52. Forni A (1994) Comparison of chromosome aberrations and micronuclei in testing genotoxicity in humans. Toxicol Lett 72: 185-190.

53. Reid TM, Feig DI, Loeb LA (1994) Mutagenesis by metal-induced oxygen radicals. Environ Health Perspect 102 Suppl 3: 57-61.

54. Wong PK (1988) Mutagenicity of heavy metals. Bull Environ Contam Toxicol 40: 597-603.

55. Izquierdo JI, Machado G, Ayllon F, d`Amico V, BalaVallarono E, et al (2003) Assessing pollution in coastal ecosystems: a preliminary survey using the micronucleus test in the mussel Mytilus edulis. Ecotoxicol. Env.Safty 55: 24-29.

56. Baillie JEM, Hilton-Taylor C, Stuart SN (2004) A global species assessment, IUCN red list of threatened species, IUCN, Cambridge, UK. p 1-217.

57. Lydeard C, Cowie RH, Ponder WF, Bogan AE, Bouchet P, et al (2004) The global decline of nonmarine mollusks. Bioscience 54: 321-330.

58. Bogan AE (1993) Freswater bivalve extinctions search for a cause (Mollusca: Unionida). Amer Zool 33: 599-609.

59. Ricciardi A, Neves RJ, Rasmussen JB (1998) Impending extinctions of North American freshwater mussels (Unionida) following the zebra mussel (Dreissena polymorpha) invasion. Journal of Animal Ecology 67: 613-619.

60. Aboul-Dahab HM (2002) Cellular responses of the molluscan host, Unio abyssinicus infected by two new haplosporidian parasites with emphasis on the morphological characteristics of the host haemolymph. Egyptian Journal of Zoology 38: 149-170.

61. Ramadan SA (2003) A comparison of the population structure of the water mite Unionicola tetrafurcatus inside two molluscan hosts from the River Nile, Egypt. Egyptian Journal of Zoology 41: 229-247.

62. Hashmi, MI, Mustafa S, Tariq SA (2002) Heavy metal concentrations in water and tiger prawn (Penaeus monodon) from grow-out farms in Sabah, North Borneo. Food Chemistry 79: 151-156.

63. World Health Organization (WHO) (1996) Guidelines for drinking-water quality,2nd edn Vol.2. Health Criteria and Other Supporting Information. Geneva, Switzerland.

64. World Health Organization(WHO) (2008) Guidelines for drinking-water quality. Geneva, 3rd edn Vol.1. Geneva, Switzerland.

Plastics: Issues Challenges and Remediation

Vipin Koushal1, Raman Sharma¹*, Meenakshi Sharma², Ratika Sharma³ and Vivek Sharma¹

¹Department of Hospital Administration, Chandigarh, India
²Senior Research fellow, School of Public health, PGIMER, Chandigarh, India
³Department of Sociology, Panjab University, Chandigarh, India

Abstract

Plastics have become a vital asset for humanity. Though extensive research and new technologies have led to invent of newer and safer plastics, but drawbacks and challenges of plastics have never been resolved and impact is on the rise. Some of the major compounds (vinyl chloride, dioxins, and plasticizers) are causative factors of hormone-disruption, reproductive dysfunction, breast growth and testicular cancers. The harmful effects are also pronounced in newborns via mothers during pregnancy or young children exposed directly.

Recycling is one of the most convenient and easiest ways. Smarter sorting, energy efficient ways, developing smarter plastics and research to develop certain fungi and bacteria that hasten degradation of conventional plastics are some of the present era needs. Source reduction (Reduce and Reuse) can occur by altering the design, manufacture or reduced use of plastic products.

Biodegradable plastics are similar to conventional plastics, with the additional quality of being able to naturally decompose and break into natural and safe byproducts. Bioplastics, nature derived plastics, are derived from biological sources such as sugar cane, cellulose etc. and these either degrade in open air or are made to compost using fungi, bacteria or enzymes.

To conclude, it is not the plastics to blame, but it is the misuse of plastics. The present time need is to look for biodegradable measures and effective policies and their implementation.

Keywords: Plastics; Impact; Prevention

Introduction

Plastic is a kind of material that is commonly known and used in everyday life in many forms. It becomes an important part of every one's life. To define plastics at molecular level, it is a kind of organic polymer, which has molecules containing long carbon chains as their backbones with repeating units created through a process of polymerization [1]. The structure of these repeating units and types of atoms play the main role in determining the characteristics of the plastic. These long carbon chains are well packed together by entanglements and Van der Waals forces forming a strong, usually ductile solid material [2]. Also, additives are usually added when manufacturing of commercial plastics is carried on, in order to improve the strength, durability or grant the plastic specific characteristics. Many of the controversies associated with plastics are associated with the additives [3].

Plastics have become a vital asset for humanity, often providing functionality that cannot be easily or economically replaced by other materials. Plastic products have brought benefits to society in terms of economic activity, jobs and quality of life. Most plastics are robust and last for hundreds of years. They have replaced metals in the components of most manufactured goods, including for such products as computers, car parts and refrigerators, and in so doing have often made the products cheaper, lighter, safer, stronger. Plastics have taken over from paper, glass and cardboard in packaging, usually reducing cost and also providing better care of the items.

Plastics also play a very vital role in hospitals and medical field. Plastics are used on large scale in hospitals. The daily plastic waste generation includes disposable syringes, I.V sets, glucose bottles, disposable plastic aprons; B.T sets; catheters and cannulas etc. are disposed of on daily basis. Plastics may be easy and convenient for everyday use, but their negative impacts on our health cannot be overlooked. Due to its non-biodegradable nature, it keeps on piling in the environment and is creating tons of trash around the world. Ian Connacher, the director of the film "Addicted to Plastics," once said in an interview with GreenMuze: "I don't think the material is to blame. I think it is our misuse of the material as consumers, the ineffective recycling policies and lack of producer [2]. In coming times also, the applications of plastics definitely are expected to increase as more new products and plastics are developed to meet demands. The increased use and production of plastic in developing and emerging countries is a particular concern, as the sophistication of their waste management infrastructure may not be developing at an appropriate rate to deal with their increasing levels of plastic waste.

Generally, there are two kinds of commercial plastics, thermoplastic (reheated, melted, and molded into different shapes) and thermosetting plastics which degrade and turn into other substances if reheated after molding [4]. Today, there are many different types of plastics manufactured in the plastic industry. The table below summarizes names of all commonly used plastics and their applications (Table 1).

New technologies and products were found after extensive research made in the field of plastics. Earlier the drawbacks of plastics were not known. The plastics are usually non-biodegradable and remain as

*Corresponding author: Dr. Raman Sharma, Department of Hospital Administration, Government Medical College & Hospital, Chandigarh – 160012, India
E-mail: drramansharmamha@rediffmail.com

Polymer type	Examples
Polyethylene Terephthalate	Fizzy drink and water bottles. Salad trays.
High Density Polyethylene	Milk bottles, bleach, cleaners and most shampoo bottles.
Polyvinyl Chloride	Pipes, fittings, window and door frames (rigid PVC). Thermal insulation (PVC foam) and automotive parts.
Low Density Polyethylene	Carrier bags, bin liners and packaging films.
Polypropylene	Margarine tubs, microwaveable meal trays, also produced as fibres and filaments for carpets, wall coverings and vehicle upholstery.
Polystyrene	Yoghurt pots, foam hamburger boxes and egg cartons, plastic cutlery, protective packaging for electronic goods and toys. Insulating material in the building and construction industry.
Unallocated References	polycarbonate which is often used in glazing for the aircraft industry

Table 1: Types of Plastics & Common uses.

waste in the environment for a very long time, thereby posing risks to human health as well as the environment. In the long run, overuse of plastics and lack of proper recycling yields many undesirable effects on our health.

The nature of traditional plastics is the reason why they cannot be biodegraded. The carbon chains of traditional plastics are too long and too well packed for microorganisms to digest, but if they are broken into small pieces the microorganisms will be able to degrade them. However, the breakdown process is too long for most of the traditional plastics, if there is no any artificial processing before being thrown in a landfill is involved. Therefore, before the plastics degrade themselves naturally, more plastics will be manufactured, causing increasing plastic pollution around the world [5]. Hence, with plastics, it is a full circle of problems and challenges that need to be resolved.

Impact

Impact on environment

Plastic is one of the major toxic pollutants of present time. Being composed of toxic chemicals and most importantly a non-biodegradable substance, plastic pollutes earth and leads to air pollution and water pollution. This also mixes with food chain effecting Environment Humans and animals. There is no safe way to dispose plastic waste and waste causes serious damage to environment during its production process, during its usage and during its disposal process.

Toxic chemicals release during manufacturing process is another significant source of the negative environmental impact of plastics. A whole host of carcinogenic, neurotoxic, and hormone-disruptive chemicals are standard ingredients and waste products of plastic production, and they inevitably find their way into our ecology through water, land, and air pollution. Some of the major compounds include vinyl chloride (in PVC), dioxins (in PVC), benzene (in polystyrene), phthalates and other plasticizers (in PVC and others), formaldehyde, and bisphenol-A, or BPA (in polycarbonate). Many of these are persistent organic pollutants (POPs)—some of the most damaging toxins on the planet, owing to a combination of their persistence in the environment and their high levels of toxicity; however, their unmitigated release into the environment affects all terrestrial and aquatic life with which they come into contact.

It is in the use phase that the benefits of plastics in durability and effectiveness are most evident. Though most plastics are benign in their intended use form, many release toxic gases in their in-place curing (such as spray foam) or by virtue of their formulation (as with PVC additives off-gassing during their use phase). Occupational exposure during installation, such as inhalation of dust while cutting plastic pipe or off-gassing vapors of curing products, is also a great concern for human health and the environment.

The disposal of plastics—the "grave" phase, is one of the least-recognized and most highly problematic areas of plastic's ecological impact. Ironically, one of plastic's most desirable traits—its durability and resistance to decomposition—is also the source of one of its greatest liabilities when it comes to the disposal of plastics. Natural organisms have a very difficult time breaking down the synthetic chemical bonds in plastic, creating the tremendous problem of the material's persistence. A very small amount of total plastic production (less than 10%) is effectively recycled; the remaining plastic is sent to landfills, where it is destined to remain entombed in limbo for hundreds of thousands of years, or to incinerators, where its toxic compounds are spewed throughout the atmosphere to be accumulated in biotic forms throughout the surrounding ecosystems.

Unfortunately, because of plastic's low density, it frequently migrates "downstream," blowing out of landfills and off garbage barges. In 1997, Captain Charles Moore discovered widespread plastic garbage contamination area, called a gyre, in the North Pacific Ocean. By 2005, the estimated area of contamination expanded to 10 million square miles. 90% of this garbage was determined to be plastic, and 80% was originally sourced from land, such as construction waste. It has been reported that there are six similar gyres across the planet's oceans, each laden with plastic refuse.

Impact on humans

The harmful effects of plastic on aquatic life are devastating, and accelerating. The impacts of plastic waste on our health and the environment are only just becoming apparent. Most of our knowledge is around plastic waste in the marine environment, although there is research that indicates that plastic waste in landfill and in badly managed recycling systems could be having an impact, mainly from the chemicals contained in plastic.

Ingestion of plastic occurs more frequently than entanglement. The MFSD has identified ingestion of waste as an indicator for monitoring environmental status. Ingestion of plastic waste has been documented in a number of species. For some species, almost all individuals contain ingested plastic [6], including sea birds, fish, turtles, mussels and mammals. Clearly different species ingest different types and sizes of plastic debris. Many animals mistake plastic waste for prey, for example, fish can confuse plastic pellets for plankton, birds may mistake pieces of plastic for cuttlefish or other prey [7,8].

There are several chemicals within plastic material itself that have been added to give it certain properties such as Bisphenol A, phthalates and flame retardants. These all have known negative effects on human and animal health, mainly affecting the endocrine system. There are also toxic monomers, which have been linked to cancer and reproductive problems. The actual role of plastic waste in causing these health impacts is uncertain. This is partly because it is not clear what level of exposure is caused by plastic waste, and partly because the mechanisms by which the chemicals from plastic may have an impact on humans and animals are not fully established. The most likely pathway is through ingestion, after which chemicals could bio accumulate up the food chain, meaning that those at the top could be exposed to greater levels of chemicals [9].

Plastic waste also has the ability to attract contaminants, such as persistent organic pollutants (POPs). Plastic could potentially transport these chemicals to otherwise clean environments and, when ingested by wildlife, plastic could cause the transfer of chemicals into the organism's system. However, in some conditions plastic could potentially act as a sink for contaminants, making them less available to wildlife, particularly if they are buried on the seafloor. With their large surface area-to-volume ratio, micro plastics may have the capacity to make chemicals more available to wildlife and the environment in comparison to larger sized plastics. However, once ingested, micro plastics may pass through the digestive system more quickly than larger plastics, potentially providing less opportunity for chemicals to be absorbed into the circulatory system [9].

Unfortunately, the properties of plastic that make it so valuable also make its disposal problematic, such as its durability, light weight and low cost. In many cases plastics are thrown away after one use, especially packaging and sheeting, but because they are durable, they persist in the environment. If plastic reaches the sea, its low density means it tends to remain on the surface. Most types of plastic are not biodegradable. Some plastics are designed to be biodegradable and can be broken down in a controlled environment, such as landfill, but it is uncertain if this will occur under other conditions, especially in oceans where the temperature is colder [10]. Even if plastic does eventually biodegrade, it will temporarily break into smaller fragments, which then produce so-called 'microplastics'. These have a specific and significant set of impacts.

The harmful effects of chemicals additives in plastics are also pronounced in newborns via mothers exposed to these toxins during their pregnancy. The second vulnerable groups are young children exposed directly to these chemicals. Since many of these chemicals (BPA and phthalates) can cross the placenta, resulting in growth retardation and neurological harm. There are also evidences to suggest hormonal derangements and cancers in children [11].

Impact of single use plastics

The single use plastics (drinking water bottles/ packing food stuff) are another issue surrounding the toxicity. Phthalates and Bisphenol A (BPA) are the two most notorious toxin which leach from plastics into food or water and when these plastic wastes are discarded improperly, they often end up in water bodies where they continue to leach these harmful chemical for an very long time.

Phthalates have been found to deposit in the fatty tissues of the body, and also causative factors of human diseases like male reproductive dysfunction, breast growth and testicular cancers [12]. BPA is often found in the food grade plastic known as polycarbonates, used in hospital disposables, has been found to have an estrogenic side effect. It is found to have detrimental effects on human placental tissues leading to premature birth, intrauterine growth retardation, preeclampsia and still birth [13]. Studies have shown that BPA may also lead to insulin resistance and diabetes also [14].

Indian Scenario

Plastic waste is a major environmental and public health problem in Indian set up particularly in the urban areas [15]. Plastic shopping or carrier bags are one of the main sources of plastic waste in our country. Plastic bags of all sizes and colors dot the city's landscape due to the problems of misuse and overuse and littering in India. Besides this visual pollution, plastic bag wastes contribute to blockage of drains and gutters, are a threat to aquatic life when they find their way to water bodies, and can cause livestock deaths when the livestock consume them. Furthermore, when filled with rainwater, plastic bags become breeding grounds for mosquitoes, which cause malaria. We have become so accustomed to the ubiquitous presence of plastic that it is difficult to envision life when woods and metals were the primary materials used for consumer products. Plastic has become prevalent because it is inexpensive and it can be engineered with a wide range of properties. Plastics are strong but lightweight, resistant when degraded by chemicals, sunlight, and bacteria, and are thermally and electrically insulating. Plastics have become a critical material in the modern economy; the annual volume of plastics produced exceeds that volume of steel [15].

The kind of recycling practiced in India is quite different from what is practiced in the rest of the world, in that state of the art technologies are not employed here. The starting point is the sorting of plastic waste (based on colour, transparency, hardness, density and opacity of the scrap). The sorted waste is then sent to the granulators to obtain granules using with the traditional mechanical and grinding techniques. The converters use these granules to make finished plastic products. The majority of such units (granulators and convertors) are often located in slums, and function single machine extruding units. Scrap storage is done in the backyards, and washing is done in open drums. These activities are often termed as backyard recycling. The technologies used in these industries are also old and local [16]. Of the types of plastics recycled in India, PVC (polyvinyl chloride) accounts for 45%, LDPE (low density polyethylene) for 25%, HDPE (high density polyethylene) for 20%, PP (polypropylene) for 7.6% and other polymers such as PS (polystyrene) for 2.4% [17].

This recycling is usually results in the down cycling of plastics into lower-quality products that have higher and more 4 leachable levels of toxic additives [18]. During recycling, the plastic scrap is cleaned to remove the dirt and foreign matter adhering to it. The wastewater generated used for this purpose is finally disposed of into open drains. This wastewater has high pollution load in terms of BOD, COD, and TSS [16].

The final stage in the life cycle of plastics is disposal. In India, there are three common ways of getting rid off plastics - by dumping them in landfills, by burning them in incinerators or by littering them. In the case of littering, plastic wastes fail to reach landfills or incinerators. It is the improper way of disposing plastics and is identified as the cause of manifold ecological problems. Incineration of plastic wastes also significantly reduces the volume of waste requiring disposal [19]. It is believed that the volume reduction brought about by incineration ranges from 80 to 95%. But the burning of these chlorine-containing substances releases toxic heavy metals and emits noxious gasses like dioxins and furans. The latter two are two of the most toxic and poisonous substances on earth and can cause a variety of health problems [18].

Preventive Measures

Recycling

Among the existing solutions recycling is one of the most convenient and easiest ways. There are various ways to participate through government programs or programs run by environmental organizations. As consumers, the recycling only requires one easy step of putting plastic wastes in right bins for disposal. Separating the plastic waste from other waste will prevent plastics to be land filled and will allow it to be recycled with other plastics of the same kind.

Recycling techniques deals with the tones of plastic waste that is choking earth. So in addition to developing smarter plastics that takes the place of conventional plastics, there is emergent need to deal with the immense quantities of toxic wastes already out there and hurting humans and the environment. Smarter sorting of plastic Wastes, energy efficient ways of getting rid of the plasticizers and increasing the scale of this entire process is very vital to overcome this challenge. Recent reports of discovery of certain fungi and bacteria that hasten degradation of conventional plastics have received a lot of scientific attention. In this process, the byproducts of this natural way of decomposition are safe for the environment and there are no hidden adverse consequences of this approach.

Reduced use of plastics

Plastic pollution can be reduced by using less plastics products and switching to alternatives. Each year, an estimated 500 billion to 1 trillion plastic bags are consumed worldwide. That comes out to over one million per minute. Billions end up as litter each year or in landfills. Now focus on another important part of eco-friendly living: reduce your use of plastic.Source reduction (Reduce and Reuse) can occur by altering the design, manufacture, or use of plastic products and materials. For example, the weight of a 2-liter plastic soft drink bottle has been reduced from 68 grams to 51 grams since 1977, resulting in a 250 million pound decrease of plastic per year in the waste stream.

Tips for safer, more sustainable use of plastics:

- Beware of cling wraps especially for microwave use.
- Avoid plastic bottled water.
- Minimize the use of canned foods and canned drinks.
- Purchase baby bottles and sippy cups or glass options.
- Bring your own cloth bags to the grocery store or any store.
- Don't buy beverages bottles in plastic.
- Carry your own reusable steel or ceramic beverage container.
- Don't buy convenience foods packages in plastic.
- Buy bread from bakeries that package in paper.
- Buy laundry detergent in boxes, not liquid in plastic containers.
- Buy farm fresh eggs in reusable paper containers.
- Package your leftovers in corning ware.
- Store all your food in glass containers instead of plastic containers.
- Buy bulk cereal; bring your own paper bags.
- Compost your trash; reduce your use of plastic trash bags.
- Line small trash bins in your house with paper bags.

- Use cloth rags for clean up around the house.
- Use matches instead of plastic encased lighters.
- Use cloth napkins. They reduce your waste and use of plastic trash bags.
- Use baby bottles made of glass.
- Use rechargeable batteries to reduce buying batteries packaged in plastic.
- Make a compost heap to reduce your food waste and put it back into the earth.
- Use a reusable cloth bag to carry your lunch to work or school.
- Spread the word, tell people about the harmful chemicals in plastic and help reduce plastic use.

Chemical decomposing

Chemical decomposing is otherwise a very effective solution to plastic pollution, since the non-biodegradable property of plastic is the main cause of plastic pollution. However, no technology has been developed yet to set up an economical and effective large-scale plastic decomposing facility. But chemical decomposing is still a field that has a great potential to develop in the future.

There are mainly two ways to decompose conventional plastics. Decomposing plastics by microorganisms is one of them. Daniel Burd, a Canadian high school student, found out that there are three kinds of microorganisms in the earth from a landfill that can break down the molecules of plastic bags. However, since this is a relatively new discovery, it is not applied industrially yet. Its economical applicability still needs to be discussed, but according to Burd, this decomposing method is possible to be applied on an industrial scale. Another way to decompose plastics is by combustion. This is a relatively easy and inexpensive way compared to using microorganisms, however, odor and toxic gases produced during combustion is a big problem. Currently, some companies have already applied this method, and Wheelabrator Technologies Inc. is one of them. In

Wheelabrator's clean energy plants, waste are burned and heat generated from combustion is turned to electricity with emission air control [20]. These waste-to-energy plants not only handle municipal waste environmentally, but also provide electricity to households and businesses.

Alternative solutions

Biodegradable Plastics (BDP): This is one of the options to the conventional plastics. One of the common constituents of BDP is polyhydroxyalkanoate (PHA). The BDP are similar to conventional plastics in all aspects with the additional quality of being able to naturally decompose and break into natural and safe byproducts. Hence if all plastics in the city waste were biodegradable, it could simply be allowed to decompose along with the food and other non-recyclable but biodegradable articles like wet paper and cotton fibers [19].

Since the technologies to manufacture BDPs are relatively new and not widely prevalent, the production cost is higher. Therefore, further research in areas of more cost effective and energy efficient manufacturing methods for biodegradable plastics is the call of the hour [19]. The incorporation of BDP is a progressive approach to a greener, healthier, and a better environment. The progressive development of several biopolymers over the years has stirred the plastic industry. The induction of biodegradable plastics is a promising and progressive prospect and will greatly reduce the dependence on fossil fuels. At the present time, it is only an option over traditional plastics, but if it is to replace traditional plastics completely, people would have no other option but to use them. Incorporating biodegradable plastics in everyday use would not only take the pressure off fossil fuels but also encourage agricultural producers who are interested in exploring and developing the natural fiber processing industry. A lot of income from agriculture can be generated if biodegradable plastics can be made mainstream. Aside from the obvious economic and environmental benefits, biodegradable plastics are progressive from scientific point of view as well. In addition to being useful for everyday life purposes, biodegradable plastics also have a great scope to be used in medicinal field.

What sets biodegradable plastics one step ahead of conventional plastics is the fact that they can be manufactured by using renewable biomass instead of biofuels. This will be of huge advantage because as "renewable biomass", will include "agro-industrial" wastes that are not only cheap, but their conversion solves another problem by turning waste materials into useful products. This makes production of biodegradable plastics possible even in the countries that lack the scope for crop expansion. In return, they are being benefitted economically and ecologically. At present biodegradable polymer technology can only offer a limited range of materials. It is due to this limitation that biodegradable plastics have not been able to go mainstream yet.

Bioplastics: A bioplastic is a plastic that is made partly or wholly from polymers derived from biological sources such as sugar cane, potato starch or the cellulose from trees, straw and cotton. Some bioplastics degrade in the open air, others are made so that they compost in an industrial composting plant, aided by fungi, bacteria and enzymes. Others mimic the robustness and durability of conventional plastics such as polyethylene or PET.

Bioplastics - partly or wholly made from biological materials and not crude oil - represent an effective way of keeping the huge advantages of conventional plastics but mitigating their disadvantages. However, that does not imply that bioplastics can naturally decompose like biodegradable plastics. The prime benefit is that it gives some respite to our depleting petroleum reserves.

Hence, further research should focus on developing bioplastics that are both biodegradable and also energy efficient to produce. Recycling is almost always more energy efficient and releases less carbon dioxide than making a new product. One major problem with efforts to recycle bioplastics is that if they become mixed with petroplastics they can contaminate the whole batch.

Polymer Blended Bitumen Roads: The non-wetting property of plastics is also being implemented successfully in road construction business. Bitumen film is often stripped off the aggregates because of the penetration of water, which results in pothole formation. When polymer (plastic waste) is coated over aggregate, the coating reduces its affinity for water due to non-wetting nature of the polymer, thereby obstructing the penetration of water. Polymers also shows higher softening temperature, thereby reduce the bleeding of bitumen during the summers [21].

To get rid of plastic waste disposal problems, Central Pollution Control Board (CPCB) has taken initiative to use the plastic waste in manufacturing units through co processing. Co-processing refers to the use of plastic waste materials in industry process such as cement, lime or steel production and power stations or any other large combustion plants. Co-processing refers to substitution of primary fuel and raw material by waste. Waste material such as plastic waste act as alternative fuels and raw material (AFR). Thus these units save fossil fuel and raw material consumption, contributing the more eco-efficient production. After getting encouraging results CPCB has granted permission to many cement plants to co-process the hazardous and non-hazardous (including plastic) waste in their kilns after trial burns [21].

1. While some people are busy developing plastic substitutes, others are bent on making conventional thermoplastics biodegradable. By throwing in additives called Prodegradant concentrates (PDCs). PDCs are usually metal compounds, such as cobalt stearate or manganese stearate. They promote oxidation processes that break the plastic down into brittle, low-molecular-weight fragments. Microorganisms gobble up the fragments as they disintegrate, turning them into carbon dioxide, water and biomass, which reportedly contains no harmful residues. When added to polyethylene (the standard plastic bag material) at levels of 3%, PDCs can promote nearly complete degradation; 95% of the plastic is in bacteria-friendly fragments within four weeks.

2. Researchers are revitalizing the idea of converting casein, the principal protein found in milk, into a biodegradable material that matches the stiffness and compressibility of polystyrene. The modern milk-based plastic doesn't crack as easily, thanks to that silicate skeleton, and they even made the stuff less toxic by substituting glyceraldehyde for formaldehyde during the process. Scientists have found a way make the protein less susceptible to cracking, thanks to a silicate clay called sodium montmorillonite.

3. Chicken feathers are composed almost entirely of keratin, a protein so tough that it can give strength and durability to plastics. It's found in hair and wool, hooves and horns -- and we can all appreciate how strong a horse's hoof can be without having the pleasure of being kicked by one. Researchers decided to tap into keratin's superstrong features by processing chicken feathers with methyl acrylate, a liquid found in nail polish. Ultimately, the keratin-based plastic proved to be substantially stronger and more resistant to tearing than other plastics made from agricultural sources, such as soy or starch, and scientists are clucking excitedly about chicken-feather plastic. After all, inexpensive, abundant chicken feathers are a renewable resource.

4. Next up is a promising new bioplastic, or biopolymer, called liquid wood. Biopolymers fake it; these materials look, feel and act just like plastic but, unlike petroleum-based plastic, they're biodegradable. This particular biopolymer comes from pulp-based lignin, a renewable resource.

5. The next three entries on this list are all biodegradable plastics called aliphatic polyesters. Overall, they aren't as versatile as aromatic polyesters such as polyethylene terephthalate (PET), which is commonly used to make water bottles. But since aromatic polyesters are completely resistant to microbial breakdown, a lot of time and effort is being pumped into finding viable alternatives in aliphatic polyesters. polycaprolactone (PCL), a synthetic aliphatic polyester that isn't made from renewable resources but does completely degrade after six weeks of composting.

6. Polyhydroxyalkanoate (PHA) polyesters, the two main members of which are Polyhydroxybutrate (PHB) and Polyhydroxyvalerate (PHV). These biodegradable plastics closely resemble man-made polypropylene. While they're still less flexible than petroleum-based plastics, you'll find them in packaging, plastic films and injection-molded bottles.

References

1. http://www.merriam-webster.com/dictionary/polymer.

2. Greenmuze (2008) Addicted to plastics.

3. Elias HG (2005) Plastics, General Survey. Ullmann's Encyclopedia of Industrial Chemistry.

4. Callister WD, Rethwisch DG (2008) Fundamentals of Materials Science and Engineering: An Integrated Approach (3rdedn).

5. Li N, Mahat D, Park S (2009) Reduce Reuse and Replace: A Study on Solutions to Plastic Wastes. An Interactive Qualifying Project Submitted to the faculty of Worcester Polytechnic Institute.

6. Ryan PG, Moore CJ, Franeker JA, Moloney CL (2009) Monitoring the abundance of plastic debris in the marine environment. Philos Trans R Soc Lond B Biol Sci B 364: 1999-2012.

7. Derraik JGB (2002) The pollution of the marine environment by plastic debris: a review. Marine Pollution Bulletin 44: 842-852.

8. Gregory MR (2009) Environmental implications of plastic debris in marine settings-entanglement ingestion smothering hangers-on hitch-hiking and alien invasions. Philos Trans R Soc Lond B Biol Sci. 364: 2013-2025.

9. Science for Environment Policy (2011) Plastic Waste: Ecological and Human Health Impacts. In-depth Reports.

10. Schuler K (2008) Smart Plastics Guide: Healthier Food Uses of Plastics. Minneapolis Minnesota.

11. Zaman T (2010) The Prevalence and Environmental Impact of Single Use Plastic Products.

12. Jobling S, Reynolds T, White R, Parker MG, Sumpter JP (1995) A variety of environmentally persistent chemicals including some phthalate plasticizers are weakly estrogenic. Environ Health Perspect 103: 582-587.

13. Benachour N, Aris A (2009) Toxic effects of low doses of Bisphenol-A on human placental cells. Toxicol Appl Pharmacol 241: 322-328.

14. Magdalena AP, Morimoto S, Ripoll C, Fuentes E, Nadal A (2006) The estrogenic effect of bisphenol A disrupts pancreatic beta-cell function in vivo and induces insulin resistance. Environ Health Perspect 114: 106-112.

15. Tammemagi HY (1999) The Waste Crisis: Landfills Incinerators and the Search for a Sustainable Future. Oxford University Press New York.

16. Narayan P (2001) Analyzing Plastic Waste Management in India: Case study of Polybags and PET bottles‖ published by IIIEE. Lund University Sweden 24-25.

17. Shah P (2001) The Plastic Devil: Ecological Menace.

18. Chaturvedi B (2002) The source of this information is a press release of No Plastics in the Environment (Nope) Imports Versus Surplus: A Glut of Plastics in India Today. Chintan Environmental Organization.

19. Begum Z (2010) Plastics and environment. Dissemination Paper for the Ministry of Environment and Forest. Government of India.

20. Wheelabrator Technologies Inc (2009) About Us.

21. Material on Plastic Waste Management (2012) Central Pollution Control Board India.

Permissions

List of Contributors

Ing-Jia Chiou and I-Tsung Wu
Department of Environmental Technology and Management, Taoyuan Innovation Institute of Technology, No. 414, Sec. 3, Jhongshan E. Rd., Jhongli, Taoyuan 320, Taiwan

Jun-Pin Su
Department of Natural Resource, Chinese Culture University, 55, Hwa-Kang Road, Yang-Ming-Shan, Taipei , 11114, Taiwan

Ching-Ho Chen
Department of Social and Regional Development, National Taipei University of Education , No. 134, Sec. 2, Heping E. Rd., Taipei City 106, Taiwan

Margarida Marchetto
Department of the Sanitary and Environmental Engineering, Federal University of Mato Grosso, Brazil

Henning Wilts and Bettina Rademacher
Wuppertal Institute for Climate, Environment,Energy GmbH, Research Group Material Flows and Resource Management, Germany

Xiao Han Sun and Kunio Yoshikawa
Department of Environmental Science and Technology, Tokyo Institute of Technology, Yokohama 226-8502, Japan

Hiroaki Sumida
Laboratory of Soil Science, Department of Chemistry and Life Science, Nihon University, Fujisawa 252-8510, Japan

Suneethi S, Sri Shalini S and Kurian Joseph
Research Scholar, Centre for Environmental Studies, Anna University, India

Abu Zahrim Yaser, Cassey TL, Hairul MA and Shazwan AS
Chemical Engineering Programme, Faculty of Engineering, Universiti Malaysia Sabah, Jalan UMS, 88400 Kota Kinabalu, Sabah, Malaysia

Duc Luong Nguyen
Department of Environmental Technology and Management, National University of Civil Engineering (NUCE), Vietnam

Xuan Thanh Bui
Faculty of Environment, Ho Chi Minh City University of Technology (HCMUT), Vietnam

The Hung Nguyen
Hanoi Urban Environment One Member Limited Company, Vietnam

Haoyi Chen
D-102, Guangzhou Boxenergytech Ltd, Guangzhou International Business Incubator, Guanagzhou City, P R China

Xiaolong Wang, Li Zhong and Shisen Xu
China Huaneng Group Clean Energy Technology Research Institute, China No. 249, Xiaotangshan Industrial Park, Changping District, Beijing, China

Tiancun Xiao
D-102, Guangzhou Boxenergytech Ltd, Guangzhou International Business Incubator, Guanagzhou City, P R China
China Huaneng Group Clean Energy Technology Research Institute, China No. 249, Xiaotangshan Industrial Park, Changping District, Beijing, China
Inorganic Chemistry Lab, Oxford University, South Parks Road, OX1 3QR, UK

Shuo Cheng and Kunio Yoshikawa
Department of Environmental Science and Technology, Tokyo Institute of Technology, Japan

Aimin Li
School of Environmental Science and Technology, Dalian University of Technology, China

Molla A, Ioannou Z, Dimirkou A and Mollas S
Soil Science Laboratory, Department of Agriculture, Crop Production & Rural Environment, School of Agricultural Sciences, University of Thessaly, Greece

E Gallego, FJ Roca and JF Perales
Laboratory Centre for Environment, Polytechnic University of Catalonia (LCMA-UPC), Spain

G Sánchez
Department of Prevention and Waste Management of Greater Barcelona, Spain

Abbas Parsaie and Amir Hamzeh Haghiabi
Department of water Engineering, Lorestan University, Khorram Abad, Iran

Mayuri Naik, Anju Singh, Seema Unnikrishnan, Neelima Naik and Indrayani Nimkar
National Institute of Industrial Engineering (NITIE), Mumbai-400087, India

Mahendra Kumar Trivedi, Alice Branton, Dahryn Trivedi and Gopal Nayak
Trivedi Global Inc., 10624 S Eastern Avenue Suite A-969, Henderson, NV 89052, USA

Ragini Singh and Snehasis Jana
Trivedi Science Research Laboratory Pvt. Ltd., Hall-A, Chinar Mega Mall, Chinar Fortune City, Hoshangabad Rd., Bhopal- 462026, Madhya Pradesh, India

Haiping Yuan
School of Environmental Science and Engineering, Shanghai Jiao Tong University, 800 Dongchuan Road, Shanghai, 200240, China

Wenxiang Yuan
Shanghai Institute for Design & Research on Environmental Engineering Co., Ltd, 11 Shilong Road, Shanghai, 200232, China

Xiaopeng Wang
School of Environmental Science and Engineering, Shanghai Jiao Tong University, 800 Dongchuan Road, Shanghai, 200240, China
Shanghai Qingcaosha Investment Construction and Development Co., ltd. 700 Jinyu Road, Shanghai, China

Queena K Qian
Endeavour Australian Cheung Kong Post-Doc Fellow, sd+b Centre, University of South Australia, Australia

Steffen Lehmann
Chair Professor of Sustainable Design; Director, sd+b Centre and CAC-SUD Centre, University of South Australia, Australia

Atiq Uz Zaman and John Devlin
PhD Candidate, sd+b Centre, University of South Australia, Australia

Dimitris P Makris
School of Environment, University of the Aegean, Mitr Ioakim Str., Myrina-81400, Lemnos, Greece

Panagiotis Kefalas
Food Quality and Chemistry of Natural Products, Mediterranean Agronomic Institute of China, China
Centre International de Hautes Etudes Agronomiques Méditerranéennes, P.O. Box 85, Chania-73100, Greece

TD Kusworo, Budiyono, J Supriyadi and DC Hakika
Department of Chemical Engineering, Diponegoro University, Indonesia

Cang Yu, Dawei Li, Zhenya Zhang and Yingnan Yang
Graduate School of Life and Environmental Science, University of Tsukuba, 1-1-1 Tennodai, Tsukuba, Ibaraki 305-8572, Japan

Qinghong Wang
State Key Laboratory of Heavy Oil Processing, China University of Petroleum, Changping, Beijing 102249, China

De Michelis I and Vegliò F
Laboratory of Integrated Treatment of Industrial Wastes and Wastewaters, Department of Industrial and Information Engineering and Economics of the University of L'Aquila, Italy

Zueva SB
Laboratory of Integrated Treatment of Industrial Wastes and Wastewaters, Department of Industrial and Information Engineering and Economics of the University of L'Aquila, Italy
Department of Environment and Chemical Engineering, Voronezh State University of Engineering Technology, Russia

Ostrikov AN and Ilyina NM
Department of Environment and Chemical Engineering, Voronezh State University of Engineering Technology, Russia

Hassan Sh Abdirahman Elmi, Muhamad Hanif Md Nor and Zaharah Ibrahim
Environmental Biotechnology, Amoud University, Borama, Somalia

Liyanage Bundunee Chanpika and Athapattu Prathapage Priyantha
Department of Civil Engineering, Faculty of Engineering Technology, The Open University of Sri Lanka

Tateda Masafumi
Department of Environmental Engineering, Toyama Prefectural University, Toyama, Japan

Sunita Chauhan and Sharma AK
Kumarappa National Handmade Paper Institute (KNHPI), Ramsinghpura, Sikarpura Road, Sanganer, Jaipur, India

Samia Bedouhene, Nassima sennani and Farida Moulti-Mati
Laboratoire de Biochimie appliquée et de biotechnologie, Faculté des Sciences Biologiques et des Sciences Agronomiques, Université M. Mammeri, BPN°17, RP 15000 Tizi-Ouzou, Algeria

Jean-Claude Marie and Jamel El-Benna
Inserm, U1149, CNRS-ERL8252, Centre de Recherche sur l'Inflammation, Paris, France

Université Paris Diderot, Sorbonne Paris Cité, Laboratoire d'Excellence Inflamex , Faculté de Médecine, Site Xavier Bichat, Paris, France

Margarita Hurtado-Nedelec
Inserm, U1149, CNRS-ERL8252, Centre de Recherche sur l'Inflammation, Paris, France
Université Paris Diderot, Sorbonne Paris Cité, Laboratoire d'Excellence Inflamex , Faculté de Médecine, Site Xavier Bichat, Paris, France
AP-HP, Unité Dysfonctionnement Immunitaire, Centre Hospitalo-Universitaire Xavier Bichat, Paris, F-75018, France

Ravindra Kumar Verma
Fellow (Doctoral Programme), Environmental Engineering & Management Group, National Institute of Industrial Engineering, Mumbai, Maharashtra - 400087, India

Shankar Murthy
Associate Professor, Environmental Engineering & Management Group, National Institute of Industrial Engineering, Mumbai, Maharashtra - 400087, India

Rajani Kant Tiwary
Senior Scientist, Water Environment Division, Central Institute of Mining and Fuel Research (CIMFR), Barwa-Road, Dhanbad, Jharkhand-826015, India

Magdy Abdelrahman, Mohyeldin Ragab and Daniel Bergerson
Department of Civil and Environmental Engineering North Dakota State University, USA

Yoshiaki Tsuzuki
Engineering, Architecture and Information Technology (EAIT), The University of Queensland, St Lucia, QLD 4072, Australia
Research Centre for Coastal Lagoon Environments (ReCCLE), Shimane University, Nishi-Kawatsu-cho, Matsue City, 690-8504, Japan

Vahidhabanu S, Abilash John Stephen, Ananthakumar S and Ramesh Babu B
CSIR- Central Electrochemical Research Institute, Karaikudi – 630 003, India

Nazile Ural
Bilecik Seyh Edebali University, Department of Civil Eng., 11210, Bilecik, Turkey

Ahsan Shamim
Associate Professor of Environmental Science, Earth and Atmospheric Sciences Department, Metropolitan State University of Denver, CO, USA

Ali Mursheda K
Adjunct Faculty, Biology Department, Metropolitan State University of Denver, CO, USA

Islam Rafiq
Soil, Water & Bioenergy Program Director, Ohio State University South Centers, Piketon, OH, USA

Vahidhabanu S and Ramesh Babu B
CSIR- Central Electrochemical Research Institute, Karaikudi- 630 003, India

Fianda Revina Widyastuti and Purwanto
Environmental Science Master's Program, Diponegoro University Semarang, Indonesia

Hadiyanto
Chemical Engineering Program, Technical Faculty, Diponegoro University, Indonesia

Tahmina Ahsan
School of Architecture and Built Environment, University of Adelaide, SA 5001, Australia

Atiq Uz Zaman
Zero Waste SA Research Centre for Sustainable Design and Behaviour, School of Art, Architecture and Design, University of South Australia, Australia

Queena K Qian
Endeavour Australia Cheung Kong Fellow, Center for Sustainable Design and Behaviour (sd+b), University of South Australia, Australia

Steffen Lehmann
Professor of Sustainable Design and Behaviour, School of Art, Architecture and Design Director, Zero Waste Research Centre, The University of South Australia, Australia

Abd Ghani Bin Khalid
Professor, Faculty of Built Environment, Universiti Teknologi Malaysia, Malaysia

Edwin HW Chan
Professor, Building and Real Estate Department, The Hong Kong Polytechnic University, Hong Kong S.A.R., China

Faiza M El Assal, Salwa F Sabet, Kohar G Varjabedian and Mona F Fol
Department of Zoology, Faculty of Science, Cairo University, Giza, Egypt

Vipin Koushal, Raman Sharma and Vivek Sharma
Department of Hospital Administration, Chandigarh, India

Meenakshi Sharma
Senior Research fellow, School of Public health, PGIMER, Chandigarh, India

Ratika Sharma
Department of Sociology, Panjab University, Chandigarh, India